SOLAR-THERMAL ENERGY SYSTEMS

SOLAR-THERMAL ENERGY SYSTEMS
Analysis and Design

John R. Howell
Professor of Mechanical Engineering
University of Texas, Austin

Richard B. Bannerot
Associate Professor of Mechanical Engineering
University of Houston

Gary C. Vliet
Professor of Mechanical Engineering
University of Texas, Austin

McGraw-Hill Book Company

New York   St. Louis   San Francisco   Auckland   Bogotá   Hamburg
Johannesburg   London   Madrid   Mexico   Montreal   New Delhi
Panama   Paris   São Paulo   Singapore   Sydney   Tokyo   Toronto

This book was set in Times Roman by Santype-Byrd.
The editors were Diane D. Heiberg and Susan Hazlett;
the production supervisor was John Mancia.
The drawings were done by ECL Art Associates, Inc.
The cover was designed by Infield D'Astolfo Associates.
Fairfield Graphics was printer and binder.

SOLAR-THERMAL ENERGY SYSTEMS
Analysis and Design

Copyright © 1982 by McGraw-Hill, Inc. All rights reserved.
Printed in the United States of America. Except as permitted under the United States
Copyright Act of 1976, no part of this publication may be reproduced or
distributed in any form or by any means, or stored in a data base or
retrieval system, without the prior written permission of the publisher.

1234567890 FGFG 898765432

ISBN 0-07-030603-6

Library of Congress Cataloging in Publication Data

Howell, John R.
   Solar-thermal energy systems.

   Includes bibliographical references and index.
   1. Solar power plants. 2. Solar energy. I. Bannerot,
Richard B. II. Vliet, Gary C. III. Title.
TK1056.H68   621.47   81-17195
ISBN 0-07-030603-6   AACR2

CONTENTS

		Preface	ix
Chapter 1		Introduction	1
	1-1	Energy: Supply and Demand	1
	1-2	The Sun as a Resource	1
	1-3	The Terminology of Solar Energy and Collector Classification	8
	1-4	Applications of Solar-Thermal Energy	11
	1-5	A Brief History of Solar Activity	12
	1-6	Conclusions	17
		References	18
Chapter 2		Thermal Conversion Systems	19
	2-1	List of Symbols	19
	2-2	Solar-Thermal Conversion Systems	21
	2-3	System Models	27
	2-4	Economic Factors	40
	2-5	Further Refinements in Modeling	43
		Problems	43
		References	44
Chapter 3		Terrestrial Solar Radiation	45
	3-1	List of Symbols	45
	3-2	The Sun-Earth Relations	46
	3-3	Measurement of Terrestrial Solar Radiation	70
	3-4	Terrestrial Insolation on Tilted Surfaces	71
		Problems	82
		References	84

Chapter 4 General Description of Solar-Thermal Collectors and Methods to Evaluate Them — 86

- 4-1 List of Symbols — 86
- 4-2 The Thermal Evaluation of Solar Energy Collectors — 87
- 4-3 Classification and General Description of Solar Energy Collectors — 91
- 4-4 Nonconcentrating Collectors — 93
- 4-5 Concentrating Collectors — 101
- 4-6 Collector Selection — 116
- 4-7 Conclusions — 119
- Problems — 119
- References — 120

Chapter 5 Analytical Design of Solar Collectors — 122

- 5-1 List of Symbols — 122
- 5-2 Optical Design — 124
- 5-3 Thermal Analysis — 133
- 5-4 Analysis of Collectors — 141
- 5-5 Application of the Analysis to the SOLSIM Program — 148
- 5-6 Collector Testing and Performance — 148
- Problems — 154
- References — 158

Chapter 6 Storage of Thermal Energy — 160

- 6-1 List of Symbols — 160
- 6-2 Methods of Storage — 161
- 6-3 Thermal Storage Requirements — 165
- 6-4 Water Energy Storage — 166
- 6-5 Pebble-Bed Storage — 173
- 6-6 Latent Heat Storage — 178
- Problems — 185
- References — 185

Chapter 7 Load Modeling — 186

- 7-1 List of Symbols — 186
- 7-2 Heating and Cooling Loads — 187
- 7-3 Domestic Hot-Water Systems — 192
- 7-4 Swimming Pool Heaters — 193
- 7-5 Industrial Energy Loads — 196
- Problems — 199
- References — 200

Chapter 8 Control of Solar-Thermal Systems — 201

- 8-1 Collector Loop Circulation — 201
- 8-2 Freeze Prevention — 204

8-3	Other Collector Loop Passive Control Components	208
8-4	Tracking Systems	209
8-5	Control of Specific Systems	210
	Problems	213

Chapter 9 Sizing Solar Energy Systems — 215

9-1	List of Symbols	215
9-2	Rules of Thumb	216
9-3	Early Optimization Studies	218
9-4	Manual Design Calculations	220
9-5	Computer Modeling	227
9-6	f-Chart Method	227
9-7	Annual f-Chart Method	233
9-8	Flat-Plate Collector Orientation and Shading	237
	Problems	239
	References	239

Chapter 10 Economics of Solar-Thermal Energy Conversion Systems — 241

10-1	General Economic Characteristics of Solar-Thermal Conversion	241
10-2	List of Symbols	242
10-3	Life-Cycle Costs of Solar Heating and Air-Conditioning Systems	242
10-4	Cost-Benefit Comparison of Energy Systems	245
10-5	Influence of Assumptions	247
10-6	Economic Analyses of Solar Systems	248
10-7	Other Economic Factors	249
	Problems	249
	References	249

Chapter 11 Case Studies — 251

11-1	Domestic Hot-Water System Design	251
11-2	Conceptual Design of a Commerical Office Heating System	255
11-3	Solar Industrial Process Heat	262
11-4	High-Temperature Applications—Solar-Driven Heat Engines	268
	Problems	271
	References	272

Appendix A Fundamental Constants and Conversion Factors — 275

A-1	Radiation Constants	275
A-2	Conversion Factors	276
A-3	Names and Symbols for Multipliers	277

Appendix B	**Insolation and Weather Data**	279
B-1	Average Daily Global Horizontal Insolation Patterns	280
B-2	Insolation and Other Data for 80 Locations in the United States and Canada	284
B-3	Ratio of Monthly Average Daily Global Insolation on a Tilted Surface Facing the Equator to That on a Horizontal Surface	294
B-4	Direct Insolation	299
B-5	Average Monthly and Yearly Degree-Days for Heating (Base 65°F) and $97\frac{1}{2}$ percent Winter Design Temperature for Selected Cities in the United States and Canada	303
Appendix C	**Heat Transfer in Solar-Thermal Applications**	309
C-1	Conduction	309
C-2	Convection	316
C-3	Radiation	321
	References	331
Appendix D	**Selected Radiative Properties of Materials**	333
D-1	Total Emissivity	334
D-2	Absorptivity for Insolation, Receiving Material at 295 K (70°F)	337
D-3	Effective Reflectances (integrated over the solar spectrum and angle of incidence)	339
	References	340
Appendix E	**SOLSIM**	341
E-1	General Information	341
E-2	Sample BASIC Program	347
E-3	Sample FORTRAN Program	373
Appendix F	**Glossary**	397
	Index	403

PREFACE

This text is aimed at providing the practicing engineer and the engineering student with a practical yet detailed understanding of systems for converting solar energy initially into useful thermal energy and ultimately into power and cooling. The authors feel that most texts for this audience concentrate only on the components and their behavior and detailed design, but not on their interrelationships or cost effectiveness as parts of a system. In the present book, therefore, we intend to present as well the economics and behavior of a complete solar system, including the collector, storage, load, and controls.

Thus both the organization of our material and its emphasis varies from those of other available works. Following an introductory chapter on the need for, origins of, and limitations on solar engery utilization, Chap. 2 concentrates on a very simple model for the transient behavior of a solar-thermal conversion system. All components of the model are discussed, but the simplest formulation is used for each (lumped storage, constant load, linear collector efficiency, on-off control, and sinusoidal solar input). A system simulation is given for the behavior of the system and a computer program is developed, which is available as an interactive program in both FORTRAN and BASIC. The simulation results are used as a tool to demonstrate the expected behavior and economics of such systems.

In succeeding chapters, each component of the system is examined, analyzed, and modeled in detail, with special emphasis on how component design affects *system* performance. Thus, for example, the ineffectiveness of using multiple covers in a collector for improving system performance in a low-temperature application is easily shown by applying the model developed in Chap. 2. And the influence of plumbing arrangements to take advantage of thermal stratification in a storage tank is shown in Chap. 6 by developing the analysis and model of a stratified storage tank, then using the model in the system program discussed earlier.

Consideration is given to thermal, optical, and mechanical design throughout the book, so that behavior as well as theory is explained to the student. By the closing chapters, therefore, a detailed model of each component has been constructed and an overall system model composed of the component models is complete. The system economics are then examined. Finally, in the last chapter, the system model is employed to analyze several practical systems, thus showing how the method outlined earlier can be applied.

Information necessary to understand the heat-transfer aspects of solar-thermal energy conversion systems is contained in App. C as a refresher and reference. For instructors more interested in practical system design, the sections on thermal design of collectors (Chap. 5) can be omitted. Problems and references are included with each chapter, and references with each appendix where appropriate.

We believe the approach used here will help the student and designer better understand the interactions of system components, and we hope that readers will forward comments to us about perceived strengths and shortcomings of the text.

John R. Howell
Richard B. Bannerot
Gary C. Vliet

CHAPTER
ONE

INTRODUCTION

1-1 ENERGY: SUPPLY AND DEMAND

In 1973, the Arab oil embargo brought attention to the fact that the dependence of any nation on oil and gas was fraught with uncertainty in the short term and must end in the long term. Many observers had been making this point for years. The so-called conventional energy resources of oil, gas, coal, and uranium are finite, and will therefore be exhausted at some time in the future. Major debate goes on as to the amount of each resource available, the projected rate of use, and therefore the time when the resources will be exhausted. No one disputes, however, that the resources *will* be exhausted.

If these consumable resources are to be exhausted in the future, where shall we find our energy? Aside from technologies not yet proven, such as fusion, the answer must lie in renewable energy sources such as geothermal energy and solar energy, including its various forms: wind, wave, offshore thermal energy conversion, biomass conversion, as well as the more direct forms of thermal and photovoltaic conversion.

The United States, for example, presently relies for its energy on a mixture of sources, the chief components of which are oil, natural gas, coal, hydropower, and nuclear energy. The historical variation of the mixture is illustrated in Fig. 1-1. One projection of available supplies of these sources, and the difference between domestic supplies and demand, is shown in Fig. 1-2. It has been generally assumed that the difference would be made up by imports of fossil fuels.

1-2 THE SUN AS A RESOURCE

The Sun

Throughout the history of humankind, the sun has been the subject of much attention and worship. The sun is, after all, the continuing source of energy for

2 SOLAR-THERMAL ENERGY SYSTEMS

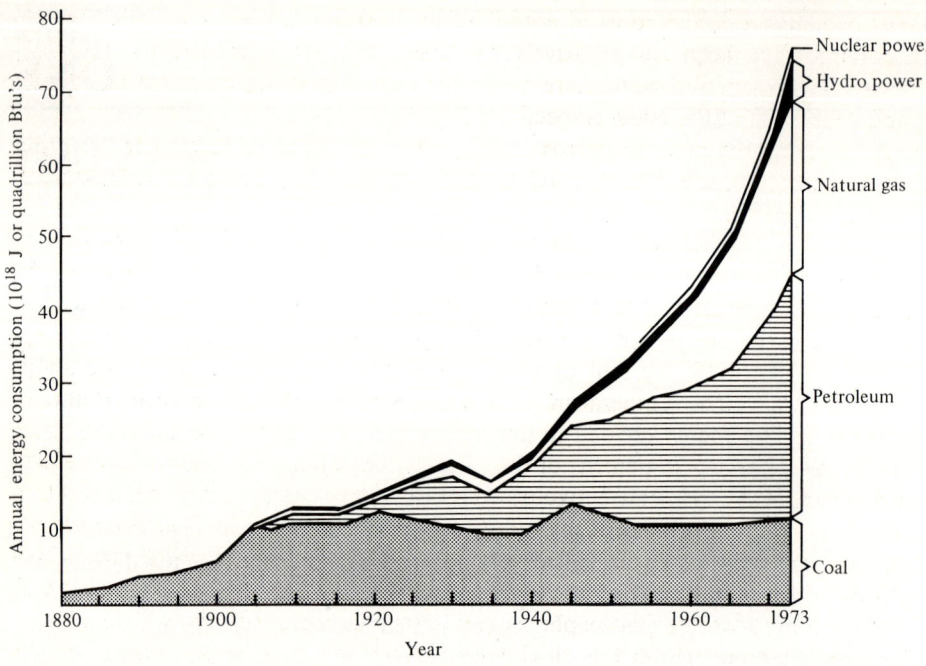

Figure 1-1 Historic energy resources for the United States.

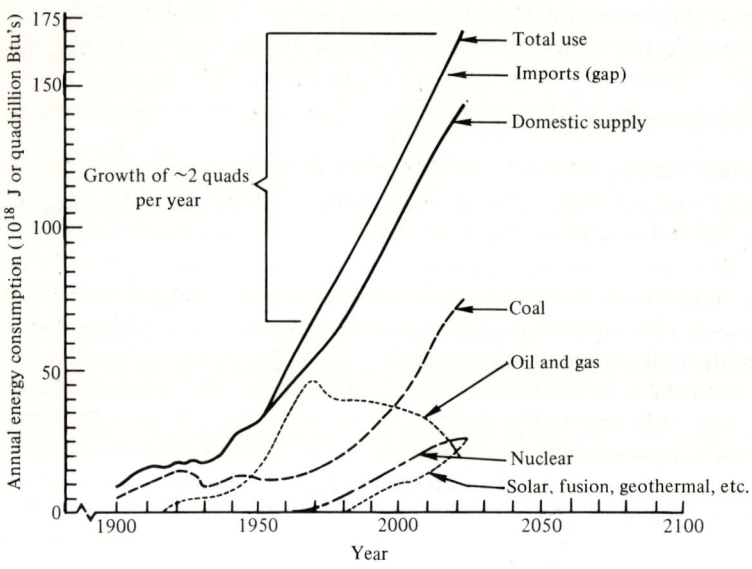

Figure 1-2 Historic and projected U.S. energy supply and demand. (*Source: DOE/EIA-0173/3 1978 Annual Report to Congress.*)

our planet. It contains most, if not all, of the elements present on earth. Over 80 elements have been quantitatively evaluated through spectroscopic measurements. Hydrogen and helium are by far the most abundant, representing over 78 and 20 percent of the mass, respectively [1].

The sun's diameter is approximately 1.39×10^9 m (864,000 mi) or 109 times the diameter of the earth. Its total mass is 1.99×10^{30} kg or about 332,000 times that of the earth.

Energy is generated within the sun in a thermonuclear fusion process in which hydrogen is transformed into helium. However, little is known about the details of the process. The process is confined to the inner core, occupying less than 2 percent of the sun's volume, yet containing 40 percent of its mass. Approximately 4×10^9 kg/s of material is converted to energy, generating about 3.7×10^{23} kW. The temperature of the core is thought to be between 10 and 20 million K. The energy created in the core is radiated outward to a distance from the center of about 70 percent of the sun's radius. The energy is then brought to the sun's surface by a convective process through a fluid that covers the sun to a depth of about 30 percent of the radius or about 0.2×10^6 km. The outer surface of the convective layer is called the *photosphere* and is essentially the direct source of all the radiative emission from the sun. Its temperature is about 6000 K (10,800°R). Above the photosphere is the chromosphere, an atmosphere of rarefied gases about 10,000 km thick. It is comprised of a large number of jets originating from the convective layer. Finally, the corona, composed largely of highly ionized gases, extends millions of kilometers into space. While the temperatures in both the chromosphere and the corona are on the order of 1 million K, these regions contribute very little to the solar emission.

Despite these impressive statistics, the sun is just an average star. Its importance to us is due to its proximity rather than to its uniqueness in our galaxy.

Extraterrestrial Solar Radiation

The earth revolves around the sun in a slightly elliptic orbit at a mean distance of 1.50×10^8 km [9.30×10^7 mi or one astronomical unit (AU)] with a variation of ± 1.7 percent. The visible disk of the sun (the photosphere) subtends an angle at the earth of 0.545° (32.7 minutes) of arc at the mean sun-earth distance. Of the 3.7×10^{23} kW of power generated by the sun, only about 1.7×10^{14} kW reaches the earth, but even this amount is approximately 5000 times the rate at which power is presently used directly by the earth's population.

The solar constant I_{sc} is the total (over all wavelengths) solar radiative energy that strikes a unit area exposed to perpendicular rays of the sun at the mean sun-earth distance. The currently accepted value of the solar constant is 1353 ± 20 W/m² [429 Btu/(ft² · h)].† The extraterrestrial radiation I_0 striking the earth varies throughout the year, primarily because of the change in the sun-earth

---

† Based on recent measurements in space, it now appears that this value will soon be corrected to 1377 W/m².

distance, but also due to sunspots, flares, and other random activity on the surface of the sun. The variation due to distance change can be expressed as

$$I_0(n) = I_{sc}\left[1 + 0.034 \cos\left(\frac{360n}{365}\right)\right] \quad (1\text{-}1)$$

where n is the day number counted from January 1.

The extraterrestrial solar spectrum in the wavelength range 0.2–2.6 μm is shown in Fig. 1-3, along with a curve indicating the fraction of the total energy occurring below specific wavelengths. Also indicated in Fig. 1-3 are the three spectral ranges making up the solar spectrum: ultraviolet (0.03–0.4 μm), visible (0.4–0.7 μm), and the infrared (0.7–1000 μm).

Example 1-1 How much extraterrestrial solar radiation occurs in the visible range on December 1?

SOLUTION From Fig. 1-3 the fractions of the solar radiation at wavelengths less than 0.40 and 0.70 μm are 0.09 and 0.47, respectively. The extraterrestrial solar radiation on December 1 is [Eq. (1-1), $n = 335$] 1393 W/m². Therefore the radiation in the visible range is

$$(0.47 - 0.09)\ 1393\ \text{W/m}^2 = 530\ \text{W/m}^2$$

The approximate distribution of the extraterrestrial solar radiation among the ultraviolet, visible, and infrared is 9, 38, and 53 percent, respectively.

The extraterrestrial solar spectrum is clearly not a smooth curve. Hence, the

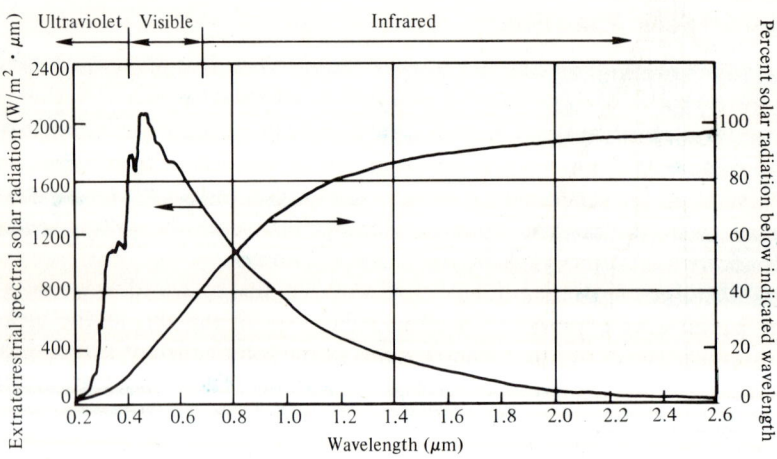

Figure 1-3 Extraterrestrial solar spectral radiation at sun-earth distance of 1 astronomical unit; NASA/ASTM standard curve; solar constant = 1353 W/m² [2].

sun does not radiate as a blackbody. However, the spectral distribution closely approximates the emission from a blackbody at 5762 K (10370°F), as illustrated in Fig. 1-4.

Terrestrial Solar Radiation

The discussion of extraterrestrial solar radiation was restricted to the sun's beam radiation and the variation was due chiefly to the change in the earth-sun distance during the year. For terrestrial solar energy applications it is essential to account for the interaction of the solar energy on passing through the atmosphere. The two primary interaction mechanisms are *absorption* and *scattering*. The results of these interactive mechanisms are both to reduce the solar radiation reaching the earth's surface and to introduce a diffuse component into the solar flux. The fraction of direct radiation is particularly important to the performance of focusing or concentrating collectors.

At the earth's surface, the sum of the incident solar radiation from all direc-

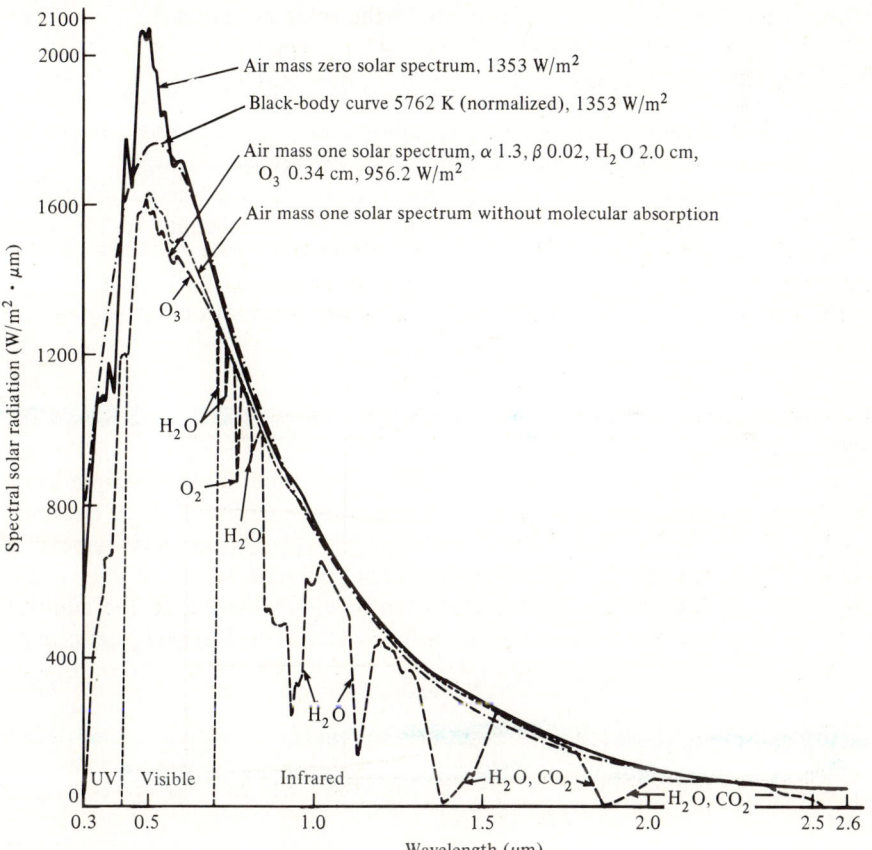

Figure 1-4 Spectral solar radiation [2].

tions is called the *global insolation*. The portion that comes directly from the sun without a change in direction (not scattered) is called the *beam* or *direct insolation*. Its value ranges from about 90 percent of the global insolation on an extremely clear day to practically zero on an overcast day. The *diffuse* or *nondirect* insolation from all directions except directly from the sun makes up the remainder of the global insolation.

The detailed analysis of the interaction of radiative energy with a partially absorbing and scattering medium, like the atmosphere, is among the most complex in all physical science. In theory the processes are well understood. The complexity, especially in the case of the atmosphere, results from the extremely large number of the interactions, the wavelength dependence of the interactions, the lack of detailed physical property data, and the lack of knowledge of the local composition of the atmosphere at any given time. Upon entering the atmosphere, the extraterrestrial beam is scattered by the air molecules, water drops, dust, and other aerosol particles, and absorbed by the atmospheric gases and water vapor.

In the atmosphere the spectral intensity of the incident beam is diminished as

$$I_\lambda = I_\lambda(0) e^{-\tau_\lambda m} \tag{1-2}$$

where I_λ = transmitted spectral intensity of the solar beam
 $I_\lambda(0)$ = extraterrestrial spectral intensity of the solar beam
 τ_λ = spectral optical thickness (dimensionless) of the atmosphere measured from the top of the atmosphere vertically downward
 m = the number of air masses which is the ratio of the actual slant path through the atmosphere to the vertical path length to the same depth

The number of air masses can be approximated as

$$m \simeq \frac{1}{\cos \zeta_s} \tag{1-3}$$

where ζ_s is the solar zenith angle, the acute angle between the direction of the solar beam and the local vertical.

The interaction between the solar radiative energy and the atmosphere has a strong spectral dependence and the total† transmitted intensity must be obtained by an integration of Eq. (1-2) over all wavelengths. This development is beyond the scope of this treatment,‡ but the results will be summarized below.

The value of spectral optical thickness can be viewed as due to the additive effects of absorption, Rayleigh scattering, and turbidity (non-Rayleigh scattering):

$$\tau_\lambda = \tau_{\lambda \text{absorption}} + \tau_{\lambda \text{Rayleigh}} + \tau_{\lambda \text{turbidity}} \tag{1-4}$$

Most of the short wavelength portion (ultraviolet, $\gtrsim 0.4$ μm) of the solar spectrum is absorbed in the upper atmosphere (ionosphere) by ozone. Throughout the atmosphere the primary absorbing constituents are H_2O and CO_2, which

---

† In this book, "total" indicates a value obtained by integration over all wavelengths, and "global" will be used to indicate a value obtained by integration over all directions. Either or both words will be deleted when the meaning is obvious.

‡ See References such as 3, 4, or 5, for details.

exhibit several absorption bands at wavelengths greater than approximately 0.7 μm, as illustrated in Fig. 1-4. The magnitude of this absorption is strongly dependent upon the amounts of CO_2 and water vapor as well as the solar zenith angle (number of air masses) (Fig. 1-5). In humid areas and in industrial areas with high CO_2 and hydrocarbon levels in the atmosphere, absorption will be greater. For all practical purposes, however, it can be considered that terrestrial solar radiation is restricted to the range $0.3 < \lambda < 1.8$ μm, because there is little extraterrestrial solar radiation at long wavelengths and what there is is strongly absorbed by the atmosphere.

Rayleigh scattering is a result of the interaction with particles which are much smaller than the wavelength of the incident radiation, i.e., particles of size much less than 0.1 μm for the case of solar scattering. It accounts for molecular scattering in the atmosphere and its effect is indicated by the general decrease in the terrestrial insolation with increasing air mass on a very clear day, as shown in Fig. 1-5. Rayleigh scattering depends on wavelength to the negative fourth power:

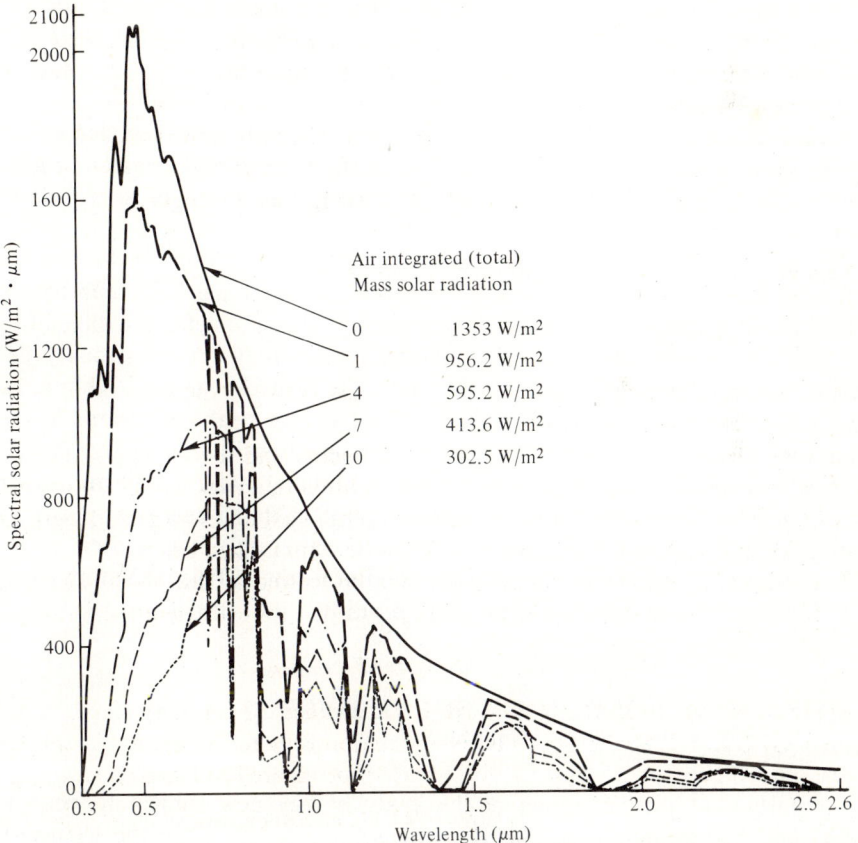

Figure 1-5 Spectral solar radiation for different air mass values, assuming U.S. standard atmosphere, 20 mm of precipitable water vapor, 3.4 mm of ozone, and very clear air [2].

$$\tau_{\lambda_{\text{Rayleigh}}} \propto \lambda^{-4} \tag{1-5}$$

This effect is evident in Fig. 1-5. Thus, in the visible spectrum, the short (blue) wavelengths are scattered more strongly, giving the sky its characteristic color. Further, at times near sunset when the air mass is large between the sun and observer, so much blue is scattered from the sun's rays that the remaining energy is at long wavelengths, giving the sun its deep red color near sunset.

Even under "clear sky" conditions, and progressively more so under hazy conditions, Rayleigh scattering and absorption by atmospheric gases and water vapor do not satisfactorily describe solar atmospheric attenuation. In the 1920s the concept of turbidity was introduced primarily to account for non-Rayleigh scattering on larger particles such as aerosols, and all other factors affecting attenuation such as haze and other pollutants. One of the most often referenced turbidity models was developed by Ångström:

$$\tau_{\lambda_{\text{turbidity}}} = \beta \lambda^{-\alpha} \tag{1-6}$$

where β is the turbidity coefficient. The wavelength exponent α has a value of about 1.3 for good atmospheric conditions. The air mass 1 curve of Fig. 1-4 includes the effects of turbidity ($\alpha = 1.3$, $\beta = 0.02$). This is a typical clear day curve. Near humid regions and/or industrial urban centers, more typical values of α and β are 0.7 and 0.2, respectively.

Cloudy atmospheres can in principle also be modeled. However, due to the general uncertainty about the presence and type of clouds, the water droplet diameter, and the thickness of the cloud structure, predictions of actual cloud attenuation are practically impossible.

As previously defined, the solar radiative energy scattered out of the beam is the nondirect or diffuse insolation. Little experimental work has been reported on the directional distribution of this radiative energy. Measurements are difficult to make and the results have limited application since they are so strongly dependent on local climatic conditions. However, the clear sky or Rayleigh atmospheric model predicts increased scattered (diffuse) energy near the horizon at the longer wavelengths but a more uniform or isotropic distribution at the shorter visible wavelengths. The turbid atmosphere, containing haze, smog, and other aerosol particles, has a much greater distribution of scattered energy at positions around the sun's position, i.e., strong forward scattering. This is seen as an apparent increased sun size. Under cloudy conditions the diffuse insolation has a cosine-like distribution centered at the sun's position.

1-3 THE TERMINOLOGY OF SOLAR ENERGY AND COLLECTOR CLASSIFICATION

It is important at the beginning of the study of any new subject to become familiar with the terminology. This is particularly true here since the science of the utilization of solar energy is young and its vocabulary is still growing. Most words have been borrowed from physics, meteorology, and engineering and are sometimes given slightly or completely different definitions. The limited number of terms discussed below is by no means the full extent of the vocabulary needed, but it does include the basic phrases upon which the text will build. They are also illustrated to the extent possible in Fig. 1-6. A glossary of these and other terms is given in App. F.

A *solar (thermal) collector* is a device which intercepts radiant energy from the sun, converts it to thermal energy, and transfers the thermal energy to a circulating fluid.

The major constituent of any collector is the *absorber*. The *absorber* or

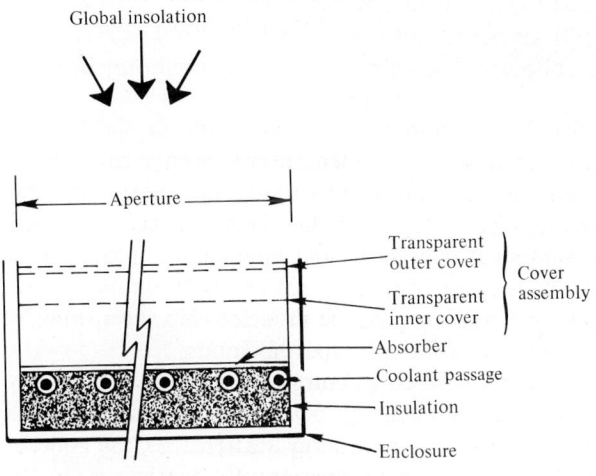

(*a*) Flat-plate collector (nonconcentrating)

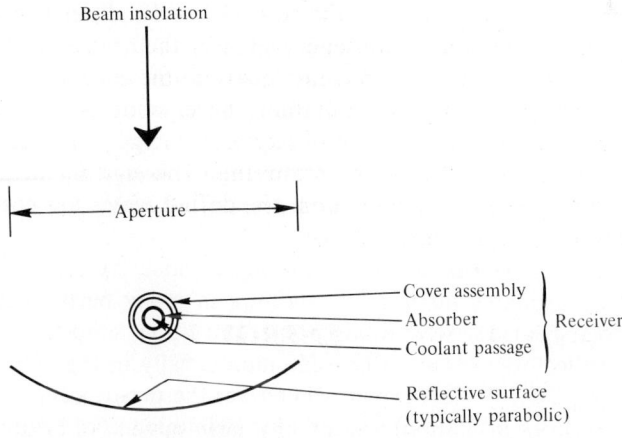

(*b*) Concentrating collector

Figure 1-6 Terminology.

absorber plate is a metal, glass, or plastic surface where the absorbed solar radiation is converted to thermal energy and transferred by thermal conduction and convection to the coolant or circulating fluid. Some collectors also operate in a batch mode. Batch heating is usually limited to small domestic systems. Also, a few collector designs depend on the direct absorption of solar energy into the fluid, e.g., the solar pond.

The *aperture* is the frontal opening of the collector which captures the sun's rays.

The *cover assembly*, *cover plate*, or simply the *cover* is the combination of glass and/or plastic sheets suspended above or around the absorber. Its two functions are to reduce the heat loss from the absorber and to protect the absorber from possible damage due to environmental exposure.

A *receiver* is an integral absorber and cover combination which is part of a larger system forming the collector. The term is commonly used only in reference to concentrating collectors.

To reduce the radiative losses from an absorber while at the same time maintaining a high solar absorptance, a *selective surface* is often applied to the absorber. These surface coatings are composed of specially formulated paints, chemical dips, or electroplated films that have the useful radiative property of high absorptance at important solar wavelengths (0.3–1.8 μm), but low emittance in the longer wavelengths where most of the radiant energy is emitted from the absorber. Hence, they act as a radiant heat trap, selectively absorbing solar energy but not reemitting significant infrared radiation.

Solar collectors are generally classified as *concentrating* or *nonconcentrating*. If the intended magnitude of solar flux (energy per unit area and time) striking the absorber has been increased above that striking the aperture, the collector is termed a "concentrating" solar energy collector. If not, it is a "nonconcentrating" collector, and the *flat-plate collector* is the most common example of the nonconcentrating designs. Concentrating collectors exhibit a wide variety of designs with the "concentration ratio" varying from as low as about two in slightly concentrating designs to several thousand in highly concentrating designs. The major advantage of concentrating systems is that, as the concentration ratio increases, it is possible to attain good collection efficiency at higher collection temperatures.

A collector can also be classified as an *air heater* or a *liquid heater*. By far the most common liquid is water. Organic liquids and pressurized water are commonly used liquids at elevated temperatures.

Flat-plate water heating collectors are by far the most widely available designs. This is due to the reasonably favorable economics in many parts of the world for year-round solar-heated water for domestic and "low temperature" commercial uses. These collectors can be purchased commercially in the United States for as little as \$40/m^2. On the less expensive end are the plastic collectors (with no cover assembly or back insulation) designed for swimming pool heating. On the expensive end, are collectors having multiple covers or using unconventional materials, special absorber coatings, and special thermal insulating techniques, where high performance is the objective.

The commercial market for flat-plate air heaters has grown steadily in the last few years but lags behind that for the liquid type because air is such a poor heat transfer fluid. Flat-plate air heaters are gaining in popularity, for space heating applications, however, because of their direct-use feature. Also, leakage, freezing, and corrosion have made maintenance a potentially expensive item for water systems. The air collectors themselves are also cheaper than comparable water heaters, although the other system costs are higher for them because of ducting, storage, and blower requirements.

Compared to flat-plate designs, there are relatively few commercially available concentrating collectors at present. Concentrating collectors normally have higher collection efficiency than flat-plate designs at temperatures above about 100°C. Concentrating collectors cost from $100/m^2 to $300/m^2 of aperture. Unlike the flat-plate type, they use only part of the available insolation since they accept only the insolation from angles within a narrow range. The higher the concentration, the narrower this acceptance angle.

Quantitative evaluation and subsequent comparison of collector performance must be made in terms of cost per unit of energy collected at a given temperature above ambient. It is very important to include the operating temperature in the analysis, for, as will be shown, the cost of solar energy is very temperature-dependent.

1-4 APPLICATIONS OF SOLAR-THERMAL ENERGY

There are very few general rules which can be used with confidence in deciding whether or not to "go solar." The key to any decision is a very careful engineering study, the necessary inputs of which include meteorological data (including insolation data), the cost and performance data for the collector system, the expected life and maintenance costs of the system, the thermal load to be supplied, and the present and projected price of the alternative fuel that the solar system is replacing.

Despite the lack of specific rules of thumb about solar's economic viability, there are some generally applicable comments: Active solar systems are usually capital-intensive and will continue to be so. To be competitive, they must exhibit sufficient savings in energy cost to offset their higher capital cost. Solar energy is most attractive for applications involving low temperatures and fairly uniform seasonal demand, locations that have good insolation and areas where conventional energy is expensive. Solar domestic water heating is often attractive, for example, particularly if electric water heating is the alternative. Space heating is currently economical where the heating season is long and the alternative fuel is expensive. On the other hand, space heating applications in climatic regions with only three or four "cold" months per year are presently difficult to justify on economic grounds.

Besides the match of load to solar availability, the use of solar energy offers the additional design challenge of a temperature match between the collector and the load. The efficiency of devices for gathering solar energy decreases with increasing temperature of collection. This decrease occurs for any given collector system because thermal losses are in approximate proportion to the temperature difference between the collecting surfaces and the surroundings. Thus, the required area of collector per unit of collected energy increases with increasing temperature.

This fact of solar life means that the unit cost of collected solar energy also tends to increase as that energy is collected at higher temperatures. However, by careful matching of particular collector types (flat plates, fixed concentrators, tracking concentrators, etc.) with the required load temperature, the cost increase with higher temperatures can be mitigated. Nevertheless, cost per unit of energy must be examined when the choice of energy source is to be made, and the cost of solar energy for medium- to high-temperature service can be high.

For low-temperature applications such as swimming pool heaters, domestic hot-water needs, and some low-temperature industrial and commercial uses such as car washes, solar heat can be provided at relatively low cost. As the required use temperature rises, the cost of solar energy becomes comparable to or higher than the cost per unit energy of conventional fuels. However, as fossil fuel prices rise because of increasing scarcity, solar heat will become competitive over an even wider range of temperatures.

Other factors also govern the possible applications of solar energy. The low energy density of sunlight means that large collector areas are required to provide high power levels, making solar energy an impractical direct energy source for certain uses such as transportation. Some solar energy might be applied to transportation through electrified rail systems or indirectly by production of synthetic fuels, but directly powered solar autos do not appear practical.

Heat engines and some other devices operate more efficiently at high temperatures. However, higher temperatures make the collectors less efficient. Costs of work output then must be optimized by choosing the temperature providing the best overall efficiency of the collector-heat engine system.

1-5 A BRIEF HISTORY OF SOLAR ACTIVITY†

An abbreviated chronology of selected solar projects is given in Table 1-1. Prior to the twentieth century (based on patent evidence) concentrating collectors were the primary interest. In fact, Tellier received the first U. S. patent for a flat-plate collector in 1890.

Until about 1850, concentrating collectors were used exclusively as furnaces in chemical and metallurgical experiments. For example, Joseph Priestley used concentrated sunlight in 1774 to liberate oxygen from mercuric oxide and thus isolated and characterized the element for the first time. With the advent of the industrial revolution, many investigators turned their attention to the development of mechanical power from solar energy. This activity continued sporadically throughout the world until World War I.

† Much of the information contained in this section is taken from Refs. 6–13.

Perhaps the most significant activity was that of John Ericsson between 1870 and 1884 in the United States and of Frank Shuman between 1906 and 1917 in the United States and Egypt. Ericsson was a Swedish-American who immigrated to the United States in 1839 at the age of 36, already a widely respected military, mechanical, and marine engineer. Besides developing a variety of solar engines, he made the most accurate measurements of insolation to date. In fact, using a series of measurements made with a portable solar calorimeter he estimated that the extraterrestrial intensity of the sun (the solar constant) was 1350 W/m^2, which is unbelievably close to the currently accepted value of 1353 W/m^2. After 15 years of work, however, his general conclusion was that solar power was too expensive and would probably be limited to remote areas of "the sun-burnt regions of our planet."

Frank Shuman utilized flat-plate collectors and a two-fluid system, water to collect the energy and ether to drive an engine, in his early work between 1906 and 1910 near Philadelphia. In 1911 the 900-m^2 array he constructed was the largest solar project to that date in the United States. (A 4700-m^2 Chilean still for desalinization of salt water was the largest in the world.) He attached flat mirrors along the north and south edges of a tiltable flat-plate array of collectors. The potential power output cost five times that of contemporary alternatives. Disappointed with the results, Shuman moved his research to Egypt, where he teamed with C.V. Boys. There they built a series of solar engines and pumps from 1912–1917. Their largest system, and the largest working solar thermal power system ever built,† was installed at Meadi in 1913. Long parabolic troughs were used to focus sunlight onto an absorbing tube. The total exposed area of the array was over 1200 m^2. The engines, removed from the Philadelphia system, developed as much as 50 kW continuously for a 5-h period. The project, however, was abandoned after a few years as the desert environment took its toll in degrading the reflective surfaces and the general maintenance costs proved to be excessive.

In the early 1900s flat-plate collectors were utilized extensively in southern California for domestic water heating. However, the solar industry came to an abrupt halt in California with the discovery in the Los Angeles area of oil and associated natural gas which provided an extremely cheap (at that time) alternative.

After a lull due to World War I and the Great Depression, interest in solar water heating returned in the late 1930s. Japan and Australia were among the leaders. The center of activity in the United States was south Florida, where many housing subdivisions were built with a solar water heater on every house. Many of the collectors in Florida, installed in the 1930s and 1940s, are still in place, but most of them have not been used since the 1950s, when again cheaper alternative sources of energy became available. However, some collectors there have seen over 40 years of operation.

---

† It will remain the largest until the mid 1980s, when a 35-MW thermal solar power tower near Barstow, CA, is scheduled to be completed.

Table 1-1 Chronology of selected solar projects

Event	Principal	Date	Remarks
Set fire to Roman fleet at Syracuse	Archimedes (Greece)	212 B.C.	Fact or myth? Proved feasible by Buffon in 1747.
Greek and Roman passive solar architecture	—	to 300 A.D.	Practice ended with Middle (Dark) Ages.
Indigenous solar passive architecture of the Americas	—	1000 A.D. to date	Cliff dwellings and pueblo structures.
Solar pump	Salomon de Caus (France)	1615 A.D.	Used air as the working fluid to pump water.
"Burning glass"	Various people (Europe)	1700s	Various refractive devices.
Solar furnace	Villette (France); others in Denmark and Persia	1750s	Polished-iron reflective solar furnaces to melt iron, copper, etc.
Powerful focusing lens	Antoine Lavoisier (France)	1770s	Disproved caloric heat theory.
Lens	Joseph Priestley (England)	1774	Discovered oxygen by heating mercuric oxide.
Photoelectric effect	Becquerel	1839	Discovery, basis of solar cells.
Distillation	Carlos Wilson (Las Salinas, Chile)	1872	4700 m² of land produced daily for 49 years over 23,000 L of fresh water for use at a nitrate mine.
Solar pump	August Mouchot (France)	1875	Tested for 6 months and then termed uneconomical.
Solar engines	John Ericsson (U.S.)	1875–1885	Built at least 8 systems, including in 1883 a 3.3 × 4.9-m parabolic collector that drove a piston with a 15-cm bore and 20-cm stroke. Claimed to deliver 2.5 kW.

6-kW solar steam engine	Adams (India)	1876	Hemispheric mirror collector with a 12-m diameter, composed of 4 × 7-cm flat mirrors.
Solar engine	Abel Pifre (France)	1880	First solar engine used in commercial venture; collector generated 1.6 kW to run a printing press.
First "flat-plate" collector	Charles Tellier (France)	1885	20-m^2 collector drove an ammonia engine.
Instrumentation	Various people notably Ångström and Abbot	1890–1940	Developed the standard instruments for measuring insolation.
Solar pump	A.G. Eneas (California and Arizona)	1901	10-m-diameter axicon (truncated conical mirror with a focal line on the axis of the cone) collector made from 1788 flat mirrors; boiler was a 4 × 0.4-m tube. Cost: $3,000. Used as an irrigation water pump.
Flat-plate solar engine	Willsie and Boyle (midwest U.S.)	1908	Several built; the largest was an engine using sulfur dioxide as the working fluid.
Solar engine	Frank Shuman (Pennsylvania)	1908	Mirrored troughs with flat-plate collector at base. Largest array had 900 m^2 of aperture.
Solar "storage"	J.A. Harrington	1910	Used a solar pump to lift water 6 m to a 19,000-L tank. Water in turn ran a turbine to produce electric lighting for a mine.
Solar pump	Shuman and Boys (Egypt)	1912	Mirrored troughs created concentration ratio of 4.5 : 1. Largest system generated 50 kW from a 1200-m^2 collector field. Used to pump irrigation water from the Nile.
Renaissance of passive solar	U.S. and Europe	early 1900s	Southern exposure with overhang.
Various solar devices	R.H. Goddard (U.S.)	1911–1930	Five patents for solar propulsion and collection in space.
Solar water heaters	California, Florida, Japan, Israel	1910–1940	Most economical application of solar to date.
1.2-kW solar engine	Charles Abbot (U.S.)	1936	Furnished power for a radio broadcast.

Table 1-1 *Continued*

Event	Principal	Date	Remarks
The demise of passive solar architecture	—	1950–1970	Due to the availability of cheap fossil-fuel-derived energy and new technology in domestic central heating and cooling systems.
Solar furnace	Felix Trombe (France)	1953	10.6-m diameter furnace made from surplus 1.5-m searchlight reflectors.
		1970	54 × 40-m stationary paraboloid with 63 heliostats (tracking plane mirrors) to achieve 1 MW at focal point (4000 K).
Silicon photovoltaic cells	Bell Laboratories (U.S.)	1954	Increased efficiency by 10× over previous solar cells.
Solar towers	Francia (Italy)	1965	50 kW; 270 heliostats of 1-m diameter.
	Atlanta, GA (U.S.)	1977	400 kW; 550 heliostats of 1.1-m diameter.
	Albuquerque, NM (U.S.)	1979	5 MW; 222 heliostats, 6 × 6 m.
	Barstow, CA (U.S.)	1983	35 MW (10 MW electric); 2000 heliostats, 6 × 6 m.

The availability of cheap fossil fuels, principally natural gas and heating oil, lessened the national interest in solar applications. Except for research projects at several universities, there was essentially no solar research or industrial activity in the 1950s and 1960s in the United States; interest in the development of solar energy slowed down worldwide in that period. The sudden impact of the Arab oil embargo in the 1970s, however, awakened many people throughout the world to the need for alternative energy sources, and the growth of solar activity was geometric in the 1970s. In the United States, encouraged somewhat by projects subsidized by the Federal government, new solar industries sprang up literally overnight and many larger established companies added "solar divisions." The major activity in the 1970s was the development of a large flat-plate collector industry in response to the consumer demands for solar water and space heating equipment. Successful higher-temperature applications and more sophisticated technologies, like solar cooling and solar power, may be just ahead. Popular interest in the subject remains high. But solar successes will come only where and when solar technology achieves an economic and/or political advantage over the alternatives.

1-6 CONCLUSIONS

The solar resource is usually viewed as low-intensity energy when compared with the conventional energy sources of today (fossil and nuclear fuels). However, when the local solar input at the ground is compared to the energy currently

Table 1-2 Examples of power use density compared to the local solar input†

	Power use density, W/m^2	Average (24-h) terrestrial solar, W/m^2
U.S. land area, 1980	0.3‡	~100
West Germany, 1980	2.0‡	~50–60
World land area, 1980	0.1‡	~60
Metropolitan areas (1965–68)		
Manhattan (N.Y.C.)	605	93
Los Angeles	21	108
Cincinnati	25	99
West Berlin	20	57
Sheffield (1952)	17	46
Moscow	123	42
Miscellaneous		
100 hp (car)	~6000	—
3 kW (house)	~15	—

† Developed from Ref. 14.
‡ Estimated from 1970 data.

generated or used locally, the solar resource provides a surprisingly good match even in today's energy-rich society, as illustrated in Table 1-2. Of course, the collection and conversion (if necessary) efficiency would reduce the amount of energy derivable from solar, but it is clear that based on land area alone solar can potentially supply a significant fraction of the energy used on the earth. If the possibility of satellite solar collection systems is accepted, then the solar resource, both in rate and longevity, becomes essentially unbounded.

REFERENCES

1. John E. Ross and Lawrence H. Aller, "The Chemical Composition of the Sun," *Science*, vol. 191, no. 4233, pp. 1223–1229, March 26, 1976.
2. M.P. Thekaekara, "Data on Incident Solar Energy," in M.P. Thekaekara (ed.), *The Energy Crisis and Energy from the Sun*, Institute of Environment Sciences, Mt. Prospect, IL, 1974.
3. Kinsell L. Coulson, *Solar and Terrestrial Radiation: Methods and Measurements*, Academic Press, New York, 1975.
4. N. Robinson (ed.), *Solar Radiation*, Elsevier Publishing Co., Amsterdam, 1966.
5. G.W. Paltridge and C.M.R. Platt, *Radiative Processes in Meteorology and Climatology*, Elsevier Publishing Co., Amsterdam, 1976.
6. F. Daniels, *Direct Use of the Sun's Energy*, Chap. 2, Ballantine Books, New York, 1953 (5th printing, 1975).
7. G. Beneveniste, "Burning Glasses: From Archimedes to Lavoisier," *Sun at Work*, vol. 1, no. 2, pp. 4–6, June 1965.
8. J.I. Yellott, "Captain John Ericsson: Pioneer in Solar Energy," *Sun at Work*, vol. 1, nos. 3 and 4, September 1956 and December 1956, and vol. 2, nos. 1 and 2, March 1957 and June 1957.
9. A.B. Meinel and M.P. Meinel, *Applied Solar Energy: An Introduction*, Chap. 1, Addison-Wesley Publishing Co., Reading, MA, 1976.
10. C.H. Pope, *Solar Heat*, Colonial Press, C.H. Simonds & Co., Boston, MA, 1903.
11. A.S.E. Ackerman, "The Utilization of Solar Energy," *Annual Report of the Board of Regents of the Smithsonian Institution*, pp. 141–165, 1915.
12. *Proceedings of World Symposium on Applied Solar Energy*, Phoenix, AZ, November 1–5, 1955.
13. Ken Butti and John Perlin, *A Golden Thread*, Van Nostrand and Reinhold Co., New York, 1980.
14. S.S. Penner and L. Icerman, *Energy, Volume 1: Demands, Resources, Impact, Technology and Policy*, pp. 298–304, Addison-Wesley Publishing Co., Inc., Reading MA, 1974.

CHAPTER
TWO

THERMAL CONVERSION SYSTEMS

The purpose of any system for converting solar energy to thermal energy is the useful application of that thermal energy.

The application can be directly in the form of heat, or indirectly by using the heat to drive a heat engine to produce useful mechanical work.

In this chapter, the general characteristics of systems for solar-thermal conversion are examined. A system model is constructed. The constraints that the second law of thermodynamics places on solar-driven heat engines are noted. Finally, some fundamental economic concepts are presented. In the following chapters, these concepts will be developed in depth and used to analyze the performance (and choose components) of practical solar-thermal systems.

2-1 LIST OF SYMBOLS

a	radiative loss parameter, $\varepsilon\sigma \bar{T}_a^4 A_e/(q_{s,\,\text{ref}} A_c)$
A	area
b	thermal loss parameter, $\bar{U} A_e \bar{T}_a/(q_{s,\,\text{ref}} A_c)$
c	fluid specific heat
F	first year fuel savings, dollars
F_R	heat removal factor, Eq. (2-4)
i	interest rate
j	annual fuel cost escalation rate
m	mass

$\dot{m}$	mass flow rate
n	amortization time, years
P	economically justifiable initial investment
q	energy per unit area per unit time
Q	energy
$\dot{Q}$	energy per unit time
t	time
T	absolute temperature
$\bar{U}$	overall heat-transfer coefficient based on absorber area
V	volume
W	work
α	absorptivity for solar energy
γ	collector flow efficiency parameter, $\delta_c \dot{m}_c c \bar{T}_a / (q_{s,\mathrm{ref}} A_c)$
δ	control function, value 0 or 1
ε	infrared emissivity
ψ	dimensionless insolation, $q_s(t)/q_{s,\mathrm{ref}}$
η	efficiency
θ	absolute temperature ratio, $T/\bar{T}_a$
ρ	density
σ	Stefan-Boltzmann constant
τ	transmissivity for solar energy

Subscripts

a	ambient
bypass	value in domestic hot-water bypass loop
c	collector or collector loop
C	Carnot cycle
d	day length
e	absorber element of collector
eff	effective value
f	fluid
in, out	inlet or outlet value
load	supplied to domestic hot-water load
L	load
min, max	minimum or maximum value
ref	reference value
rise	sunrise
s	solar
st	storage
stag	stagnation
sup	supply value
t	thermodynamic
u	usable

Superscript

$-$	average value

2-2 SOLAR-THERMAL CONVERSION SYSTEMS

Many solar-thermal conversion systems can be represented by the same block diagram which is shown in Fig. 2-1. The system consists of a solar collector and a storage device that supply thermal energy to a load. This simple schematic, when augmented by a control system, represents many solar domestic hot-water systems, domestic space-heating systems, and industrial process heat systems. If the load $\dot{Q}_L$ is taken to be the heat required for driving an absorption cooling device, then the schematic can represent solar-driven air-conditioning systems as well. If the storage device is taken to be a swimming pool and the load is taken to be the pool heat losses, then the diagram also applies to solar pool-heating systems. Finally, if the load is the input to a heat engine, then solar-driven power production and mechanically driven air-conditioning systems can be included. Other uses of solar energy could also fit into the block diagram.

Some hot-water systems require slightly different modeling, and the required modifications in the simple system model are discussed in Sec. 2-5.

Solar Collectors—Simple Models

The concepts to be developed in Chaps. 4 and 5 show that the useful energy collected by a solar collector $Q_u(t)$ is influenced by three major factors: (1) the ability of the absorbing element to absorb the available insolation $q_s(t)$ that is incident on the element after being reflected from or transmitted through other collector components; (2) the magnitude of thermal losses due to convection to the ambient air; and (3) the magnitude of thermal losses due to radiative exchange with the surroundings. These can be expressed as

$$\dot{Q}_u(t) = q_u(t)A_c = (\tau\alpha)_{\text{eff}} q_s(t) A_c - \bar{U}A_e(\bar{T}_e - T_a) - \varepsilon_{\text{eff}} \sigma A_e(\bar{T}_e^4 - T_a^4) \qquad (2\text{-}1)$$

$$\begin{pmatrix}\text{usable energy}\\\text{collected}\end{pmatrix} = \begin{pmatrix}\text{energy}\\\text{absorbed}\end{pmatrix} - \begin{pmatrix}\text{convective}\\\text{losses}\end{pmatrix} - \begin{pmatrix}\text{radiative}\\\text{losses}\end{pmatrix}$$

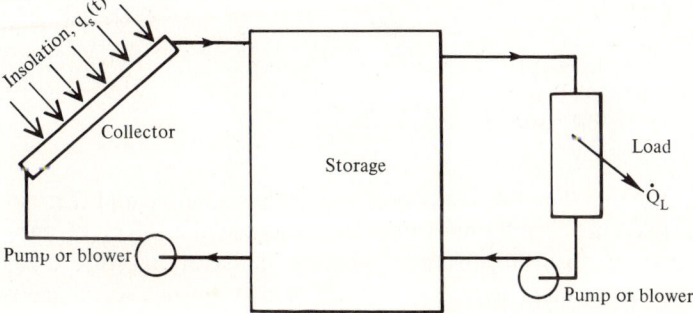

Figure 2-1 System schematic.

In this relation, $(\tau\alpha)_{\text{eff}}$ is the fraction of the insolation striking the collector aperture that is absorbed by the absorbing element of the collector, and accounts for all effects of transmission, absorption, and reflection losses because of cover plates, reflectors, lenses, or other optical elements in the collector. A typical value for $(\tau\alpha)_{\text{eff}}$ is about 0.8 or less. The term $q_s(t)A_c = \dot{Q}_s(t)$ represents the instantaneous insolation per unit of aperture area striking the collector aperture, $q_s(t)$, times the collector aperture area A_c.

Thus, the first term on the right of Eq. (2-1) is the solar energy absorbed by the absorbing element of the collector.

The second term on the right represents losses due to convection and conduction from the collector in terms of an average overall heat-transfer coefficient $\bar{U}$ times the absorbing element area A_e times the difference between the average absorber surface temperature $\bar{T}_e$ and the ambient temperature T_a. Both $\bar{T}_e$ and T_a may be time-dependent. The absorber area A_e and aperture area A_c will be equal for certain collectors such as flat plates, but different for concentrating collectors.

The final term on the right of Eq. (2-1) accounts for the exchange of infrared radiation between the collector and the surroundings, and normally is small for a collection temperature below 100°C. It is assumed here that the radiating temperature of the surroundings can be taken equal to the ambient temperature T_a. Often, the radiating temperature on clear days is lower than T_a. The value of ε_{eff} is not the absorber plate emittance, but is a more complex function that includes the effect of the cover glass. It may approach a value of unity for an unglazed collector, but normally is less than unity for better designs.

A measure of the collector performance is the ratio of the useful collected energy $\dot{Q}_u(t)$ to the available incident energy $q_s(t)A_c$. This ratio is called the *collector efficiency* η. Equation (2-1) can be rearranged to obtain:

$$\eta = \frac{\dot{Q}_u(t)}{q_s(t)A_c} = (\tau\alpha)_{\text{eff}} - \frac{\bar{U}A_e}{q_s(t)A_c}(\bar{T}_e - T_a) - \frac{\varepsilon_{\text{eff}}\sigma A_e}{q_s(t)A_c}(\bar{T}_e^4 - T_a^4) \qquad (2\text{-}2a)$$

in dimensional form or:

$$\eta = \frac{\dot{Q}_u(t)}{q_s(t)A_c} = (\tau\alpha)_{\text{eff}} - [b(\bar{\theta}_e - \theta_a) + a(\bar{\theta}_e^4 - \theta_a^4)]/\psi(t) \qquad (2\text{-}2b)$$

in dimensionless form where:

$$b = \bar{U}A_e \bar{T}_a/(q_{s,\text{ref}} A_c)\,;\; a = \varepsilon_{\text{eff}} \sigma \bar{T}_a^4 A_e/(q_{s,\text{ref}} A_c)$$

$$\bar{\theta}_e = \bar{T}_e/\bar{T}_a\,;\; \theta_a = T_a/\bar{T}_a\,;\; \psi(t) = \frac{q_s(t)}{q_{s,\text{ref}}}$$

The parameters a and b are related to the magnitude of the radiative and thermal losses relative to a reference insolation value, and include the effect of concentration in the ratio of absorbing element area to collector aperture area, A_e/A_c. The $q_{s,\text{ref}}$ is any convenient reference value of insolation such as the peak clear-day value, and $\bar{T}_a$ is the daily average ambient temperature.

Example 2-1 For a nonconcentrating collector with $(\tau\alpha)_{\text{eff}} = 0.8$, operating at a daily average ambient temperature of 285 K and using $q_{s,\text{ref}} = 0.8 \text{ kW/m}^2$, find the constants in Eq. (2-2b) for a collector with $\bar{U} = 2 \text{ W/(m}^2 \cdot \text{K)}$ and $\varepsilon_{\text{eff}} = 0.2$. Write the resulting expression for collector efficiency.

SOLUTION For this collector

$$a = \frac{\varepsilon_{\text{eff}}\,\sigma \bar{T}_a^4 (A_e/A_c)}{q_{s,\text{ref}}} = \frac{(0.2)[5.67 \times 10^{-11} \text{kW/(m}^2 \cdot \text{K}^4)](285 \text{ K})^4(1)}{0.8 \text{ kW/m}^2}$$

$$= 9.4 \times 10^{-2}$$

$$b = \frac{\bar{U}\bar{T}_a(A_e/A_c)}{q_{s,\text{ref}}} = \frac{[2 \text{ W/(m}^2 \cdot \text{K})](285 \text{ K})(1)(1 \text{ kW/1000 W})}{0.8 \text{ kW/m}^2} = 0.71$$

Thus,

$$\eta = 0.8 - [0.71(\bar{\theta}_e - \theta_a) + (9.4 \times 10^{-2})(\bar{\theta}_e^4 - \theta_a^4)]/\psi(t)$$

Note that unless $(\bar{\theta}_e^4 - \theta_a^4) \gg (\bar{\theta}_e - \theta_a)$, the radiation loss term will not be significant.

The efficiency of a collector used in low-temperature service (up to about 50°C above ambient) can be expressed in a linear relation with the difference between the average absorber temperature and the ambient temperature (see Chap. 5). This linear form results because the final (nonlinear) term in Eq. (2-2) is negligible at low temperatures. Physically, this means that the net radiative exchange between the collector and the surroundings is small compared with the convective exchange. Thus, for a low-temperature collector, Eq. (2-2b) becomes

$$\eta = (\tau\alpha)_{\text{eff}} - b(\bar{\theta}_e - \theta_a)/\psi(t) \tag{2-3}$$

To simplify both the analysis and the testing of collectors, it is desirable to put Eq. (2-3) in terms of the dimensionless fluid inlet temperature $\theta_{f,\text{in}} = T_{f,\text{in}}/\bar{T}_a$ rather than in terms of the average dimensionless surface temperature $\bar{\theta}_e$. If a relationship of the form

$$F_R\left[(\tau\alpha)_{\text{eff}} - \frac{b(\theta_{f,\text{in}} - \theta_a)}{\psi(t)}\right] = \left[(\tau\alpha)_{\text{eff}} - \frac{b(\bar{\theta}_e - \theta_a)}{\psi(t)}\right] \tag{2-4}$$

is assumed, then Eq. (2-3) becomes in dimensional form:

$$\eta = F_R\left[(\tau\alpha)_{\text{eff}} - \frac{\bar{U}A_e(T_{f,\text{in}} - T_a)}{q_s(t)A_c}\right] \tag{2-5a}$$

or in dimensionless form:

$$\eta = F_R[(\tau\alpha)_{\text{eff}} - b(\theta_{f,\text{in}} - \theta_a)/\psi(t)] \tag{2-5b}$$

The F_R relates the average collector temperature to the more easily measured fluid inlet temperature, and the collector efficiency can now be found in terms of the fluid inlet temperature. This concept of introducing an effectiveness factor that relates actual performance to a reference performance is common in engineering. A common example is the effectiveness used in heat exchanger design. Methods of determining F_R by analysis and by experiment are discussed in Chap. 5. Figure 2-2 illustrates the collector efficiency as given by Eq. (2-5).

For a well-designed flat-plate collector, $(\tau\alpha)_{\text{eff}}$ will have values in the range of 0.75 to 0.95. The value of $(\tau\alpha)_{\text{eff}}$ depends on the optical design of the collector. The values above 0.9 exist for simpler unglazed collectors such as those designed for swimming pool heating, while values of 0.75 to 0.85 apply to glazed collectors used in space heating and domestic hot-water service. For more complex collectors using multiple covers, moderate concentration, or honeycomb convection suppression which may reduce optical transmission to the absorber, the values of $(\tau\alpha)_{\text{eff}}$ will be lower.

The value of bF_R depends on the thermal design of the collector, and it determines the change of the collector efficiency with temperature. A swimming-pool collector, designed for operation near ambient temperature, could have a value of bF_R as large as 15, while an evacuated-tube collector with concentration and a selective surface will have a much smaller bF_R value. In the latter case, values of 0.5 or smaller are attainable.

In Eq. (2-5), the values of $\theta_{f,\text{in}}$, θ_a, and $\psi(t)$ all may be time-dependent, and the efficiency η would therefore also be a function of time. The $(\tau\alpha)_{\text{eff}}$, which depends on the incident angle of insolation, is also time-dependent. However, it is relatively constant over the important hours for solar collection, and is taken as constant here.

The net enthalpy gain $\dot{Q}_u(t)$ of the fluid flowing through the collector is given

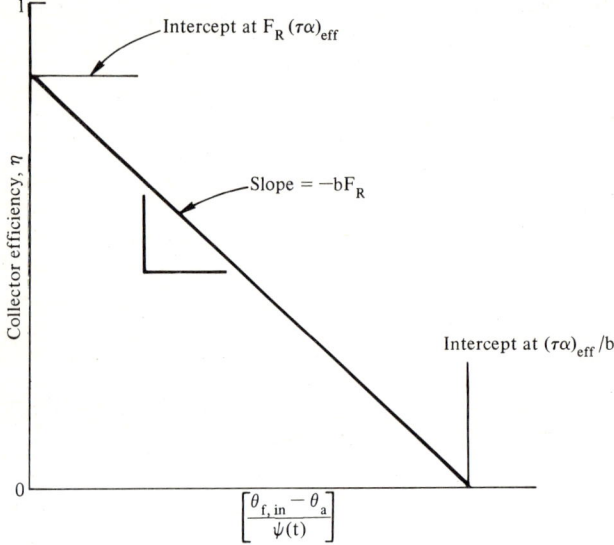

Figure 2-2 Efficiency curve for low-temperature collector.

by

$$\dot{Q}_u(t) = \delta_c \dot{m}_c c(T_{f,\text{out}} - T_{f,\text{in}}) \tag{2-6}$$

where $\dot{m}_c$ is the mass flow rate of fluid through the collector, $T_{f,\text{in}}$ and $T_{f,\text{out}}$ are the inlet and outlet temperatures of the collector fluid, respectively, and c is the specific heat capacity of the fluid. The factor δ_c is a control variable for the system model. It will have the value 1 when the collector loop pump is running, and zero when it is not.

The value of $\dot{Q}_u(t)$ is related to the insolation $q_s(t)$ through the definition of collector efficiency by

$$\dot{Q}_u(t) = \eta q_s(t) A_c \tag{2-7}$$

where A_c is the aperture area of the collector. Equations (2-6) and (2-7) neglect the effect of the heat capacities of the collector and collection fluid. This can be an important effect under transient conditions, especially during morning startup.

Substitution of Eqs. (2-5) and (2-6) into Eq. (2-7) gives, after solving for $T_{f,\text{out}}$,

$$T_{f,\text{out}} = T_{f,\text{in}}\left(1 - \frac{F_R b}{\gamma}\right) + \frac{F_R \bar{T}_a}{\gamma}[(\tau\alpha)_{\text{eff}} \psi(t) + b\theta_a] \tag{2-8}$$

where

$$\gamma = \delta_c \dot{m}_c c \bar{T}_a / q_{s,\text{ref}} A_c$$

The parameter γ is a measure of the ratio of the ability of the working fluid to remove energy from the collector to the reference solar energy. The value of γ can range from zero to very large values depending on the relative values of the mass flow rate through the collector, the collector area, and the reference insolation. To illustrate the point, as γ approaches zero ($\dot{m}_c \to 0$, near stagnation conditions), Eq. (2-8) predicts [using Eq. (2-4)] that the *no-flow* or *stagnation temperature* of the plate is given by

$$T_{\text{stag}}(t) = T_a(t) + \frac{(\tau\alpha)_{\text{eff}} \psi(t)}{b} \quad \bar{T}_a = T_a(t) + \frac{(\tau\alpha)_{\text{eff}} A_c q_s(t)}{\bar{U} A_e} \tag{2-9}$$

The average plate temperature at stagnation is thus predicted to reach a value of $(\tau\alpha)_{\text{eff}} q_s(t) A_c / \bar{U} A_e$ degrees above the ambient value of T_a. The term $(\tau\alpha)_{\text{eff}} q_s(t) A_c / \bar{U} A_e$ is simply the absorbed solar energy divided by the convective loss per degree above ambient.

If $\gamma \to \infty$ (very high flow rate per unit area and/or little insolation) then Eq. (2-8) predicts $T_{f,\text{out}} = T_{f,\text{in}}$ as expected.

Example 2-2 A flat-plate collector has values of $b = 1.2$, $(\tau\alpha)_{\text{eff}} = 0.8$, and $F_R = 1$, and is operating on a day when the daily average temperature $\bar{T}_a$ is 25°C. The mass flow rate per unit area through the collector is 0.31 kg/(m² · min), and the fluid specific heat capacity is 4.2 kJ/(kg · °C). At a particular time, the inlet temperature of the fluid to the collector is 80°C, the insolation is 800 W/m², and the ambient air is at 30°C. All collector constants are based on a reference insolation of $q_{s,\text{ref}} = 800$ W/m². What are the values of (a) the collector efficiency, (b) the outlet fluid temperature, and (c) the stagnation temperature.

SOLUTION
(a) Using Eq. (2-5),

$$\eta = F_R[(\tau\alpha)_{\text{eff}} - b(\theta_{f,\text{in}} - \theta_a)/\psi(t)]$$

$$= (1)\left[0.8 - 1.2\left(\frac{80+273}{25+273} - \frac{30+273}{25+273}\right)\frac{800}{800}\right]$$

$$= 0.8 - 1.2(1.184 - 1.017) = 0.60 \text{ or } 60\%$$

(b) Note that

$$\gamma = (\dot{m}_c/A_c)c\bar{T}_a/q_{s,\text{ref}} = \frac{(0.31)(4.2 \times 10^3)(273+25)}{(800)(60)} = 8$$

Using Eq. (2-8),

$$T_{f,\text{out}} = T_{f,\text{in}}\left(1 - \frac{F_R b}{\gamma}\right) + \frac{F_R \bar{T}_a}{\gamma}[(\tau\alpha)_{\text{eff}}\psi(t) + b\theta_a]$$

$$= (80+273)\left(1 - \frac{1.2}{8}\right) + \frac{(273+25)}{8}\left[0.8\left(\frac{800}{800}\right) + 1.2\left(\frac{273+30}{273+25}\right)\right]$$

$$= 300.2 + 37.2[0.8 + 1.2(1.016)] = 375 \text{ K or } 102°\text{C}$$

(c) Using Eq. (2-9),

$$T_{\text{stag}} = T_a + \frac{(\tau\alpha)_{\text{eff}}}{b}\psi(t)\bar{T}_a = (273+30) + \frac{0.8(800)}{1.2(800)}(273+25)$$

$$= 496 \text{ K or } 219°\text{C}$$

Storage—Simple Modeling

In the simplest case, the storage medium can be assumed to be completely mixed at uniform temperature T_{st} and then treated as a lumped thermal capacitance which does not change phase. The rate of change in the amount of energy stored is related to the temperature change of the storage medium by

$$\dot{Q}_{\text{st}} = \rho c_v V_{\text{st}}\frac{dT_{\text{st}}}{dt} = m_{\text{st}}c_v\frac{dT_{\text{st}}}{dt} \tag{2-10}$$

For solids and liquids, $c_p = c_v$. Gases are not used for practical energy storage because of the excessive volume required. Therefore, c_v will be replaced by c. The value of $\dot{Q}_{\text{st}}$ is the rate of energy collection less the rate of energy delivery to the load and the rate of energy loss from the storage tank. More sophisticated modeling is of course required if the effects of stratification (layering of liquid into zones of differing temperatures) in liquid storage are to be accounted for, or if the very important effects of temperature fronts in rock-bed storage are to be ad-

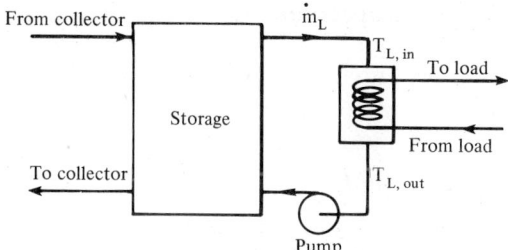

Figure 2-3 Load schematic.

equately studied (Chap. 6). However, for our purpose here of studying the behavior of a simple system, Eq. (2-10) will be adequate.

Load Models

It is paradoxical, but the apparently simplest solar systems for residential use have difficult thermal loads to model, while large-scale industrial or utility loads are relatively simple to model. The latter loads tend to be constant with time over a day or at least to have known variations with time. Residential loads, however, closely follow the vagaries of domestic hot-water demand and of the weather with all of its hourly, daily, and seasonal fluctuations. The coupling of the highly variable solar input with varying residential load requirements thus presents an interesting challenge to the designer and modeler.

One useful load model for many purposes is to assume that the load is provided through heat exchange from hot fluid taken from storage (Fig. 2-3). Then the rate at which energy is taken from storage and provided to the load, $\dot{Q}_L$, is

$$\dot{Q}_L = \delta_L \dot{m}_L c(T_{L,\text{in}} - T_{L,\text{out}}) \tag{2-11}$$

where δ_L is the control variable which, for on-off control, has a value of unity when the load pump is on and zero otherwise. For our simple model, $T_{L,\text{in}}$ is equal to the storage temperature. The value of $\dot{Q}_L$ must be provided by the modeler, either in the form of time-varying loads or as a constant value. In sophisticated residential or commercial heating and cooling models, engineering or architectural heat load programs can be used to predict heating or cooling loads based on individual building characteristics.

2-3 SYSTEM MODELS

Basic System Model

The basic system, with the equations derived from the behavior of each component, can now be put together. The system is shown schematically in Fig. 2-4, and the conditions at various points in the system are shown. The relations among the conditions are given in Table 2-1.

With these relationships established, the transient behavior of the system can be modeled numerically. A flow chart of such a model is shown in Fig. 2-5. A straightforward approach is to provide the required system configuration parameters: collector area and type as defined by values for A_c, b, F_R, and $(\tau\alpha)_{\text{eff}}$; flow loop specification in terms of $\dot{m}_c$, $\dot{m}_L$, c; storage size as either volume V_{st} or mass m_{st} and $U_{st} A_{st}$, the overall thermal conductance between the storage and the surroundings; and load and insolation data $\dot{Q}_L(t)$, $q_s(t)$, and $q_{s,\,\text{ref}}$; and the daily average ambient temperature $\bar{T}_a$ and instantaneous ambient temperature $T_a(t)$. By setting the single initial storage temperature condition of $T_{st}(t = 0)$, the transient behavior of the system can be computed by simply choosing a time interval Δt, working through the temperatures from positions 1 through 9 at $t = 0$, calculating a new value of $T_{st}(t = t + \Delta t)$ at position 9, and repeating the process.

Some important factors that have been omitted so far will soon become apparent if this is done. As with a real system, computed temperatures will soon reach values that cause practical difficulties (freezing or boiling, for example). Thus, a system control strategy must be employed. The values δ_L and δ_c can be set to provide flow or no flow in the load and collector loops, respectively. Temperature is the most easily measured variable, so on-off control is usually done in a real system by monitoring collector outlet fluid temperature and storage temperature at the coldest point in the storage tank. The set of criteria used in the program described here is

- If $T_{\text{stag}}(t) - T_{f,\,\text{in}}(t) \geq 6°C$, and

  $T_{f,\,\text{out}}(t) - T_{f,\,\text{in}}(t) \geq 2°C$, set $\delta_c = 1$.

- If $T_{f,\,\text{out}}(t) - T_{f,\,\text{in}}(t) < 2°C$, or

  $T_{st}(t) > T_{st,\,\text{max}}$, set $\delta_c = 0$.

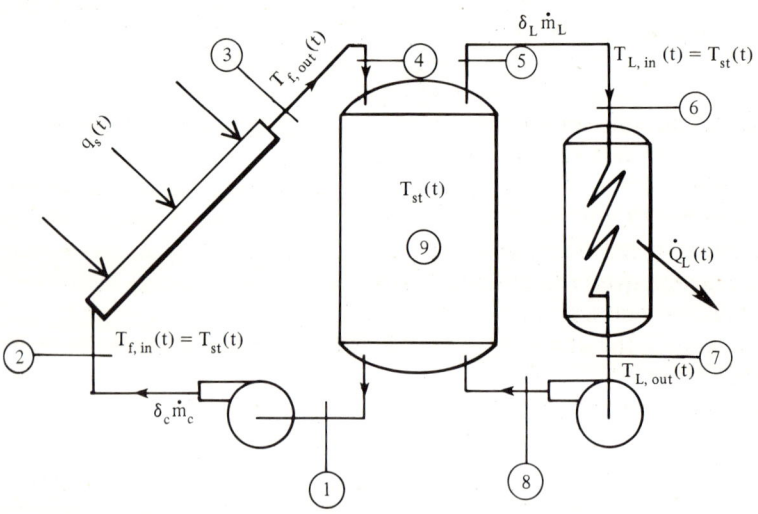

Figure 2-4 Schematic of solar-thermal system.

Table 2-1 System conditions and relations

Position	Temperature	Flow rate	Relation or comment
1	$T_{st}(t)$	$\delta_c \dot{m}_c$	—
2	$T_{f,\,in}(t)$	$\delta_c \dot{m}_c$	Could include blower or pump work effect on T_f, heat losses in piping: otherwise $T_{f,\,in}(t) = T_{st}(t)$
3	$T_{f,\,out}(t)$	$\delta_c \dot{m}_c$	Eq. (2-8)
4	$T_{f,\,out}(t)$	$\delta_c \dot{m}_c$	Again, could modify by including heat losses in pipe
5	$T_{st}(t)$	$\delta_L \dot{m}_L$	—
6	$T_{L,\,in}(t)$	$\delta_L \dot{m}_L$	Could include piping heat losses: otherwise $T_{L,\,in}(t) = T_{st}(t)$
7	$T_{L,\,out}(t)$	$\delta_L \dot{m}_L$	Eq. (2-11)
8	$T_{L,\,out}(t)$	$\delta_L \dot{m}_L$	Could include blower or pump work and piping losses
9	$T_{st}(t)$	—	Eq. (2-10) using $\dot{Q}_{st}(t) = \dot{Q}_c(t) - \dot{Q}_L(t) - \bar{U}_{st} A_{st}[T_{st}(t) - T_a(t)]$

Similarly, the load control functions of

- If $\dot{Q}_L(t)/\dot{m}_L c < 2°C$, or $T_{st}(t) < T_{st,\,min}$, set $\delta_L = 0$.
- If $\dot{Q}_L(t)/\dot{m}_L c > 2°C$, set $\delta_L = 1$.

These control functions turn on the collector loop pump or blower whenever the collector stagnation temperature exceeds the storage temperature by 6°C or more and the collector fluid outlet temperature exceeds the storage (collector inlet) temperature by 2°C or more, and turns it off otherwise. Similarly, if the storage temperature exceeds some predetermined maximum, the pump or blower is turned off to avoid overpressures (or boiling in a liquid system). A further check on the collector temperature could be made so that, if $T_{f,\,out}(t)$ drops below some present minimum, the pump is turned on to circulate warm storage fluid to avoid freezing. This is not done in the simple model.

On the load side, the circulating pump or blower is turned on whenever the load is sufficient to cause a fluid temperature drop of more than 2°C, and turned off otherwise. In addition, circulation is stopped whenever the storage temperature drops below some preset minimum $T_{st,\,min}$. For residential heating, for example, the load might only be satisfied as long as the temperature provided to

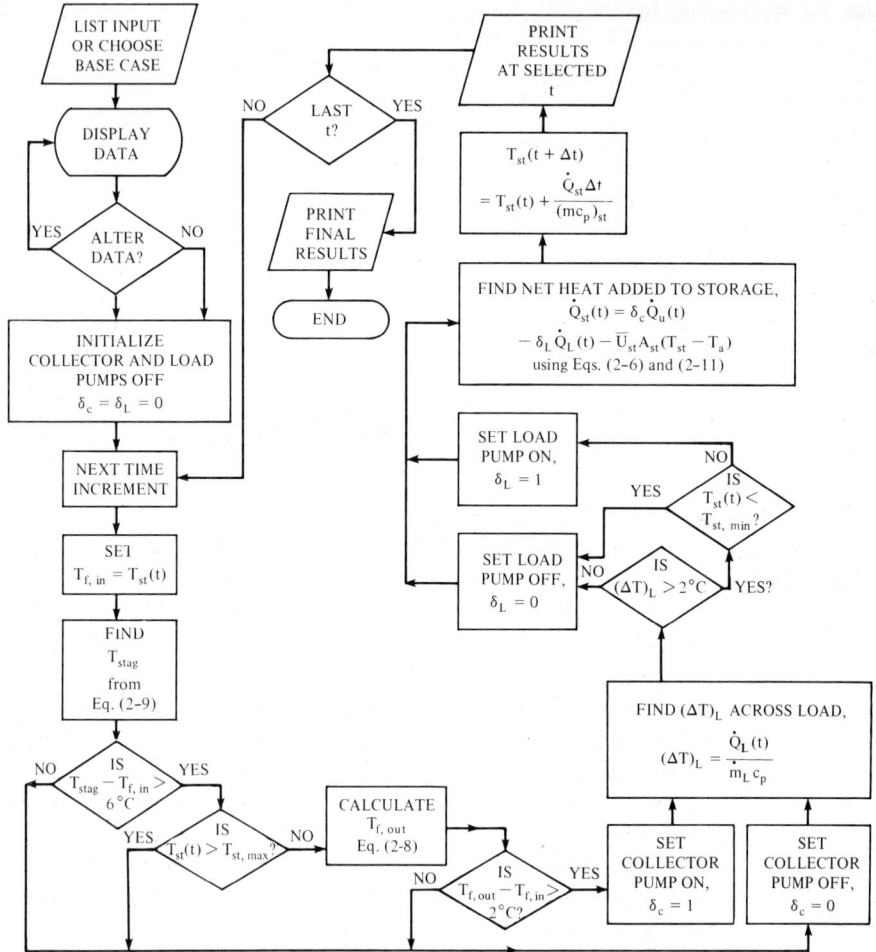

Figure 2-5 Computer flow chart for basic system.

the load heat exchanger from storage exceeded 50°C. Further discussion on control systems and strategies is given in Chap. 8.

Results

The system model outlined just above has been programmed in both FORTRAN and BASIC for examining the behavior of solar-thermal systems. The program is called SOLSIM and both versions are provided in App. E. It is instructive to examine the results for a particular problem.

A base case is chosen for study to illustrate some fundamental behavior patterns of solar-thermal conversion systems. For this case, a constant thermal load of 4.2 kW is to be supplied at a temperature of 300 K or above. The collector efficiency is related to the collector temperature by

$$\eta = 0.95\left[0.8 - \frac{1.2(\theta_{f,\text{in}} - \theta_a)}{\psi(t)}\right]$$

The insolation between sunrise at time t_{rise} and sunset is given by

$$\psi(t) = \frac{q_s(t)}{q_{s,\text{max}}} = \sin\frac{\pi(t - t_{\text{rise}})}{t_d}$$

where t_d is the maximum sunshine time in minutes and $q_{s,\text{max}} = q_{s,\text{ref}}$ is the insolation available on the collector at solar noon. With these values, $\psi(t)$ is never negative.

The ambient temperature is modeled as

$$T_a = \bar{T}_a + (\Delta T)_a \sin\frac{2\pi[t - (t_{\text{rise}} + 360)]}{1440}$$

where $(\Delta T)_a$ is the temperature swing around the average, and the phase shift in the sine term is chosen to cause the minimum daily temperature to occur at sunrise and the maximum to occur at 12 h after sunrise. (In the computer simulation, $t = t_{\text{rise}}$ occurs at sunrise, taken to be $t_d/2$ before solar noon.)

The other model inputs are shown in Table 2-2, and the insolation incident on the collector is shown in Fig. 2-6. The storage size and the flow rates through the collector and load loop were chosen to give behavior typical of good designs, i.e., about one day of storage, an average 5 K temperature rise through the collector, and a 5 K temperature drop through the load heat exchanger.

Figure 2-6 also shows the storage temperature as it increases from its initial value of 300 K to 340 K and then begins to decline as the load heat removal rate from storage exceeds the heat addition rate from the collectors. The maximum positive slope in the curve of storage temperature versus time occurs at solar noon, the time of maximum rate of heat addition (since the heat extraction rate is constant in this example).

In Fig. 2-7, the behavior of the storage temperature is shown over a 3-day period for two cases. The curve labeled 30 m² is for the system shown in Fig. 2-6, and the other curve is for a system with the collector area doubled to 60 m² but with all other characteristics remaining the same.

The base case storage temperature is seen to return to its initial value each night, indicating that the collector area is too small to supply 100 percent of the load. The energy collected is less than the load on a daily basis. If more realistic insolation is considered, as it is in Chap. 3, then even more collector area will be required.

When the collector area is doubled, enough solar energy is collected to drive the storage to the limiting maximum storage temperature of 373 K. The control system then turns off the collector pump, and the peaks on the storage temperature curves are chopped. (Actually, the chopped peaks are a series of small jagged peaks oscillating around the maximum of 373 K. This is caused

Table 2-2 Inputs to "base case" for SOLSIM

Input Data in SI Units

Time of simulation is 2880 min
Calculation interval is 15 min
Frequency of output is 4 intervals
$q_s(t) = 800 \sin(\pi t/600)$ W/m^2
$q_{s,\text{ref}} = 800$ W/m^2
Daily mean temp. is 285 K
$(\Delta \bar{T})_a$ is 10 K
$\tau\alpha_{\text{eff}}$ is 0.8
Value of b is 1.2
Value of F_R is 0.95
Collector mass flow is 20 kg/min
Specific heat of fluid is 4183 J/kg · °C
Collector area is 30 m^2
Lumped liquid storage
Storage capacity is 5,000,000 J/°C
U_{st} value of storage tank is 0.28 W/m^2 · °C
Length/diameter of storage is 3†
Initial storage temperature is 300 K
Minimum usable temperature is 300 K
Maximum safe temperature is 373 K
Load is constant at 250,000 J/min
Load mass flow is 10 kg/min
Expected system life is 10 years‡
Interest rate is 12 %‡
Fuel escalation rate is 15 %‡
Fuel cost is now 0.05 $/kWh‡

† Used to compute tank area for heat losses.
‡ See Sec. 2-4.

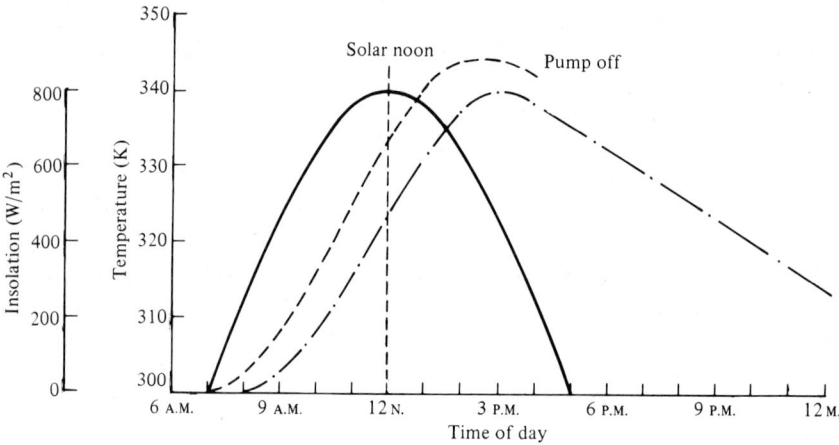

Figure 2-6 Insolation and storage temperature—base case. —— Insolation; —·— storage temperature; – – –collector outlet temperature.

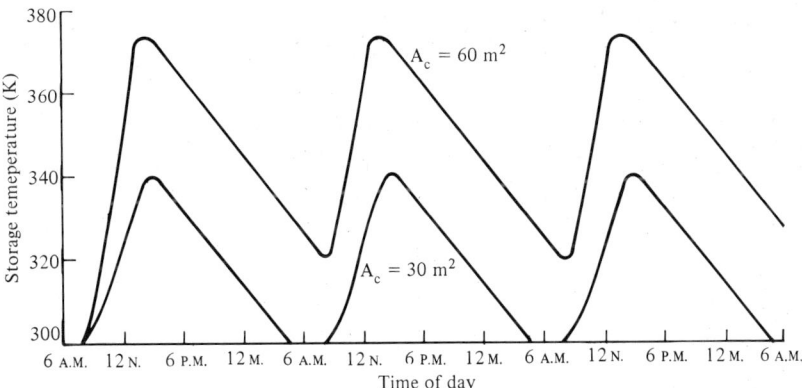

Figure 2-7 Storage temperature versus time.

by the on-off pump control that senses the storage temperature as being above or below 373 K and turns the pump off or on accordingly. These fluctuations are not plotted, since their exact shape depends to some extent on the time interval Δt chosen in the simulation.)

The shape of the curves for the larger collector area also becomes repetitive after the first day because the system reaches the peak storage temperature. Some available solar energy is not collected with this system because the storage reaches its maximum temperature over part of the collection period. The system is thus less efficient than the base system, not only because it is shut off for a portion of the day but also because the average collector and storage temperatures are higher. The latter two effects cause increased thermal losses. Using more storage capacity would improve the relative performance of this system. Further, because the total flow rate through the collector, $\dot{m}_c$, was not changed when the collector area was doubled, the temperature increase through the collector is roughly doubled so that the collector efficiency is somewhat reduced because of the higher average collector temperature. As a result, there is a shorter period of time when the storage "bottoms out," and there is a greater fraction of load provided by solar energy.

Figure 2-8 shows the effect of doubling the thermal capacity of the storage system. The solid line is a replotting of the base case. The dashed line shows the effect of doubling the storage with no change in collector area. The peak temperature is reduced as are the slopes of storage temperature versus time for both the heat addition and removal periods. The efficiency for the system with doubled storage capacity will be slightly greater than for the base system, since the collector will operate at a lower inlet temperature.

Doubling both the storage capacity and the collector area results in the upper (dot-dashed) curve. Because of the larger storage and collector, one observes that a condition of exactly repeating daily storage temperature profiles takes three or more days. If the simulation had been extended beyond 3 days, the maximum storage temperature would have been reached as observed in Fig. 2-7.

	Collector area, m²	Storage capacity, kJ/K	Percentage of load supplied	Collection efficiency over three-day period
——	30	5000	85.1	56.5
-----	30	10000	93.8	62.7
—·—	60	10000	97.9	46.4

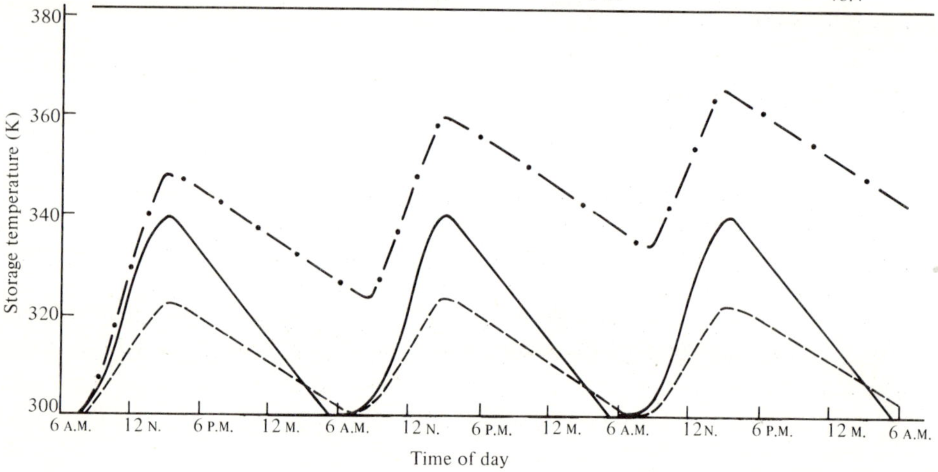

Figure 2-8 Effect of storage capacity on system behavior.

In such a case, the system is oversized, since the storage is driven to temperatures greater than needed. This results in less efficient collector operation as well as discard of part of the potential energy to be collected.

A secondary effect of these changes in storage capacity and collector area is to change the rate of heat losses to the surrounding environment. Whether the losses will increase or decrease with an increase in storage volume is not obvious, since the area for heat loss will increase but the average storage temperature will decrease.

Domestic Hot-Water Systems

To model a simple domestic hot-water (DHW) system, some changes must be made to the simple model shown in Fig 2-1. The collector and storage can be modeled as before, but as shown in Fig. 2-9, the load portion of the model must be changed. In the DHW case, it is generally required that water be supplied to the user at a fixed temperature T_{load} [usually about 330 K (140°F) for typical use in dishwashers and clothes washers, less for showers and baths] and at a prescribed flow rate $\dot{m}_{load}$. This results in a prescribed energy rate $\dot{Q}_{load}(t)$. The total mass flow to load, $\dot{m}_{load}$, can thus be found from

$$\dot{Q}_{load}(t) = \dot{m}_{load}(t)c(T_{load} - T_{sup}) \qquad (2\text{-}12)$$

However, water from the storage tank may be available at any given time at a temperature $T_{st}(t)$ above the required load temperature T_{load}. In that case, some

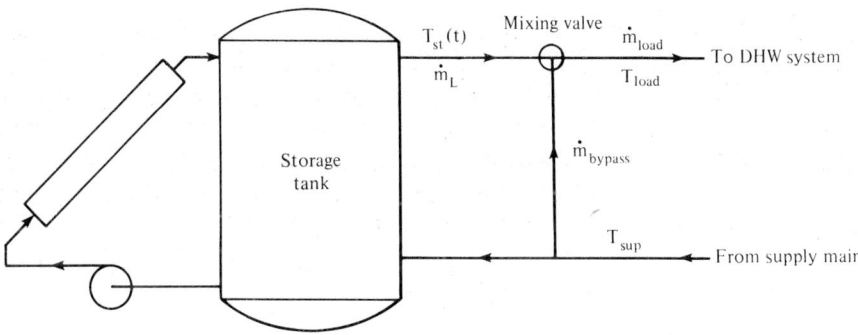

Figure 2-9 Model of simple domestic hot-water system.

unheated water $\dot{m}_{bypass}$ from the supply source at temperature T_{sup} is mixed with water $\dot{m}_L$ at $T_{st}(t)$ by using the mixing valve. An energy balance around the mixing valve gives

$$\dot{m}_L c T_{st} + \dot{m}_{bypass} c T_{sup} = \dot{m}_{load} c T_{load} \tag{2-13}$$

and a mass balance gives

$$\dot{m}_L + \dot{m}_{bypass} = \dot{m}_{load} \tag{2-14}$$

Combining Eqs. (2-12), (2-13), and (2-14) to eliminate $\dot{m}_{bypass}$ and $\dot{m}_{load}$ yields

$$\dot{m}_L(t) = \frac{\dot{Q}_{load}(t)}{c[T_{st}(t) - T_{sup}]} \tag{2-15}$$

Thus, the mass flow rate of fluid taken from storage at any time, $\dot{m}_L(t)$, can be calculated. The rate at which energy is removed from storage is then

$$\dot{Q}_L(t) = \dot{m}_L(t) c [T_{st}(t) - T_{sup}] \tag{2-16}$$

It is seen that, unlike the model shown in Fig. 2-1, the value $\dot{m}_L(t)$ varies with the storage temperature and not simply with whether the load pump is on or off. If the storage temperature falls below T_{load}, then only enough auxiliary heat must be supplied to raise T_{st} to T_{load}.

The program SOLSIM offers the user the option of modeling a DHW system using the approach outlined. In addition to the usual inputs, the water supply temperature T_{sup} must be specified.

High-Temperature Solar-Thermal Systems

Collector modeling In systems where energy to the load must be supplied at high temperatures, the model proposed in Fig. 2-1 is still valid. However, the nonlinear radiative loss term in Eq. (2-2) cannot be neglected. Solving for the collector outlet temperature by substituting Eqs. (2-2) and (2-6) into Eq. (2-7) gives, using Eq. (2-4),

$$T_{f,\text{out}} = T_{f,\text{in}}\left(1 - \frac{bF_R}{\gamma}\right) + \frac{F_R \bar{T}_a}{\gamma}[(\tau\alpha)_{\text{eff}}\psi(t) + b\theta_a] - \frac{\bar{T}_a a}{\gamma}(\bar{\theta}_e^4 - \theta_a^4) \quad (2\text{-}17)$$

This equation as it stands requires an iterative solution because $\bar{\theta}_e$, the nondimensional average absorber temperature, depends on $T_{f,\text{out}}$. Because the term involving $\bar{\theta}_e$ is usually small except near stagnation temperatures, it is usually acceptable to replace $\bar{\theta}_e$ with $\theta_{f,\text{in}}$, the nondimensional temperature of the inlet fluid to the collector. For a collector operating at 300 K and with a 5 K temperature rise, the approximation of $\bar{\theta}_e$ is in error by 3.4 percent, and the error decreases at higher temperatures. The error in the radiative loss caused by this approximation also depends on the value of θ_a. The collector outlet temperature is then given by

$$T_{f,\text{out}} = T_{f,\text{in}}\left(1 - \frac{bF_R}{\gamma}\right) + \frac{F_R \bar{T}_a}{\gamma}[(\tau\alpha)_{\text{eff}}\psi(t) + b\theta_a] - \frac{\bar{T}_a a}{\gamma}(\theta_{f,\text{in}}^4 - \theta_a^4) \quad (2\text{-}18)$$

The stagnation temperature can be found by setting $\eta = 0$ in Eq. (2-2). However, the equation remains nonlinear in form:

$$T_{\text{stag}} = \theta_a \bar{T}_a + \frac{(\tau\alpha)_{\text{eff}} \bar{T}_a q_s(t)}{b q_{s,\text{ref}}} - \frac{a}{b}(\theta_{\text{stag}}^4 - \theta_a^4)\bar{T}_a$$

$$= T_a + \frac{(\tau\alpha)_{\text{eff}} q_s(t) A_c}{\bar{U} A_e} - \frac{\varepsilon_{\text{eff}}\sigma}{\bar{U}}(T_{\text{stag}}^4 - T_a^4) \quad (2\text{-}19)$$

and again must be solved iteratively. The physical interpretation of Eq. (2-19) is similar to that for Eq. (2-9), except that the additional radiative loss term lowers the predicted stagnation temperature from that computed from Eq. (2-9).

Simple storage modeling The storage system is modeled here as it was done for the low-temperature system.

Load modeling The higher-temperature thermal energy from the collectors can be used directly as heat as in the low-temperature system. However, at increased temperatures, power production by use of a heat engine becomes feasible. To maximize the work output of a given heat engine, it is desirable to operate at or near the temperature that maximizes the product of collector efficiency (which decreases with increasing temperature) and heat engine efficiency (which increases with increasing temperature). This temperature may be tied to the rate of work output or power of the heat engine, $\dot{W}_C$, through the thermodynamic efficiency η_t, which for the special case of the Carnot cycle operating between the load inlet temperature T_L and the ambient temperature T_a, is

$$\eta_t \equiv \frac{\dot{W}_C}{\dot{Q}_L} = 1 - \frac{T_a}{T_L} \equiv \eta_C \quad (2\text{-}20)$$

or

$$\dot{Q}_L(t) = \frac{\dot{W}_C(t)}{1 - T_a(t)/T_L(t)} \quad (2\text{-}21)$$

If the value of $T_L(t)$ is assumed to be at the inlet temperature to the load heat exchanger shown in Fig. 2-3 and the desired work output of the heat engine is assumed constant, then

$$\dot{Q}_L(t) = \frac{\dot{W}_C}{1 - T_a(t)/T_{L,\,\text{in}}(t)} \tag{2-22}$$

System modeling The system model shown in Figs. 2-4 and 2-5 applies equally well to the solar heat engine system if Eq. (2-18) is used for the collector and Eq. (2-21) or some other thermal load function is used to compute the load. The control strategy is also valid, although it is not optimum for a real situation.

Heat Engine Systems

The United States is presently embarked on a program to demonstrate the feasibility of large-scale conversion of solar-thermal energy to electrical power (100 MW electrical from about 250 MW of collected solar thermal input). The leading concept is currently the "solar tower," which uses a large array of individually tracking mirrors (heliostats) to reflect beam insolation onto a boiler at a central location (Fig. 2-10). The concept is described further in Chap. 4. References 1 through 3 give details of the system.

Characteristic	100-MWe module	Pilot plant, 10 MWe
Field dimensions	1720 × 1896 m	527 × 527 m
Number of heliostats	29300	2290
Heliostat size	6.1 m	6.1 m
Tower height	305 m	95 m
Peak receiver temperature	516°C	516°C
Storage temperature	316°C	302°C
Turbine inlet temperature	510°C	510°C
Turbine throttle pressure	12.5 × 10³ kPa	10.1 × 10³ kPa
Net plant efficiency		
Direct	33%	30%
From storage	27%	23.6%

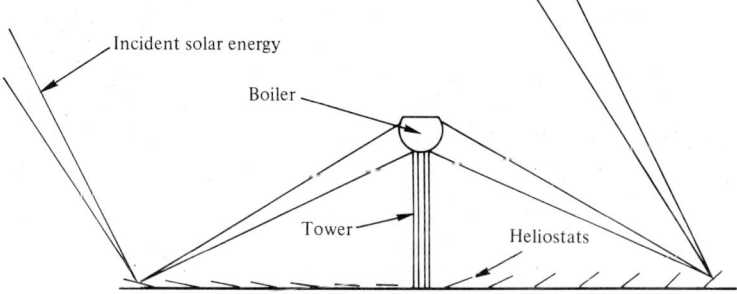

Figure 2-10 Characteristics of commercial solar power plant.

At present, a 5-MW (thermal) tower-heliostat test facility in Albuquerque, NM, is being used to test components for a 10-MW solar-electrical demonstration tower system now under construction near Barstow, CA. If the 10-MWe system proves successful, full-scale systems will be considered.

As an example of how these highly concentrating systems affect collector performance, consider the 5-MW test facility. In early construction stages it had 72 multifaceted heliostats of 36-m^2 reflector area each that provided an effective collection area A_c of about 2600 m^2. The absorber (or receiver) target area A_e was roughly 2.5 m^2. Thus, the thermal loss parameter b for any set of conditions is very small. (For $A_e/A_c = 1/900$ to account for mirror blocking and shading, $\bar{U} = 0.01$ kW/(m$^2 \cdot$ K), $\bar{T}_a = 300$ K, $q_{s,\,\text{ref}} = 0.83$ kW/m^2, gives $b = 4.04 \times 10^{-3}$.) Similarly, the radiative parameter a becomes

$$a = \frac{\varepsilon \sigma T_a^4}{q_{s,\,\text{ref}}} \left(\frac{A_e}{A_c}\right) = \frac{(1)\{5.67 \times 10^{-11} \text{ [kW/(m}^2 \cdot \text{K}^4)]\}(300 \text{ K})^4}{0.83 \text{ kW/m}^2} \cdot \left(\frac{1}{900}\right)$$

$$= 6.2 \times 10^{-4}$$

For this collector, Eq. (2-2) gives, for $(\tau\alpha)_{\text{eff}} = 0.90$,

$$\eta = 0.90 - \frac{[(4.04 \times 10^{-3})(\bar{\theta}_e - \theta_a) + (6.2 \times 10^{-4})(\bar{\theta}_e^4 - \theta_a^4)]}{\psi(t)}$$

Note that this efficiency is based on beam insolation alone. The small values of a and b assure that this collector will have high efficiency even at large temperatures. For example, at $T_e = 1200$ K, $T_a = \bar{T}_a = 300$ K, and $q_s(t) = q_{s,\,\text{ref}} = 0.83$ kW/m^2,

$$\eta = 0.90 - \left\{(4.04 \times 10^{-3})\left(\frac{1200}{300} - 1\right)\right.$$
$$\left. + (6.2 \times 10^{-4}) \cdot \left[\left(\frac{1200}{300}\right)^4 - 1\right]\right\} \frac{0.83}{0.83} = 0.73$$

It is seen from this admittedly rough analysis that the great advantage of the tower-heliostat system is its very high concentration ratio, which leads to radiative and convective losses that are extremely small in comparison with the collected energy (which causes a and b to be very small). These small losses in turn allow good collection efficiency at high temperatures. The stagnation temperature of this collector (no useful energy to the fluid loop so that $\eta = 0$) can be calculated from Eq. (2-19):

$$T_{\text{stag}} = 300 + \frac{(0.9)(0.83 \text{ kW/m}^2)(900)}{0.01 \text{ kW/(m}^2 \cdot \text{K})}$$

$$- \frac{(1)[5.67 \times 10^{-11} \text{ kW/(m}^2 \cdot \text{K}^4)][T_{\text{stag}}^4 - (300)^4]}{0.01 \text{ kW/(m}^2 \cdot \text{K})}$$

$$= 300 + (7.05 \times 10^4) - (5.67 \times 10^{-9})[T_{\text{stag}}^4 - (81 \times 10^8)]$$

Solving by trial and error for the result gives $T_{\text{stag}} = 1868$ K, or 2900°F!

The SOLSIM computer model will now be used to examine the behavior of a conceptual 100-MWe solar power station that approximates the behavior of contemplated designs. Let the working fluid have the properties of liquid water, transferring its energy to a storage system of large capacity. We will neglect the actual heat-transfer problem in the high-temperature boiler, where the working fluid (usually steam) undergoes a change of phase. We will try for a 100-MWe output over a period of 10 h on each clear day, using a Carnot engine to provide the shaft work to the generator.

A typical system of this type might have approximately 36,000 heliostats, each 4 × 4 m² for a total collector mirror area of 576,000 m². About 27,000 kg/min (1000 lb$_m$/s) of working fluid will pass through the absorber on the tower, which might have an area for energy collection of about 500 m². After accounting for mirror shading and blocking and other optical effects, the concentration ratio of the mirror–field–absorber system will be about 1000. Note that the system is defocused by design to keep temperatures below material limits. The absorber efficiency can be approximated by Eq. (2-2):

$$\eta = (\tau\alpha)_{\text{eff}} - \frac{b(\bar{\theta}_e - \theta_a) + a(\bar{\theta}_e^4 - \theta_a^4)}{\psi(t)}$$

where $(\tau\alpha)_{\text{eff}}$ in this case accounts for reflective losses at the mirrors and the absorptance of the absorber surface. A typical value is 0.9. Values of a and b for the absorbers used with these systems are quite small, as noted before, and might be near $a = 0.0006$, $b = 0.004$.

If we design for a 750 K maximum storage temperature, and a minimum turbine inlet temperature of 600 K with exhaust to the daily average ambient temperature of 285 K, then the storage capacity necessary to provide 100 MWe for 4 h (to provide operation on stored energy past an assumed 6-h period of storage charging and operation from collected solar energy), assuming a maximum of about a 60 percent Carnot heat engine efficiency, is

$$(mc)_{\text{st}} = \frac{\Delta t}{\Delta T} \cdot \frac{W_C}{\eta_C} = \frac{(4)(3600)}{(750 - 600)} \cdot \frac{100 \times 10^6 \text{ W}}{0.6} = 16 \text{ GJ/K}$$

A heat balance across the load heat exchanger gives a reasonable mass flow rate [if $c \sim 4.2$ kJ/(kg · K)] of $\dot{m}_L = 10{,}000$ kg/min for a fluid loop temperature drop of 200 K.

These are "first-try" design numbers, and must be varied to obtain good system behavior. For the computer program, the input will now be as shown in Table 2-3.

The computer model predicts the mean storage temperature, collector efficiency, and Carnot engine efficiency to be as shown in Fig. 2-11. Note that the Carnot engine can be run for about 8 h on the startup day, and then for about 11 h/day by the third day of clear-day operation. However, the storage system has reached the maximum allowable value of 750 K early in the second and third days, so that we should either reduce collector area, increase storage capacity, or

Table 2-3 Input values to model 100-MWe solar power system†

Input data in SI units	Input Data in SI Units
Time of simulation is 4320 min	Storage capacity is 1.6×10^{10} J/°C
Calculation interval is 15 min	$\bar{U}_{st}$ value of storage tank is 0.28 W/m² · °C
Frequency of output is 4 intervals	Length/diameter of storage is 3
$q_s(t) = 800 \sin[\pi(t - 420)/600]$ W/m²	Initial storage temperature is 300 K
$q_{s,\,ref}$ is 800 W/m²	Minimum usable temperature is 600 K
Daily mean temperature is 285 K	Maximum safe temperature is 750 K
$(\Delta T)_a$ swing is 10 K	Load is a Carnot cycle to deliver 1×10^8 W
$(\tau\alpha)_{eff}$ is 0.9	Value of a is 6.2×10^{-4}
Value of b is 0.00404	Load mass flow is 10000 kg/min
Collector mass flow is 27,000 kg/min	Expected system life is 10 years
Specific heat of fluid is 4183 J/kg · °C	Interest rate is 12 %
Collector area is 576,000 m²	Fuel escalation rate is 15 %
Lumped liquid storage	Fuel cost is now 0.05 \$/kWh

do both to better optimize the system and avoid discarding energy that could otherwise be collected.

The system as designed does provide 100 MW of shaft work from the Carnot engine; however, for an actual thermodynamic cycle (Rankine for steam cycles or Brayton for gas cycles) the engine efficiency will be much lower, and requirements for solar heat from the tower correspondingly higher.

Note that an actual power-tower system would be designed to feed energy directly from the tower to the heat engine to take advantage of the higher enthalpy during times of available sunlight. Only a fraction of the tower energy would be used to charge the storage system. Mixing the tower energy into storage, and then drawing energy from storage to the heat engine, introduces a major inefficiency into the system modeled here.

2-4 ECONOMIC FACTORS

In determining the economical feasibility of any system alternative, the designer must know the present cost of money (interest rate) and desired payback period for the equipment as well as the cost of equipment, fuels, labor, maintenance, etc. In most cases these figures are available. However, when evaluating a technically new product such as a solar conversion system, cost data may be neither readily available nor accurate. In performing the analysis, two questions of importance have to do with what cost per unit area of solar collector can be economically justified and whether collectors are available at that price. These factors may govern all others because solar collector costs are often the single most expensive item in the system. If the present trend continues, and there is much evidence that in an era of fuel shortages it will, then the cost of conventional fuels may continue to escalate at 10 to 15 percent/year and perhaps at even higher rates. General curves have been developed which show the economically justifiable initial invest-

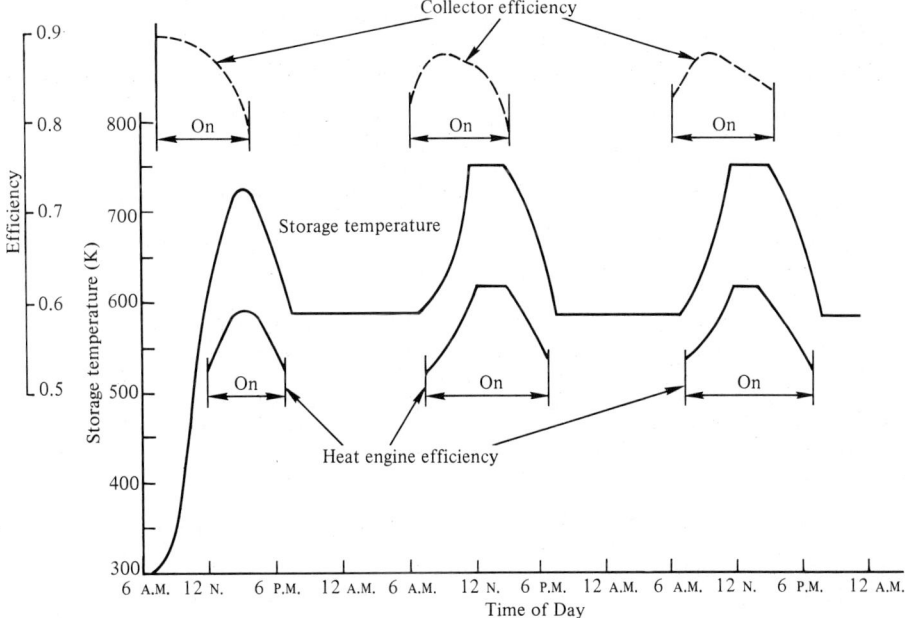

Figure 2-11 Performance of a model 100-MWe solar power system.

ment P versus the life of the solar system for various escalation rates. These curves, shown in Fig. 2-12, are based on the method given in Ref. 4 for heating systems with increasing fuel costs. The equation used is

$$P/F = \frac{(1 + i_{\text{eff}})^n - 1}{i_{\text{eff}}(1 + i_{\text{eff}})^n} \qquad (2\text{-}23)$$

where $i_{\text{eff}} = (i - j)/(1 + j)$, F is the fuel cost saved in the first year, and n is the amortization period (or sometimes useful life) of the system. The i_{eff} is the effective interest rate, which depends on the annual interest rate i and the annual rate of fuel cost escalation, j.

These curves may be used in conjunction with the annual fuel savings calculated from SOLSIM to determine an economically justifiable system based on a rate of fuel cost escalation.

The SOLSIM program includes the calculation of the value for P if the operator inputs the values for i, j, n, and the cost per kilowatt of the energy replaced by solar. This calculation is based on finding F, the fuel savings in the first year by the solar system by using the fraction of required load provided by the solar equipment.

Example 2-3 A solar energy system saves $600 in fuel costs the first year and has a system life of 10 years. If the interest paid on the capital cost of the

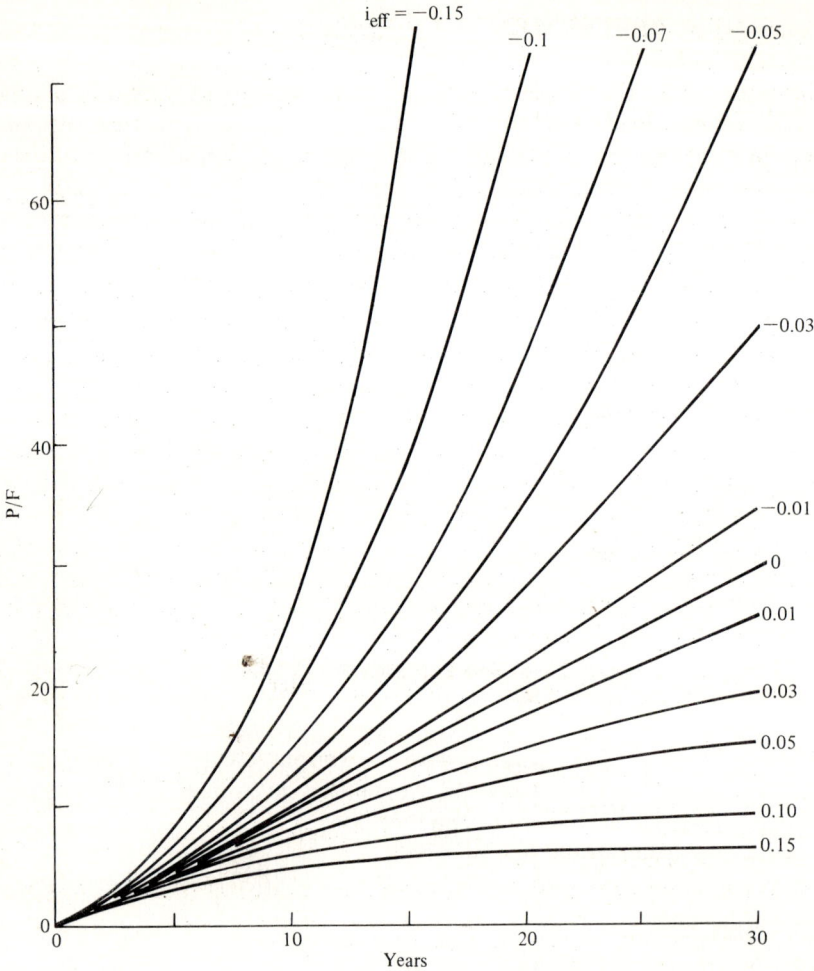

Figure 2-12 Economically justifiable investment versus system life for various effective interest rates. F = initial annual savings from solar system; P = present justifiable capital expenditure. $i_{\text{eff}} = (i - j)/(1 + j)$ where i = annual interest rate and j = rate of fuel price escalation.

system is 12 percent/year, and fuel escalates in cost at 15 percent/year, then determine the maximum economically justifiable investment.

SOLUTION

$$i_{\text{eff}} = \frac{0.12 - 0.15}{1.15} = -0.026$$

and, from Fig. 2-12 or Eq. (2-23), $P/F = 11.6$ or $P = \$6960$. Thus, for these conditions, the buyer could afford to invest up to \$6960 for the solar energy system, and in 10 years the fuel savings would be equal to or greater than the cost of the system, including interest.

2-5 FURTHER REFINEMENTS IN MODELING

In this chapter, only the simplest model has been defined to introduce a solar collection, storage, load, and control system. To accurately model the performance of a real system, a number of additional factors must be included and the effect of some of the assumptions in the simplified model must be assessed.

The additional factors to be considered include the use of more realistic insolation data; the more complete thermal, optical, and fluid mechanical modeling of the collector; the modeling of the detailed transient behavior of the storage system; the detailed transient behavior of the system load; comparison of various control strategies; and the effect of all these factors on system performance and economics.

These subjects are considered in later chapters.

PROBLEMS

2-1 For a collector with efficiency given by

$$\eta = 0.79 - 1.05 \frac{(\theta_{f,\,in} - \theta_a)(q_{s,\,ref})}{q_s(t)}$$

plot the efficiency curve. Also, find the efficiency of the collector when operating at 200°F at an average daily ambient temperature of 80°F, at an instantaneous insolation value of 300 Btu/(h · ft²). Assume

$$q_{s,\,ref} = q_s(t) \quad \text{and} \quad T_a(t) = 90°F$$

2-2 For the base case solar thermal model inputs shown in Table 2-2 run SOLSIM and graph the results to find the following:
(a) Storage tank temperature for a two-day period
(b) Collector efficiency and outlet temperature versus time during operation of the collector pump
Use 1-h time intervals for the printout.

2-3 For the SOLSIM base case, vary the collector area from 5 to 50 m² (take 5, 10, 20, 30, 40, and 50 m²) for a 3-day period and plot to find
(a) Allowable capital cost versus collector area
(b) The fraction of load provided by solar versus collector area
To reduce runtime, print out results only once each day (that is, for $\Delta t = 15$ min, let the printout occur for each 96 time intervals). Discuss why the curves have the shapes that you find. If installed collector costs are about \$300/m² and the storage, controls, piping, pumps, etc., cost about \$2000 for the system modeled, plot the actual total system capital cost on the same graph as for part (a). Discuss how you would choose an economical size for the system modeled as a result.

2-4 Repeat Prob. 2-3 for alternative fuel costs of \$0.20/kWh. (If you have done Prob. 2-3, then this can be done without using the computer.)

2-5 A storage tank is to hold sufficient water to provide 2 days' storage to provide a load of 30,000 Btu/h. The maximum storage temperature is to be 210°F and energy to the load must be provided at or above 120°F. What volume (ft³) and mass (lb$_m$) must the tank hold?

2-6 A solar collector has efficiency of

$$\eta = 0.8 - \frac{1.2(T_{f,\,in} - 80°F)}{q_s(t)}$$

where $q_s(t)$ is in Btu/(h · ft²). The insolation normal to the collector is approximately

$$q_s(t) = 240 \sin\left[\frac{\pi(t - 420)}{600}\right] \text{Btu/(h · ft}^2\text{)}$$

and $T_{f,\text{in}} = 180°\text{F}$ and t is in minutes.

(a) How much useful energy is collected over the 10-h period of sunshine by 1 ft² of collector if the collector is run continuously during the sunshine period?

(b) What is the collector efficiency at solar noon?

(c) What is the average collector efficiency over the 10-h sunshine period?

(d) Discuss your results.

Hint: Since $q_s(t)$ is an analytic function, you can carry out this solution by direct integration.

REFERENCES

1. V.A. Baum, R.R. Aparissi, and B.A. Garf, "High Power Solar Installations," *Solar Energy*, vol. 1, p. 6, 1957.
2. L.L. Vant-Hull, and A.F. Hildebrandt, "Solar Thermal Power System Based on Optical Transmission," *Solar Energy*, vol. 18, no. 1, pp. 31–39, 1976.
3. A. Sobin, W. Wagner, and C.R. Easton, "Central Collector Solar Energy Receivers," *Solar Energy*, vol. 18, no. 1, pp. 21–30, 1976.
4. Jan F. Kreider, and Frank Kreith, *Solar Heating and Cooling: Engineering, Practical Design and Applications*, Hemisphere Publishing Co., New York, 1975.

CHAPTER
THREE

TERRESTRIAL SOLAR RADIATION

The prediction of the available solar energy at a given location and at a given time is necessary for the accurate design of any system for providing useful energy from the sun. Measurement of available solar energy has been carefully done at some locations, but not at enough to give the engineer the necessary data for careful design at any arbitrary site.

This chapter presents the basic information available to the engineer along with the methods of measurement commonly used and the methods of calculating or approximating site-specific data from the sparse information available.

3-1 LIST OF SYMBOLS

ET   equation of time, Eq. (3-6), Fig. 3-2 and Table 3-1
H total, global insolation on a horizontal surface
$\bar{H}$ monthly average daily total global insolation on a horizontal surface
H_b beam component of insolation on a horizontal surface
$\bar{H}_b$ monthly average daily beam component of insolation on a horizontal surface
H_d diffuse component of insolation on a horizontal surface
$\bar{H}_d$ monthly average daily diffuse component of insolation on a horizontal surface
H_0 extraterrestrial insolation on a horizontal surface, Eq. (3-16)
$\bar{H}_0$ daily average extraterrestrial insolation on a horizontal surface, Eq. (3-17), *or* monthly average daily extraterrestrial insolation on a horizontal surface for the typical day
I_b solar beam radiative flux (terrestrial)

I_0 extraterrestrial solar radiative flux
I_{sc} solar constant
K_T hourly clearness factor, Eq. (3-26)
$\bar{K}_T$ monthly average daily clearness factor, Eq. (3-29)
n day number counted from January 1 each year, Table 3-1
q_b component of the solar beam on a tilted surface, Eq. (3-7)
q_s global insolation on a tilted surface, Eq. (3-19)
$\bar{q}_s$ monthly average daily global insolation on a tilted surface, Eq. (3-31)
r $H/\bar{H}$, Eq. (3-39)
r_d $H_d/\bar{H}_d$, Eq. (3-38)
R q_s/H, Eq. (3-25)
$\bar{R}$ $\bar{q}_s/\bar{H}$, Eq. (3-34)
R_b ratio of beam insolation on a tilted surface to that on a horizontal surface, Eq. (3-22)
$\bar{R}_b$ monthly average daily ratio of beam insolation on a tilted surface to that on a horizontal surface, Eq. (3-35)
R_d ratio of the diffuse insolation on a tilted surface to that on a horizontal surface, Eq. (3-23)
R_r the radiative configuration factor from the ground and surroundings to a tilted surface, Eq. (3-24)
t_d day length, Eq. (3-5)
α_s solar altitude, Eq. (3-2), Fig. 3-3
β surface tilt from horizontal, Fig. 3-9
γ_c surface azimuth, Fig. 3-9
γ_s solar azimuth, Eq. (3-3), Fig. 3-3
δ declination, Eq. (3-1), Table 3-1, Fig. 3-2
θ solar beam incident angle relative to a tilted surface, Eqs. (3-8) to (3-12)
λ local meridian
λ_{st} standard meridian for a specified time zone
ρ_r effective diffuse ground reflectance, App. D
ϕ latitude
ω hour angle
ω_{ss} sunset or sunrise hour angle, Eq. (3-4)
ω'_{ss} surface sunset or sunrise hour angle, Eq. (3-13)
ζ_s solar zenith, $90° - \alpha_s$, Fig. (3-3)
ζ_{sp} solar zenith projected into the north-south plane

3-2 THE SUN-EARTH RELATIONS

Sun Angles

As the earth rotates about the sun, it spins about an axis which points to the North Star and is inclined at 23°27'8.2" (approximately 23.45°) to the orbital plane, as illustrated in Fig. 3-1. Therefore, the angle between the earth's equa-

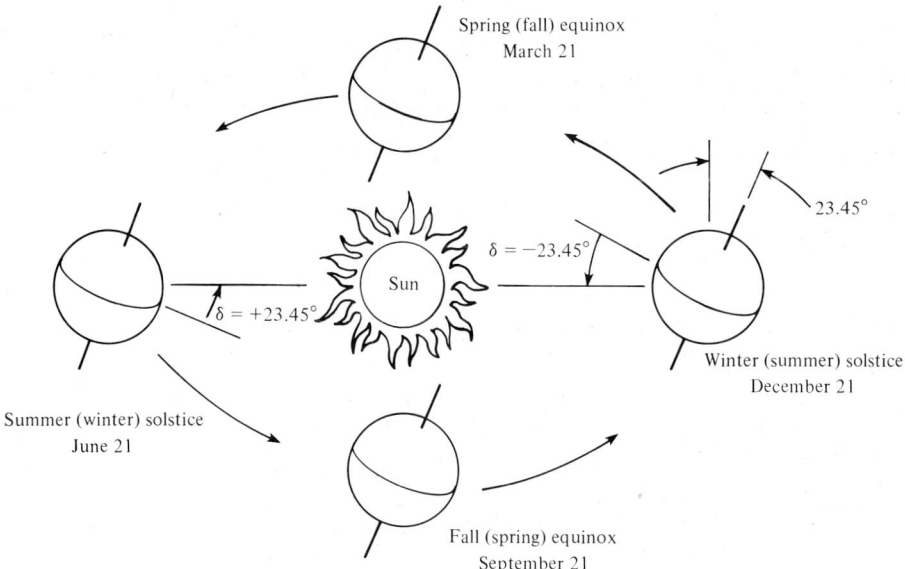

Figure 3-1 Northern (southern) hemisphere earth-sun relationship.

torial plane and the earth-sun line varies between ±23.45° throughout the year. This angle is called the *declination*, δ. Declinations north of the equator are positive (summer in the northern hemisphere); those south are negative. The declination is tabulated in Table 3-1 and represented graphically in Fig. 3-2. It can be approximated by

$$\delta \cong 23.45° \sin\left[360°\left(\frac{284+n}{365}\right)\right] \tag{3-1}$$

where n is the day of the year (Table 3-1). The value of δ calculated from Eq. (3-1) will be correct within $+0.37°$ (with maximum positive deviation on May 1) and $-1.70°$ (with maximum negative deviation on October 9) [2].

The sun's location in the sky relative to a point on the surface of the earth can be defined with two angles, the solar altitude α_s and the solar azimuth γ_s, as illustrated in Fig. 3-3. The solar altitude at a point on the earth is the angle between the line passing through the point and the sun and the line passing through the point tangent to the earth and passing below the sun. The solar azimuth is the angle between the line under the sun and the local meridian pointing to the equator, or due south in the northern hemisphere. It is positive measured to the east and negative to the west (in both hemispheres). The solar zenith angle ζ_s, defined in Chap. 1 as the angle between a solar ray and local vertical direction, is the complement of α_s.

The sun's location in the sky is a function of the location on the earth, the time of year, and the time of day. The location on the earth is specified by the

48 SOLAR-THERMAL ENERGY SYSTEMS

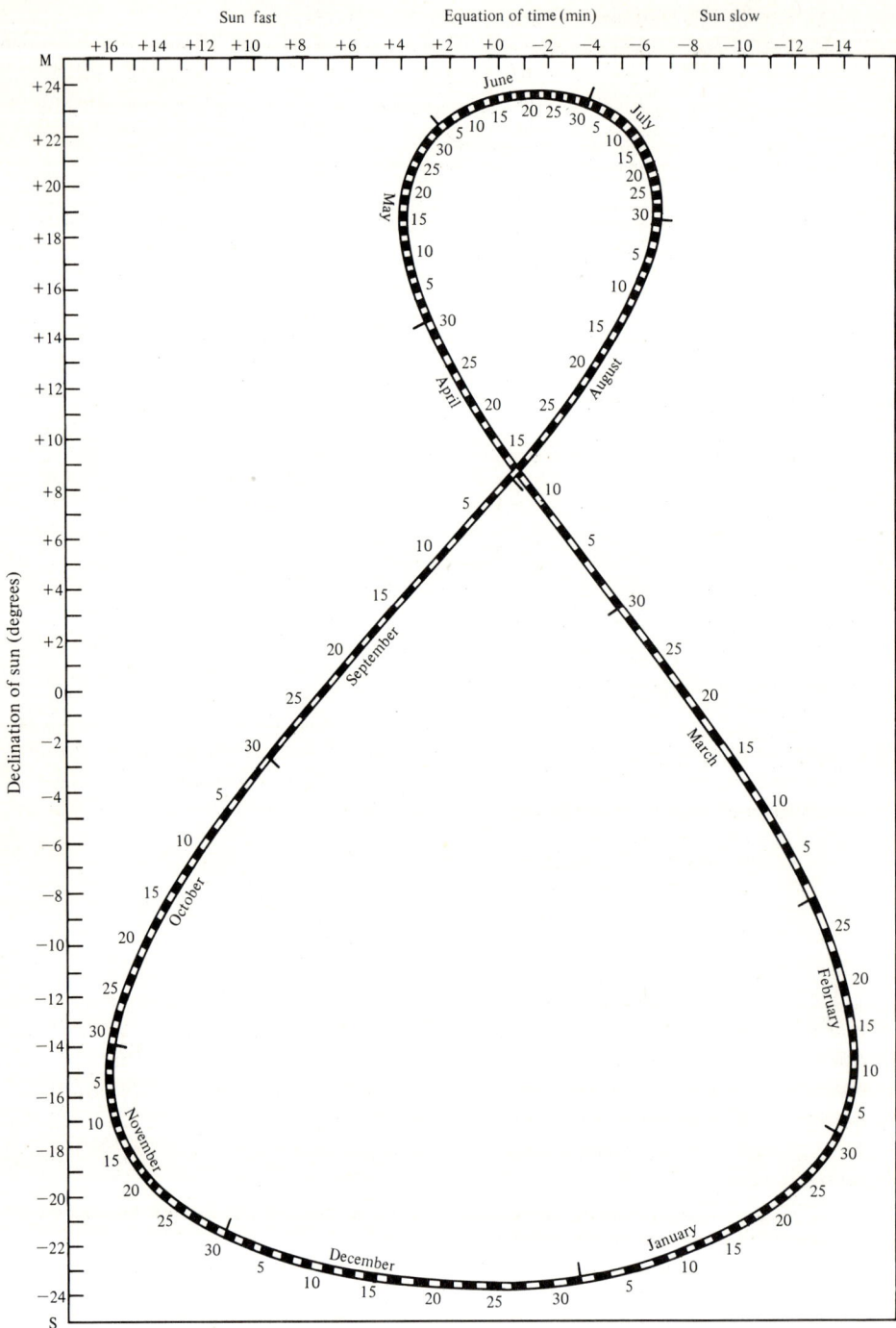

Figure 3-2 The analemma: a graphical representation of the declination and equation of time for both hemispheres. (Equation of time shown in minutes.)

Table 3-1 Declination and equation of time for 1975†

Date	Day n	Declination δ, deg†	ET, minutes of time §‡	Date	Day n	Declination δ, deg‡	ET, minutes of time §‡
Jan. 1	1	−23.0	−3.1	July 9	190	+22.9	−5.2
10	10	−22.0	−7.6	19	200	20.9	−6.2
20	20	−20.2	−11.1	29	210	18.8	−6.4
30	30	−17.8	−13.3				
				Aug. 8	220	16.2	−5.6
Feb. 9	40	−14.8	−14.3	18	230	13.2	−3.8
19	50	−11.4	−13.9	28	240	9.8	−1.2
Mar. 1	60	−7.7	−12.4	Sept. 7	250	6.2	+2.0
11	70	−3.9	−10.0	17	260	2.4	5.5
21	80	+0.1	−7.2	27	270	−1.5	9.0
31	90	4.0	−4.2				
				Oct. 7	280	−5.4	12.6
Apr. 10	100	7.8	−1.3	17	290	−9.1	14.6
20	110	11.4	+1.1	27	300	−12.7	16.1
30	120	14.7	2.8				
				Nov. 6	310	−15.9	16.3
May 10	130	17.5	3.6	16	320	−18.7	15.2
20	140	19.9	3.5	26	330	−20.9	12.7
30	150	21.7	2.5				
				Dec. 6	340	−22.5	8.9
June 9	160	22.9	0.8	16	350	−23.3	4.4
19	170	23.4	−1.3	26	360	−23.4	−0.6
29	180	23.3	−3.4	31	365	−23.1	−3.0

† Since each year is 365.25 days long, the precise daily value varies from year to year. The table was developed from Reference 1.
‡ At 12 N. solar time, Greenwich meridian.
§ Equation of time [see Eq. (3-6)].

latitude ϕ. At the equator, $\phi = 0$. North of the equator, latitudes are positive; south, negative. The time of year is specified by the solar declination δ, previously defined. The time of day is specified by the hour angle ω. The hour angle is defined as zero at local solar noon ($\gamma_s = 0$), and increases by 15° for each hour before local solar noon [e.g., for 8 A.M. (solar time), $\omega = 60°$] and decreases by 15° for each hour after solar noon [e.g., at 3 P.M. (solar time, $\omega = -45°$) in both hemispheres].

With the help of spherical geometry, expressions for the solar altitude and the solar azimuth can be developed in terms of ϕ, δ, and ω. Thus,

$$\sin \alpha_s = \sin \phi \sin \delta + \cos \phi \cos \delta \cos \omega \qquad (3\text{-}2)$$

and

$$\sin \gamma_s = \frac{\cos \delta \sin \omega}{\cos \alpha_s} \qquad (3\text{-}3)$$

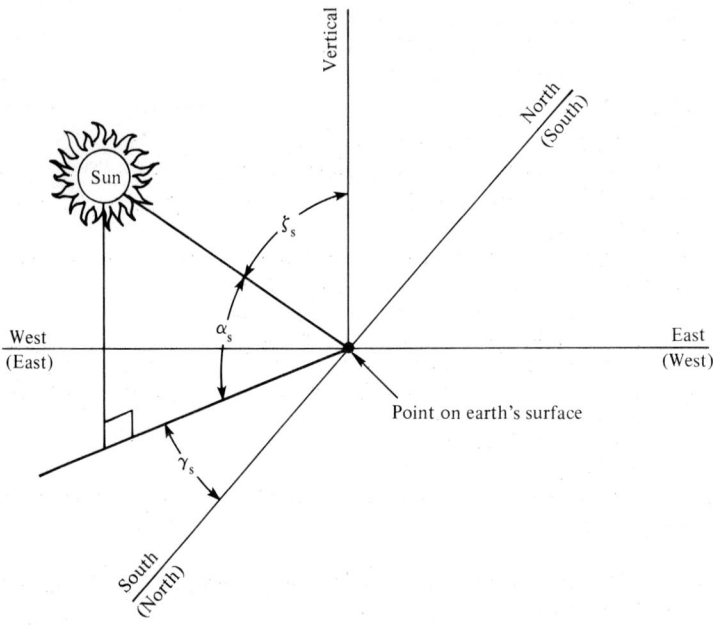

Figure 3-3 Solar altitude, zenith, and azimuth for northern (southern) hemisphere.

Example 3-1 Determine the location of the sun at 9 A.M. and at 2 P.M. solar time at Houston, TX, $\phi = 30$, and at Rio de Janeiro, Brazil, $\phi = -23°$, on February 1.

SOLUTION From Eq. (3-1),

$$\delta = 23.45 \sin \frac{360(284 + 32)}{365} = -17.5°$$

or from Table 3-1, $\delta = -17.2°$ (Fig. 3-2, $\delta = -17.3°$). From Eq. (3-2), for Houston ($\phi = 30°$) and $\omega = +45°$ (9 A.M.):

$$\sin \alpha_s = \sin (30) \sin (-17.5) + \cos (30) \cos (-17.5) \cos (45)$$

so that

$$\alpha_s = 25.7°$$

From Eq. (3-3)

$$\gamma_s = \sin^{-1} \frac{\cos (-17.5) \sin (45)}{\cos (25.7)} = 48.4°$$

The results of all the calculations are shown in the accompanying table.

	Houston		Rio de Janeiro	
	9 A.M.	2 P.M.	9 A.M.	2 P.M.
δ	−17.5	−17.5	−17.5	−17.5
ϕ	30	30	−23	−23
ω	45	−30	45	−30
α_s	25.7	34.4	47.6	88.7
γ_s	48.4	−35.4	61.4	−84.5

The rapid change in the solar azimuth between 9 A.M. and 2 P.M. at Rio de Janeiro is due to the fact that the sun is very nearly overhead at solar noon $[\alpha_s(\omega = 0) = 84.5°]$ so that it moves from "east" to "west" quickly at noon.

The sunrise time, sunset time, and the day length can be determined using Eq. (3-2). At sunrise or sunset

$$\alpha_s = 0 = \sin \phi \sin \delta + \cos \phi \cos \delta \cos \omega_{ss}$$

where ω_{ss} is the sunset (or sunrise) hour angle, which may then be expressed as:

$$\omega_{ss} = \cos^{-1}(-\tan \phi \tan \delta) \qquad (3\text{-}4)$$

The day length is twice the sunset hour angle. From Eq. (3-4) the day length can be expressed in hours as

$$t_d = \tfrac{2}{15}\omega_{ss} = \tfrac{2}{15} \cos^{-1}(-\tan \phi \tan \delta) \qquad (3\text{-}5)$$

when ω_{ss} is expressed in degrees. The actual time during which the sun is visible is somewhat longer than that indicated by Eq. (3-5). First, the refractive effect of the atmosphere allows one to see "around" the earth's horizon. Second, the sun is not a point source of light as assumed in the development of Eq. (3-4) but it has a finite size so that a "limb" of it is visible before its center appears.

The path of the sun on any given day is approximately in a plane tilted at an angle equal to the local latitude from the vertical. Relative to a given point on the earth's surface, the plane moves north (summer in the northern hemisphere) and south (winter in the northern hemisphere) throughout the year. At equinox, the plane contains the point. It is only at equinox that every point on the earth's surface moves essentially in the relative orbit plane of the sun. At equinox the sun rises due east and sets due west for every point on the earth. These statements can be better understood by examining Figs. 3-4 and 3-5.

Figure 3-4a illustrates sun-path diagrams for a location at 30°N latitude at the equinoxes and at the solstices. The view from the east (Fig. 3-4b) clearly shows the sun's relative orbit plane and its constant tilt.

Example 3-2 For 30°N latitude determine the sunrise time, the day length, the sunrise solar azimuth angle, and the solar noon zenith angle for the solstices and equinox and check the solutions with Fig. 3-4.

52 SOLAR-THERMAL ENERGY SYSTEMS

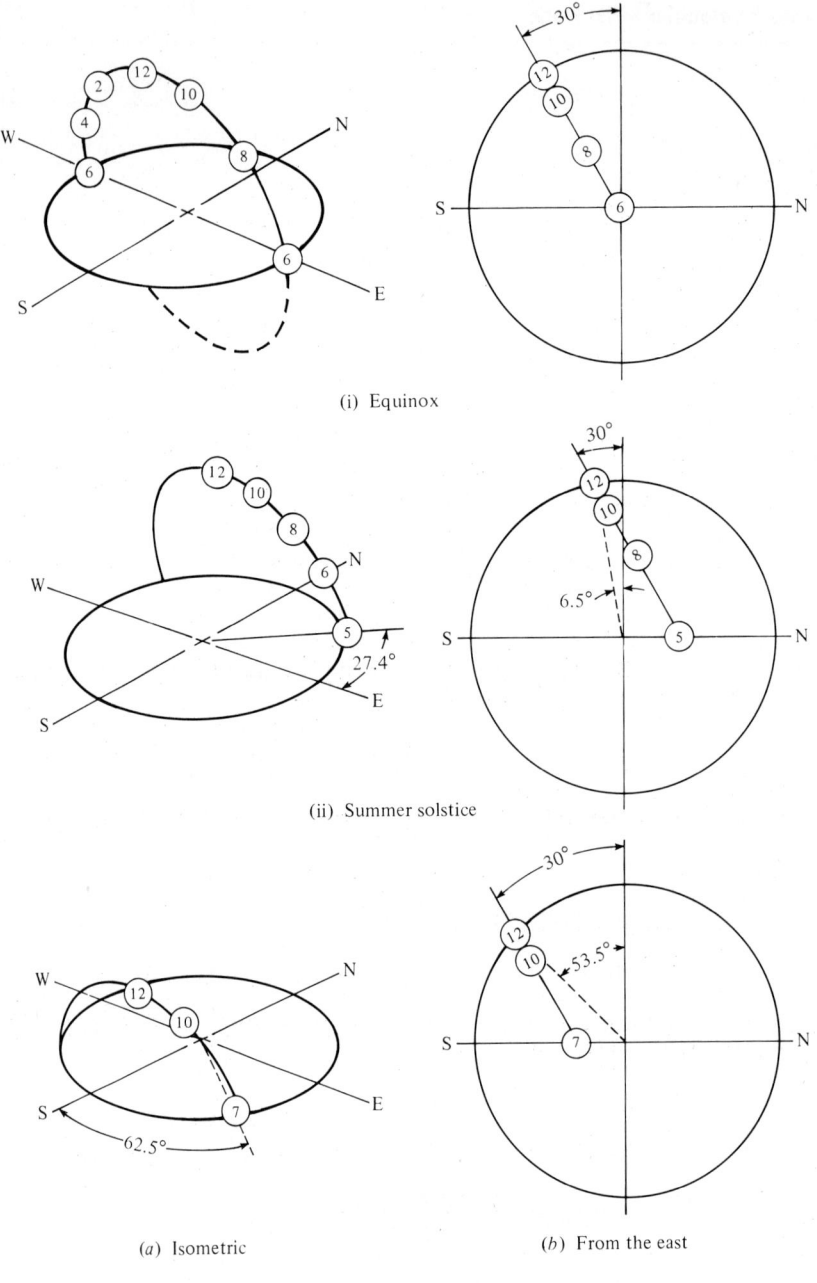

(a) Isometric   (b) From the east

(iii) Winter solstice

Figure 3-4 Sun paths for equinox (i) and summer (ii) and winter (iii) solstices (northern hemisphere) for a site at 30°N latitude. Solar time is shown by the circled numbers [3].

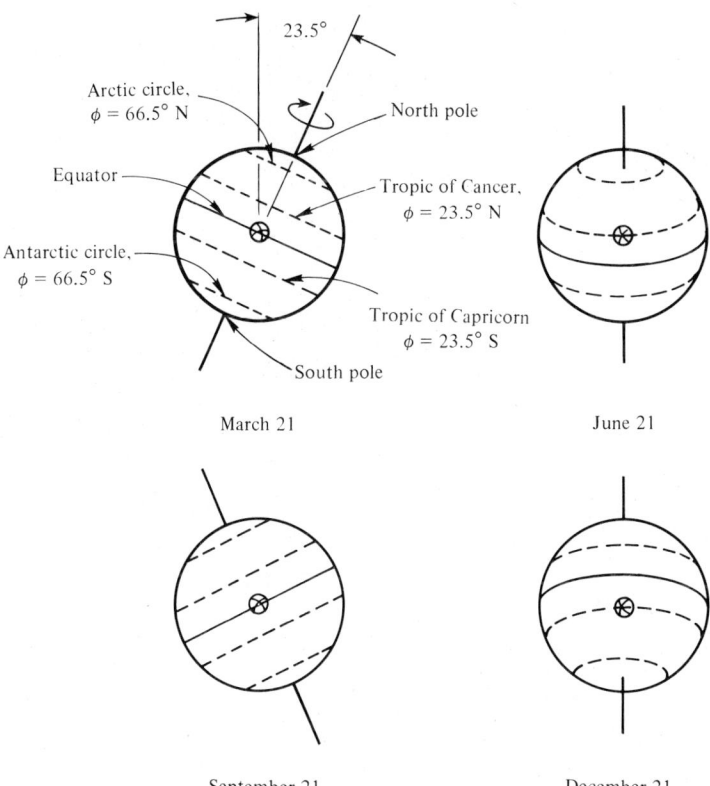

Figure 3-5 The earth viewed from the sun. ⊗ Location at which sun is directly overhead at local solar noon.

SOLUTION For equinox [with Eq. (3-4)]

$$\omega_{ss} = \cos^{-1}(-\tan 30 \tan 0) = \pm 90° = \pm 6 \text{ h}$$

or sunset and sunrise at 6 P.M. and 6 A.M. solar time (as expected). The daylength [from Eq. (3-5)] is:

$$t_d = \tfrac{2}{15} \cos^{-1}(-\tan 30 \tan 0) = \tfrac{2}{15}(90) = 12 \text{ h}$$

as expected. For the sunrise solar azimuth [from Eq. (3-3)]

$$\sin \gamma_{ss} = \frac{\cos(0)\sin(\pm 90°)}{\cos(0)} = \pm 1$$

so that

$$\gamma_{ss} = \pm 90°$$

That is, the sun rises 90° east of south (due east) and sets 90° west of south (due west). The solar noon zenith ζ_{sn} can be determined from Eq. (3-2):

$$\cos \zeta_{sn} = \sin \alpha_{sn} = \sin \alpha_s(\gamma_s = 0) = \sin \phi \sin \delta$$
$$+ \cos \phi \cos \delta \cos (0) = \cos (\phi - \delta)$$

which implies that

$$\zeta_{sn} = \phi - \delta = \phi$$

since $\delta = 0$ at equinox.

This procedure can be repeated for the other cases, with the results being summarized in the accompanying table.

	Equinox	Summer solstice	Winter solstice
ϕ	30°	30°	30°
δ	0	23.45°	−23.45°
ω_{ss}	6 h	6.97 h	5.03 h
t_d	12 h	13.93 h	10.07 h
γ_{ss}	90°	117.4°	62.5°
ζ_{sn}	30°	6.5°	53.5°

Both γ_{ss} and ζ_{sn} are shown in Fig. 3-4.

Figure 3-5 depicts views of the earth as seen from the sun for the equinox and the solstice times. The fact that lines of constant latitude are straight at equinox illustrates that sunrise and sunset are, respectively, due east and due west for every point on the surface of the earth. The earth's "inward" tilt on June 21 indicates that sunrise and sunset are both north of due east and due west, respectively, in both hemispheres. Similarly on December 21, the sunrise and sunset are south of due east and due west, respectively.

The tropic of Cancer at approximately 23.5° and the tropic of Capricorn at approximately −23.5° are the most northerly and southerly latitudes, respectively, at which the sun reaches a zero zenith angle (is directly overhead). The arctic and antarctic circles are defined as those latitudes above which the sun does not rise above the horizon plane at least once per year.

From Eqs. (3-2) and (3-3) it is clear that at a given location (latitude) the solar altitude and azimuth are functions of only two variables. Therefore, they can be represented on a two-dimensional diagram as functions of δ and ω. One such diagram is called a sun chart and is illustrated in Fig. 3-6 for 40°N latitude. Sun charts are also available at other latitudes. The sun chart is used as follows: The solar azimuth is plotted along the horizontal axis from east to west. The solar altitude is plotted along the vertical direction. The sun's position is determined by locating the interception of the appropriate day and time lines. Additional information about the use of the sun chart and other sun diagrams can be found in many architectural reference books (e.g., Refs. 4 and 5).

Example 3-3 From the sun chart, determine the solar azimuth and altitude at 40°N latitude at the times indicated in the table below.

Date	Time (solar)
June 11	2:00 P.M.
March 11	8:00 A.M.
October 15	3:30 P.M.
November 5	9:45 A.M.

Check the results with Eqs. (3-2) and (3-3).

SOLUTION The table below lists the value of α_s and γ_s as determined from Fig. 3-6 and compares them with the results of using the equations.

		Sun chart		Equations	
Date	Time	α_s	γ_s	α_s	γ_s
June 11	2:00 P.M.	59°	−64°	59.5°	−65.1°
Mar. 11	8:00 A.M.	20°	67°	19.8°	66.7°
Oct. 15	3:30 P.M.	21°	−58°	21.5°	−57.5°
Nov. 5	9:45 A.M.	26°	37°	26.1°	36.6°

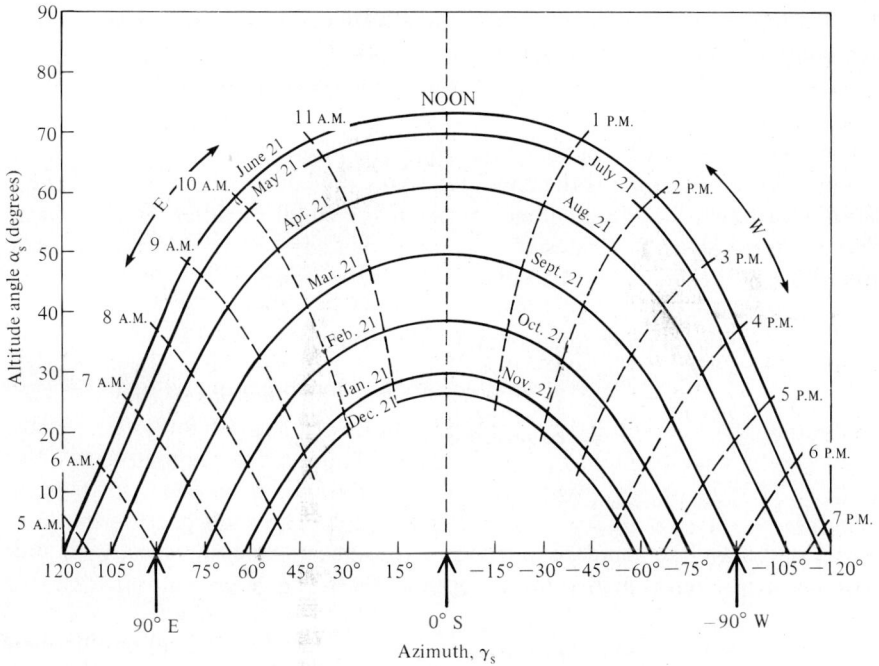

Figure 3-6 Sun chart for 40°N latitude [4].

Time

So that all "days" can start at approximately the same local time, a worldwide system of 24 time zones has been established by international agreement. The earth revolves 360° in 24 h, or 15° per hour. Therefore, each of the 24 time zones represents a different hour and physically consists of the land contained between two meridians 15° apart, although the latter is often approximate since various political and natural boundaries cause some modification of the time zone boundaries. The time zones are measured westward (positive) from the prime meridian through Greenwich, England. In each time zone the zone mean time (ZMT) or clock time is Greenwich time less the number of hours equal to the time zone meridian divided by 15. In the United States, the time zone meridians associated with the various time zones are eastern, 75°W; central, 90°W; mountain, 105°W; and pacific, 120°W; Alaska and Hawaii are in the time zone associated with the 150°W meridian. When it is noon (local time) in Greenwich, the zone mean time (local time) in the U.S. central time zone is 6 A.M. (of the same day).

A *solar day* at a given point on the earth is defined as the length of time between successive solar noons. However, because of the eccentricity of the earth's orbit and the inclination of the earth to the earth-sun orbit plane, the length of a solar day is not constant and in fact varies by up to about 30 s between successive days at certain times of the year. The cumulative effect of the phenomenon is a shift in the time scale of up to 15 min from the mean time. This shift (the difference between true solar time and mean solar time), called the *equation of time* (ET), is shown in Fig. 3-2 and tabulated in Table 3-1. Lamm [6] has developed a five-term series for ET that is accurate within less than 3.6 s for any day of the year.

The relationship between true solar time and zone (clock) time can now be developed. True solar time is the zonal time plus the equation of time correction plus an adjustment for the difference between the zonal meridian and the local meridian. Since the earth rotates 15° per hour, this adjustment is 4 min for each degree of longitude. Solar time can therefore be determined as

Solar time = zonal time or clock time

+ 4(zonal meridian − local meridian)

+ equation of time

or $$\text{Solar time} = \text{ZMT} + 4(\lambda_{st} - \lambda) + \text{ET} \tag{3-6}$$

where λ_{st} and λ are the standard and local meridian in degrees, respectively, and the two correction terms are in minutes of time.

Example 3-4 Determine the sunrise and sunset time (local time) on October 30 at Phoenix, AZ (33°N, 112°W).

SOLUTION For October 30 and from Fig. 3-2 (or Table 3-1)

$$\delta = -13.4° \qquad ET = +16.2 \text{ min}$$

From Eq. (3-4)

$$\omega_{ss} = \cos^{-1}[\tan(33)\tan(-13.4)] = 81.1°$$

Sunrise, therefore, occurs at

$$\tfrac{81.1}{15} = 5.41 \text{ h before solar noon or 6:35 A.M. solar time}$$

Sunset occurs at 5:25 P.M. solar time. From Eq. (3-6), sunrise occurs at

$$\text{Local time} = 6\text{:}35 \text{ A.M.} - 4(105 - 112)\text{min} - 16 \text{ min}$$

$$= 6\text{:}47 \text{ A.M. mountain standard time (MST)}$$

Similarly, sunset is at $5\text{:}25 + 26 - 16 = 5\text{:}37$ P.M. MST. Note that most states in the United States switch to daylight saving time the last Sunday in April and change back to standard time the last Sunday in October. Arizona, however, does not. In a state that does change to daylight saving time, the June 1 local time would be advanced 1 h, e.g., 7 A.M. MST becomes 8 A.M. MDT. The local time on October 30 may be daylight or standard time, depending upon the year.

Solar Angles Relative to Tilted Surfaces

Probably the most common calculation made in any solar design problem is to determine the angle of incidence of the solar beam to an arbitrarily oriented surface like a window, a skylight, or other solar collector. The sun's position is established by the solar altitude and the solar azimuth. The orientation of the irradiated surface is defined by its azimuth γ_c, measured from the local meridian toward the equator (positive to the east and negative to the west in both hemispheres), and a tilt angle β, measured relative to the horizontal. These angles are illustrated in Fig. 3-7. The solar incident angle relative to the surface is designated θ. The quantity $\cos \theta$ is the fraction of the illuminated surface that is projected into the solar beam direction. Thus, the amount of the solar beam flux striking a unit area of the surface is

$$q_b = I_b \cos \theta \tag{3-7}$$

where I_b is the solar beam intensity.

It can be shown that

$$\cos \theta = \cos(\gamma_s - \gamma_c) \cos \alpha_s \sin \beta + \sin \alpha_s \cos \beta \tag{3-8}$$

or in terms of the latitude, hour angle, and declination [with Eqs. (3-2) and (3-3)]

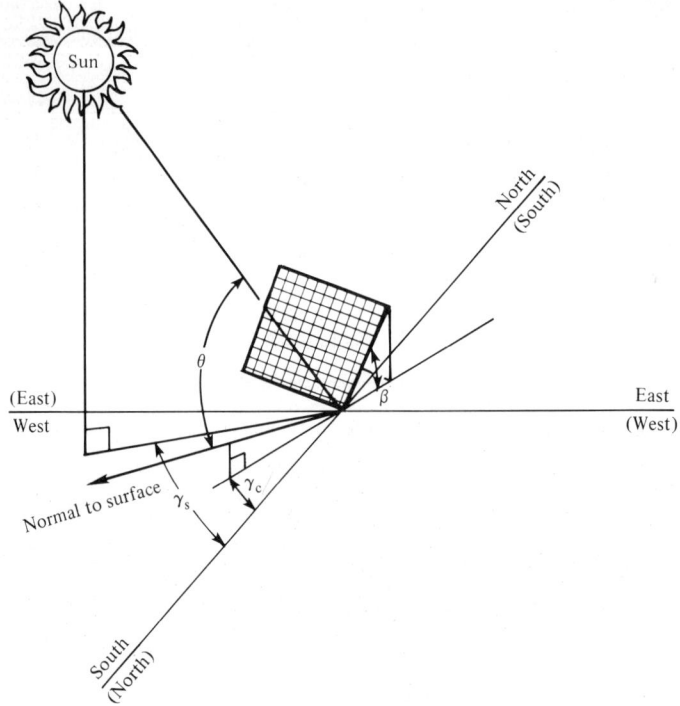

Figure 3-7 Orientation angles for tilted surface in northern and southern hemispheres.

that

$$\cos \theta = \sin \delta (\sin \phi \cos \beta - \cos \phi \sin \beta \cos \gamma_c)$$
$$+ \cos \delta \cos \omega (\cos \phi \cos \beta + \sin \phi \sin \beta \cos \gamma_c)$$
$$+ \cos \delta \sin \beta \sin \gamma_c \sin \omega \qquad (3\text{-}9)$$

For special cases of interest these expressions can be simplified:

For a surface facing the equator, for which $\gamma_c = 0$,

$$\cos \theta = \cos \gamma_s \cos \alpha_s \sin \beta + \sin \delta \cos \beta$$
$$= \sin (\phi - \beta) \sin \delta + \cos (\phi - \beta) \cos \delta \cos \omega \qquad (3\text{-}10)$$

For a vertical surface facing the equator, for which $\gamma_c = 0$ and $\beta = 90°$,

$$\cos \theta = \cos \gamma_s \cos \alpha_s = -\sin \delta \cos \phi + \cos \delta \sin \phi \cos \omega \qquad (3\text{-}11)$$

For a horizontal surface, for which $\beta = 0$,

$$\cos \theta = \sin \alpha_s = \sin \phi \sin \delta + \cos \phi \cos \delta \cos \omega \qquad (3\text{-}12)$$

Equation (3-10) for the surface facing the equator is plotted as a function of $(\phi - \beta)$, δ, and ω in Fig. 3-8. Note that these curves can also be used to determine the solar altitude α_s by selecting $\beta = 0$ $[(\phi - \beta) = \phi]$ and evaluating $\cos \theta = \sin \alpha_s$. Values of $\cos \theta$ ($\gamma_c = 0$) for $(\phi - \beta) < 0$ can also be obtained from Fig. 3-8 by reversing the sign for δ [see Eq. (3-10)].

Many times it is necessary to determine the total time during a day that a particular surface is illuminated by the sun. As in Eq. (3-4), this time can be expressed in terms of the effective surface sunrise hour angle ω'_{ss}. The angle is the smaller of (1) the actual sunrise hour angle (relative to the horizontal) or (2) the sunrise hour angle relative to the tilted surface. In turn, the sunrise hour angle relative to the tilted surface can be determined from Eq. (3-9) by setting $\cos \theta = 0$ and solving for ω. In general, this cannot be done explicitly. However, for a surface facing the equator [Eq. (3-10) with $\cos \theta = 0$] there results

$$\omega_{\text{facing equator}} = \cos^{-1} \left[-\tan (\phi - \beta) \tan \delta \right]$$

Therefore, the effective surface sunrise hour angle for a surface facing the equator is the minimum of

$$\omega'_{ss} = \cos^{-1} (-\tan \phi \tan \delta)$$

or
$$\omega'_{ss} = \cos^{-1} (-\tan (\phi - \beta) \tan \delta) \quad (3\text{-}13)$$

When δ and ϕ have the same sign (summer in either the northern or southern hemisphere),

$$(\omega'_{ss})_{\text{summer}} = \cos^{-1} \left[-\tan (\phi - \beta) \tan \delta \right] \quad (3\text{-}14)$$

When δ and ϕ have opposite signs (winter in either hemisphere) (unless β negative),

$$(\omega'_{ss})_{\text{winter}} = \cos^{-1} (-\tan \phi \tan \delta) \quad (3\text{-}15)$$

Extraterrestrial Total Beam Solar Radiation on a Horizontal Surface

The amount of beam insolation that reaches the top of the earth's atmosphere was given in Eq. (1-1). This is the radiative energy per unit time per unit area perpendicular to the beam. Also of interest is the amount of beam energy that strikes an extraterrestrial horizontal surface

$$H_0 = I_0(n) \sin \alpha_s = I_{sc}\left(1 + 0.034 \cos \frac{360n}{365}\right) \sin \alpha_s \quad (3\text{-}16)$$

where α_s is the solar altitude in Eq. (3-2). The symbol H indicates insolation on a horizontal surface. The subscript "0" is used to indicate "extraterrestrial."
terrestrial." The daily average extraterrestrial insolation on a horizontal surface can be obtained by integrating Eq. (3-16) from sunrise to sunset. The result is

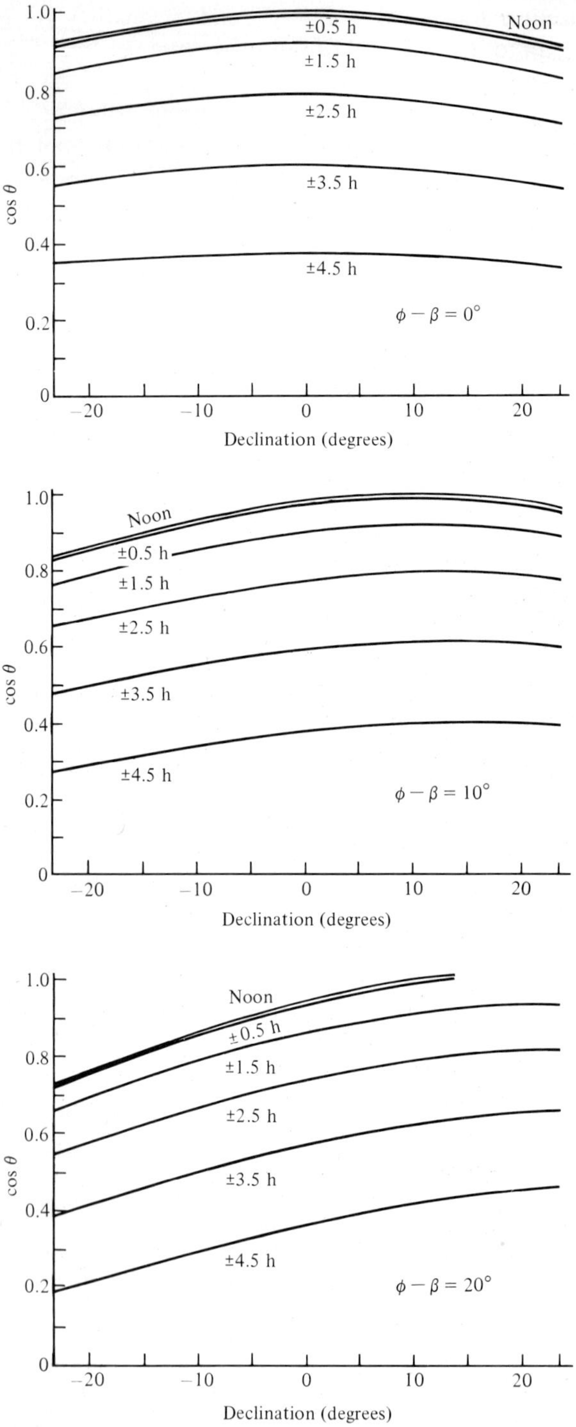

60 SOLAR-THERMAL ENERGY SYSTEMS

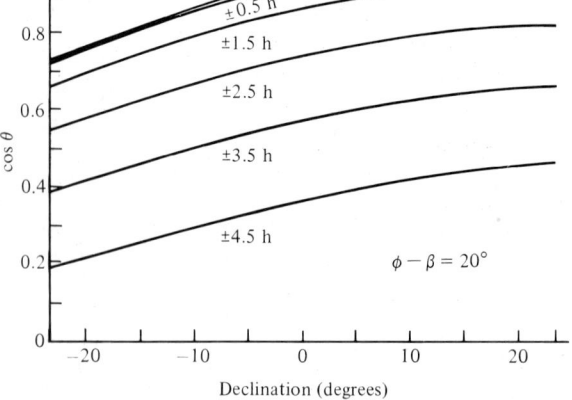

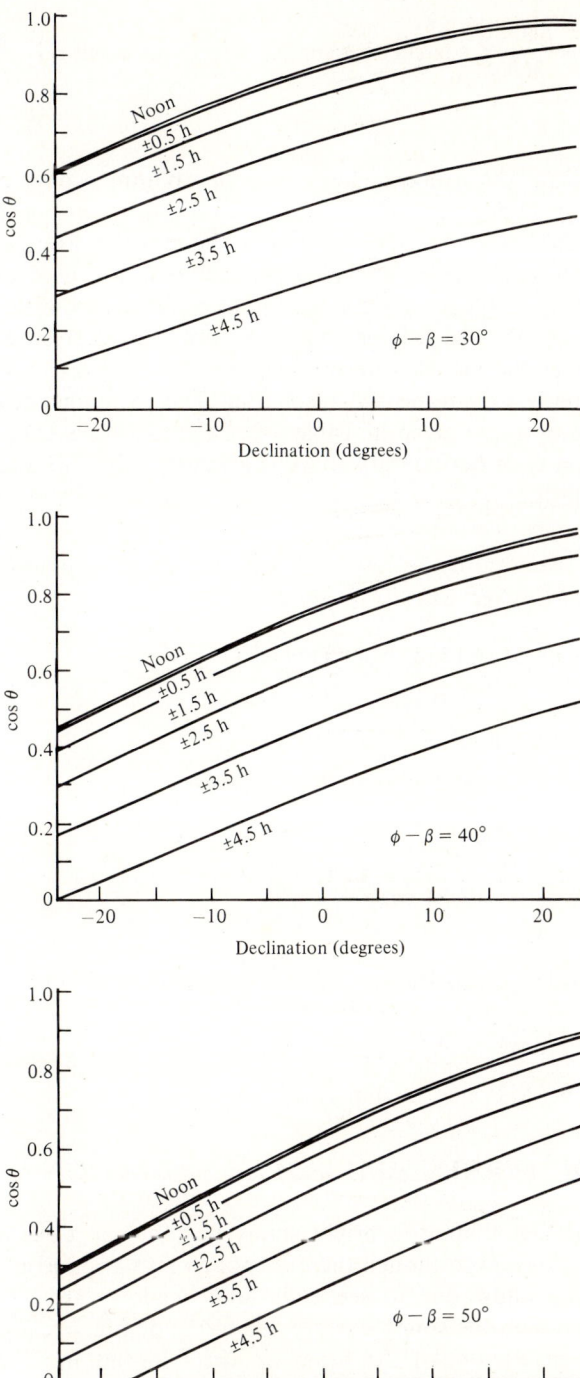

Figure 3-8 Cosine of angle of incidence of beam radiation on surfaces tilted toward the equator for various values of $\phi - \beta$ [7].

$$\bar{H}_0 = \frac{24}{\pi} I_{sc} \left(1 + 0.034 \cos \frac{360n}{365}\right)\left(\cos \phi \cos \delta \sin \omega_{ss} + \frac{\pi \omega_{ss}}{180} \sin \phi \sin \delta\right)$$

(3-17)

The overbar is used to indicate the average daily value in kWh/(m² · day) if I_{sc} is in kW/m². Note that ω_{ss} is in degrees.

In many of the correlations used to predict insolation, the monthly average daily extraterrestrial insolation on a horizontal surface for each month is used as a reference. Klein [8] suggests that this reference value can be estimated from Eq. (3-17) by selecting, for each month, the day for which the daily extraterrestrial insolation is nearly the same as the monthly mean value. In this case, H_0 indicates the monthly average daily extraterrestrial insolation. The recommended characteristic days for each month are given in Table 3-2. The monthly average daily extraterrestrial insolation on a horizontal surface is given in Table 3-3 and in Fig. 3-9 as a function of latitude.

Example 3-5 Determine the monthly average daily extraterrestrial insolation on a horizontal surface for Denver, CO (40°N), for May.

SOLUTION From Eq. (3-1) on May 15 (or from Table 3-1)

$$\delta \cong 18.8°$$

Then from Eq. (3-4) the sunset hour angle is

$$\omega_{ss} = \cos^{-1}[-\tan(40)\tan(18.6)] = 106.4°$$

From Eq. (3-17), for $n = 135$ (May 15),

$$\bar{H}_0 = \frac{24}{\pi} 1.353 \text{ kW/m}^2 \left(1 + 0.034 \cos \frac{360(135)}{365}\right)$$

$$\times \left[\cos(40)\cos(18.8)\sin(106.4) + \frac{106.4(2\pi)}{360} \sin(40)\sin(18.6)\right]$$

$$= 10.88 \text{ kWh/(m}^2 \cdot \text{day}) = 39.15 \text{ MJ/(m}^2 \cdot \text{day})$$

compared to a value of 39.2 MJ/(m² · day) in Table 3-3.

3-3 MEASUREMENT OF TERRESTRIAL SOLAR RADIATION

Everyone is familiar with the fact that on a bright sunny day the sun casts a sharp shadow. On an overcast day, even though there is enough insolation available to illuminate objects and allow one to see, only faint shadows are distinguishable. The sharp shadows on the sunny day are due to the dominance (up to 90 percent of the global insolation) of the beam or direct insolation. The uniform illumination of the overcast day is due to the low level of the direct insolation and the dominance of the "diffuse"† insolation. As defined in Chap. 1,

† Diffuse in this sense does not necessarily imply a uniform distribution over direction.

Table 3-2 Recommended characteristic day for each month.†

Month	Day of the year, n	Date
January	17	January 17
February	47	February 16
March	75	March 16
April	105	April 15
May	135	May 15
June	162	June 11
July	198	July 17
August	228	August 16
September	258	September 15
October	288	October 15
November	318	November 14
December	344	December 10

† Reference 8.

the direct or beam insolation is the sunlight that reaches the surface of the earth directly from the sun without being scattered in the earth's atmosphere. The diffuse component of the insolation is the sunlight that reaches the earth's surface after being scattered or diverted from its original path in the atmosphere. As defined in Chap. 1, the term "global insolation" is used in this book to indicate the sum of all solar radiation on a surface. This includes not only the beam and diffuse insolation but also any insolation reflected from surrounding objects.

Over the past 150 years many different types of radiometers have been developed to measure the intensity of insolation. They have ranged from sophisticated absolute instruments based on the calorimeter principle to simple lens devices used to burn sun paths on paper. Descriptions of these instruments can be found

Table 3-3 Monthly average daily extraterrestrial insolation on a horizontal surface in the northern hemisphere.†

Latitude	Jan.	Feb.	Mar.	Apr.	May	June	July	Aug.	Sept.	Oct.	Nov.	Dec.
20°	26.7	30.2	34.4	37.5	38.9	39.1	38.9	37.8	35.2	31.3	27.4	25.5
25°	23.9	28.1	32.8	37.1	39.4	40.0	39.6	37.8	34.2	29.4	24.9	22.7
30°	21.0	25.7	31.1	36.5	39.6	40.7	40.1	37.5	32.9	27.2	22.2	19.7
35°	18.1	23.1	29.2	35.5	39.5	41.1	40.3	37.0	31.3	24.8	19.3	16.7
40°	15.4	20.3	27.0	34.3	39.2	41.3	40.3	36.2	29.5	22.3	16.3	13.6
45°	12.0	17.4	24.7	32.9	38.7	41.3	40.1	35.1	27.5	19.5	13.3	10.6
50°	9.0	14.5	22.1	31.2	38.0	41.1	39.6	33.8	25.2	16.7	10.3	7.6
55°	6.1	11.5	19.4	29.3	37.2	40.9	39.1	32.4	22.9	13.8	7.4	4.8
60°	3.5	8.3	16.8	27.4	36.3	40.6	38.4	30.5	20.2	10.6	4.4	2.2
65°	1.2	5.4	13.9	25.2	35.5	40.6	37.9	28.8	17.4	7.6	2.0	0.3

† All values in MJ/(m² · day). Based on $I_{sc} = 1353$ W/m².

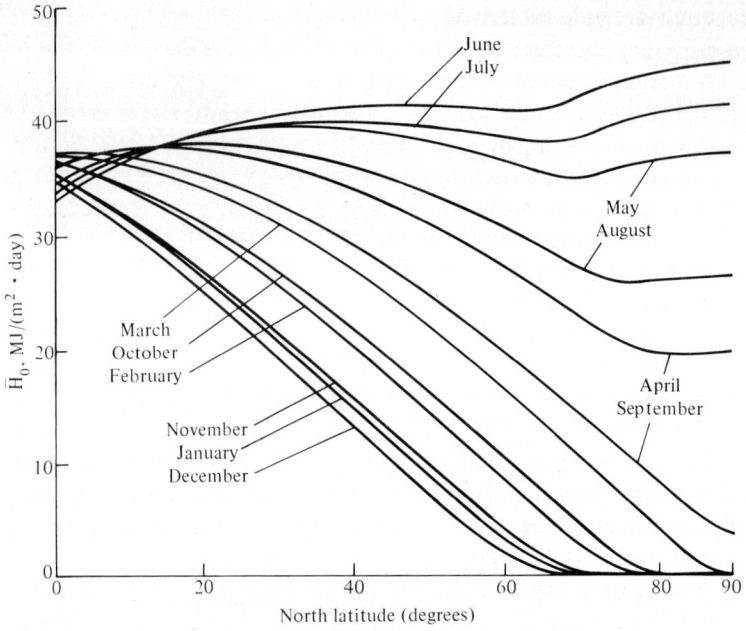
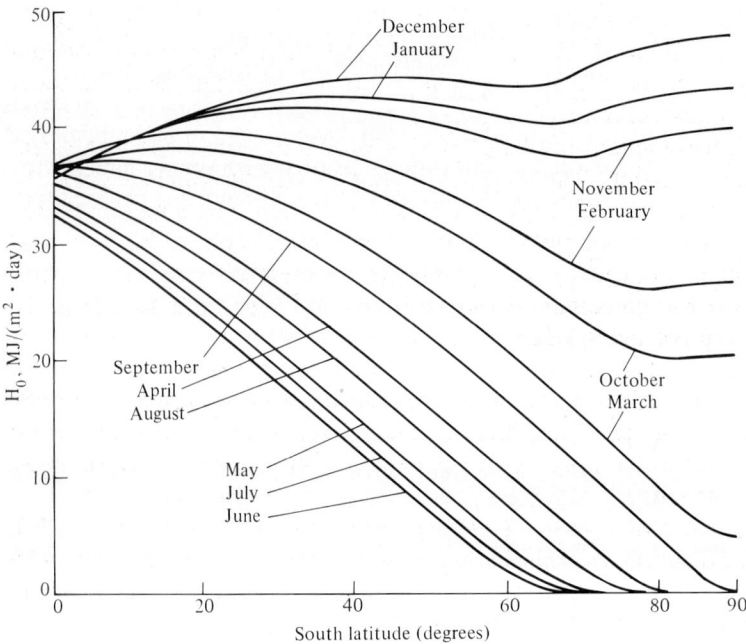

Figure 3-9 Monthly average daily extraterrestrial insolation on a horizontal surface (based on $I_{sc} = 1353$ W/m^2) [9].

in several references, e.g., Refs. 10–13. Most of the commonly used instruments today are based on either the thermoelectric effect (using a thermopile: thermocouples in series) or the photoelectric effect (solar cell).

The thermoelectric effect is utilized by attaching the hot junction of the thermopile to one side of a thin metallic plate. The other side of the plate is "blackened" to be highly absorptive of solar radiation and "gray" (i.e., constant absorptivity over the significant solar wavelength band) and is exposed to the sun's radiation. The cold junction of the thermopile is exposed to a cavity within the instrument. The output is compensated electrically for the cavity temperature. The elevated temperature achieved by the hot junction and the resulting electromotive force (emf) generated is related to the amount of insolation. The response is linearized and calibrated, so that the output voltage can be readily converted to the radiative flux. Since the sensor is a thermal device, it measures the total (integrated over all wavelengths) insolation and its output is influenced by many factors. These include (if not properly compensated for) the ambient temperature, the heat-transfer patterns near the sensor (especially convection), and the radiative directional properties of the sensor itself as well as the instrument as a whole. Also, response times can be excessive.

The photoelectric sensors are simpler and cheaper, have essentially instantaneous response, and have a good overall stability. Their major disadvantage is that their spectral response is not uniform in the solar band. Silicon cells, for example, respond primarily in the 0.6- to 1.0-μm wavelength band while the bulk of the solar energy is in the 0.3- to 0.7-μm band. If the spectral distribution of the terrestrial insolation were constant, this mismatch would not effect its operation since it could be calibrated. However, the spectral distribution is not constant and is dependent on the site altitude, solar altitude, cloudiness, and the water vapor content and the turbidity of the atmosphere.

There are three standard insolation measurements that are made with essentially two basic instruments. The three measurements are of the direct (beam), the nondirect (diffuse), and the global (hemisphere) insolation. The direct insolation measurement is made with a pyrheliometer [or as it is officially called, a normal incidence pyrheliometer (NIP)]. The global measurement is made with a pyranometer (or, as it was commonly called until 1965,† a 180° pyrheliometer). The difference between the two instruments is the field of view; the pyrheliometer is restricted to a narrow field of view while the pyranometer views a half-space or a hemisphere of solid angle. Either sensor, the thermopile or the solar cell, can be used with either instrument. The nondirect measurements are made with a pyranometer over which a shadow band is deployed to continuously shade the beam insolation from the sensor.

In pyrheliometers, the sensor is located at the base of a tube fitted with a series of annular diaphragms such that only nearly normal incident radiation can reach the base, as illustrated in Fig. 3-10. The early pyrheliometers had rectangular openings of various fields of view but standard practice now is to use a

† Based on recommendations of the World Meteorological Organization, the term "pyranometer" has been established for the instrument measuring global insolation.

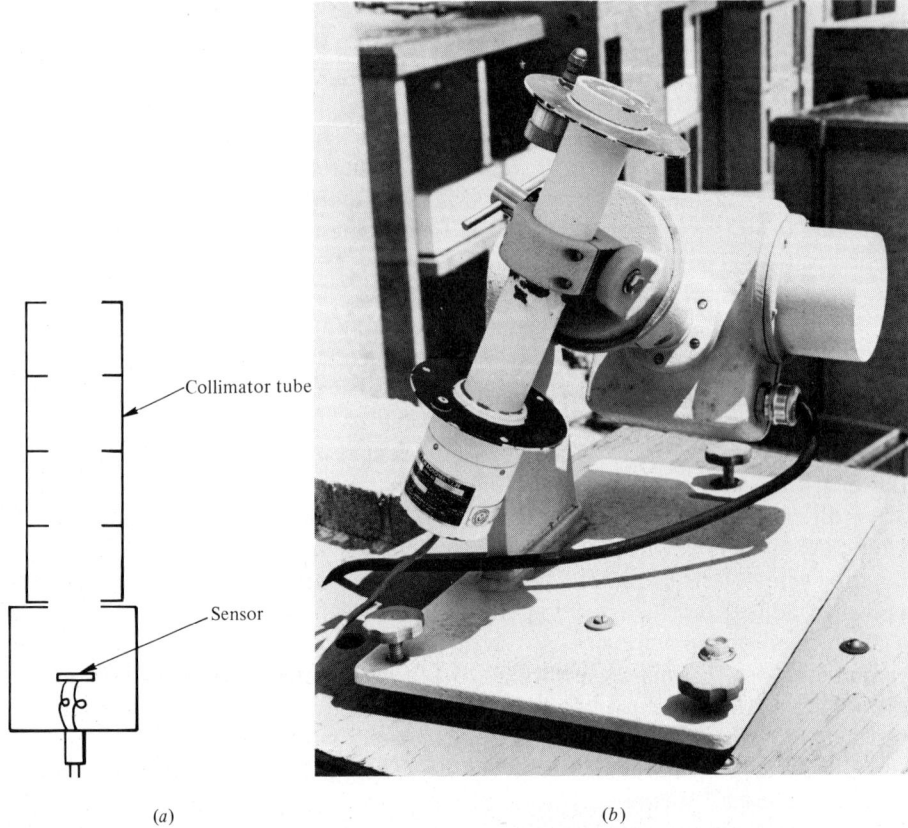

Figure 3-10 Pyrheliometer for the measurement of beam insolation (a) Schematic. (b) Eppley pyrheliometer with tracker.

circular opening with a full conical acceptance angle of about 6° (5.7° for the most commonly used instrument, the Eppley NIP).

The pyreheliometer must track the sun. Since the solar disk subtends an arc of only about 0.5°, precise tracking is not necessary. A single-axis clock drive (with tilt adjustments every day or so depending on the season, i.e., an equatorial mount†) is sufficiently accurate. While most of the diffuse insolation is blocked from the sensor, the field of view includes a significant region around the sun [14]. This fact has led to concern by some that a correction should be made for the additional "circumsolar" radiation when reporting direct insolation measurements.

In the pyranometer, one or two glass hemispheres enclose the sensor, as illustrated in Fig. 3-11. This cover protects and helps to thermally isolate the sensor. The pyranometer is intended for use in a permanently mounted horizontal position and this is the position in which it is calibrated.

There are two main sources of error unique to the pyranometer. First, the directional characteristics of the sensor surface coating and the reflections within

† The entire tracking device is tilted at the local latitude. Manual adjustments correct for the declination. The clock drive follows the hour angle (one revolution in 24 h).

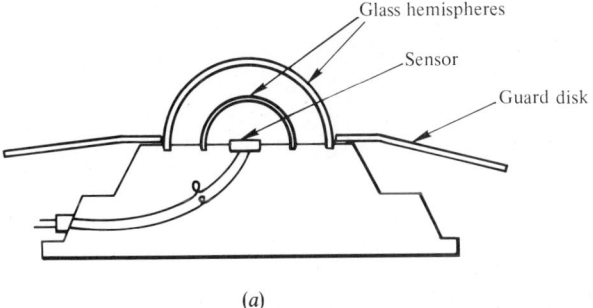

Figure 3-11 Pyranometer for the measurement of global insolation. (*a*) Schematic. (*b*) Eppley precision pyranometer.

the glass hemisphere are a function of the incident angle or altitude angle of the incoming radiation. Errors due to this problem (see Ref. 10, pp. 106–107, and Ref. 15) (the so-called "cosine error") of less than 5 percent can be expected for incident angles up to 60° from the vertical. Beyond 60°, errors as high as 40 percent have been reported for selected instruments. Second, tilting of the instrument can significantly change the convective heat transfer patterns within the hemisphere [15, 16]. This fact can lead to errors of up to 10 percent if a thermoelectric sensor is used. Most errors due to tilting, however, are less than 5 percent. Frequently, one needs to determine the amount of insolation on a tilted surface such as a window, a skylight, or a solar collector. Pyranometers are commonly

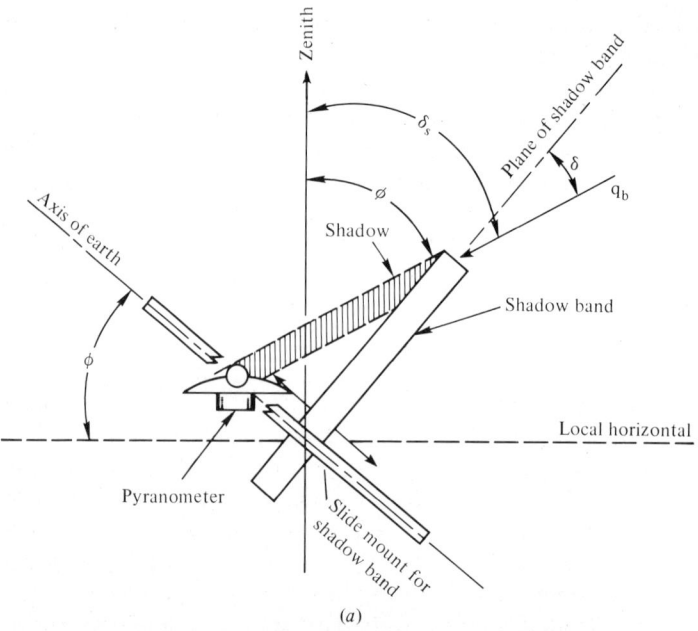

Figure 3-12 (*a*) Geometry of the shadow band. (*b*) Pyranometer with shadow band for measurement of diffuse insolation.

mounted on these tilted surfaces. If proper corrections are made to the resulting data, this is an acceptable practice.

A pyranometer with a shadow band is illustrated in Fig. 3-12. The band is adjusted back and forth along the direction parallel to the earth's axis to keep the entire glass dome of the pyranometer shaded from direct sunlight over the entire day. The frequency of the adjustment depends on the rate of change of the sun's declination (season), the size of the pyranometer dome, and the width of the band. Corrections [17] must be made for the blockage of the diffuse insolation from the sensor by the band. These corrections range between 5 and 25 percent, depending on the season, the local latitude, and the condition of the sky. The correction for the latitude and the season can be determined theoretically. Unfortunately, the local sky condition, or more precisely the directional distribution of the diffuse insolation, cannot be determined theoretically and can lead to significant errors if not properly accounted for. Recently it has been demonstrated [17] that reflections from the underside of the shadow band due to direct irradiation from the sun can lead to a significant overestimate of the diffuse component early and late in the day. Generally, one must determine the shadow band corrections experimentally by comparing the measurements of a pyrheliometer, an unshaded pyranometer, and a shaded pyranometer. Despite some uncertainty in the correction factors, pyranometers with shadow bands are popular since they can be used with an unshaded pyranometer to estimate the direct insolation without a tracking pyrheliometer and hence with no electric power requirement. However, corrections for the pyranometer cosine error and the shadow-band-induced errors must be made [18]. In many areas of the word (for example, in parts of Europe), there is a significant amount of diffuse insolation and its direct measurement is preferable to an indirect determination from pyrheliometer and pyranometer measurements.

By far the most common insolation measurements are made with a horizontally mounted pyranometer. Over 100 sites take such data in the United States and at least that many more throughout the world, mostly in Europe, Japan, and the southern half of Africa. Only a small fraction of these sites also record direct insolation with a pyrheliometer or diffuse insolation with a shaded pyranometer. Information is sparse on other than horizontal global insolation.

The optimal orientation of a solar collector is rarely horizontal so that the actual insolation incident on the collector is seldom directly available. Thus the horizontal global data must often be used to estimate insolation on other surfaces. If either the beam or the diffuse insolation is known in addition to the horizontal global insolation, it is a relatively straightforward and accurate calculation to determine the insolation on other arbitrarily oriented surfaces. However, without this information, the calculation depends on a general correlation between the measured horizontal global insolation and the calculated extraterrestrial insolation on a horizontal surface, as will be illustrated in Sec. 3-4. Note that all such calculations are based on the assumption that the so-called "diffuse" component is indeed isotropic. This is generally not true, and the only accurate way is to measure insolation with a properly calibrated pyranometer tilted in the correct direction.

70 SOLAR-THERMAL ENERGY SYSTEMS

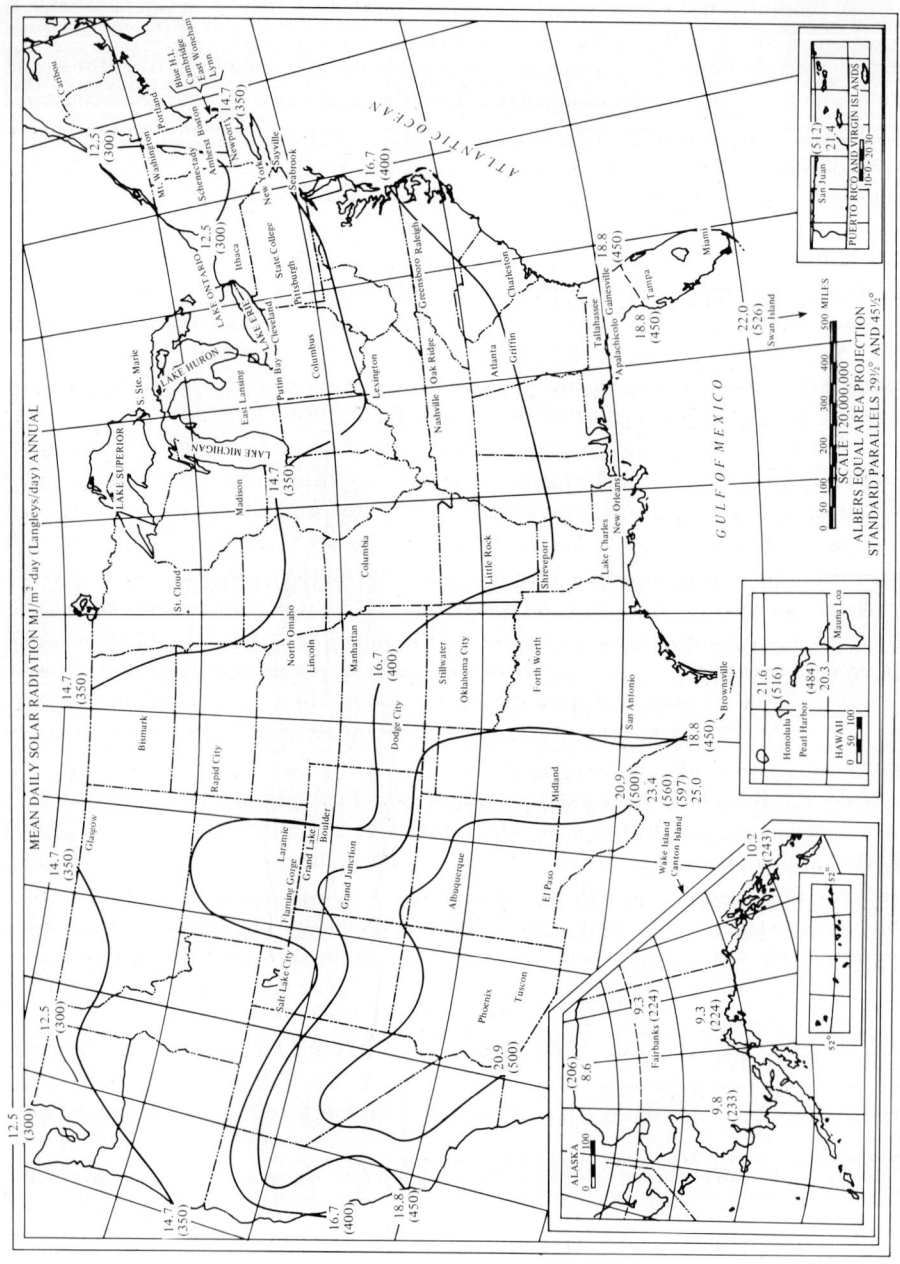

Figure 3-13 Annual average daily horizontal global terrestrial insolation.

The measured horizontal global insolation at selected sites can be used to construct iso-insolation maps like the one illustrated in Fig. 3-13 in which the annual average daily, horizontal, global terrestrial insolation is given. Annual average values actually have little use in detailed solar design since insolation levels and sun angles vary so much throughout the year. However, some general correlations, useful for preliminary design, have been developed based on annual average insolation. More useful, though, is the monthly average daily horizontal global terrestrial insolation. Maps illustrating these values are included for the United States in App. B. Maps of the monthly average daily beam insolation can also be found in App. B.

3-4 TERRESTRIAL INSOLATION ON TILTED SURFACES

An important question that the solar designer may ask† is, "How much sunlight strikes this collector, window, or skylight on the average each hour, or each day, or each month, or each year?"

To obtain an answer, two main problems must be faced: First, except at specific sites, there is little data available, and second, average hours, days, months, and years are difficult to determine. However, just as significant is the fact that in many design situations one does not really want to determine the "average" insolation. For example, it should be clear from the system studies of Chap. 2 that a solar water heater requires a minimal level of insolation even to begin to operate. For a given temperature above ambient, enough insolation must be available to overcome the thermal losses from the collector at that temperature. Available insolation above this threshold level can be used to heat the water. Therefore, the energy collected in a 2-day period during which one day is clear and the other is completely overcast is greater than that collected in another 2-day period during which both days experience intermittent clouds or consistently reduced insolation even though the sum of the insolation during each 2-day period is the same.

Although detailed design and/or performance studies rely on hourly insolation data, preliminary designs can be developed using correlations based on daily or monthly average daily insolation. In the sections that follow, the methods for estimating the hourly and monthly average daily insolation on tilted surfaces will be presented. There is a limited amount of information available on daily average insolation on tilted surfaces. However, this information is not particularly useful in design calculation.

Instantaneous or Hourly Insolation on Tilted Surfaces

Methods will be given to determine the instantaneous insolation available on tilted surfaces. But one may use the same procedures to estimate hourly average

---

† Approximately 20 percent of the articles in *Solar Energy*, the technical journal of the International Solar Energy Society, for the 1978–1979 period have been on the measurement and/or prediction of insolation.

insolation by performing the same calculations at the midpoint of each hour interval. The particular calculations required are dependent upon the availability of information. Commonly, the available information is (1) local horizontal global (pyranometer) and beam (pyrheliometer) terrestrial insolation or (2) local horizontal terrestrial insolation only. Without at least the local measured horizontal global terrestrial insolation, it is extremely difficult to accurately predict the insolation on tilted surfaces. Many papers have appeared over the last 60 years proposing various formulas to estimate the horizontal global terrestrial insolation based on various climatological data including sunshine hours, cloud cover, ambient temperature, and humidity. Many of the correlations can be successfully applied locally. However, there are so many factors, and their interactions are potentially so complex, that no general formulation of good accuracy has been developed.

It is assumed in the formulations to follow that the surface orientation (β, γ_c) (see Fig. 3-7), the time (δ, ω), and the place (ϕ) are all known. The basic procedure is as follows:

1. Determine the horizontal global insolation.
2. Determine the beam insolation.
3. Determine the diffuse insolation. (The order of 2 and 3 will depend on the type of correlation used.)
4. Assume that the diffuse insolation is truly diffuse (i.e., isotropic).†
5. Determine the angle between the beam insolation and the tilted surface.
6. Determine the separate contributions of the beam, diffuse, and reflected (also assumed isotropic) insolation on the tilted surface.

Sample calculations using the two types of commonly available information are now presented.

Case 1 The horizontal total global terrestrial insolation H and the total beam terrestrial insolation I_b are known.

The diffuse insolation on the horizontal surface is

$$H_d = H - H_b = H - I_b \sin \alpha_s \tag{3-18}$$

where $H_b =$ solar beam contribution to the global insolation on a horizontal surface
$\alpha_s =$ the solar altitude [Eq. (3-2)]

The insolation on the tilted surface [19] is therefore

$$q_s = I_b \cos \theta + H_d \cos^2 \frac{\beta}{2} + H\rho_r \sin^2 \frac{\beta}{2} \tag{3-19}$$

where $\theta =$ the angle between the solar beam and the normal to the tilted surface [Eqs. (3-9)–(3-12)]

† This model was first suggested by Liu and Jordan [19]. Other models have since been suggested which can account for increased circumsolar effects as well as other anisotropic effects, e.g., Refs. 20–24.

$\cos^2 \frac{\beta}{2}$ = the radiative configuration factor from the tilted surface to the sky (it is the fraction of the sky "seen" by the surface and represents the fraction of the diffuse insolation that strikes the surface)

$\sin^2 \frac{\beta}{2}$ = the radiation configuration factor from the tilted surface to the ground and surroundings (it is one minus the surface-to-sky configuration factor)

ρ_r = effective diffuse ground reflectance of the diffuse plus beam insolation on a horizontal surface (values are given in App. D [Ref. 25]).

Equation (3-19) is commonly written as:

$$q_s = H_b R_b + H_d R_d + H \rho_r R_r \tag{3-20}$$

or simply

$$q_s = HR \tag{3-21}$$

where

$$R_b = \frac{\cos \theta}{\sin \alpha_s} \tag{3-22}$$

$$R_d = \cos^2 \frac{\beta}{2} \tag{3-23}$$

$$R_r = \sin^2 \frac{\beta}{2} \tag{3-24}$$

$$R = \left(1 - \frac{H_d}{H}\right) R_b + \frac{H_d}{H} R_d + \rho_r R_r \tag{3-25}$$

Case 2 Only the horizontal total global terrestrial insolation H is known.

A way must be found to estimate either the beam or the diffuse insolation. Two correlations are described below. Both are based on the parameter

$$K_T = \frac{H}{H_0} \tag{3-26}$$

where H_0 is the extraterrestrial insolation on a horizontal surface [Eq. (3-16)]. The quantity K_T is given various names in the literature including the "cloudiness index," the "clearness index," and the "hourly percent sunshine." Here it will be called the "hourly clearness index." Its value ranges from about 0.8 under very clear conditions to nearly zero in severe overcast.

Based on beam and horizontal global terrestrial insolation measurements in the United States, Boes et al. [26] have developed a very simple correlation between K_T and I_b,

$$\begin{aligned} I_b &= -520 + 1800 K_T \; (\text{W}/\text{m}^2) & 0.85 > K_T \geq 0.30 \\ I_b &= 0 & 0.30 > K_T \end{aligned} \tag{3-27}$$

Alternatively, based on a limited amount of data, Orgill and Hollands [27] have developed a correlation between K_T and the ratio of the hourly diffuse to the hourly global terrestrial insolation on a horizontal surface,

$$\frac{H_d}{H} = \begin{cases} 1.0 - 0.249 K_T & K_T < 0.35 \\ 1.557 - 1.84 K_T & 0.35 \le K_T \le 0.75 \\ 0.177 & 0.75 < K_T \end{cases} \quad (3.28)$$

With H and H_o known, then either Eq. (3-27) or Eq. (3-28) could be used with Eq. (3-18); the results would then be used in Eq. (3-20). Alternatively, the results of both Eqs. (3-27) and (3-28) could be used directly in Eq. (3-20).

Example 3-6 The following insolation measurements were made at Houston, TX (30°N), at 10 A.M. (solar time) on April 20:

$$\text{Horizontal pyranometer:} \quad 750 \text{ W/m}^2$$

$$\text{Pyrheliometer:} \quad 650 \text{ W/m}^2$$

Determine the insolation on a surface facing 45° east of south and tilted 20° from the horizontal at the same time. The surface is located in an urban institutional environment.

SOLUTION From Eq. (3-2)

$$\sin \alpha_s = \sin (30) \sin (11.1) + \cos (30) \cos (11.1) \cos (30)$$

which implies the solar altitude

$$\alpha_s = 56.3°$$

and from Eq. (3-3)

$$\gamma_s = \sin^{-1} \frac{\cos (11.1) \sin (30)}{\cos (56.3)} = 62.2°$$

From Eq. (3-8)

$$\cos \theta = \cos (62.2 - 45) \cos (56.3) \sin (20)$$
$$+ \sin (56.3) \cos (20) = 0.963$$

so that

$$\theta = 15.6°$$

From Eq. (3-18)

$$H_d = 750 - 650 \sin (56.3) = 209 \text{ W/m}^2$$

From Eq. (3-19)

$$q_s = 650 \cos (15.6) + 209 \cos^2 (10) + 750(0.38) \sin^2 (10) = 874 \text{ W/m}^2$$

Example 3-7 Rework Example 3-6 but assume that only the pyranometer datum is available.

SOLUTION With the actual beam insolation unknown, one of the correlations presented above must be used. There are three options:

1. Use both Eqs. (3-27) and (3-28).

2. Use only Eq. (3-27) with Eq. (3-18).
3. Use only Eq. (3-28) with Eq. (3-18).

All three methods are used below. The calculation of K_T is common to all three. From Eq. (3-16)

$$H_0 = 1353\left[1 + 0.034 \cos \frac{2\pi(100)}{365}\right] \sin(56.3) = 1120 \text{ W/m}^2$$

From Eq. (3-26)

$$K_T = \frac{750}{1120} = 0.670$$

Method 1: From Eq. (3-27)

$$I_b = -520 + 1800(0.670) = 685 \text{ W/m}^2$$

From Eq. (3-28)

$$H_d = 750[1.557 - 1.84(0.670)] = 243 \text{ W/m}^2$$

and with Eq. (3-18)

$$H = 243 + 685 \sin(56.3) = 813 \text{ W/m}^2$$

From Eq. (3-19),

$$q_s = 685 \cos(15.6) + 243 \cos^2(10) + 813(0.38) \sin^2(10) = 904 \text{ W/m}^2$$

Note the calculated value of H compared to the measured insolation of 750 W/m². The fact that this calculated value of H is not the same as that originally given is not unexpected. The correlations, for example, are based on different data bases. Equation (3-27) is an average for the United States based on a limited number of stations (about 30) not including Houston. Equation (3-28) is based on many years' data but from only one site, Toronto, Canada. Even at a given site there is considerable variation among the beam, diffuse, and global insolation from day to day and the equation must be considered to correlate the average relations among the insolation types.

Method 2: If the beam correlation alone is used, Eqs. (3-18) and (3-19) yield

$$H_d = 750 - 685 \sin(56.3) = 180 \text{ W/m}^2$$

$$q_s = 685 \cos(15.6) + 180 \cos^2(10)$$
$$+ 750(0.38) \sin^2(10) = 844 \text{ W/m}^2$$

Method 3: If the diffuse insolation correlation is used, Eqs. (3-18) and (3-19) yield

$$I_b = \frac{H - H_d}{\sin \alpha_s} = \frac{750 - 243}{\sin(56.3)} = 609 \text{ W/m}^2$$

$$q_s = 609 \cos(15.6) + 243 \cos^2(10) + 750(0.38) \sin^2(10)$$
$$= 831 \text{ W/m}^2$$

Even though the two correlations lead to considerable difference in the breakdown between the beam and the diffuse components, the difference in the predictions for the insolation on the tilted surface is less than 2 percent. All predictions based on the correlations are within 5 percent of the results for q_s from Example 3-6, which is based on the pyrheliometer and pyranometer measurements.† The fact that one correlation yields an answer closer to the answer based on the measured data has no particular significance at this point, since this calculation represents only one check. However, the procedure used in Method 1 using two independent correlations is suspect, and is discouraged.

Monthly Average Daily Insolation on Tilted Surfaces

The instantaneous or hourly calculations are used for detailed performance studies. If long-term performance evaluation is required, a computer simulation of the system's thermal characteristics and available insolation can be developed. However, such detailed evaluation is often not justified. Further, the detailed (hourly) insolation data is rarely available. An alternative is to use performance evaluations based on a monthly average daily insolation model. Monthly average hourly insolation data can also be developed from this model. The model is based on the definition of a monthly average daily clearness index $\bar{K}_T$ analogous to the hourly clearness index defined in Eq. (3-26),

$$\bar{K}_T = \frac{\bar{H}}{\bar{H}_0} \qquad (3\text{-}29)$$

where $\bar{H}$ is the monthly average daily global terrestrial insolation on a horizontal surface and $\bar{H}_0$ is the monthly average daily extraterrestrial insolation on a horizontal surface. The $\bar{H}_0$ is determined by evaluating the daily average extraterrestrial insolation on the appropriate day of the month, as explained earlier in this chapter [Eq. (3-17)]. Over the last 20 years many researchers have developed correlations between $\bar{K}_T$ and the ratio of the monthly average daily diffuse (terrestrial) insolation on a horizontal surface, $\bar{H}_d$, and the monthly average daily global (terrestrial) insolation on a horizontal surface, $\bar{H}$. One of the simplest correlations, yet one of satisfactory accuracy, is due to Page [28]:

$$\frac{\bar{H}_d}{\bar{H}} = 1.00 - 1.13 \bar{K}_T \qquad (3\text{-}30)$$

Other correlations can be found in References 20 and 29–32. The successful use of this correlation depends upon the development of an expression for the monthly average daily insolation on a tilted surface, $\bar{q}_s$, analogous to Eqs. (3-19)–(3-21) for the instantaneous (hourly) insolation on a tilted surface. The development is as follows:

---

† The actual error in general depends on many factors, including the solar angles, atmospheric conditions, surface orientation, and local meteorological conditions.

The monthly average daily beam insolation on a horizontal surface can be expressed as

$$\bar{H}_b = \bar{H} - \bar{H}_d \tag{3-31}$$

The monthly average daily insolation on a tilted surface is

$$\bar{q}_s = \bar{H}_b \bar{R}_b + \bar{H}_d R_d + \bar{H}\rho_r R_r \tag{3-32}$$

or

$$\bar{q}_s = \bar{H}\bar{R} \tag{3-33}$$

where $\bar{R}_b$ = monthly average value of R_b
R_d, R_r = same quantities defined in Eqs. (3-23) and (3-24), respectively

and

$$\bar{R} = \left(1 - \frac{\bar{H}_d}{\bar{H}}\right)\bar{R}_b + \frac{\bar{H}_d}{\bar{H}} R_d + \rho_r R_r \tag{3-34}$$

Actually the beam contribution $\bar{H}_b \bar{R}_b$ should be calculated by integrating or summing the $H_b R_b$ product over the month. However, without the hourly values of H_b, this is not possible. It is an accepted practice to make the approximation [from Eq. (3-22)]

$$\bar{R}_b \simeq \frac{\overline{\cos \theta}}{\overline{\sin \alpha_s}} \tag{3-35}$$

that is, the monthly average daily value of $\cos \theta$ divided by the monthly average daily value of $\sin \alpha_s$. In theory this calculation would be performed only for the representative day of the month (Table 3-2). However, even with this simplification the evaluation is in general long and tedious. However, for surfaces facing the equator

$$\bar{R}_b(\gamma_c = 0) = \frac{\cos(\phi - \beta)\cos\delta \sin\omega'_{ss} + (\omega'_{ss}\pi/180)\sin(\phi - \beta)\sin\delta}{\cos\phi\cos\delta\sin\omega_{ss} + (\omega_{ss}\pi/180)\sin\phi\sin\delta} \tag{3-36}$$

(evaluated on the representative day) where ω_{ss} is the sunset hour angle in degrees and ω'_{ss} is the effective sunset hour angle in degrees for the surface [Eq. (3-13)]. Figure 3-14a through d presents the results of this calculation for selected values of $(\phi - \beta)$. Reference 33 recommends the use of these results for surfaces oriented within 15° from directly facing the equator (due south in the northern hemisphere), that is, $-15° \leq \gamma_c \leq 15°$. Hence, $\bar{R}_b$ for surfaces facing the equator is a function of ϕ, $(\phi - \beta)$, and δ. For a given value of $\bar{K}_T$, we can determine $\bar{H}_d/\bar{H}$. Then, if a value of $\rho_r(=0.10)$ is assumed, $\bar{R}$ can be calculated from Eq. (3-34). Such values are given in Table 3 of App. B.

In their classic paper on the interrelationship among direct, diffuse and global insolation, Liu and Jordan [19] assumed and then demonstrated with insolation data from several sites that

$$\frac{\bar{H}_d}{\bar{H}_0} \simeq \frac{H_d}{H_0} \tag{3-37}$$

that is, the long-term (monthly) average daily ratio of the diffuse to the global insolation on a horizontal surface is approximately equal to the instantaneous

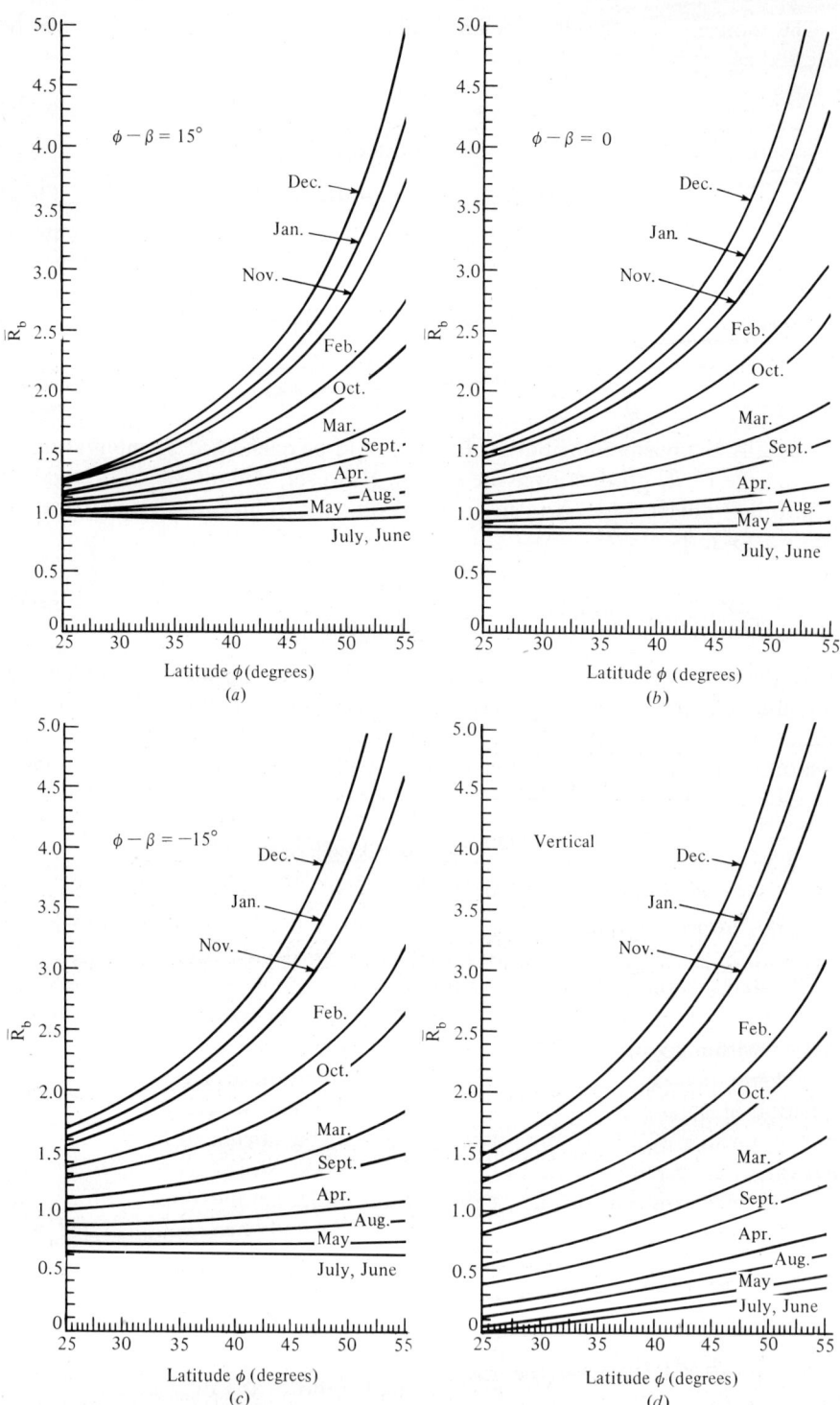

Figure 3-14 $\bar{R}_b$ for south-facing surfaces [33].

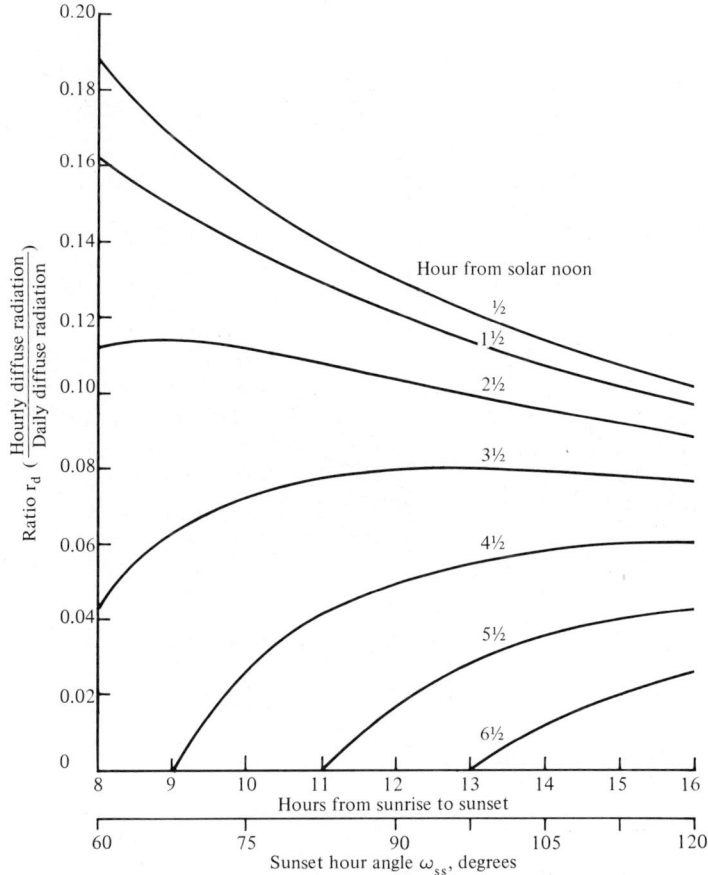

Figure 3-15 Ratio of the hourly diffuse to the daily diffuse insolation on a horizontal surface [19].

(hourly) ratio of the same quantity. Equation (3-37) leads to the result that

$$r_d = \frac{H_d}{\bar{H}_d} = \frac{\pi}{24} \frac{\cos\omega - \cos\omega_{ss}}{\sin\omega_{ss} - \omega_{ss}\cos\omega_{ss}} \tag{3-38}$$

This relationship is plotted in Fig. 3-15, and has been compared with actual data with good results.

A similar but not so exact correlation was shown to exist for

$$r = \frac{H}{\bar{H}} \tag{3-39}$$

Liu and Jordan [19] presented a plot of Eq. (3-39) which appears here as Fig. 3-16.

Example 3-8 Determine the monthly average daily global insolation in October for a south-facing surface tilted at 40° from the horizontal in Houston, TX. The surface is located in a residential area. Then estimate the

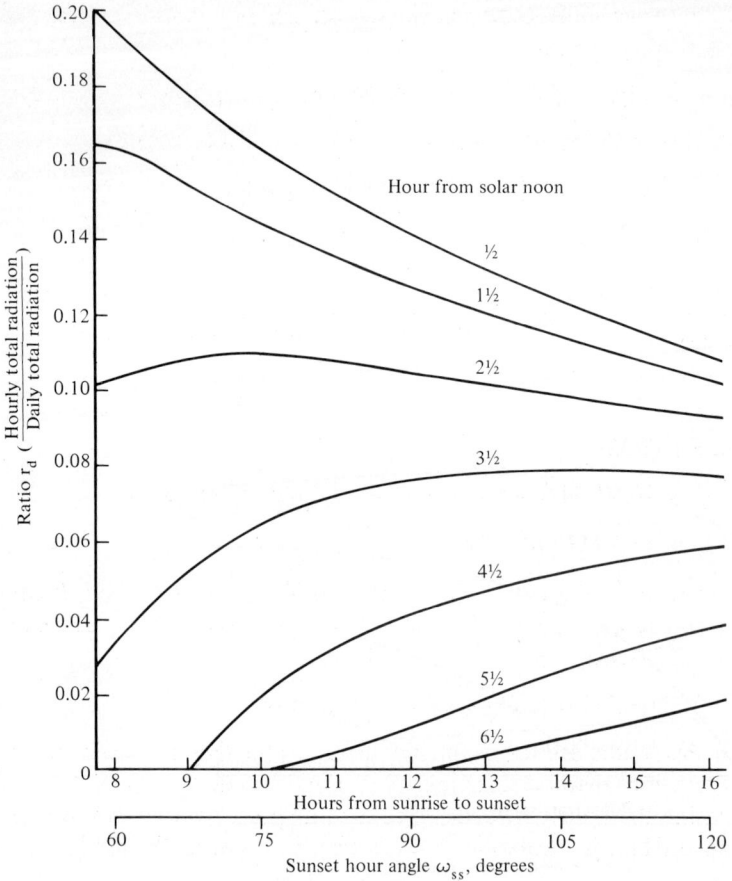

Figure 3-16 Ratio of the hourly global to the daily global insolation on a horizontal surface [19].

hourly average global insolation on the surface between 11 A.M. and 12 N. local time [Central daylight time (CDT)].

SOLUTION From the insolation maps in App. B, the monthly average daily global insolation on a horizontal surface is approximately 15.8 MJ/(m² · day). The monthly average daily extraterrestrial insolation on a horizontal surface in October at 30° N (from Table 3-3) is 27.2 MJ/(m² · day). Then from Eq. (3-29)

$$\bar{K}_T = \frac{\bar{H}}{\bar{H}_0} = \frac{15.8}{27.2} = 0.581$$

and from Eq. (3-30)

$$\bar{H}_d = \bar{H}(1.00 - 1.13\bar{K}_T) = 15.8 \text{ MJ/(m}^2 \cdot \text{day)} [1.00 - 1.13(0.581)]$$
$$= 5.4 \text{ MJ/(m}^2 \cdot \text{day)}$$

The monthly average daily beam insolation on the horizontal surface, Eq. (3-31), is

$$\bar{H}_b = 15.8 - 5.4 = 10.4 \text{ MJ/(m}^2 \cdot \text{day)}$$

For the mean day of the month, October 15, $n = 288$ (Table 3-2), then

$$\omega_{ss} = \cos^{-1}[-\tan(30)\tan(-8)] = 85.3°$$

From Eq. (3-36) with $\omega'_{ss} = \omega_{ss}$,

$$\bar{R}_b = \frac{\cos(30-40)\cos(-8.0)\sin(85.3) + (85.3\pi/180)\sin(30-40)\sin(-8.0)}{\cos(30)\cos(-8.0)\sin(85.3) + (85.3\pi/180)\sin(30)\sin(-8.0)}$$

$$= 1.342$$

Finally, from Eq. (3-32)

$$\bar{q}_s = 10.4(1.342) + 5.4\cos^2(20) + 15.8(0.3)\sin^2(20)$$

$$= 19.3 \text{ MJ/(m}^2 \cdot \text{day)}$$

Alternatively, after the calculation of $\bar{K}_T$ above, Table 3 of App. B can be used to estimate $\bar{q}_s$. For $\phi = 30°$, $(\phi - \beta) = -10°$, $\bar{K}_T = 0.581$, and, for October, $\bar{R} = 1.25$. Then,

$$\bar{q}_s = \bar{R}H = (1.25)(15.8) = 19.7 \text{ MJ/(m}^2 \cdot \text{day)}$$

Even though the results are for a different value of ρ_r, they are close to the results of the previous method since the effect of ρ_r is small.

To determine the hourly insolation, solar time must be used as the time frame. To convert 11 A.M. local time to solar time, Eq. (3-6) can be used:

$$\text{Solar time} = 11 \text{ A.M.} - 1\text{ h} + 4(90 - 95) \text{ min} + 13.7 \text{ min}$$

$$= 10 \text{ A.M.} - 20 \text{ min} + 13.7 \text{ min}$$

$$\cong 9{:}54 \text{ A.M.}$$

Thus, the insolation between 9:54 A.M. and 10:54 A.M. solar time is required. From Figs. 3-15 and 3-16 ($\omega_{ss} = 85.3°$, $t_d = 11.4$ h) at 1 h 26 min from solar noon:

$$r_d = \frac{H_d}{H_d} = 0.127$$

$$r = \frac{H}{H} = 0.133$$

Then between 11 A.M. and 12 N. CDT

$$H_d = 0.127(5.4) = 0.69 \text{ MJ/(m}^2 \cdot \text{h)}$$

$$H = 0.133(15.8) = 2.10 \text{ MJ/(m}^2 \cdot \text{h)}$$

The problem can now be treated as if the hour data were given. From Eq. (3-18)

$$H_b = H - H_d = 1.41 \text{ MJ/(m}^2 \cdot \text{h)}$$

At 10:24 A.M. solar time, Eq. (3-2) gives

$$\sin \alpha_s = \sin (30) \sin (-8.0) + \cos (30) \cos (-8.0) \cos (1\tfrac{36}{60})(15)$$
$$= 0.714$$

From Eq. (3-10)

$$\cos \theta = \sin (30 - 40) \sin (-8) + \cos (30 - 40) \cos (-8) \cos (24)$$
$$= 0.915$$

Note: $\cos \theta$ and $\sin \alpha_s$ can be determined directly from Fig. 3-8:

$$\phi - \beta = -10° < 0$$

so that $\cos \theta$ is found on the $\theta - \beta = +10°$ line where $\delta = -(-8) = 8°$. Therefore at 10:24 A.M. solar time

$$\cos \theta = 0.92$$
$$\sin \alpha_s = \cos \theta = 0.72 \quad \text{for} \quad (\phi - \beta = \phi = 30°)$$

Then from Eq. (3-20),

$$q_s = 1.41 \frac{0.915}{0.714} + 0.69 \cos^2 (20) + 2.10(0.3) \sin^2 (20)$$
$$= 2.49 \text{ MJ/(m}^2 \cdot \text{h)}$$

PROBLEMS

3-1 Determine the location of the sun (α_s, γ_s) at 10 A.M. solar time in Denver, CO ($\phi = 40°$N) on July 4 and on Christmas Day.

3-2 Determine the length of the longest and the shortest days of the year at 45°N, 30°N, and 10°N latitudes and at the equator.

3-3 Repeat Prob. 3-2 for 40°S and 20°S latitudes.

3-4 Determine the sunrise azimuth angles for 30°N, 50°N, and 70°N latitudes on May 15.

3-5 Determine the local time when solar noon occurs for Houston, TX, on the summer and winter solstices and on both equinoxes. Houston is at 95°W longitude. Texas imposes daylight saving time.

3-6 Repeat Prob. 3-1 at 10 A.M. local time. Denver is at 105°W longitude.

3-7 Determine the sunrise or sunset azimuth angles at your location for the date this assignment is due and check it out with a compass. (You must also correct for the difference between true north and magnetic north.)

3-8 Determine the sunrise and sunset local time today at your location and compare it to the times given in the local weather report.

3-9 Determine the sun's location in your city at 9 A.M. and 2 P.M. local time today.

3-10 At 80°N latitude, during what dates is the sun (*a*) never visible and (*b*) always visible? How many days a year is the sun not visible?

3-11 Describe the sun's apparent motion on the summer solstice as viewed from the north pole.

3-12 Determine the incident solar angle on a south-facing vertical surface at 10 A.M. solar time on October 15 at 40°N latitude.

3-13 Repeat Prob. 3-12 for a surface inclined by 30° from the vertical.

3-14 Determine the incident solar angle on a vertical surface facing southeast at 10 A.M. solar time on October 15 at 40°N latitude.

3-15 Determine the incident solar angle on a south-facing surface tilted at 30° from the horizontal at 30°N latitude on equinox at 8 A.M. solar time.

3-16 Develop an expression for the incident solar angle on a surface facing the equator, tilted at an angle β from the horizontal at latitude ϕ. Under what condition does the incident solar angle change exactly as a sine function as assumed in SOLSIM?

3-17 For a south-facing surface at 30°N latitude, tilted at 30° from the horizontal, compare the cosine of the incident solar angle on summer solstice with that assumed in SOLSIM by sketching both on the same graph.

3-18 Compute the monthly average daily extraterrestrial insolation on a horizontal surface for January, March, and June at your location. Compare your results to those given in Table 3-3.

3-19 Calculate the hourly values of R_b for a surface located at 40°N latitude, facing south, and tilted at 20° from the horizontal on August 16.

3-20 From the results of Prob. 3-19 determine a value of R_b on August 16. Compare the value with that calculated from Eq. (3-36).

3-21 For values of the hourly clearness index of 0.75 and 0.50, determine the hourly values of R [Eq. (3-25)] for the conditions of Prob. 3-19. Assume $\rho_r = 0.10$.

3-22 For assumed values of the monthly average daily clearness index of 0.75 and 0.50, compare values of $\bar{R}$ calculated from the results of Prob. 3-21 with values calculated from Eq. (3-34) and from Table B-3.

3-23 Determine the monthly average daily global insolation in December for a south-facing surface tilted at 55° from the horizontal in Columbia, MO. The surface is located on a house in an open field. Assume first that the field is snow-covered. Repeat the calculation for no snow.

3-24 Estimate the hourly average global insolation on the surface described in Prob. 3-23 between 1 and 2 P.M. local time (in December). Assume there is snow (see Table B-2).

3-25 For your location, assume a collector is mounted facing the equator with a tilt equal to the local latitude. Determine the monthly average daily global insolation on the collector for January, April, July, and October.

3-26 For the conditions given in Prob. 3-25, estimate the hourly average global insolation on the collector for one day during each of the months. Plot your results. Fit a sine curve through the data suitable for use in the SOLSIM simulation.

3-27 Use insolation data developed in Prob. 3-26 as input to SOLSIM and estimate the yearly performance (based on three-day runs at selected times during the year) of the SOLSIM base system at your location.

3-28 Obtain local insolation data (both global horizontal and beam if possible). Compare it to insolation predicted with techniques developed in this chapter.

3-29 Place a pyranometer on a tilted surface and determine the hourly distribution of global insolation on the surface. (Try to do this on a clear day.) Compare these results with those obtained from the methods developed in this chapter.

3-30 For March 12, 1981, in Austin, TX (30°N latitude, 97.7°W longitude), find (a) the time of sunrise, Central standard time, and (b) the day length.

3-31 For January in San Antonio, TX (latitude 29.5°N), find
    (a) The monthly average daily extraterrestrial insolation on a horizontal surface, MJ/(m² · day).
    (b) The monthly average daily global insolation on a horizontal surface, MJ/(m² · day).
    (c) The monthly average daily global insolation on a south-facing surface tilted at the latitude (i.e., $\beta = \phi = 29.5°$).

(d) The hourly insolation values expected on the tilted surface (monthly average hourly insolation). Fit a sine curve of the form $q_s(t) = q_{s,\,max} \sin[\pi(t - t_{rise})/t_{day}]$ to the result, assuming sunrise is at the expected local time for January 17.

3-32 For $q_{s,\,max} = 800$ W/m^2, vary the day length in the SOLSIM base case over 6, 8, and 10 h and plot the fraction of load and collection efficiency versus day length. Use mixed storage and a 3-day run.

3-33 For the SOLSIM base case with mixed storage, let $q_{s,\,max}$ be 600, 800, and 1000 W/m^2 for a 3-day run each. Plot the fraction of load supplied and collection efficiency versus $q_{s,\,max}$.

REFERENCES

1. *The American Ephemeris and Nautical Almanac for the Year 1975*, U.S. GPO Publication No. 0854-00046, U.S. Government Printing Office, Washington, DC, 1976.
2. Enrico Coffari, "The Sun and the Celestial Vault," Chap. 2 in A.A.M. Sayigh (ed.), *Solar Energy Engineering*, Academic Press, New York, 1977.
3. H. Tabor, "Stationary Mirror Systems for Solar Collectors," *Solar Energy*, vol. 2, pp. 27–33, 1958.
4. Edward Mazria, *The Passive Solar Energy Book*, Rodale Press, Emmaus, PA, 1979.
5. Bruce N. Anderson, *Solar Energy: Fundamentals in Building Design*, McGraw-Hill Book Co., New York, 1977.
6. H.C. Hottel and B.B. Woertz, "Performance of Flat-Plate Solar-Heat Collectors," *Trans. ASME*, vol. 64, p. 91, 1942.
7. L.O. Lamm, "A New Analytical Expression for the Equation of Time," *Solar Energy*, vol. 26, no. 5, p. 465, 1981.
8. S.A. Klein, "Calculation of Monthly Average Insolation on Tilted Surfaces," *Solar Energy*, vol. 19, no. 4, pp. 325–329, 1977.
9. J.A. Duffie and W.A. Beckman, *Solar Energy Thermal Processes*, John Wiley & Sons, New York, 1974.
10. Kinsell L. Coulson, *Solar and Terrestrial Radiation: Methods and Measurements*, Academic Press, New York, 1975.
11. N. Robinson (ed.), *Solar Radiation*, Elsevier Publishing Co., Amsterdam, 1966.
12. M.P. Thekaekara, "Solar Radiation Measurement: Technique and Instrumentation," *Solar Energy*, vol. 18, no. 4, pp. 309–329, 1976.
13. Byard D. Wood, "Solar Energy Measuring Equipment," Chap. 19 in *Solar Energy Engineering* (see Ref. 2 above).
14. Thomas H. Jeys and Lorin L. Vant-Hull, "The Contribution of the Solar Aureole to the Measurement of the Pyrheliometer," *Solar Energy*, vol. 18, no. 4, pp. 343–347, 1976.
15. A. Jeffrey Mohr, Duane A. Dahlberg, and Inge Dirmhirn, "Experiences with Tests and Calibration of Pyranometers for a Mesoscale Solar-Irradiance Network," *Solar Energy*, vol. 22, no. 3, pp. 197–203, 1979.
16. D.J. Norris, "Calibration of Pyranometers in Inclined and Inverted Positions," *Solar Energy*, vol. 16, no. 1, pp. 53–55, 1974.
17. B.A. LeBaron, W.A. Peterson, and I. Dirmhirn, "Corrections of the Diffuse Irradiance Measured with Shadow-bands," *Solar Energy*, vol. 25, no. 1, pp. 1–13, 1980.
18. Jeffrey A. Secrest and Inge Dirmhirn, "Accuracies Achievable with Indirect Measurements of Direct Solar Irradiance Component," *Solar Energy*, vol. 23, no. 6, pp. 509–512, 1979.
19. B.Y.H. Liu and R.C. Jordan, "The Interrelationship and Characteristic Distribution of Direct, Diffuse and Total Solar Radiation," *Solar Energy*, vol. 4, no. 3, pp. 1–19, 1960.
20. John E. Hay, "Calculation of Monthly Mean Solar Radiation for Horizontal and Inclined Surfaces," *Solar Energy*, vol. 23, no. 4, pp. 301–307, 1979.
21. R.C. Tempo and K.L. Coulson, "Solar Radiation Incident Upon Slopes of Different Orientations," *Solar Energy*, vol. 19, no. 2, pp. 179–184, 1977.

22. V.M. Puri, R. Jimenez, M. Menser, and F.A. Costello, "Total and Non-Isotropic Diffuse Insolation on Tilted Surfaces," *Solar Energy*, vol. 25, no. 1, pp. 85–90, 1980.
23. T.A. Weiss and G.O.G. Löf, "The Estimation of Daily Clear-Sky Solar Radiation Intercepted by a Tilted Surface," *Solar Energy*, vol. 24, no. 3, pp. 287–294, 1980.
24. T.M. Klucher, "Evaluation of Models to Predict Insolation on Tilted Surfaces," DOE/NASA Report No. 1022-78/28 and NASA Technical Memorandum No. TM-78842, March 1978.
25. B.D. Hunn and D.O. Calafell, II, "Determination of Average Ground Reflectivity for Solar Collectors," *Solar Energy*, vol. 19, no. 1, pp. 87–89, 1977.
26. E.C. Boes, I.J. Hall, R.R. Prairie, R.P. Stromberg, and H.E. Anderson, "Distribution of Direct and Total Solar Radiation Available for the USA," *Proceedings of the 1976 Annual Meeting of the Am. Sec. of ISES, Sharing the Sun*, vol. 1, Winnipeg, August 15–20, 1976, pp. 238–263.
27. J.F. Orgill and K.G.T. Hollands, "Correlation Equation for Hourly Diffuse Radiation on a Horizontal Surface," *Solar Energy*, vol. 19, no. 4, pp. 357–359, 1977.
28. J.K. Page, "The Estimate of Monthly Mean Values of Daily Total Short-Wave Radiation on Vertical and Inclined Surfaces from Sunshine Records for Latitudes 40°N to 40°S," Paper No. 35/5/98, *Proc. UN Conference on New Sources of Energy*, Rome, 1961.
29. N.K.O. Choudhury, "Solar Radiation at New Delhi," *Solar Energy*, vol. 7, no. 2, pp. 44–52, 1963.
30. G. Stanhill, "Diffuse Sky and Cloud Radiation in Israel," *Solar Energy*, vol. 10, pp. 96–101, 1966.
31. D.J. Norris, "Solar Radiation on Inclined Surfaces," *Solar Energy*, vol. 10, pp. 72–77, 1966.
32. Manuel Collares-Pereira and Ari Rabl, "Simple Procedure for Predicting Long Term Average Performance of Nonconcentrating and of Concentrating Solar Collectors," *Solar Energy*, vol. 23, no. 3, pp. 235–253, 1979.
33. William A. Beckman, Sanford A. Klein, and John A. Duffie, *Solar Heating Design by the f-Chart Method*, John Wiley & Sons, New York, 1977.

CHAPTER
FOUR

GENERAL DESCRIPTION OF SOLAR-THERMAL COLLECTORS AND METHODS TO EVALUATE THEM

An introduction to the thermal performance of solar collectors is presented, including the development and discussion of the concept of the collector's instantaneous efficiency. This introduction is followed by a general description of solar collector designs beginning with the nonconcentrating flat-plate collectors and continuing through tracking concentrating collectors. The chapter ends with a discussion of the tradeoffs associated with collector selection. Detailed evaluation and thermal analysis are not discussed here, but are addressed in Chap. 5. The purpose of this chapter, however, is to present sufficient information so that the system designer not particularly interested in the detailed thermal analysis of the collector, will still become familiar with the concepts, the terminology, and the selection tradeoffs.

4-1 LIST OF SYMBOLS

A_c area of collector aperture, m² (collector area)
A_e area of collector absorber, m²
C_g equal to A_c/A_e; collector geometric or ideal concentration ratio
F_R heat removal factor, Eq. (4-2)
q_s incident solar flux (insolation), W/m²

q_u usable energy collected per unit area, W/m²
T_a ambient temperature, K
$T_{f,\,in}$ collector fluid inlet temperature, K
T_e temperature of absorber plate, K
$\bar{T}_e$ integrated average temperature of collector absorber plate, K
$\bar{U}$ collector thermal loss coefficient, W/(m² · K)
α absorber solar absorptance
η collector (instantaneous) thermal efficiency, Eq. (4-1)
ϵ absorber infrared emittance

4-2 THE THERMAL EVALUATION OF SOLAR ENERGY COLLECTORS

The purpose of a solar thermal collector is to absorb the sun's radiative energy and to transfer the resultant thermal energy to a fluid which, in turn, delivers the energy to storage or directly to the ultimate use (load). Therefore, in the evaluation of a solar thermal collector, the determination of the collector's thermal performance (as described above) is the primary objective. However, a complete evaluation should include considerations of cost and durability or expected useful life.

Of these last aspects, only cost has received much consideration in most analyses and it will be addressed specifically in the collector selection section at the end of this chapter. The remaining aspect, durability, is very important but often overlooked in system evaluation. This is due largely to a general lack of information on which to base a judgment. Most collectors and indeed systems are just too new to have developed a "track record."

A standard method for evaluating the thermal performance of a solar collector has been more or less accepted in the industry. The actual testing procedure is described in a Standard [1] published by the American Society of Heating, Refrigerating, and Air Conditioning Engineers (ASHRAE) based on an NBS report [2]. These procedures are discussed at the end of Chap. 5.

There are three main objectives in the thermal design of a solar energy collector [refer to Eq. (2-5)]:

1. Capture or absorb as much as possible of the available solar energy (high optical efficiency): $(\tau\alpha)_{eff}$ approaching 1.
2. Retain as much of this received energy as possible (low thermal loss coefficient): b as small as possible.
3. Transfer as much of this retained energy as possible into the coolant (high effectiveness or heat removal factor): F_R close to 1.

It is usually a mistake to try to optimize collector thermal performance by independently considering only one or two of the three objectives listed above. Design changes which address one objective (e.g., decrease the thermal loss coefficient) may also inadvertently affect another (e.g., decrease the optical efficiency). While not specifically addressed here, it is obvious that incremental cost also plays an important role in the decision to improve the thermal performance by changing the design. Hence, a good design will strike a balance among the three design objectives, the cost, and the durability.

There are two techniques or testing procedures used to evaluate the collector thermal performance. Both yield figures of merit or efficiencies which are the ratio of the thermal energy collected to available insolation. The main difference is in the time period over which the measurements are made. Thermal performance measured over a time period measured in minutes at a specific collector temperature is termed the "instantaneous collector efficiency," usually called simply "collector efficiency." Thermal performance measured over a much longer time period, e.g., for an entire day, for a range of operating conditions and collector temperature is termed the "all-day collector efficiency."

All-Day Collector Efficiency

As the name implies, all-day collector efficiency is the ratio of the total thermal energy collected in 1 day (or other suitable time interval) to the available insolation at the collector aperture over the same day (or same time interval). In a properly run experiment, the results of this type of evaluation can be very useful. However, many of the factors which influence all-day performance are impossible to control and/or difficult to monitor—for example, the nature and intensity of the insolation, wind speed and direction, ambient temperature, and piping and storage thermal losses. All-day performance testing actually results in an evaluation of system performance rather than collector performance. The results are also limited to the geographical and climatic region of the test. Therefore, as a universal method to evaluate and compare collector performance in a wide variety of uses, the all-day technique is not satisfactory. However, for side-by-side comparison it is the most realistic and useful testing technique since many operating system effects, such as the incident sun angle, interaction with storage, and transients, are included.

Instantaneous Collector Efficiency

The currently accepted method for the evaluation of solar collectors apart from a complete system is based on their instantaneous efficiency. Short-term steady state collector performance is obtained under rigidly controlled conditions. The instantaneous thermal efficiency is the same quantity defined in Eq. (2-2) as collector efficiency and is the ratio of the rate at which the coolant is heated (the product of the mass flow rate, the specific heat, and the temperature rise across the collector) to the rate at which solar energy strikes the aperture of the collector. Using Eq. (2-5) at low collection temperatures the instantaneous collector efficiency is

$$\eta = F_R \left[(\tau\alpha)_{\text{eff}} - \frac{\bar{U} A_e (T_{f,\text{in}} - T_a)}{A_c q_s} \right]$$

$$= F_R \left[(\tau\alpha)_{\text{eff}} - \frac{\bar{U}}{C_g} \left(\frac{T_{f,\text{in}} - T_a}{q_s} \right) \right] \qquad (4\text{-}1)$$

where $\quad C_g = A_c/A_e$

is the collector geometric or ideal concentration ratio and

$$F_R = \left[(\tau\alpha)_{\text{eff}} - \frac{\bar{U}}{C_g} \left(\frac{\bar{T}_e - T_a}{q_s} \right) \right] \Big/ \left[(\tau\alpha)_{\text{eff}} - \frac{\bar{U}}{C_g} \left(\frac{T_{f,\text{in}} - T_a}{q_s} \right) \right] \qquad (4\text{-}2)$$

is the heat removal factor. The heat removal factor is the ratio of the actual thermal energy removed from the collector by the coolant to the thermal energy that would have been removed if the entire absorber were at the fluid inlet temperature. The fluid inlet temperature is the lowest possible temperature associated with the absorber under operating conditions. Therefore, energy losses evaluated while assuming that the average absorber temperature is equal to the fluid inlet temperature $[\bar{U}(T_{f,\text{in}} - T_a)]$ would always underestimate the true thermal losses from the collector. In turn, the resulting net collected energy

$$Q_u = q_s (\tau\alpha)_{\text{eff}} A_c - \frac{\bar{U} A_c}{C_g} (T_{f,\text{in}} - T_a)$$

would be overestimated. The heat removal factor is therefore always less than 1 when $\bar{T}_e > T_a$ (which is usually the case), but approaches 1 as the temperature differences between the absorber and the fluid (good thermal transfer within the collector) and between the coolant inlet and outlet (high coolant flow rate or low q_s) approach 0. The fluid flow rate is specified by the system designer but is constrained by the pressure drop in the system, which of course increases as the flow rate increases (and results in increased pumping power requirement). Clearly, low values of q_s are not desirable for good overall system performance. The collector designer can produce a collector with good thermal performance ($T_e \simeq T_{f,\text{in}}$), but good collector performance within the overall system depends on good system design (proper fluid flow rates and collector orientation). Hence, "reasonable," not maximum, values of F_R are desirable. "Reasonable values" are, of course, defined by a system analysis, but for nonconcentrating collectors usually range between 0.85 and 0.90 for a liquid heater and between 0.6 and 0.7 for an air heater.

The instantaneous thermal efficiency is a function of the collector design [F_R, $(\tau\alpha)_{\text{eff}}$, $\bar{U}$, and C_g] and the quantity $(T_{f,\text{in}} - T_a)/q_s$. In dimensional form, efficiency is usually represented graphically as shown in Fig. 4-1 (in analogy to the dimensionless form of Fig. 2-2). The collector efficiency is represented on the vertical or y axis and the quantity $(T_{f,\text{in}} - T_a)/q_s$ on the horizontal or x-axis. A second horizontal axis may be included which allows a direct determination of

90 SOLAR-THERMAL ENERGY SYSTEMS

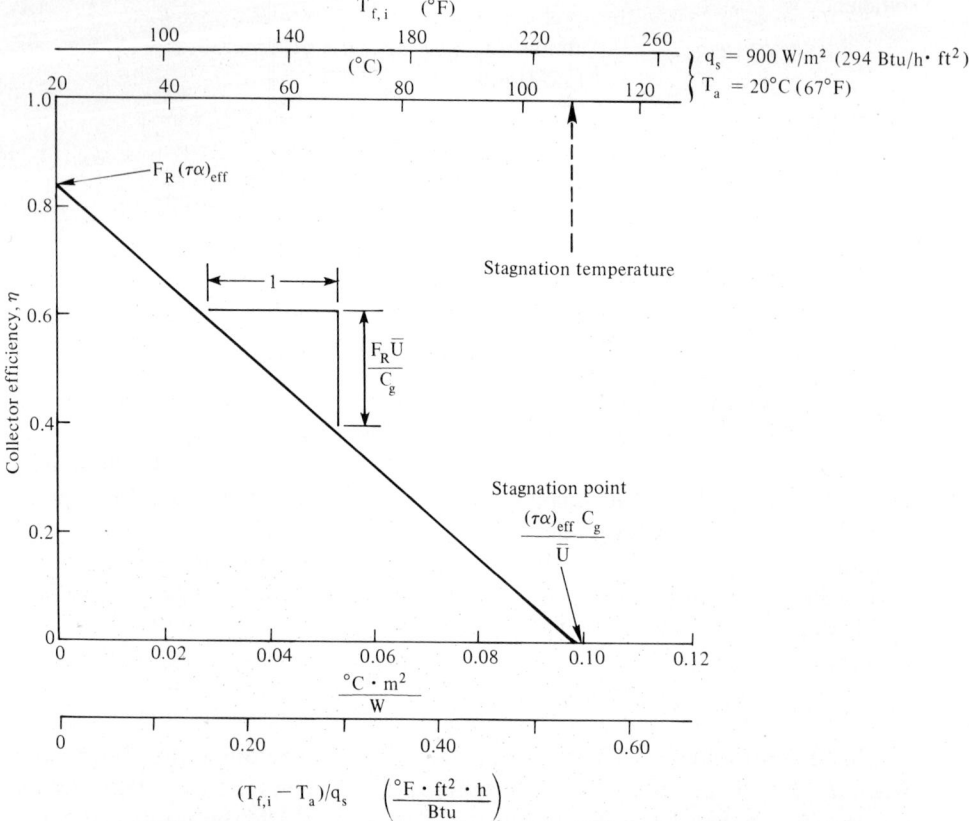

Figure 4-1 Instantaneous efficiency curve for a typical flat-plate collector.

collector efficiency as a function of $T_{f,\text{in}}$ for a fixed level of insolation and ambient temperature. {In the figure, $q_s = 900$ W/m² [294 Btu/(ft² · h)], and $T_a = 20°C$ (67°F).} The illustrated performance is that of a single-glazed (one glass cover), nonselective ($\alpha = \epsilon$), flat-plate ($C_g = 1$) collector. The y intercept of the performance curve is the optical efficiency times the heat removal factor, $F_R(\tau\alpha)_{\text{eff}}$. The magnitude of the slope of the curve is the product of the heat removal factor and the thermal loss coefficient divided by the geometric concentration ratio, $F_R \bar{U}/C_g$ ($= F_R \bar{U}$ for a flat-plate collector with $C_g = 1$). The stagnation temperature [from Eq. (2-9)] for the collector is defined as the maximum equilibrium temperature attainable and therefore corresponds to operation at no flow, when no energy is withdrawn from the collector by the coolant ($\eta = 0$). The stagnation point is the x intercept of the curve at $x = C_g(\tau\alpha)_{\text{eff}}/\bar{U}$. As discussed in Chap. 2, the thermal loss coefficient $\bar{U}$ is not constant but is a function of temperature (due primarily to the radiative transfer effects). For the sake of illustration it has been assumed to be constant in Fig. 4-1. In fact, the efficiency curves should be slightly concave down because the loss coefficient increases with increasing temperature.

Example 4-1: Write the expression for the instantaneous efficiency for the collector described in Fig. 4-1. Determine the rate at which useful energy can be extracted from the collector if $q_s = 850$ W/m², $T_a = 25°C$, and $T_{f,\text{in}} = 75°C$.

SOLUTION From Fig. 4-1 we have at the y intercept

$$F_R(\tau\alpha)_{\text{eff}} = 0.83$$

and at the x intercept

$$\frac{(\tau\alpha)_{\text{eff}}}{\bar{U}} = 0.098 \ (°C \cdot m^2)/W \qquad (C_g = 1)$$

Then

$$\frac{F_R(\tau\alpha)_{\text{eff}}}{(\tau\alpha)_{\text{eff}}/\bar{U}} = F_R \bar{U} = \frac{0.83}{0.098} = 8.47 \ W/(m^2 \cdot °C)$$

From Eq. (4-1)

$$\eta = 0.83 - 8.47\left(\frac{T_{f,\text{in}} - T_a}{q_s}\right)$$

At the given condition

$$\eta = 0.83 - 8.47 \ W/(m^2 \cdot °C) \left[\frac{(75-25)°C}{850 \ W/m^2}\right] = 0.32$$

$$q_u = \eta q_s = 282 \ W/m^2$$

4-3 CLASSIFICATION AND GENERAL DESCRIPTION OF SOLAR ENERGY COLLECTORS

The relative thermal performances of hypothetical flat-plate, moderately concentrating, and highly concentrating collectors are shown in Fig. 4-2. This comparison is made based on global insolation, assuming a 90 percent beam component, and on the assumption that each collector has the same aperture and the same materials for the absorber (receiver) and cover assemblies. From Eq. (4-1) and under the assumption that $\bar{U}$ (based on absorber area) is approximately the same for each collector, the three curves represent collectors with geometric concentration ratios of approximately 1, 3, and 15. The slopes of the curves are inversely proportional to the concentration ratios. The y intercept or $F_R(\tau\alpha)_{\text{eff}}$ for each collector was chosen somewhat arbitrarily but is intended to qualitatively illustrate the expected drop in optical efficiency $(\tau\alpha)_{\text{eff}}$ and moderate increase in

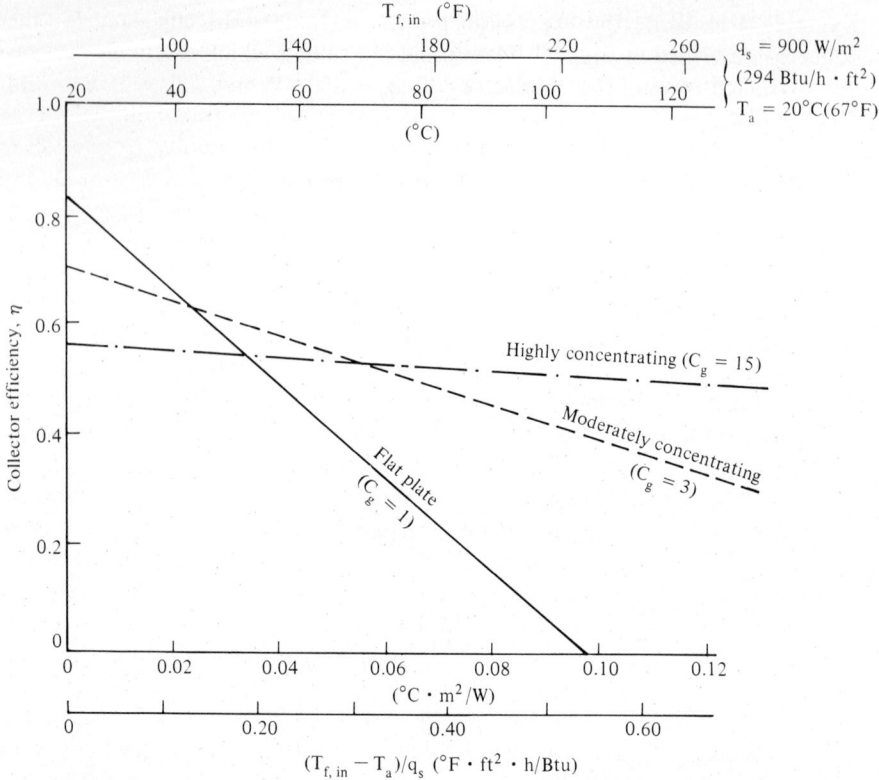

Figure 4-2 Relative performances for flat-plate, moderately concentrating, and highly concentrating solar collectors.

F_R for concentrating collectors. The decrease in optical efficiency is due primarily to two effects:

1. *Acceptance angle effect.* Only the direct or beam insolation can be used by a highly concentrating or focusing collector. That is, its optical acceptance angle is small. As the concentration ratio decreases, the acceptance angle increases, but only when the concentration ratio reaches 1 (a flat plate) does it reach 180°.† Therefore, if performance is based on the global insolation striking the aperture, the concentrating collector takes an immediate loss. This loss shows up as a decreased optical efficiency. Incidentally, if misalignment occurs during operation (due to a tracking error or wind loading, etc.), the performance can suffer significantly.
2. *Internal optical effects.* Concentration can be achieved only through reflection or refraction of the incident beam. Either process will result in a decrease in the amount of energy in the solar beam due to absorption (and/or scattering) in the lens or mirror.

† Actually, due to the effects of refractive optics at interfaces between different materials (like air and glass), the useful acceptance angle for a flat-plate collector is considerably less than 180°. This effect is discussed further in Chap. 5.

In concentrating collectors, these two effects (particularly the second) cause the image size to be increased. If the concentration ratio is increased by decreasing the receiver size until the receiver is smaller than the image, then a significant fraction of the radiation may be lost. The losses are often accounted for by multiplying the $(\tau\alpha)_{\text{eff}}$, which is primarily a function of the radiative properties of the optical components, by an "intercept factor," which is primarily a function of the geometry (see Chap. 5).

The heat removal factor F_R may be larger in concentrating collectors, since the smaller absorber reduces or eliminates the "fin" efficiency effects (the temperature drop due to conduction along the absorber panel to the fluid tube) and the bond resistance between the absorber panel fins and the tube. These effects are analyzed in detail in Chap. 5, but the impact can be understood through the discussion of collector hardware to follow in this chapter. In summary, the value of $F_R(\tau\alpha)_{\text{eff}}$ depends on the thermal design and the optical quality of the collector, its alignment, and the angular distribution of the insolation.

Due to their higher optical efficiency, most flat-plate collectors outperform the concentrators at low temperature. However, as the collector temperature rises, the smaller ratio of absorber to aperture area characteristic of the concentrators, results in a less rapid decrease in efficiency with temperature. At sufficiently elevated temperatures, a properly designed concentrator will outperform a typical flat-plate collector. The actual crossover in performance depends on the details of the various designs and the relative amounts of the beam and diffuse insolation. In practice, the choice between a concentrating and a flat-plate collector is based on cost difference (both initial and operating) and local insolation characteristics as well as the load temperature requirements and the performance of the particular collectors.

It is a common belief that the only objective of concentration is to achieve a high temperature. In fact, the major objective of concentration is to achieve the desired high temperature while maintaining an acceptable efficiency. Through concentration, absorber (receiver) losses are reduced by reducing absorber area, which generally permits maintenance of a better efficiency than for a flat plate at moderate to high temperatures.

A general discussion of collector designs now follows. First, the nonconcentrating collectors are covered and then the concentrating. Details of the thermal analysis of these collectors will follow in Chap. 5.

4-4 NONCONCENTRATING COLLECTORS

Liquid Heaters

The flat-plate design is by far the most common type of nonconcentrating collector. The basic designs for liquid- and air-heating flat-plate collectors are illustrated in Figs. 4-3 to 4-7. An absorbing surface with means for efficiently

94 SOLAR-THERMAL ENERGY SYSTEMS

(*a*) Liquid heater

(*b*) Air heater

Figure 4-3 Photographs of flat-plate collectors.

GENERAL DESCRIPTION OF SOLAR-THERMAL COLLECTORS 95

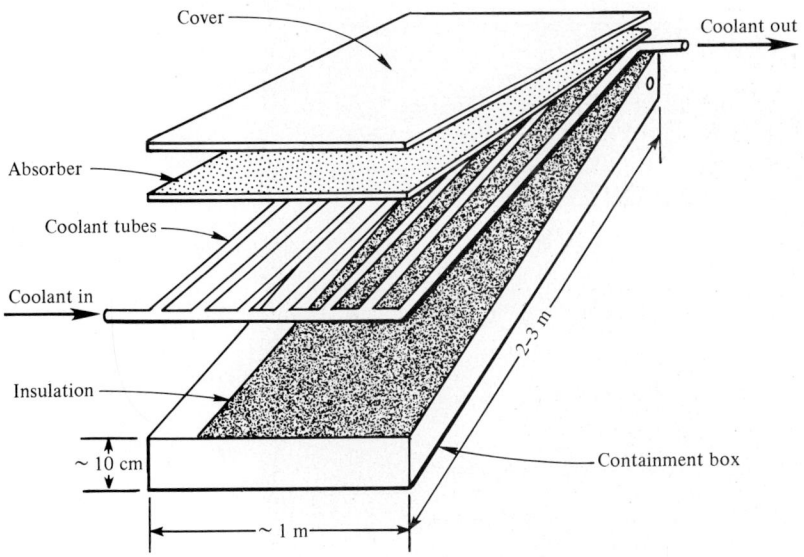

(a) Single-cover liquid heater

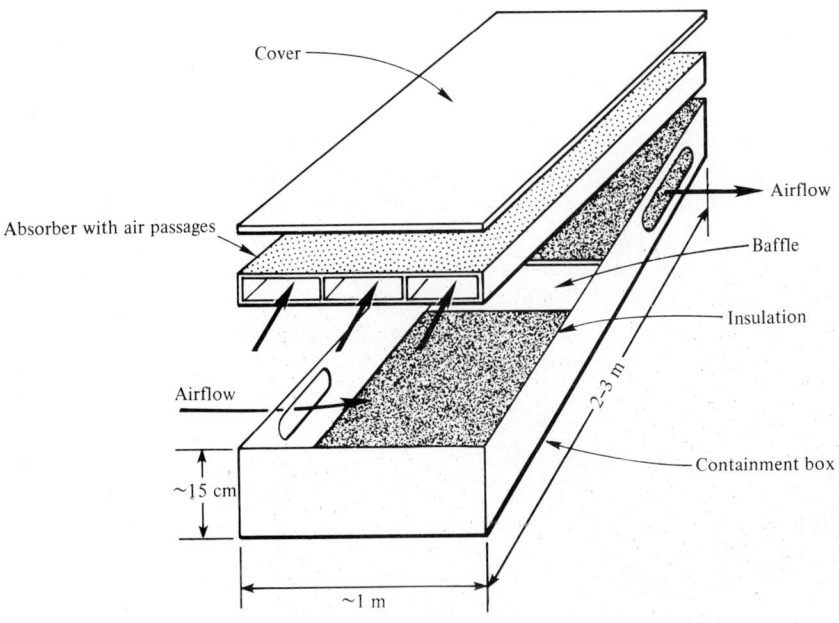

(b) Single-cover air heater

Figure 4-4 Flat-plate collectors.

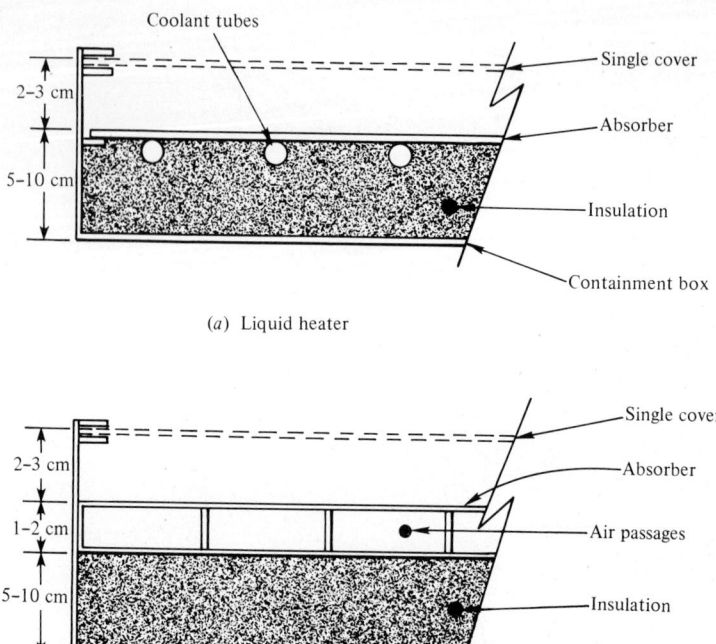

Figure 4-5 End view cutaway of flat-plate collectors.

transferring the absorbed solar energy to a fluid is the heart of the collector. The materials used for this absorber include copper, aluminum, steel, glass, and plastic, with copper and aluminum being by far the most common. In liquid heaters, the liquid usually flows in tubes which are attached to or integral with a fin or sheet that serves as the absorber, as illustrated in Fig. 4-6. Copper is the most popular material for the tubes and fins because of its good thermal conductivity and corrosion resistance. However, an aluminum absorber with copper tubes attached by a forced fit combines the desirable features of copper with the econ-

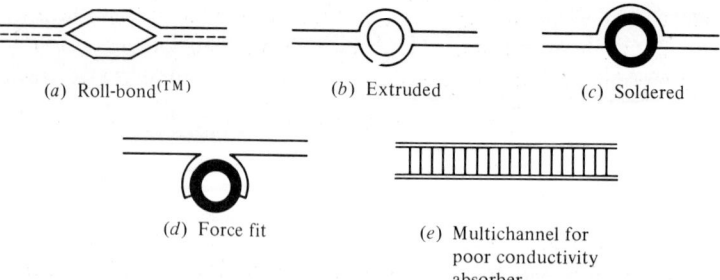

Figure 4-6 Liquid heater absorber—tube details.

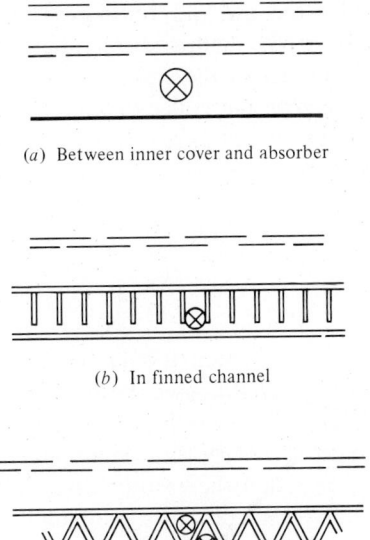

(a) Between inner cover and absorber

(b) In finned channel

(c) Around corrugated insert

Figure 4-7 Air heater absorber detail. ⊗ Designates air flow.

omy of aluminum (Fig. 4-6d). The absorber is usually coated with a material with large solar absorptivity. The coating may also be selective, that is, it may have a low infrared emissivity.

The cover above the absorber surface reduces the convective and radiative heat loss from the collector. Many materials are used for covers, including glass, plastic sheet and film, and glass-reinforced plastic. Glass is the most commonly used. The absorber, with insulation on the unirradiated surface, is enclosed by the glazing and a weatherproof container. Fiberglass is the most common insulation because it will withstand flat-plate stagnation temperatures. In low-temperature applications such as for heating swimming pools, the absorber plate is used without glazing, insulation, or enclosure.

Almost all commercially available liquid-heating collectors use parallel flow through the collector, as illustrated in Fig. 4-4a. The individual tubes connect into headers at each end. Each collector then has two external connections—one for inlet and one for outlet which preferably should be at diagonally opposite corners of the collector to enhance even flow distribution through the parallel tubes as well as to facilitate plumbing of collector banks. To prevent air entrapment and facilitate fill and drain, the flow through the collector is from bottom to top.

Air Heaters

In air heaters the air flow is usually between two metallic surfaces which separate the air from the cover and the insulation. However, other designs have been used, as illustrated in Fig. 4-7. The fins and corrugations are used to both increase the heat-transfer coefficient between the surface and the air and to increase the surface area in contact with the air. This reduces the temperature difference between the surface and the air, which, in turn, improves the collector thermal efficiency. This also increases the pressure drop across the collector and consequently increases blower power. The collector design is optimized in terms of its thermal efficiency and the blower power requirements. Except for the size of the ducting, a flat-plate air heater is similar in appearance to a flat-plate liquid heater, as illustrated in Fig. 4-3.

Evacuated Tubular Collectors

Another type of nonconcentrating collector is the evacuated tubular collector. Several designs are shown in Figs. 4-8 and 4-9. They are used both as nonconcentrating collectors and as receivers for linear concentrating systems. The cylindrical geometry affords the structural strength necessary to accommodate a vacuum. The major interest in this type of collector is that the vacuum essentially eliminates conduction and convection losses. Thus these collectors, coupled with a radiatively selective absorber, provide a sufficiently low overall thermal loss coefficient such that efficient high-temperature operation is possible. All the designs in Fig. 4-8 have "plug-in" capability, as illustrated in Fig. 4-9. The inlet and outlet are at the same end and are fitted to an external manifold. One of the designs (Fig. 4-8d) incorporates an internal mirror to provide some concentration.

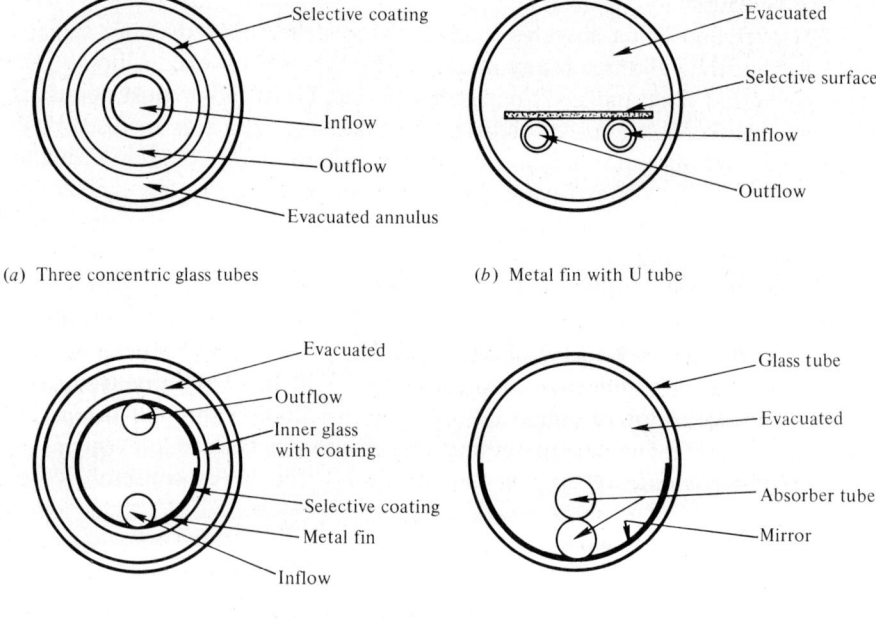

(a) Three concentric glass tubes
(b) Metal fin with U tube
(c) Cylindrical metal fin
(d) Half silvered outer glass

Figure 4-8 Evacuated tubular collectors—cross section.

Figure 4-9 An array of tubular collectors.

Low Temperature Collectors

For heat collection at 10–15°C above ambient, a much simpler collector can be used. The principal application here is for swimming pools. For these collectors, no cover, insulation, or box is used, leaving only the absorber. The absorber must be able to withstand the exposure to the environment. Cost is reduced substantially. At these low temperatures heat losses are low enough that the cost of glazing is not justified in terms of the gain in efficiency. The collectors are relatively efficient at low temperatures but their performance degrades rapidly if operation at higher temperatures is attempted. The collectors are commonly in the form of either a plastic mat (~ 1 cm high) with parallel flow channels (Fig. 4-6e) and large tubular headers or, simply, the metal absorber plates described under "Liquid Heaters" (page 96).

Performance Comparison

The relative performances of the nonconcentrating collectors just discussed are illustrated in Fig. 4-10. The previously discussed one-cover, nonselective absorber collector (D) is shown again for comparison. The unglazed collector (E) is clearly useful only at temperatures about 20°C or less above ambient. The relatively poor performance of the air heater (F) is due primarily to its low value of F_R. Near stagnation (where efficiency does not depend strongly on F_R), its performance is similar to the comparable liquid heater. A second cover and/or selective absorbers (C, B) clearly enhance performance. Beyond 100°C above ambient, the evacuated collectors (A) appear to have achieved a significant performance advantage.

The relatively poor optical performance [i.e., small $(\tau\alpha)_{\text{eff}}$] of the evacuated tube collector is due to the unfavorable optics of the concentric cylinder geometry and the problem of "coverage," i.e., the gaps between adjacent tubes. Further details can be found in Ref. 3.

Of course, each improvement in performance is usually associated with an increase in cost. This fact will be discussed further at the end of this chapter.

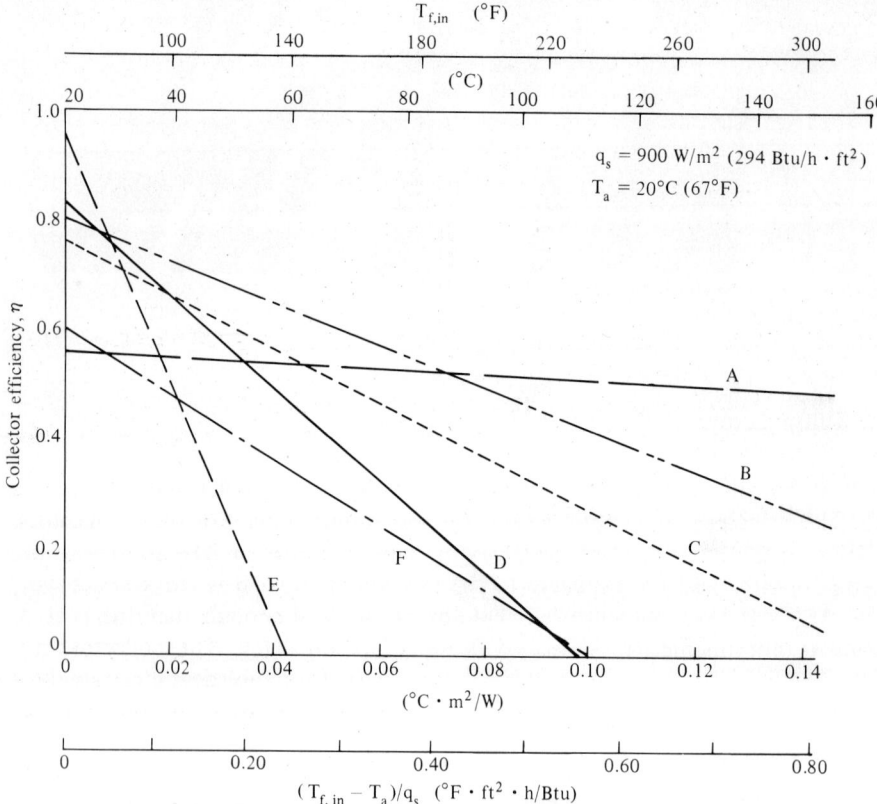

Figure 4-10 Relative performance of nonconcentrating collectors.

Collector Orientation

Nonconcentrating collectors are usually mounted in a stationary array with an orientation optimized for a particular region and load. For year-round operation a south-facing collector in northern latitudes (north-facing in the southern hemisphere), tilted from the horizontal at an angle approximately equal to the local latitude plus 10°, is common practice. For winter use, it would be tilted higher. System performance is a weak function of azimuthal orientation around the optimal. Fifteen to twenty degrees either side of south can cause up to 5 percent degradation in annual performance. Tilt angles are somewhat more critical, but $\pm 5°$ is an allowable tolerance. Orientation is discussed further in Chap. 9.

4-5 Concentrating Collectors

Concentrating solar collectors can be loosely classified in two ways. One classification describes the geometry of the absorber (linear or point concentration). The other indicates the means of concentration (reflection or refraction). A given collector generally falls into one category in each classification. Hence, a concentrating collector could, for example, be identified as a "linear reflecting" or a "point refracting" collector. However, not all concentrators fall neatly into these classifications. For example, two-stage concentration may be achieved by a combination of reflection and refraction.

Linear concentrators will be discussed next: First reflecting collectors, then refracting. Details of the thermal analysis of the linear concentrators will be given in Chap. 5. Brief comments on point concentration will appear near the end of this chapter.

Linear Concentration by Reflection

Most linear reflective concentrators use parabolic or troughlike reflector geometries with tubular absorbers. (However, linear concentration can also be achieved utilizing a spherical- or conical-shaped reflector.) For purposes of discussion, the linear concentrators will be subdivided in two classes: nontracking and tracking.

Nontracking concentrators The concept of flat "side" mirrors to boost the performance of an otherwise nonconcentrating collector was first used about 1910. Fixed-geometry groove configurations have been studied, both as thermal collectors [4] and as solar cell concentrators [5]. The reflectors can be large and placed beside a more or less conventional flat-plate collector or they can be smaller and formed as an integral part of the collector. In this latter case, finned tubes can be used as the absorber and the cover assembly of the collector can also enclose and protect the reflectors.

The grooves are aligned east-west so that the sun "moves" approximately along the groove during the day. Seasonal tilt adjustments help to keep the groove aligned. However, it is only at equinox that it is possible to perfectly align the groove for the full day's operation due to the north-south diurnal swing of the sun, which progressively increases from equinox to solstice (see Fig. 3-4).

These grooved collectors are not focusing devices, even though they do concentrate. Therefore, they can have a rather large acceptance angle. That is, they can concentrate radiation over a wide angle of incidence. It is this characteristic, of course, that makes it possible to use them in the nontracking mode.

The simplest groove design is a trapezoid [6] shown in Fig. 4-11a. Planar reflectors form the side walls and the absorber is the base. Optimal designs utilize a total opening angle of approximately 30° with various depth-to-base ratios depending on the desired concentration ratio and the acceptable frequency for collector tilt adjustment. Annual concentration ratios up to about 1.5 are possible without tilt adjustment for a depth-to-base ratio of about 1. The better the solar alignment, maintained by more frequent tilt adjustments, the deeper the groove can be and the higher the resulting concentration ratios. The practical limit for

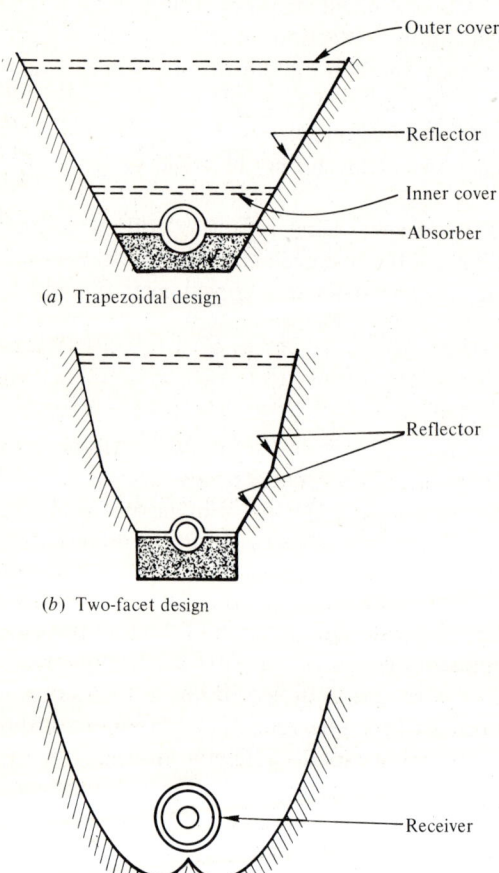

(a) Trapezoidal design

(b) Two-facet design

(c) One of many CPC-cusp designs

Figure 4-11 Nontracking concentrators.

the concentration ratio for the trapezoidal design is a little above 2. Higher concentration ratios can be achieved by utilizing multiple plane facets to replace the straight walls (Fig. 4-11b). Concentration ratios of 3 and above can be achieved with two facets.

Practical concentration ratios up to 6 can be achieved in troughlike non-tracking collectors if curved walls are used. Cylindrical and several parabolic cusp mirror geometries have been investigated. A particularly interesting geometry is the Winston or compound parabolic concentrator (CPC) [7–9]. This design is shown schematically in Fig. 4-11c. Two symmetrical half-parabolic cylinders form the reflecting sidewalls. The walls are designed to deliver to the absorber all incident rays within a given angle (the acceptance angle) and at the same time achieve the maximum concentration possible for the given acceptance angle. From a practical viewpoint this so-called "full" CPC design requires an excessive reflector area per unit of aperture area. For example, a concentration ratio of 9.4 can be attained with a total acceptance angle of 12°, but the reflector area is approximately six times larger than the aperture area. However, the mirror area can be reduced by a considerable amount without significantly degrading performance. A maximum concentration ratio of 4.2 is possible with a total acceptance angle of 27.5°. The reflector area can be reduced by 50 percent by removing the portion of the reflector farthest from the absorber, and the concentration ratio drops only to 3.6. Hence, the "truncated" CPC design has more practical value. Several CPC-like designs are commercially available. Anodized aluminum sheets are used as the reflectors and tubular collectors (some evacuated) are used as receivers, as illustrated in Fig. 4-11c.

Tracking concentrators There are three tracking concepts used for linear concentration. These are (1) total or fully tracking (the entire collector moves as a unit), (2) fixed-reflector, tracking receiver and (3) fixed-receiver, tracking reflector.

There are a large number of tracking collectors, but the most common is a parabolic or cylindrical reflective surface which is illustrated in Fig. 4-12. While simple in concept, the optics of such systems can become quite complex and they are discussed further in Chap. 5. It should be noted, however, that a two-dimensional parabolic reflector is the only surface which will generate a line focus. Small cylindrical sectors only approximate this behavior. For one-axis tracking, the receiver is usually aligned along an east-west line and tracking occurs about that axis. The "cosine losses" (the loss of insolation normally available to a fully tracking system) and off-axis aberrations reduce the performance of the collector but the mounting is more convenient than a north-south system. North-south alignment is less common primarily because of the limits to the trough length imposed by this mounting arrangement.

The difficulties in designing a total tracking collector go beyond the optics. Consideration must be given to structural strength, wind loading, drive power, and construction tolerances. Blockage and shading of insolation by adjacent collectors can be significant in an array located in a limited space and using east-west orientation. Some of the difficulties can be alleviated by reducing the structure that must track. There are a number of companies currently in com-

Figure 4-12 Parabolic tracking collector (Courtesy Sandia National Laboratory).

mercial production of tracking linear systems. True two-axis tracking is uncommon for linear concentration.

Two types of large-scale fixed-mirror tracking receiver collector systems have been built. One is a troughlike design in which a Fresnel, or faceted, mirror replaces the parabolic reflector [10]. The other uses a fixed spherical mirror to achieve linear concentration [3, 11]. Each design is discussed below.

In the trough design, flat, slatlike mirrors are mounted on a fixed cylindrical structure in a stairlike arrangement, as illustrated in cross section in Fig. 4-13a. The mirrors are tilted toward the middle of the trough at an angle equal to one-fourth their angular position around the circumference. The receiver tube is mounted on a support pivoted about the axis of the cylindrical mirror array. As the sun angle changes (in the plane of the figure), the position of the receiver shifts, as illustrated in Figs. 4-13b and c. The major disadvantages of the concept are the cosine losses, a severe end effect in a short design, and aberrations due to "out-of-plane" insolation which change the focal length and defocus the system. A photograph of one concept of this design is presented in Fig. 4-14. The mirror base is cast cement; the slat reflectors are back-surface glass mirrors; the receiver is a pipe at the base of an inverted trapezoidal groove (Fig. 4-11a) so that a secondary concentration is achieved. Another prototype has been built using stamped metal parts fastened together like a grandstand.

The fixed spherical mirror concentrator is illustrated in Fig. 4-15. The absorber tube swings on a rod pivoted about a fixed point at the center of curvature. The details of the optics can be found in Ref. 3. When the absorber is aligned with the beam radiation, the mirror reflects all the sunlight onto the absorber whose length is approximately one-half the mirror radius. At least three prototypes of this system have been constructed: two in the United States, one in France. These designs are described below.

An aluminum mirror 9.5 m (31 ft) in diameter was built into the roof of a house in Boulder, CO, in 1975 [3, 11]. A 20 m (65 ft) in diameter dish composed of small flat glass mirrors was constructed in west Texas (Crosbyton) in 1979 (Fig. 4-16) [12]. The absorber, which is 14-cm (5.5-in) in diameter, and 5.6 m ($18\frac{1}{2}$ ft) long, consists of helically wound stainless steel tubing. Ten 61-m (200-ft) dishes are planned for a 5-MW solar thermal electric power plant at the Crosbyton site. In France, a 40-m (130-ft) spherical dish composed of reflective petals was constructed in 1978. Each petal is composed of up to twenty-four 0.90-m (3-ft) "diameter" hexagonal-shaped segments. The design features both a high- ($C_g \cong 250$) and a low-pressure ($C_g \cong 50$) boiler.

One major disadvantage of the tracking receiver is that a slip or flexible coupling is required for the circulation of the heat-transfer fluid. As a practical matter this has been a problem. However, it can be overcome with a fixed-receiver tracking mirror design. One such design which is commercially available looks like the fully tracking parabolic trough except that the reflector moves about a fixed receiver. The reflector rotates on the receiver axis. In another design the parabolic reflector is replaced with a Fresnel approximation and is made up of individually rotatable slats. These are positioned to redirect the beam insolation onto the receivers as indicated in Figs. 4-17 and 4-18 [13].

Linear Concentration by Refraction

Compared to concentration by reflection there are relatively few practical designs which concentrate solar energy by refraction. For these few, ordinary convex lenses are seldom used due to their high cost and weight.

The imaging characteristics of a lens depend primarily on the surface curvature. Therefore, essentially the same optical performance can be achieved if the center of the lens is removed, the surface is divided into small segments which maintain their original slope, and these segments are displaced to the center line of the lens. Such a lens is called a Fresnel lens and is illustrated in Fig. 4-19. They are in common use in noncritical focus applications and in the vast majority of the cases where solar energy is concentrated by refraction.

Fresnel lenses for solar applications are generally made of plastic. The plastic is light weight and can be easily extruded through or cast against a die.

A few linear focus collectors are available which use Fresnel lenses, a representative design of which is also shown in Fig. 4-19. The grooves may either face toward or away from the sun, but the latter is preferred because the facet edges are protected and it is easier to keep the lens clean. These collectors can be used in a non-tracking mode aligned east-west, so long as north-south tilt adjust-

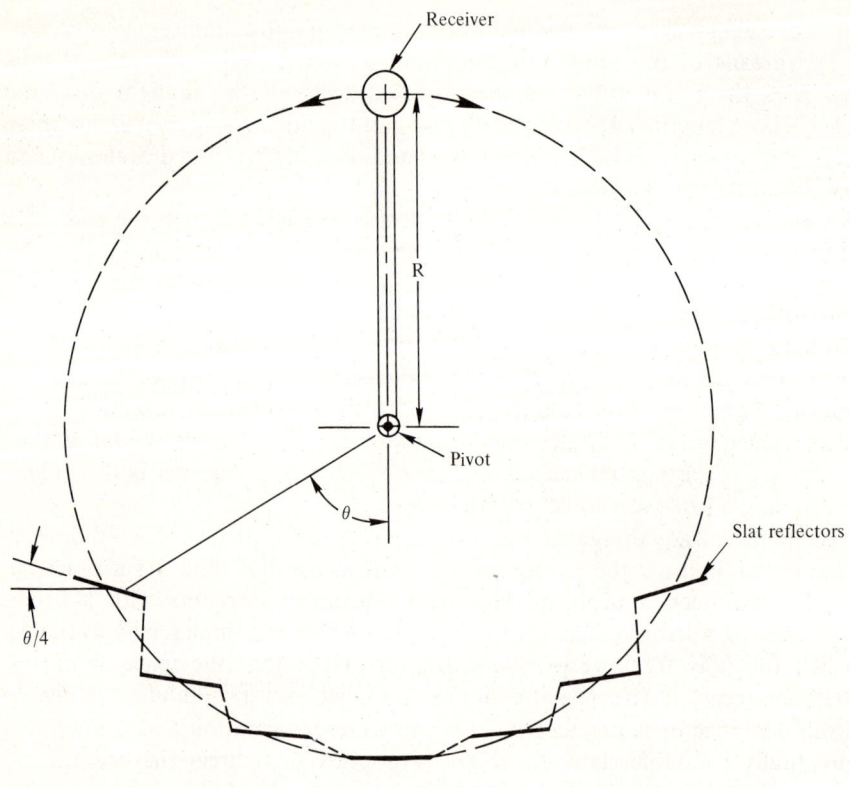

(a) Design

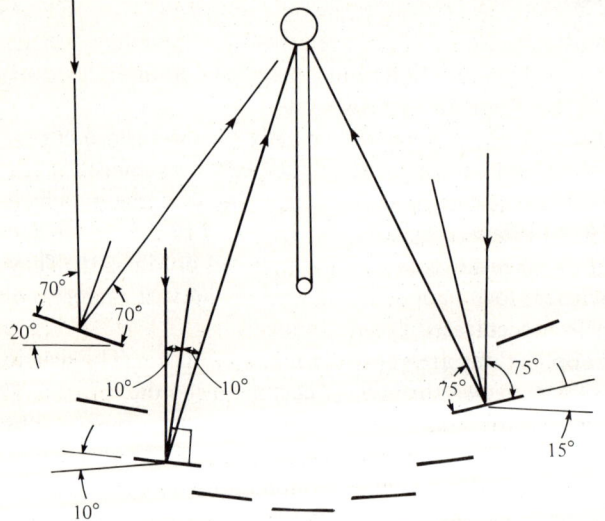

(b) Normal-incidence ray trace

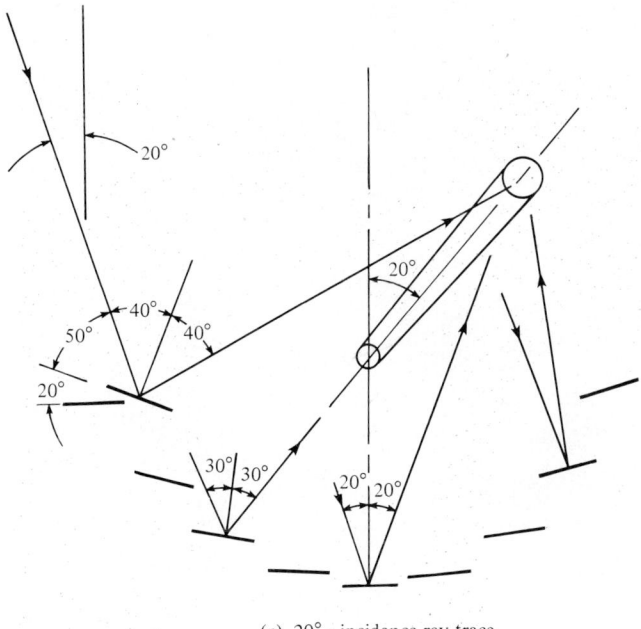

(c) 20° - incidence ray trace

Figure 4-13 Fixed-mirror tracking-receiver troughlike collector (reflectors enlarged and reduced in number for clarity).

ments are made. However, off-axis aberrations experienced early and late in the day cause severe defocusing. Because of this fact a north-south alignment with east-west one-axis tracking, possibly with seasonal tilt adjustments, is the preferred design.

Point Concentration

Point focus concentration is achieved using an axisymmetric reflective or refractive concentrator, requiring two-axis tracking.

In practice, "point" focus concentration is not truly attainable (due to the finite size of the sun) nor is it usually desired. Extremely high temperatures are theoretically attainable with an even moderate concentration ratio leading to possible damage of the absorber. The extremely large energy fluxes associated with point focusing can exceed the values that can be transferred from the absorber using available heat-transfer technology. However, the receiver can be placed on-axis but away from the focal point to realize acceptable flux levels.

These considerations help to define mechanical tolerances designed into the collector structure and tracking mechanism. There are few commercially available point focus systems. Most use a segmented reflector (for ease of manufacture,

108 SOLAR-THERMAL ENERGY SYSTEMS

Figure 4-14 Photograph of fixed-mirror tracking receivers, troughlike design (Courtesy Sandia National Laboratory).

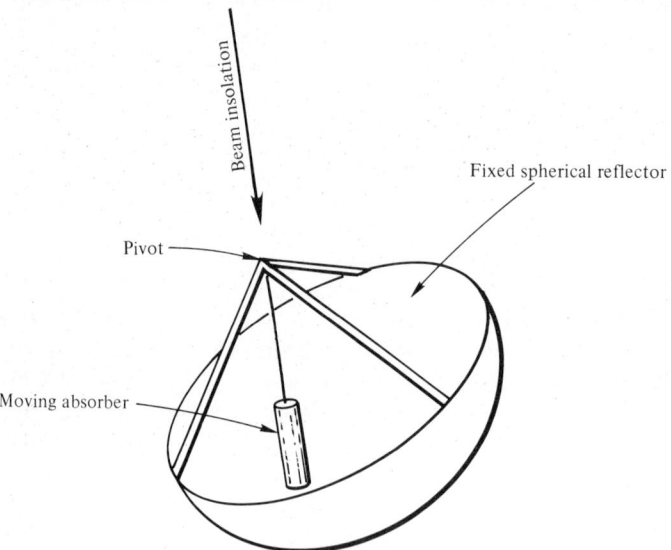

Figure 4-15 Stationary-reflector tracking-absorber spherical mirror collector.

Figure 4-16 Photograph of the Crosbyton spherical reflector collector (Courtesy Texas Solar Energy Society).

assembly, and transportation) on the order of a few meters in diameter, as illustrated in Fig. 4-20.

An extension of the CPC concept can be made to three dimensions in an axially symmetric geometry. This concept has limited practical application but has been suggested as a secondary concentrator in a larger point concentrator system.

Three particularly interesting large-scale point focus systems are the 1-MW (thermal) solar furnace in the Pyrenees in France, a 400-kW (thermal) solar tower in Atlanta, GA, and a 5-MW (thermal) solar tower in Albuquerque, NM.

The 1-MW system, built in 1970, consists of a stationary, vertical paraboloidal mirror [39.6 m (129 ft) high by 53.3 m (173 ft) wide] containing 9500 individual mirrors within a total reflecting area of 1920 m^2 (21,000 ft^2). This mirror forms one wall for a 14-story building. Beam insolation is directed to this mirror by 63 heliostats [individually tracked mirrors; total area of 2840 m^2

Figure 4-17 Photograph of fixed-receiver rotating-slat reflector (Courtesy Sandia National Laboratory).

(31,000 ft^2)] which are mounted on a terraced hill. The paraboloidal mirror in turn focuses the insolation through a window in the solar laboratory (Fig. 4-21).

The 400-kW tower is located at the Advanced Components Test Facility on the campus of Georgia Tech in Atlanta. It is powered by a hexagonal array of 550 circular heliostats [1.1-m (3.6-ft) diameter] with a total reflecting area of 532 m^2 (5720 ft^2). The heliostats are driven by a master mechanical drive train which maintains the entire field in focus. The peak radiative flux delivered to the receiver suspended from a tower 21 m (69 ft) above the field is about 200 W/(cm)^2 [630 × 10^6 Btu/(ft^2 · h)].

The 5-MW Sandia Central Receiver Test Facility was built in 1977 to test components (receivers) for larger central tower systems expected to be developed in the future. The Sandia facility has 222 individually tracked heliostats in a north field array but this number can be increased. Each heliostat consists of

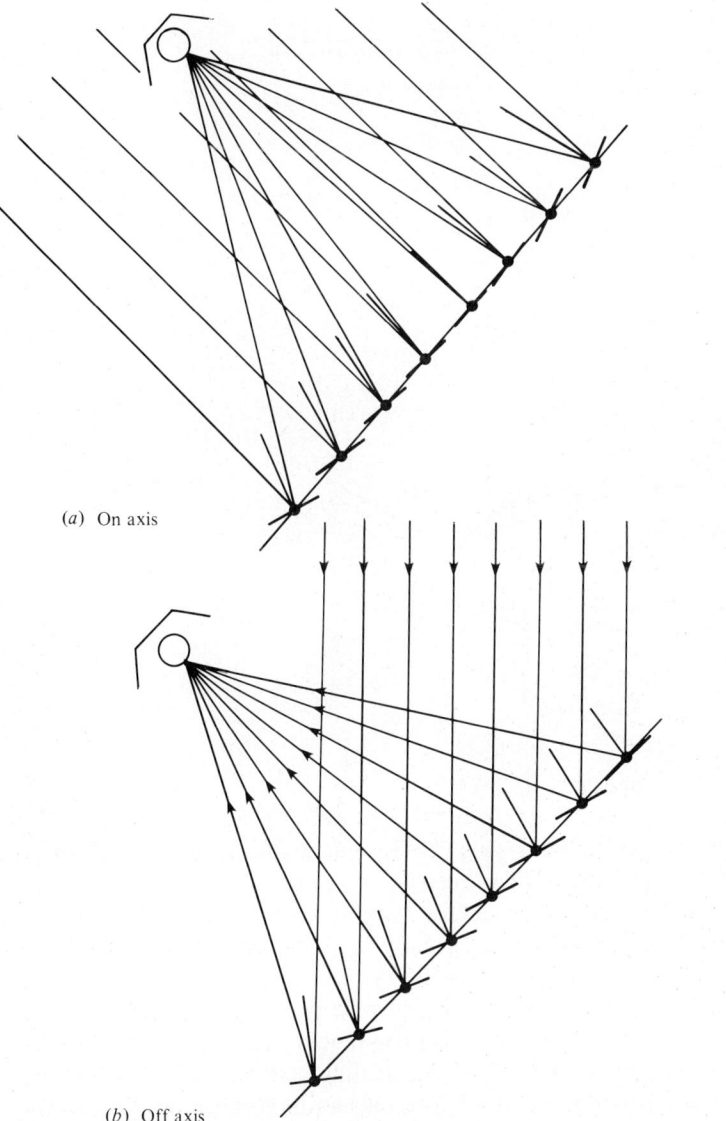

(a) On axis

(b) Off axis

Figure 4-18 Optics of fixed-receiver rotating-slat receiver.

twenty-five 1.2-m (4.0-ft) square flat mirrors individually aimed. The total reflecting area is 8250 m² (88,000 ft²). A concrete tower 61 m (200 ft) tall by 15 m (49 ft) in diameter has test platforms at 36 m (118 ft), 43 m (141 ft), and 49 m (160 ft) above ground level, facing the north field (Fig. 4-22).

Performance Comparison for Concentrating Collectors

The comparisons among concentrating collectors should be examined with caution. First, there are more operational problems associated with tracking systems

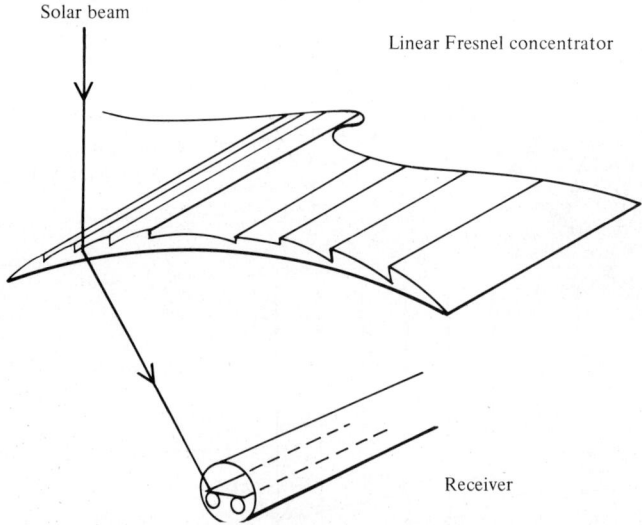

Figure 4-19 Fresnel lens and collector.

than for nontracking systems, and second, the extent of environmental degradation of exposed reflector surfaces has been shown to be potentially severe. Both of these factors could contribute to widely varying cost claims. The desire for increased performance at reduced cost usually leads to extending materials to their limit. This is especially true for focusing systems where absorber materials, absorber coatings, and heat-transfer fluids must be properly specified. In addition, even a momentary failure in the controls could result in significant deterioration if not destruction of a highly concentrating system.

Even thermal performance results themselves can be misleading if not properly interpreted. The major source of uncertainty in evaluating concentrating collectors is the nature of the insolation during the test. A side-by-side test of collector types is, of course, desired. However, if test results are furnished for a variety of collectors, each tested at a different site under different insolation, there is considerable uncertainty in comparisons unless the angular distribution of the beam insolation is completely specified. The problem that can develop is due to the fact that, the higher the concentration ratio achieved in the collector, the more limited is the available insolation under a given sky condition since the acceptance angle or "field of view" of the collector decreases with increasing concentrating ratio. Therefore, the primary comparison parameter, collector efficiency, is subject to considerable variation depending upon how q_s is defined.

Figure 4-20 Commercially procured point focus concentrator (Courtesy Sandia National Laboratory).

For example, a standard pyrheliometer has an acceptance angle of 5.5°. However, for a linear concentration ratio of 50, the acceptance angle is about 1°. This implies that the pyrheliometer would indicate more insolation than would in fact be available to the collector. Hence, the collector's performance would be downgraded depending upon the amount of circumsolar radiation, i.e., the insolation that appears to come from the region around, but outside of, the solar disk and is due to forward scattering in the atmosphere as discussed in Chap. 3. On the other hand, if the collector is tested at a "clear-sky site" with little circumsolar radiation but is to be used at a different site with significant circumsolar radiation, the use of pyrheliometer data for system design and evaluation leads to a considerable overprediction of performance at the second site.

Other performance uncertainties may be related to deformation of the reflector due to wind loading and/or inadequate support structure, and the ability of the tracking system to keep the absorber at the focal line or point of the reflector or lens.

These comments are not meant to indicate that useful comparisons and analysis cannot be made, but are simply to point out that there usually are a large number of uncertainties in the analysis.

Figure 4-23 illustrates a comparison among a variety of concentrating collectors, all of which have been previously discussed. As can be seen by the range of performance of even the same type of collector, these results should not be used to rank-order the collector concepts. The figure presents experimental data on specific collectors. Even these results do not give a fair comparison since only "peak noon efficiency" is given. Reduction in all-day (or all-year) performance due to off-axis aberrations, cosine effects, end effects, and degradation will affect

Figure 4-21 1-MW solar furnace, Odeillo, France.

GENERAL DESCRIPTION OF SOLAR-THERMAL COLLECTORS **115**

Figure 4-22 Sandia 5-MW central receiver test facility, Albuquerque, NM.

116 SOLAR THERMAL ENERGY SYSTEMS

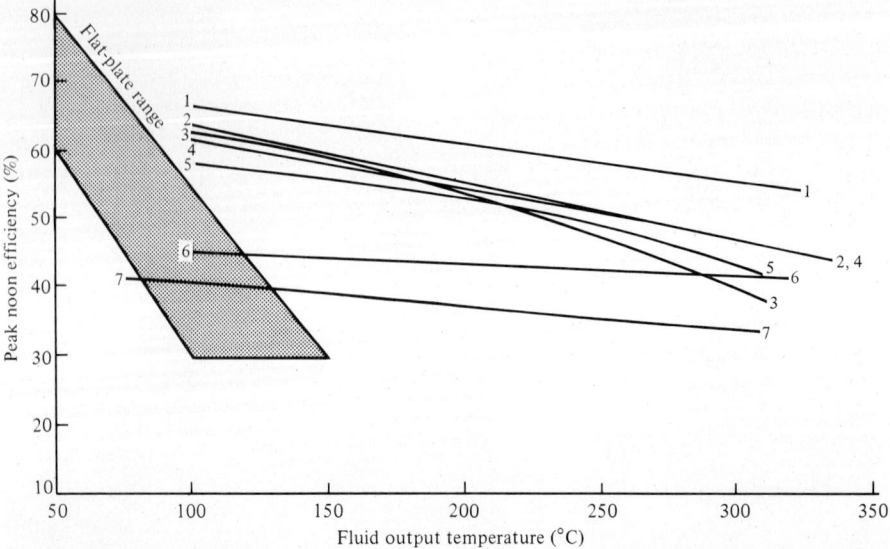

Figure 4-23 Linear concentrating collector efficiency [14]. 1: Parabolic trough, manufacturer A; 2: segmented reflector, stationary absorber; 3: parabolic trough, manufacturer B; 4: parabolic trough, manufacturer C; 5: linear Fresnel refractor; 6: segmented reflector, moving absorber; 7: segmented reflector, moving absorber.

each design differently. Also, tracking sophistication and cost have significantly different impact on the different designs.

4-6 COLLECTOR SELECTION

The collector manufacturer should supply, in addition to a warranty, at least two items of information with the collector literature: (1) the collector thermal efficiency curve (from tests performed by a certified laboratory) and (2) the installed (or at least delivered) cost of the collector per unit of aperture area. Standards for durability, maintenance, or degradation are currently being developed but, because they do not now exist for collectors, the buyer's best judgment on these factors must be used.

Collector thermal performance curves for some typical collector types have been previously presented in Figs. 4-10 and 4-23. However, as pointed out above, thermal performance alone is not enough information for collector selection. The most desirable collector will be the one which delivers the required amount of energy at the specified temperature at the lowest cost, based on life cycle cost analysis. The most efficient collector may not be the most economical if it costs considerably more than a less efficient one. To demonstrate this fact, the efficiency curves for the five liquid-heating collectors shown in Fig. 4-10 are divided into their respective estimated but realistic costs. The results are plotted in Fig. 4-24 along with the cost data used. Each curve represents a measure of the relative cost per unit energy delivered by each collector per unit area as a function of the usual collector efficiency variable, $(T_{f,\,in} - T_a)/q_s$. For given values of the insolation, ambient temperature, and the required operating temperature, the collector with the lowest cost per unit of energy delivered can be determined. The specification of the required temperature is clearly important since, as is characteristic of the thermal performance curves, these cost curves also experience crossovers as the temperature changes. The results illustrated in Fig. 4-24 can be used to determine that part of the initial system cost (installed cost) due to the col-

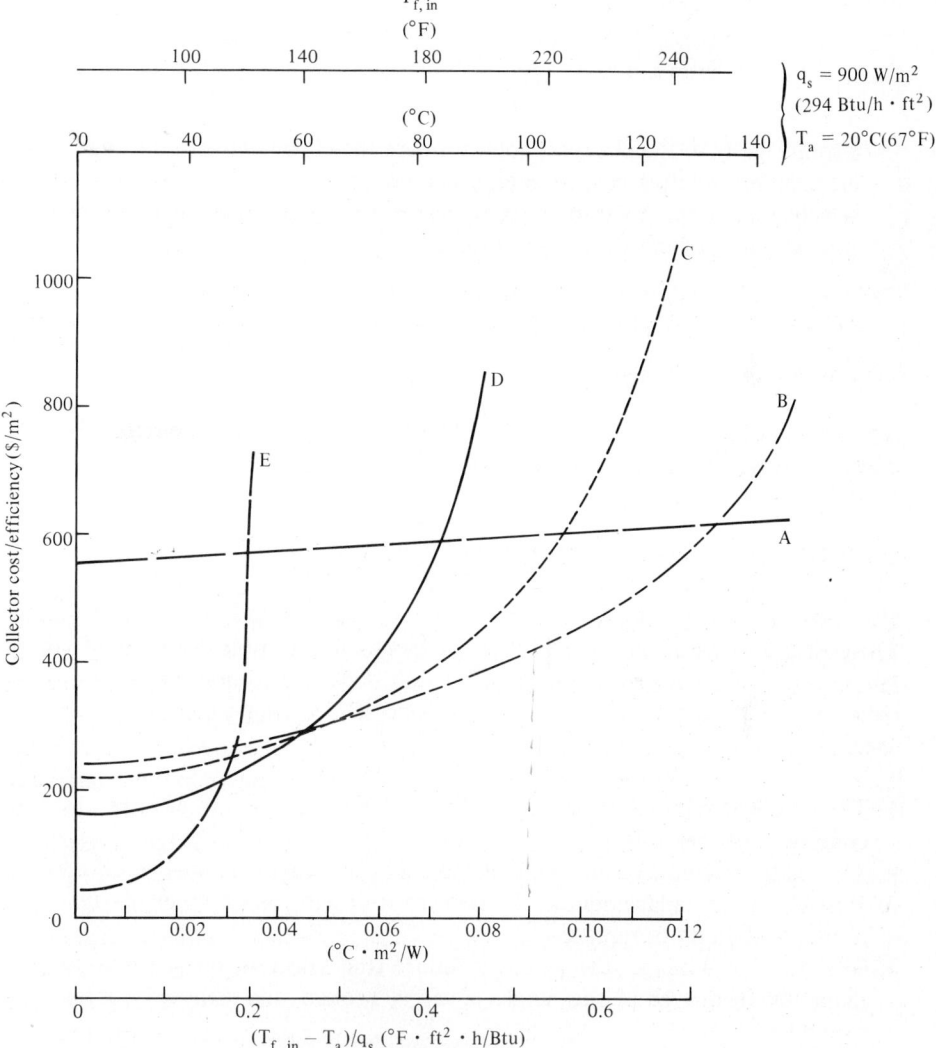

Figure 4-24 Cost per unit area divided by efficiency for collectors shown in Fig. 4-10. A: Evacuated tubular, $300/m^2; B: one glass cover, selective absorber, $180/m^2; C: two glass covers, nonselective absorber, $150/m^2; D: one glass cover, nonselective absorber, $130/m^2; E: unglazed, nonselective absorber, $40/m^2.

lectors alone. Of course, the additional costs of storage, distribution, pumps, controls, structure, etc., must be added to it.

Example 4-2 Determine the cheapest collector to meet the following requirements:

$$T_{f,\,in} = 90°C$$
$$T_a = 30°C$$
$$q_s = 900 \text{ W/m}^2$$

SOLUTION

$$\frac{T_{f,\,in} - T_a}{q_s} = \frac{(90 - 30)°C}{900 \text{ W/m}^2} = 0.0667 \; (°C \cdot m^2)/W$$

From Fig. 4-24, the cheapest collectors are the (1) one-cover selective absorber, $330/m^2$ relative cost of energy per unit area and (2) the two-cover nonselective absorber, $370/m^2$ relative cost of energy per unit area. The installed price per unit of energy delivery rate is

$$\frac{\text{Cost per unit area}}{\text{Collected energy rate per unit area}} = \frac{\text{Cost}}{\eta q_s}$$

For the one-cover selective type,

$$\frac{\text{Cost}}{\eta q_s} = \frac{\$330/m^2}{900 \text{ W/m}^2} = \frac{\$367}{kW}$$

For the two-cover type,

$$\frac{\text{Cost}}{\eta q_s} = \frac{\$370/m^2}{900 \text{ W/m}^2} = \frac{\$411}{kW}$$

The cost difference between the two best designs in Example 4-2 is slight (about 10 percent). In all cases, but especially where cost differences are small, other collector design features should be considered in deciding which to use. In this specific instance, one would consider

1. The expected relative durability of the selective and the nonselective absorber coating
2. The relative weights of the two collectors (i.e., one cover versus two covers)
3. Possible cover replacement (if the two-cover design has an inner plastic cover, it may be difficult to replace)
4. The possible damage due to cover failure (the selective surface may be permanently destroyed if the single cover is broken and exposed to rain, for example)
5. The overall weatherability of the collectors
6. The possible effect of a prolonged stagnation condition (the selective absorber collector has a significantly higher stagnation temperature)

4-7 CONCLUSIONS

Readers should now be in a position to make reasonable and educated decisions about the selection of a solar collector for a given application. Based on supplied efficiency curves and cost data for input to SOLSIM, realistic system simulations can now be performed with confidence. However, much of the detail of the analysis is not included in this chapter. If one wishes to design a collector, the details of the optics and heat transfer associated with collector performance must be known. These are presented in Chap. 5. Those satisfied with the collector overview presented in Chap. 4 may wish to skip Chap. 5 and move on to Chap. 6.

PROBLEMS

4-1 From Fig. 4-10 determine $(\tau\alpha)_{\text{eff}}$ and $\bar{U}$ for the six collectors shown. Fill in the results in the table (note F_R is given for each collector below). Calculate the stagnation temperature for each collector with $q_s = 800 \text{ W/m}^2$, $T_a = 25°\text{C}$.

	F_R	$(\tau\alpha)_{\text{eff}}$	$\bar{U}$	T_{stag}
Evacuated	0.96			
1 cover, selective	0.91			
2 cover, nonselective	0.90			
1 cover, nonselective	0.88			
Air	0.70			
Unglazed	0.95			

4-2 Using the data in Fig. 4-10, find the constants in the collector equation [Eq. (2-2)]

$$\eta = (\tau\alpha)_{\text{eff}} - [b(\theta_e - 1)q_{s,\text{ref}}/q_s(t)]$$

for all collectors if $\bar{T}_a = 27°\text{C}$ and $q_{s,\text{ref}} = 1000 \text{ W/m}^2$.

4-3 A collector designer calculates that a particular flat-plate collector will operate at 50 percent efficiency when exposed to insolation of 800 W/m² at an operating (fluid inlet) temperature of 95°C and ambient temperature of 25°C.

(a) What minimum stagnation temperature would you predict for this collector if the designer's calculations are correct?

(b) If one glass cover and a selective absorber were used, $(\tau\alpha)_{\text{eff}} \cong 0.83$, and if we assume $F_R = 0.95$, what efficiency would you expect if the insolation were 400 W/m² and the other conditions remained the same? What are the estimated stagnation temperatures at insolations of 400 and 800 W/m²?

4-4 Assume that the test conditions for the results presented in Fig. 4-23 were

$$q_s = 950 \text{ W/m}^2$$
$$T_a = 30°\text{C}$$
$$T_{f,\text{out}} - T_{f,\text{in}} = 5°\text{C}$$

Calculate $F_R(\tau\alpha)_{\text{eff}}$ and $F_R\bar{U}/C_g$ for collectors 1, 2, 5, and 6.

4-5 Plot on a single graph η versus $(T_{f,\,\text{in}} - T_a)/q_s$ the performances of the following collectors:
  (a) Evacuated tube (Fig. 4-10)
  (b) One-glass cover selective absorber (Fig. 4-10)
  (c) One-glass cover nonselective absorber (Fig. 4-10)
  (d) Parabolic trough—"manufacturer A" (Fig. 4-23)
  (e) Linear Fresnel refractor (Fig. 4-23)
  Use the conditions of Prob. 4-4 for the collectors of Fig. 4-23, and base all efficiencies on the given value of q_s.

4-6 Determine the cheapest collector from Fig. 4-24 to meet the following requirements:

$$T_{f,\,\text{in}} = 120°C$$

$$T_a = 30°C$$

$$q_s = 1000 \text{ W/m}^2$$

What is the installed collector cost per kilowatt of capacity?

4-7 Repeat Prob. 6 above but also consider two new evacuated collector arrays. The first is a less expensive version of the one illustrated in Fig. 4-10. Its instantaneous efficiency is given by

$$\eta = 0.58 - 0.8 \left(\frac{T_{f,\,\text{in}} - T_a}{q_s} \right)$$

and its cost is $200/m². The second utilizes moderate concentration with the evacuated tubes as the receivers. Based on the collector aperture, it also costs $200/m² with an instantaneous efficiency of

$$\eta = 0.50 - 0.42 \left(\frac{T_{f,\,\text{in}} - T_a}{q_s} \right)$$

4-8 Repeat Prob. 7 at 300°C and include in your candidates the two concentrating systems discussed in Prob. 5. Assume the parabolic trough is available at $250/m² and the Fresnel refractor at $175/m², and that of the global insolation, 85 percent is available in the beam component.

REFERENCES

1. "Method of Testing Solar Collectors Based on Thermal Performance," ASHRAE Standard 93-77, ASHRAE, New York, January 1977.
2. J.E. Hill et al., "Development of Proposed Standards for Testing Solar Collectors and Thermal Storage Devices," National Bureau of Standards Report No. NBS-TN 899, February 1976.
3. F. Kreith and J.F. Kreider, *Principles of Solar Engineering*, pp. 286–294, McGraw-Hill Book Co., New York, 1978.
4. H. Tabor, "Stationary Mirror Systems for Solar Collectors," *Solar Energy*, vol. 2, pp. 27–33, 1958.
5. R.G.T. Hollands, "A Concentrator for Thin-Film Cells," *Solar Energy*, vol. 13, pp. 149–163, 1971.
6. Richard B. Bannerot and John R. Howell, "The Effect of Non-Direct Insolation on the Radiative Performance of Trapezoidal Grooves Used as Solar Energy Collectors," *Solar Energy*, vol. 19, no. 5, pp. 539–545, 1977.
7. Roland Winston, "Principles of Solar Concentrators of a Novel Design," *Solar Energy*, vol. 16, no. 2, pp. 89–94, 1974.
8. Ari Rabl, "Optical and Thermal Properties of Compound Parabolic Concentrators," *Solar Energy*, vol. 18, no. 6, pp. 497–511, 1976.
9. W.W. Schertz, "Nonimaging Concentrators Deliver High Temperatures for Industry," *Solar Engineering*, pp. 28–29, July 1977.

10. J.R. Schuster, G.H. Eggers, and J.L. Russell, Jr., "Operating Experience with the General Atomic Fixed Mirror Solar Concentrator," *Proceedings of the 1978 Annual Meeting of the Am. Sec. of ISES*, Denver, CO, 1978, pp. 863–870.
11. J.F. Kreider, "Thermal Performance Analysis of the Stationary Reflector/Tracking Absorber Solar Concentrator," *J. Heat Transfer*, vol. 97, pp. 451–456, August 1975.
12. *Solar Engineering*, December 1980.
13. R.L. French, L.G. Mooney, J.H. McDowell, and R.B. Useton, "Performance Testing of the SLAT™ Concentrating Collector"; Part I: "Equipment and Procedures," and Part II: "Results and Analysis," *Proceedings of the 1978 Annual Meeting of the Am. Sec. of ISES*, Denver, CO, 1978, pp. 871–884.
14. Charles Wyman, James Castle, and Frank Kreith, "A Review of Collector and Energy Storage Technology for Intermediate Temperature Application," *Solar Energy*, vol. 24, no. 6, pp. 517–540, 1980.

CHAPTER
FIVE

ANALYTICAL DESIGN OF SOLAR COLLECTORS

Two general problems may face engineers who work with solar collectors. These are the problems connected with the analysis and design of the solar collector itself and problems associated with choosing a collector and sizing a collector or array for a particular application. In the former category, engineers are interested in predicting the performance of a given collector design and in determining possible performance changes that might occur because of modifications in a given design. In choosing and sizing collectors for a given application, on the other hand, they work with given performance and cost data as well as a careful model of overall solar energy system performance, for example, SOLSIM. They then choose collectors based on this overall cost/performance comparison.

Both of these problem categories are addressed in this chapter. First, the optical and thermal behavior of solar collectors is analyzed, and then the design considerations are examined for collector selection and performance.

5-1 LIST OF SYMBOLS†

a	extinction coefficient, 1/m
a, a_1	collector dimensionless coefficients, Eqs. (5-13) and (5-14)
A	area, m^2
B	constant in free convection correlations, $Nu = B(Gr)^n$
$b, b_1,$ b_2, b_3	collector thermal loss coefficients

† Terms are dimensionless unless noted.

c	specific heat capacity, J/(kg · °C)
C_g	geometric concentration ratio, (A_c/A_e)
C_R	concentration ratio, $(A_c/A_e)I = C_g I$
d	projected width of absorber element, m
D	distance between surfaces of cover plates, or between cover and absorber, m
D_i	thickness of insulation, m
f	focal length, m
F	collector heat loss efficiency factor, Eq. (5-21)
F_R	heat removal factor
g	standard gravitational acceleration, 9.8 m/s^2
G	mass flow rate per unit of collector absorber area, $\dot{m}/A_e$, kg/(m^2 · s)
Gr	Grashof number, $g\beta(\Delta T)D^3/v^2$
h	convective heat-transfer coefficient, W/(m^2 · °C)
I	intercept factor
k	thermal conductivity, W/(m · °C)
L	thickness of a single cover plate, cm
$\dot{m}$	mass flow rate, kg/s
n	simple refractive index, or exponent in free convection correlations
Nu	Nusselt number
q	energy per unit time per unit area, W/m^2
$\dot{Q}$	energy per unit time, W
r	distance from focal point to rim, m
R	thermal resistance, °C/W
T	absolute temperature, K
$\bar{U}$	overall thermal loss coefficient, W/(m^2 · °C)
W	width of collector aperture, m
x	axial position in tube, m
α	absorptivity: fraction of insolation absorbed
β	temperature coefficient of volume expansion [(1/K)] or dimensionless group, $\bar{U}[T_{f,\text{in}} - \bar{T}_a]/[(\tau\alpha)_{\text{eff}} q_s C_R]$
γ	flow parameter, $\mu\bar{U}/Gc$
ϵ	infrared emissivity: energy emitted divided by energy emitted from blackbody at the same absolute temperature.
η	collector efficiency
η_{fin}	fin efficiency
θ	angle with respect to collector normal; angle of incidence
Θ	ratio of absolute temperature, $T/\bar{T}_a$
Φ	rim angle of concentrating collector
λ	wavelength, m
μ	thermal resistance ratio, Eq. (5-18)
v	kinematic viscosity, m^2/s
ψ	ratio of instantaneous to reference insolation, $q_s(t)/q_{s,\text{ref}}$
ρ	reflectivity: fraction of incident energy reflected
σ	Stefan-Boltzmann constant, 5.667×10^{-8} W/(m^2 · K^4)

τ transmissivity; fraction of incident energy transmitted
$(\tau\alpha)_{\text{eff}}$ effective absorptance of cover-absorber assembly, Eq. (5-6)
χ angle of refraction
Ω collector acceptance angle

Subscripts

a	ambient
abs	absorbed
b	beam
back	back of collector
c	collector or collector aperture or fin base
conv	convective
e	absorber element
eff	effective value
f	fluid
in, out	inlet, outlet
i	insulation or inside of tube
ir	infrared
n	cover plate n
r	reflected
rad	radiative
ref	reference
s	solar
stag	stagnation
u	useful
w	wind
λ	wavelength-dependent
I, II, ...	number of plates in cover assembly
1, 2, ...	plate number index in multicover assembly

Superscripts

—	value for an assembly of cover plates $(1, 2, \ldots, n)$ or average over collector surface or daily average

5-2 OPTICAL DESIGN

Aside from the tracking and alignment errors associated with concentrating collectors, the optical design of a collector centers on two factors: First, minimizing optical losses caused by the reflection of insolation from collector elements as well as by the absorption of insolation in collector elements other than the absorber itself; and second, on maximizing absorption of the available insolation by the absorber. Loss of energy by radiative emission from the absorber must also be minimized, but this effect is treated as a thermal loss and will be discussed separately.

Cover Assembly Properties

First, let us examine the transmission of radiant energy through the cover-plate assembly of a flat-plate collector. Most good cover-plate materials transmit over 90 percent of the radiant energy in the visible spectrum, but are nearly opaque to radiation at infrared wavelengths. It is therefore convenient to separately analyze the energy in the solar spectrum (chiefly visible) from that emitted by the absorber (infrared). Figure 5-1 shows this typical behavior for glass. Note that Fig. 5-1 gives $\bar{\tau}_\lambda$, which is the fraction of energy incident on a glass sheet that emerges from the other side. The $\bar{\tau}_\lambda$ thus accounts for all interreflections that occur from the first and second surfaces of the glass as well as for absorption of radiation within the glass. The $\bar{\tau}_\lambda$ of two types of glass are shown in Fig. 5-1 and it is seen that the transmittance increases sharply at a wavelength of 0.2–0.4 μm and then drops sharply at about 2.5 μm, depending on the glass. Table 5-1 gives $\bar{\tau}_s$, the total solar transmittance for some typical single cover-plate materials.

For a single plate, the effective reflectance $\bar{\rho}_1$ is (including reflections from the front and rear surface and internal absorption)

$$\bar{\rho}_1 = \rho\left[1 + \frac{(1-\rho)^2 \tau^2}{1 - \rho^2 \tau^2}\right] \tag{5-1}$$

where ρ is the first surface reflectance and τ the transmittance of the glass which accounts only for the internal absorption. Both ρ and τ are strongly dependent

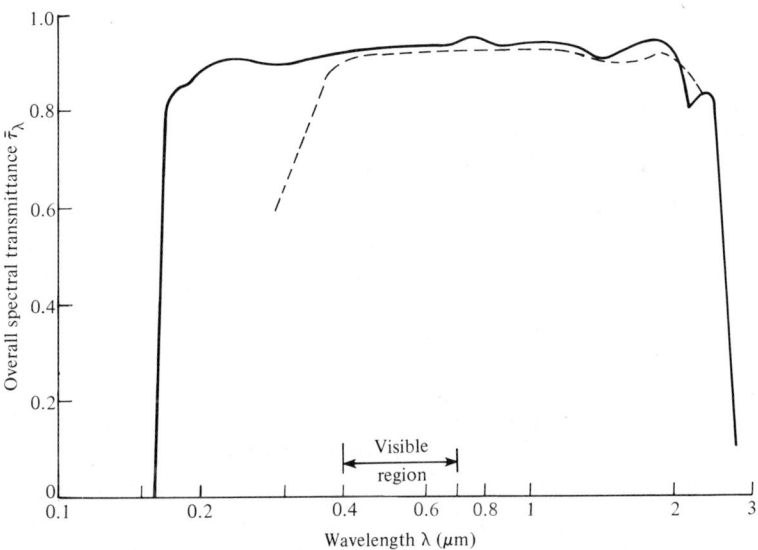

Figure 5-1 Normal-incidence overall spectral transmittance of glass plate (including surface reflections) at 298 K. ---- Borosilicate glass, 0.476 cm (3/16 in) thick; —— fused-silica glass, 1.27 cm (0.5 in) thick. (*Replotted from Ref. 1.*)

on the angle of incidence to the plate [3]. The effective transmittance is

$$\bar{\tau}_I = \tau\left(\frac{1-\rho}{1+\rho}\right)\left(\frac{1-\rho^2}{1-\rho^2\tau^2}\right) \quad (5\text{-}2)$$

and the effective absorptance

$$\bar{\alpha}_I = 1 - \bar{\tau}_I - \bar{\rho}_I \quad (5\text{-}3)$$

For multiple plates, Ref. 3 shows that, for two plates,

$$\bar{\rho}_{II} = \bar{\rho}_I + \frac{\bar{\rho}_I \bar{\tau}_I^2}{1-\bar{\rho}_I^2}$$

$$\bar{\tau}_{II} = \frac{\bar{\tau}_I^2}{1-\bar{\rho}_I^2} \quad (5\text{-}4)$$

$$\bar{\alpha}_{II} = 1 - \bar{\tau}_{II} - \bar{\rho}_{II}$$

and for three plates

$$\bar{\rho}_{III} = \bar{\rho}_I + \frac{\bar{\rho}_{II} \bar{\tau}_I^2}{1-\bar{\rho}_I \bar{\rho}_{II}}$$

$$\bar{\tau}_{III} = \frac{\bar{\tau}_I \bar{\tau}_{II}}{1-\bar{\rho}_I \bar{\rho}_{II}} \quad (5\text{-}5)$$

$$\bar{\alpha}_{III} = 1 - \bar{\tau}_{III} - \bar{\rho}_{III}$$

Table 5-1 Properties and comments on selected cover-plate candidates

Material	$\bar{\tau}_s$†	Advantages	Disadvantages
Glass (125 mil) Low-iron glass	0.79–0.85 0.91	Self-supporting, weathers well, moderate price	Weight, poor impact properties, breakage due to thermal expansion
Etched glass	0.96	Self-supporting	Cost, weathering, weight, impact
Lexan© (125 mil)	0.65–0.75	Shatter-resistant	Cost
Tedlar© (4 mil)	0.88–0.92	Moderate price	Limited temperature range, weathering uncertain
Teflon© (1 mil)	0.93–0.95	Good quality	Cost
Mylar© (5 mil)	0.80–0.85	Inexpensive	Ultraviolet and thermal degradation

† Solar transmittance.

Relations are given in Ref. 3 for any number of identical plates and for sets of n plates of one type and m plates of another material.

Considering the visible plus near-infrared portion of the spectrum, Fig. 5-2 shows the overall transmittance of a set of up to three identical glass plates [2, 3].

At angles of incidence up to about $\theta = 60°$, it is possible to simplify Eqs. (5-1) and (5-2) by expanding the $(1 - \rho)^2$ terms, simplifying, and then dropping any remaining ρ^2 terms, since ρ is typically less than 0.1. The simplified forms are given in many references, since important insolation is often incident only for $\theta < 60°$. At greater θ angles, however, the simplification may lead to significant errors. The equations given here are quite simple in form and are recommended.

When the solar energy passes through a series of n identical plates and strikes an absorber with absorptivity α, a fraction $(1 - \alpha)$ is reflected back into the cover-plate assembly where again a complicated series of interreflections occurs. Some of the reflected energy returns to the absorber, and the pattern is repeated.

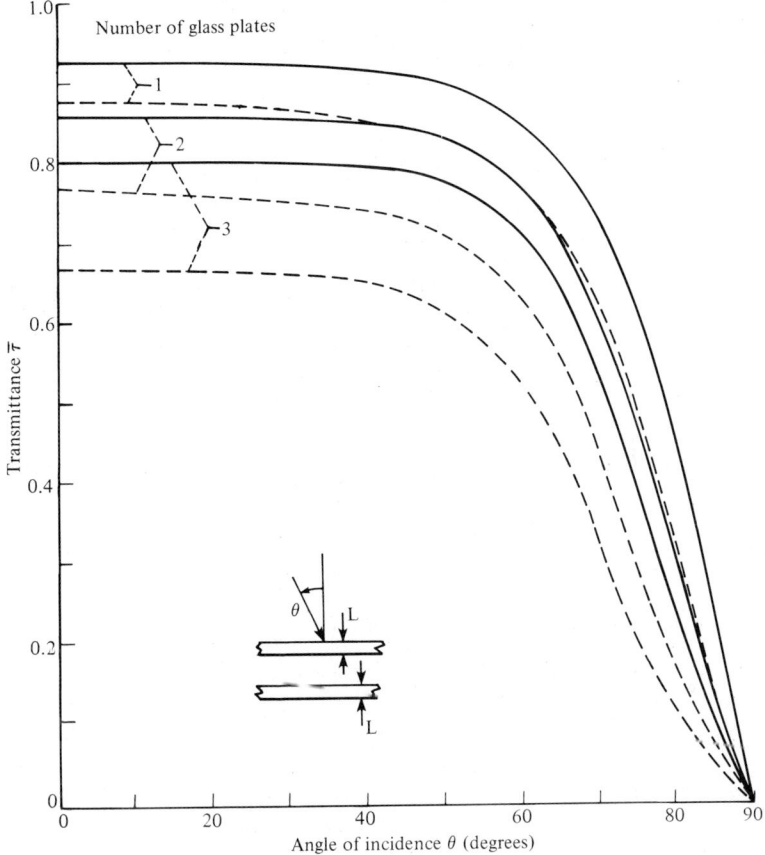

Figure 5-2 Effect of incidence angle and absorption on transmittance of multiple-glass plates; $n = 1.5$. $aL = :$ ———— 0 per plate [4]; – – – – 0.0542 per plate [2].

The fraction of the energy incident on the cover-plate assembly of n identical plates that is absorbed by the absorber with absorptivity α is

$$(\tau\alpha)_{\text{eff}} = \frac{\alpha \bar{\tau}_n}{1 - (1-\alpha)\bar{\rho}_n} \tag{5-6}$$

and the notation $(\tau\alpha)_{\text{eff}}$ indicates the effective value of the transmission through the cover-plate assembly times the absorptance of the absorber. The $(\tau\alpha)_{\text{eff}}$ value includes all interreflections among the absorber and cover plates, and all effects of absorption by the covers. The chief error in using Eq. (5-6) for the solar $(\tau\alpha)_{\text{eff}}$ value is the neglect of shading by the collector sides.

Example 5-1 The transmittance τ of a glass plate is given by

$$\tau = e^{-0.165 L / \cos \chi}$$

where L is the thickness of the plate (cm), and χ is the angle of travel of the solar energy in the glass relative to the normal to the glass surface.

The reflectance at the interface between two materials (in this case, air and glass) for any incident angle θ can be approximated by Fresnel's equation [3], which, for unpolarized radiation, is

$$\rho = \frac{1}{2} \frac{\sin^2(\theta-\chi)}{\sin^2(\theta+\chi)} \left[1 + \frac{\cos^2(\theta+\chi)}{\cos^2(\theta-\chi)} \right]$$

Here, χ is the angle of refraction, related to the angle of incidence by the refractive index n,

$$n = \frac{\sin \theta}{\sin \chi}$$

For normal incidence, $\chi \to \theta \to 0$, and ρ becomes

$$\rho(\theta = 0) = [(n-1)/(n+1)]^2$$

Consider a solar collector which has a single thick glass cover plate [0.32 cm ($\frac{1}{8}$ inch) thick, $n = 1.5$] over an absorber with $\alpha = 0.92$. Using the above information find

(a) $(\tau\alpha)_{\text{eff}}$ for normally incident beam insolation ($\theta = 0$)
(b) $(\tau\alpha)_{\text{eff}}$ for beam insolation at $\theta = 60°$

SOLUTION (a) For this case,

$$\tau = e^{-0.0524} = 0.949 \quad \text{and} \quad \rho = \left(\frac{1.5-1}{1.5+1}\right)^2 = 0.04$$

From Eqs. (5-1) and (5-2),

$$\bar{\rho}_1 = \rho \left[1 + \frac{(1-\rho)^2 \tau^2}{1 - \rho^2 \tau^2} \right] = 0.04 \left[1 + \frac{(1-0.04)^2(0.949)^2}{1 - (0.04)^2(0.949)^2} \right] = 0.0732$$

$$\bar{\tau}_1 = \tau \left(\frac{1-\rho}{1+\rho}\right)\left(\frac{1-\rho^2}{1-\rho^2 \tau^2}\right) = 0.876$$

(which agrees with Fig. 5-2 for $aL = 0.165L = 0.0524$ for this example) and, from Eq. (5-6),

$$(\tau\alpha)_{\text{eff}} = \frac{\alpha\bar{\tau}_I}{1 - (1-\alpha)\bar{\rho}_I} = \frac{(0.92)(0.876)}{1 - (1-0.92)(0.0732)} = 0.811$$

(b) For this case,

$$\chi = \sin^{-1}\left(\frac{\sin\theta}{n}\right) = \sin^{-1}\left(\frac{\sin 60°}{1.5}\right) = 35.3°$$

and

$$\tau = e^{-0.0524/\cos 35.3} = 0.938$$

$$\rho = \frac{1}{2}\frac{\sin^2(24.7)}{\sin^2(95.3)}\left[1 + \frac{\cos^2(95.3)}{\cos^2(24.7)}\right] = 0.0889$$

Now

$$\bar{\rho}_I = 0.0889\left[1 + \frac{(1-0.0889)^2(0.938)^2}{1-(0.0889)^2(0.938)^2}\right] = 0.154$$

$$\bar{\tau}_I = 0.938\left(\frac{1-0.089}{1+0.089}\right)\left[\frac{1-(0.089)^2}{1-(0.089)^2(0.938)^2}\right] = 0.783$$

(which again agrees with Fig. 5-2), and

$$(\tau\alpha)_{\text{eff}} = \frac{0.92(0.783)}{1-(1-0.92)(0.154)} = 0.729$$

The fraction of incident direct insolation that is absorbed by the absorber plate thus drops from 81 percent at normal incidence to 73 percent at an incident angle of 60°. The dropoff is much larger at greater incidence angles.

In most cases, the infrared portion of the spectrum can be conveniently treated by assuming that the cover plates are perfect absorbers so that the radiative transfer between the absorber and cover or between two covers can be treated as the exchange between parallel black plates. However, if the cover has one or more spectral "windows" in the infrared region, as is the case with plastic films, then the infrared radiation exchange should be treated more exactly.

The properties of some typical absorber surfaces for flat-plate collectors are given in Table 5-2.

Table 5-2 Properties of some absorber surfaces

Surface	α	ϵ	(α/ϵ)
Black nickel	0.87–0.96	0.07–0.12	8–14
Copper oxide	0.81–0.93	0.11–0.17	5–8.5
Black chrome	0.90–0.95	0.15–0.25	3–6
Lead sulfide (crystals)	0.89	0.20	4.5
Black paints	0.90–0.98	0.90–0.98	~1

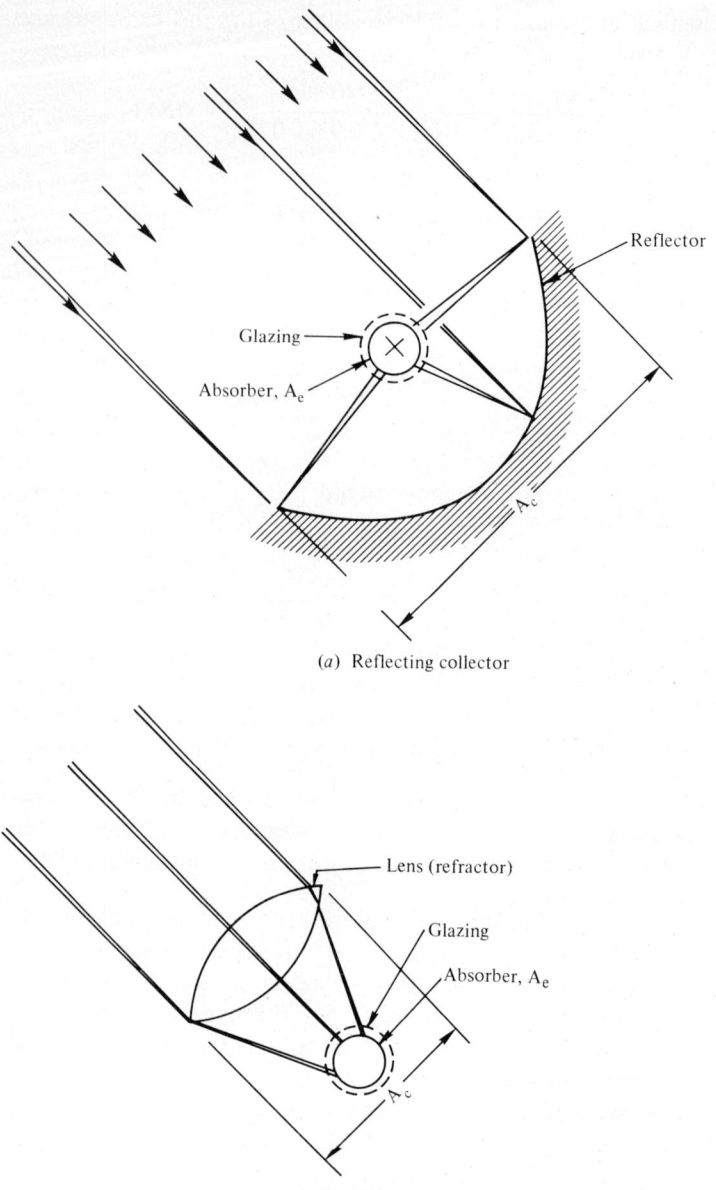

(a) Reflecting collector

(b) Refracting collector

Figure 5-3 Reflecting and refracting focusing collectors (receiver oversized for clarity).

Concentrating Collectors

Concentrating collectors use an optical element (lens or reflector) to focus or intensify (concentrate) the beam component of solar energy onto a receiving area (absorber) smaller than the intercepting area (aperture) of the optical element. Figure 5-3 illustrates the basic reflective (mirrorlike) and refractive (lens) types of concentrating collectors. The geometric or ideal concentration ratio (C_g) is generally defined as the aperture area A_c divided by the entire exposed surface area A_e of the absorber, $C_g = A_c/A_e$. The advantage of concentration is that, since the thermal losses from the absorber are proportional to the absorber area, the reduced absorber area usually permits maintaining a higher efficiency at high temperatures than for nonconcentrating collectors. Disadvantages of concentrating collectors are that, except for designs with low concentration ratios, only the beam component of radiation can be collected (the diffuse component is lost); there is a need for accurate tracking and relative positioning of the optical elements; and there can be significant losses due to the loss of the nonspecularly reflected (or refracted) radiation by the internal optical elements. However, if high temperatures are required, the single advantage of concentrating collectors outweighs the disadvantages.

Figure 5-4 illustrates a ray trace diagram and some common terminology for a simple reflective optical system of the planar-parabolic (trough) or axisymmetric parabolic types. The focal length f is the distance from the reflector apex to the focal point (line) for on-axis parallel incoming light rays. The acceptance angle Ω is the maximum angle over which incident radiation can be reflected to the absorber. For an ideal concentrator, the acceptance angle could be set equal to the angle subtended by the solar disk, approximately 32 minutes ($\sim 0.5°$) of arc. In that case, the limiting size of the image at the focal point is specified by $r \tan \Omega$, where r is the distance from the absorber to the reflector rim (the furthest distance from the absorber to any point on the reflector). The angle between the optical axis and the reflected ray from the rim is designated the rim angle Φ, and

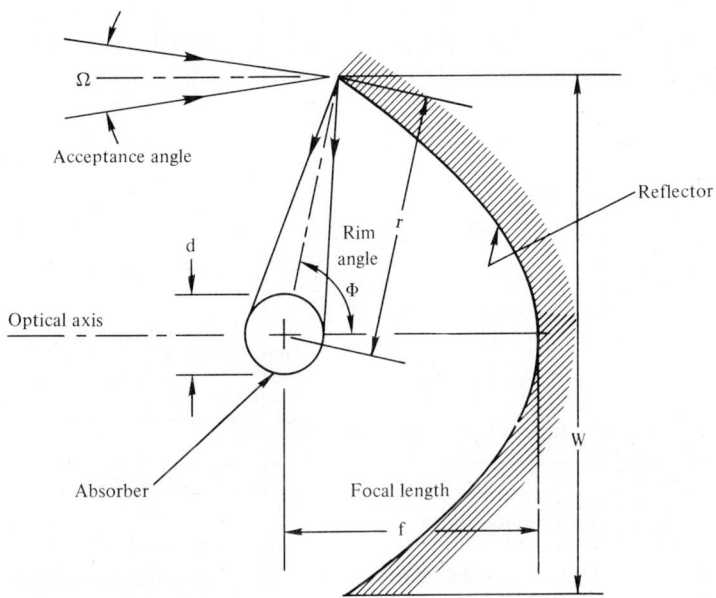

Figure 5-4 Nomenclature for reflective optical system (receiver oversized for clarity).

is related to f and r by

$$r = \frac{2f}{1 + \cos \Phi}$$

Since r, Φ, and W, the width of the aperture, are related by

$$\frac{W}{2} = r \sin \Phi$$

the projected (intercepting) width d at the focal plane is

$$d = 2r \tan \left(\frac{\Omega}{2}\right) = 2r \tan (16') = \frac{4f \tan (16')}{1 + \cos \Phi} = \frac{W \tan (16')}{\sin \Phi}$$

However, due to diffraction effects in rear surface mirrors, macroscopic imperfections in the optical quality of the mirror, misalignment, or other aberrations, the image at the focal plane will be spread beyond this width. Figure 5-5 illustrates a typical intensity at the focal plane relative to that for perfect optics. The receiver (absorber) width d is selected to intercept some moderately high fraction of this energy, but not so large as to cause the absorber thermal losses (which are in proportion to the absorber area) to become excessive.

Concentrating collector performance analysis The analysis of concentrating collectors is complicated by the difficulty of precisely characterizing the optical properties of the mirror or lens (optical element), the accuracy of positioning the receiver relative to the reflector, and the tracking accuracy of the collector (reflector/absorber) relative to the beam radiation. These uncertainties become more important as one attempts to collect at higher temperature, i.e., for higher concentration ratios. The optical quality of the lens or reflector must be considered in terms of its microscopic and macroscopic characteristics, which will now be

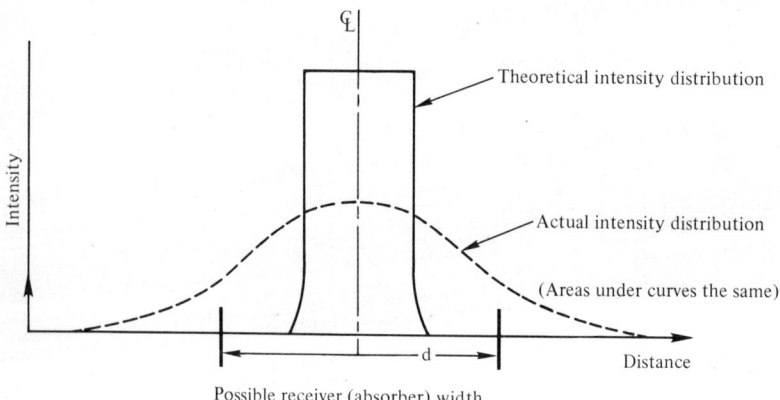

Figure 5-5 Relation between theoretical and actual intensity for planar optics, assuming the same internal optical losses.

discussed. For the following discussion, reference will be made to reflective optics, though similar comments apply to refractive optics. The specular reflectivity ρ' of a surface represents the fraction of the incident energy that is reflected in a specular (or mirrorlike) manner and is the portion of the incident beam energy that could strike the absorber. However, due to the macroscopic imperfections in the reflector as well as to improper size and positioning of the absorber and lack of perfect tracking accuracy of the collector, only some fraction of the specularly reflected energy will be intercepted by the absorber. This fraction is designated as the intercept factor I. It is quite difficult to predict I accurately, but its deviation from unity can represent a very important loss in a concentrating system. Figure 5-5 illustrates the intensity of radiation across the focal plane of the optical element and the projected width d of the absorber for perfect positioning and tracking. The intercept factor is thus a function of the width d of the receiver.

The net energy absorbed by the absorbing element in a perfectly tracking concentrating collector is then

$$\dot{Q}_{abs} = A_e q_{abs} = (\tau\alpha)_{eff} I q_s A_c \qquad (5\text{-}7)$$

or

$$q_{abs} = (\tau\alpha)_{eff} I C_g q_s = (\tau\alpha)_{eff} C_R q_s \qquad (5\text{-}8)$$

where C_R is the concentration ratio of the collector (equal to IC_g) and q_{abs} means an average absorbed energy flux over the absorber area, A_e. The heat flux on the absorber is thus equal to the insolation times the geometric concentration ratio times the intercept factor times an effective transmission-absorption factor that accounts for optical losses in the optical elements and on the absorber itself. For a flat-plate collector, $C_g = 1$ and the intercept factor is near unity (varying only because of possible shading by side walls).

5-3 THERMAL ANALYSIS

In this section, an overall thermal analysis is presented, including the effects of multiple covers and concentrator optics on collector performance. To simplify, however, we consider a two-cover flat-plate collector.

Figure 5-6 describes the energy flow in a two-cover flat-plate collector. The visible (solar) radiation has been separated from the infrared radiation and convection energy transfer modes. For an incident solar flux q_s, individual cover transmittance $\bar{\tau}_1$ and reflectance $\bar{\rho}_1$, and absorber plate absorptance α, the transmission through two covers would be $q_s \bar{\tau}_{II}$ [Eq. (5-4)] and the energy absorbed by the absorber plate would be $q_s(\tau\alpha)_{eff}$ [Eq. (5-6)]. The solar radiation interaction with the collector can be assumed to be independent of the collector component temperatures and the only energy terms resulting from solar radiation which contribute to the overall energy balance (infrared + convective) on the collector are the solar absorptance in the cover glasses, $\bar{\alpha}_1$ and $\bar{\alpha}_2$, and the solar absorptance by the absorber itself, α.

134 SOLAR-THERMAL ENERGY SYSTEMS

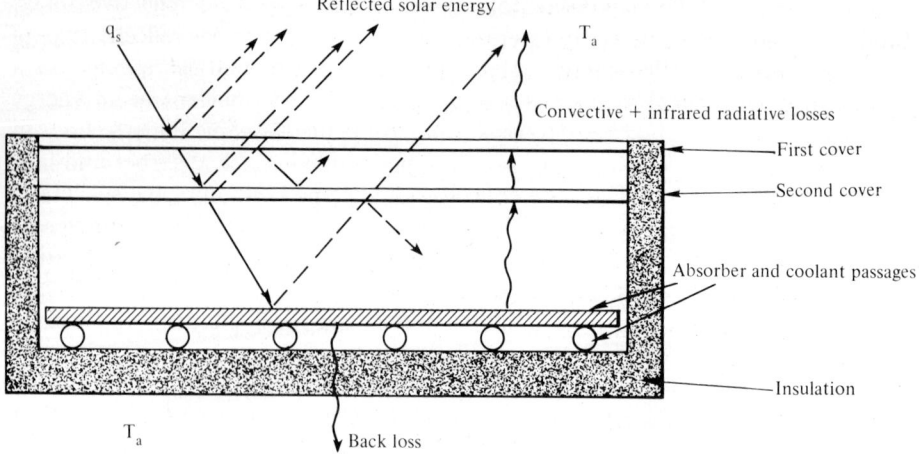

(a) Schematic

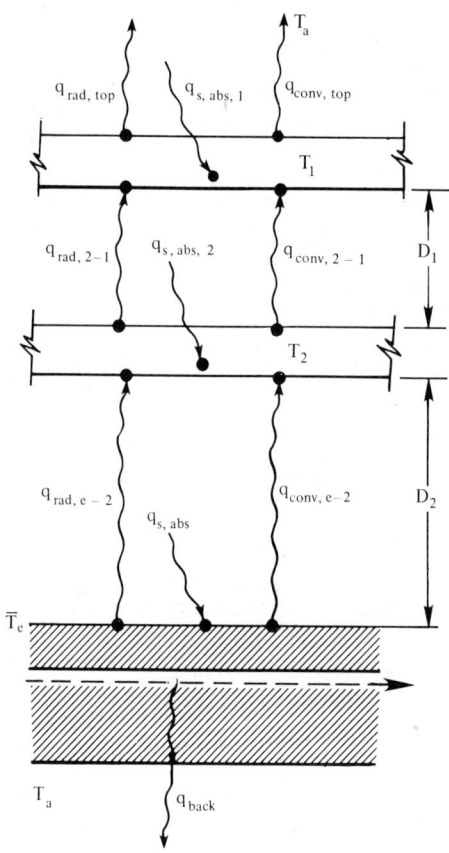

(b) Nodal representation

Figure 5-6 Model for flat-plate collector analysis.

The analysis of the performance of the collector begins by writing energy balances on each of the covers for an assumed temperature of the absorber, and solving for the cover temperatures. As will be seen, this requires an iterative solution. A subsequent energy balance on the absorber will permit solution of the net useful energy absorbed by the absorber and carried away by the coolant, $\dot{Q}_u$, which when divided by the incident solar energy rate $\dot{Q}_s$, provides the collector efficiency η at the assumed absorber temperature. In the analysis to follow, it is assumed: the surfaces are gray for the infrared, the covers are opaque to the infrared, the outer cover loses energy by both radiation and convection to T_a, through a lower effective temperature than T_a could be used for the radiation sink.

For steady state an energy balance on each glazing is† (Fig. 5-6)

$$q_{\text{in, rad}} + q_{\text{in, conv}} + q_{s,\text{abs}} = q_{\text{out, rad}} + q_{\text{out, conv}} \tag{5-9}$$

For cover 1 (outer cover)

$$\frac{\sigma}{1/\epsilon_1 + 1/\epsilon_2 - 1} (T_2^4 - T_1^4) + \frac{kB}{D_1} \text{Gr}_{D,1}^n (T_2 - T_1) + q_s \bar{\alpha}_1$$
$$= \varepsilon_1 \sigma (T_1^4 - T_a^4) + h(T_1 - T_a) \tag{5-10}$$

(Infrared radiation: cover 1 to cover 2) + (convection: cover 1 to cover 2) + (solar absorption in cover 1) = (infrared radiation: cover to ambient) + (convection: cover to ambient)

where $\text{Gr}_{D,1} = g\beta(T_2 - T_1)D_1^3/\nu^2$ is the Grashof number for the space between the covers.

For cover 2

$$\frac{\sigma}{1/\epsilon + 1/\epsilon_2 - 1} (\bar{T}_e^4 - T_2^4) + \frac{kB}{D_2} \text{Gr}_{D,2}^n (\bar{T}_e - T_2) + q_s \bar{\alpha}_2$$
$$= \frac{\sigma}{1/\epsilon_1 + 1/\epsilon_2} (T_2^4 - T_1^4)$$
$$+ \frac{kB}{D_1} \text{Gr}_{D,1}^n (T_2 - T_1) \tag{5-11}$$

Infrared radiation: absorber to cover 2) + (convection: absorber to cover 2) + (solar absorption) = (infrared radiation: cover 1 to cover 2) + (convection: cover 1 to cover 2)

where $\text{Gr}_{D,2} = g\beta(\bar{T}_e - T_1)D_2^3/\nu^2$. Note that $\bar{\alpha}_1$ and $\bar{\alpha}_2$ are the absorptance of plates 1 and 2. These are different quantities from $\bar{\alpha}_\text{I}$ and $\bar{\alpha}_\text{II}$ of Eqs. (5-3) and (5-4), which are for *sets* of one and two plates, respectively.

The constants B and n in the second term of each equation depend on the

---

† For the relations used here in calculating the various q values, see App. C.

Grashof number for the natural convection between the surfaces. These two equations can be solved iteratively for T_1 and T_2 for a given value of collector surface temperature $\bar{T}_e$ and then the temperature T_2 may be used in the equations below to determine the net useful energy absorbed by the absorber and carried away by the fluid.

Absorber energy balance

$$q_u = q_s(\tau\alpha)_{\text{eff}} - \frac{\sigma}{1/\epsilon + 1/\epsilon_2 - 1}(\bar{T}_e^4 - T_2^4) - \frac{kB}{D_2}\text{Gr}_{D,2}^n(\bar{T}_e - T_2)$$

$$- \frac{1}{1/h + D_i/k_i}(\bar{T}_e - T_a) \tag{5-12}$$

(Useful energy) = (absorbed solar) − (radiation: absorber cover 2) − (convection: absorber cover 2) − backloss where q_u is the net useful energy absorbed by the absorber per unit area and time.

Dividing the above equation by q_s gives the collector efficiency η as

$$\eta = \frac{q_u}{q_s} = (\tau\alpha)_{\text{eff}} - \frac{\sigma}{1/\epsilon + 1/\epsilon_2 - 1}\left(\frac{\bar{T}_e^4 - T_2^4}{q_s}\right) - \frac{kB}{D_2}\text{Gr}_{D,2}^n\left(\frac{\bar{T}_e - T_2}{q_s}\right) \tag{5-13}$$

$$- \frac{1}{1/h + D_i/k_i}\left(\frac{\bar{T}_e - T_a}{q_s}\right) = (\tau\alpha)_{\text{eff}} - \frac{b_1(\bar{\Theta}_e - \Theta_2) + b_2(\bar{\Theta}_e - \Theta_a) + a_1(\bar{\Theta}_e^4 - \Theta_2^4)}{\psi(t)}$$

where $\Theta = \dfrac{T}{T_a}$

$$\psi(t) = \frac{q_s(t)}{q_{s,\text{ref}}}$$

$$b_1 = \frac{kB\bar{T}_a}{D_2 q_{s,\text{ref}}}\text{Gr}_{D,2}^n$$

$$b_2 = \frac{\bar{T}_a}{q_{s,\text{ref}}(1/h + D_i/k_i)}$$

$$a_1 = \frac{\sigma\bar{T}_a^4}{q_{s,\text{ref}}(1/\epsilon + 1/\epsilon_2 - 1)}$$

Now the collector efficiency from Eq. (5-13) can be calculated by substituting T_2 as found by simultaneous solution of Eq. (5-10) and (5-11). Because the equations are nonlinear in T, a closed-form solution for efficiency in terms of $\bar{T}_e$ and T_a is generally not possible.

To simplify the form of the efficiency equation, it is usually written in terms of a single overall loss coefficient $\bar{U}$ between the absorber-plate surface temperature and the ambient temperature, so that the efficiency can be given by

$$\eta = (\tau\alpha)_{\text{eff}} - \frac{b(\bar{\Theta}_e - \Theta_a) + a(\bar{\Theta}_e^4 - \Theta_a^4)}{\psi(t)} \tag{5-14}$$

where b is proportional to $\bar{U}$. This form is the same as Eq. (2-2) used in Chap. 2.

Comparison of Eq. (5-14) with Eq. (5-13) shows that the form of Eq. (5-14) includes the back-loss coefficient b_2 with a front-loss coefficient in the factor b.

If radiation is small as is usual for the operating temperatures of flat-plate collectors, then the $(\bar{\Theta}_e^4 - \Theta_a^4)$ term can be linearized by

$$(\bar{\Theta}_e^4 - \Theta_a^4) = (\bar{\Theta}_e + \Theta_a)(\bar{\Theta}_e^2 + \Theta_a^2)(\bar{\Theta}_e - \Theta_a) \simeq 4\Theta_a^3(\bar{\Theta}_e - \Theta_a)$$

and Eq. (5-14) then becomes

$$\eta = (\tau\alpha)_{\text{eff}} - \frac{b_3(\bar{\Theta}_e - \Theta_a)}{\psi(t)} \qquad (5\text{-}15a)$$

where $b_3 = b + 4a\Theta_a^3$.

It is assumed that b_3 can be taken as proportional to $\bar{U}$, that is,

$$b_3 = \frac{\bar{U}\bar{T}_a}{q_{s,\text{ref}}}$$

and this is the form used in SOLSIM for flat-plate and other low-temperature collectors.

Expanding Eq. (5-15a) into dimensional variables then gives

$$\eta = (\tau\alpha)_{\text{eff}} - \frac{\bar{U}(\bar{T}_e - T_a)}{q_s(t)} \qquad (5\text{-}15b)$$

The efficiencies of various collectors depend on many quantities, including

- Climatic conditions: ambient temperature, wind, insolation
- Number of covers and their radiative properties (reflectance, absorptance)
- Incident solar angle
- Radiative properties of the absorber plate (α and ϵ), and its construction
- Spacing of covers and absorber (i.e., space may be air, partially evacuated, or may have a honeycomb structure to reduce convection)
- Fluid type, flow rate and fluid passage area
- Insulation of collector enclosure and degree of infiltration

If one considers Eq. (5-14), it is seen that the collector efficiency consists of the term $(\tau\alpha)_{\text{eff}}$ less terms involving temperature differences divided by the solar flux. That is, for a given collector, the efficiency is degraded by temperature differences between the absorber and surroundings and enhanced by increasing q_s. Figure 5-7a and b explains this pictorially for a single-glazed and a double-glazed collector. Note that, while the single-glazed collector will exhibit better performance at lower temperatures than the double-glazed type, the convection and radiation losses for the double-glazed collector are less, and thus it performs better at higher collection temperatures.

The performance of a double-glazed flat-plate collector under different insolation levels and ambient temperatures was determined by solving Eqs. (5-10) to (5-12) on a computer. For these calculations it was assumed that the absorber had $\alpha = 0.9$ and $\epsilon = 0.5$ (slightly selective). The covers were assumed to have

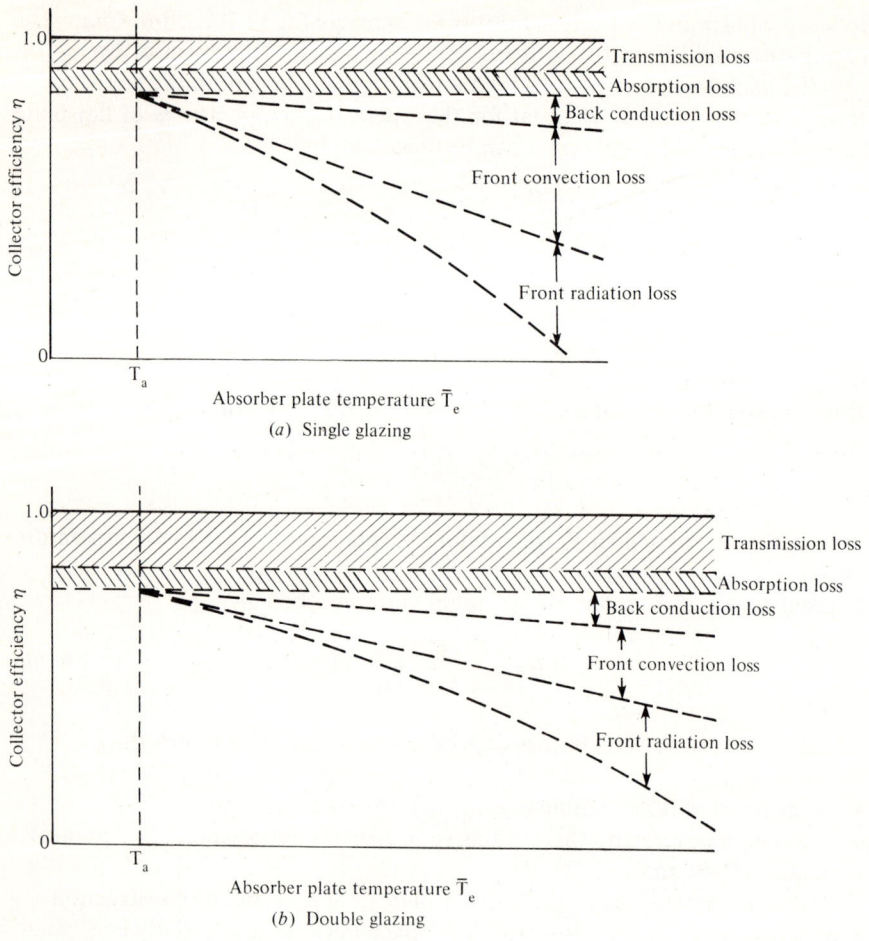

Figure 5-7 Losses for flat-plate collectors.

reflectivities at each surface of 0.04 and bulk absorption of 0.02, i.e., an overall transmittance of 90 percent (low-iron glass). Air spaces of 1.25 and 5 cm (0.5 and 2 in) between the covers and between second cover and absorber were used. The enclosure was insulated with a foam insulation of $k_i = 0.035$ W/(m · °C) [0.02 Btu/(h · ft · °F)] and thickness $D_i = 5$ cm (2.0 in). A wind velocity of 4.4 m/s (10 mi/h) was used to compute the convection coefficient to ambient from the first cover and the enclosure. Figures 5-8a and b show the performance results for ambient temperatures of 7 and 35°C (45 and 90°F), respectively. These results thus represent a typical double-glazed flat-plate collector with a slightly selective surface and low-iron glass. Note that the performance could be improved by

1. Treating the glass to reduce reflective losses. Overall transmittance values of >95 percent can be achieved.

ANALYTICAL DESIGN OF SOLAR COLLECTORS **139**

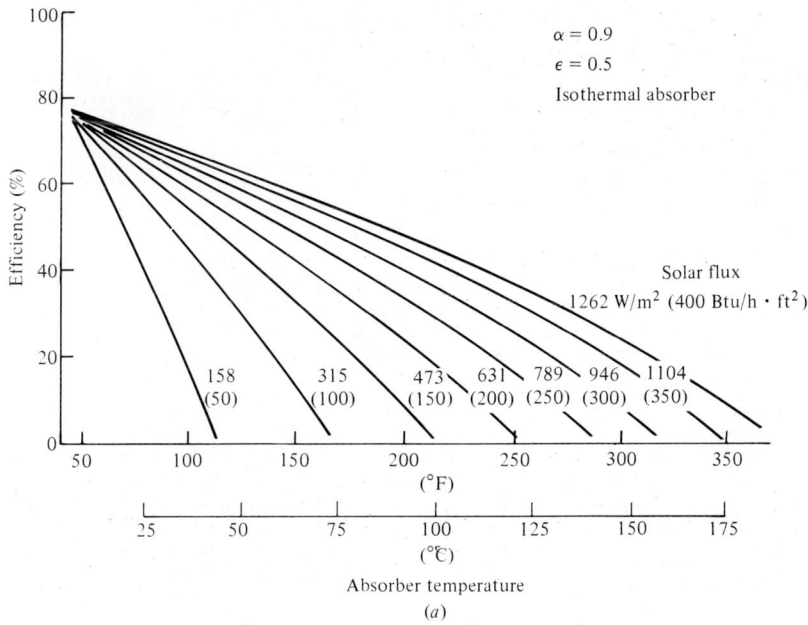

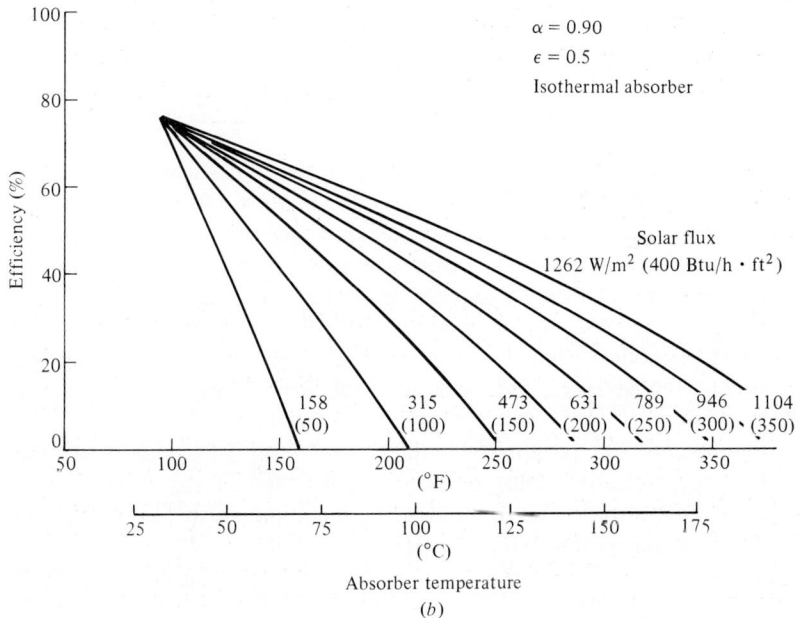

Figure 5-8 Double-glazed collector efficiency for (a) 7°C (45°F) and (b) 35°C (95°F) ambient temperature.

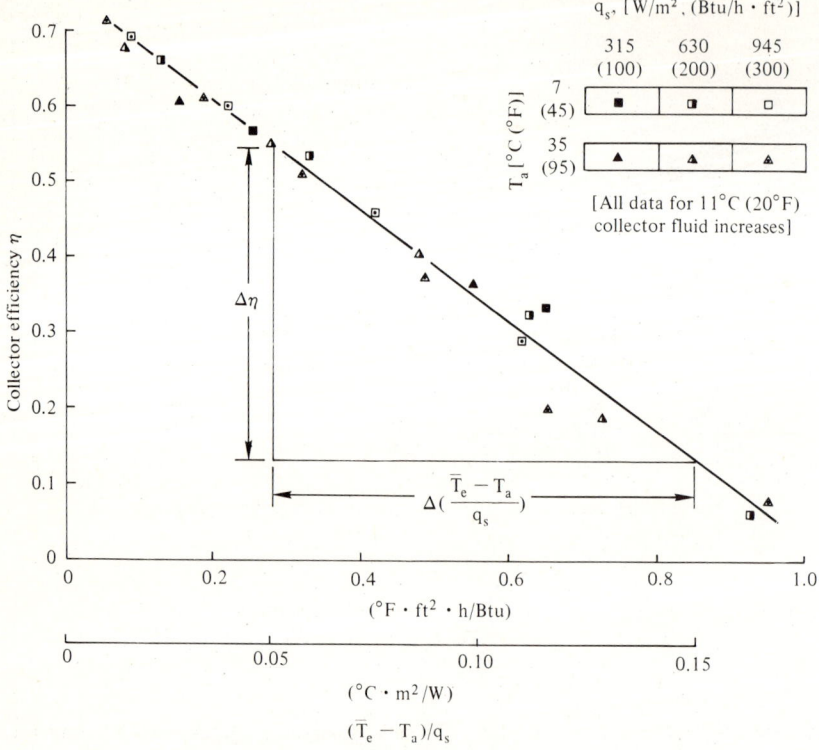

Figure 5-9 Correlation of predicted collector efficiency data in terms of the parameter $(T_{f,\text{in}} - T_a)/q_s$.

2. Applying a better selective coating to the collector absorber. Selective coatings of copper, nickel, or chromium oxide will provide values of $\alpha/\epsilon \simeq 6$.
3. Partially evacuating the spaces between covers and absorber. This significantly reduces convective losses (but not conductive loss unless a hard vacuum is achieved). However, evacuation introduces severe structural design problems.

It is seen that the collector efficiency η decreases with decreasing insolation, decreasing ambient temperature, and increasing absorber temperature, as one would expect.

Figure 5-9 presents the performance data from Fig. 5-8a and b for the double-glazed slightly selective collector at two ambients, in the form of $\eta = f_n[(T_{f,\text{in}} - T_a)/q_s]$. To transform the data into the form shown in Fig. 5-9 it was assumed that there is negligible difference between the fluid and absorber temperatures† and that the flow rates are such that there is approximately an 11°C

---

† Note that a temperature difference between fluid and absorber will exist, is a function of the absorber/fluid flow passage geometry and materials, and may be very critical in the resulting collector performance.

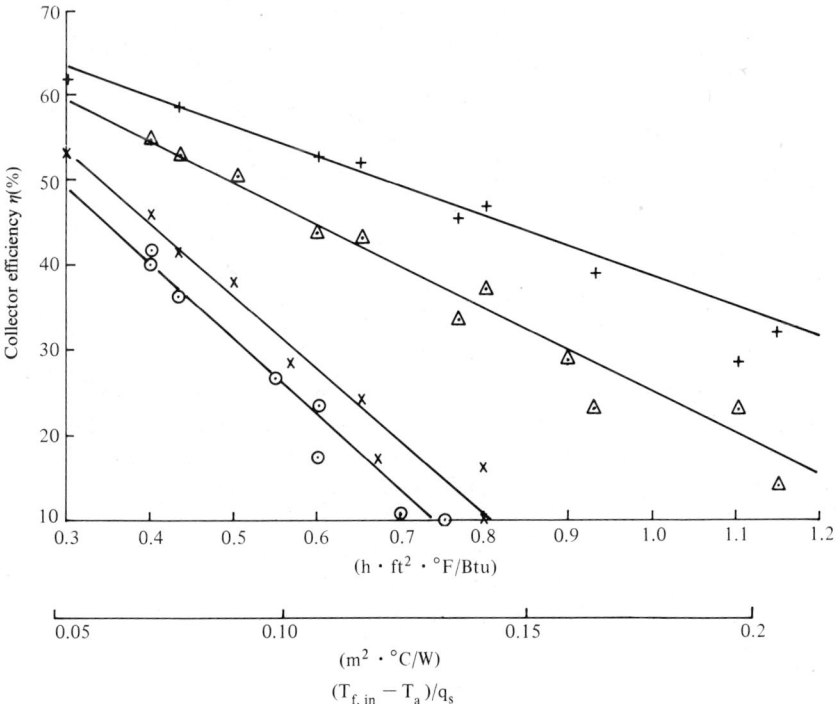

Figure 5-10 Predicted performance of flat-plate collectors for $\alpha = 0.9$; ($\bigcirc$)$\epsilon = 0.9$, ($\triangle$)$\epsilon = 0.15$ for convection; ($\times$)$\epsilon = 0.9$, ($+$)$\epsilon = 0.15$ for no convection, partially evacuated.

(20°F) fluid rise along the collector.† At various conditions of q_s and $\bar{T}_e$ for Fig. 5-8a and b an efficiency was chosen where $T_{f,\,in}$ would thus be $\bar{T}_e - 5.5$°C. It is seen that the predictions of Fig. 5-8a for a wide range of q_s, $\bar{T}_e$, and T_a correlate well in the form of $\eta = f_n[T_{f,\,in} - T_a)/q_s]$.

Figure 5-10 presents similar data for a double glazed collector but with grey ($\alpha = \epsilon = 0.9$) and selective ($\alpha = 0.9$, $\epsilon = 0.15$) absorbers and with nonevacuated and partially evacuated (conduction-limited) collectors. The importance of these factors is evident. It is seen that a selective surface results in a large increase in performance and that partial evacuation results in a moderate improvement. Note that all performance curves show scatter even in predicted performance due to the nonlinear behavior of the heat-transfer effects.

5-4 ANALYSIS OF COLLECTORS

The analysis presented in Sec. 5-3 is correct but cumbersome because of the iterative nature of the solution. A practical simplified method of analyzing the performance of a flat-plate collector was proposed by Hottel and Woertz [5], and

† The fluid temperature rise will be $\Delta T_f = \eta q_s A_c/(\dot{m}_c c)$ where $\dot{m}_c/A_c$ is the fluid mass flow rate per unit of collector aperture area and c is fluid specific heat.

developed by Hottel and Whillier [6] and Bliss [7]. The method can be extended to include concentrating collectors. The method neglects the effects of thermal capacitance in the collector; that is, transient effects induced by storage of energy within the collector itself are ignored. Klein, Duffie, and Beckman [8] have shown that ignoring these effects leads to negligible errors in applications to real collectors, and that the so-called Hottel-Whillier-Woertz-Bliss or HWWB method is often preferable to more complex formulations unless detailed meteorological information is available. Lunde [9] and Howell [10] have given other approaches and interpretations to the HWWB method.

The HWWB method is developed below.

Analysis

For a conductive fin attached to an absorber fluid passage, the fin efficiency or fin effectiveness η_{fin} is defined as the actual energy gained or lost by a fin-tube assembly divided by the energy gained or lost if the assembly were uniformly at the temperature of the fin base, T_c. For an elemental section of a solar collector (Fig. 5-11), an energy balance can be written, using the fin efficiency and neglecting thermal capacitance and longitudinal (x-direction) conduction [11], as

$$\begin{aligned} q_u(x)\, dA_e &= \dot{m}c\, dT_f \\ \text{(Useful energy collected)} &= \text{(change in fluid enthalpy)} \\ &= \eta_{\text{fin}}\{(\tau\alpha)_{\text{eff}} I q_s\, dA_c - \bar{U}[T_c(x) - T_a]\, dA_e\} \\ &= \text{(net collected energy)} \end{aligned} \qquad (5\text{-}16)$$

Here, $q_u(x)\, dA_e$ is the useful energy collected from the absorber area element dA_e at position x along the flow path, $\bar{U}$ is the overall thermal loss coefficient between

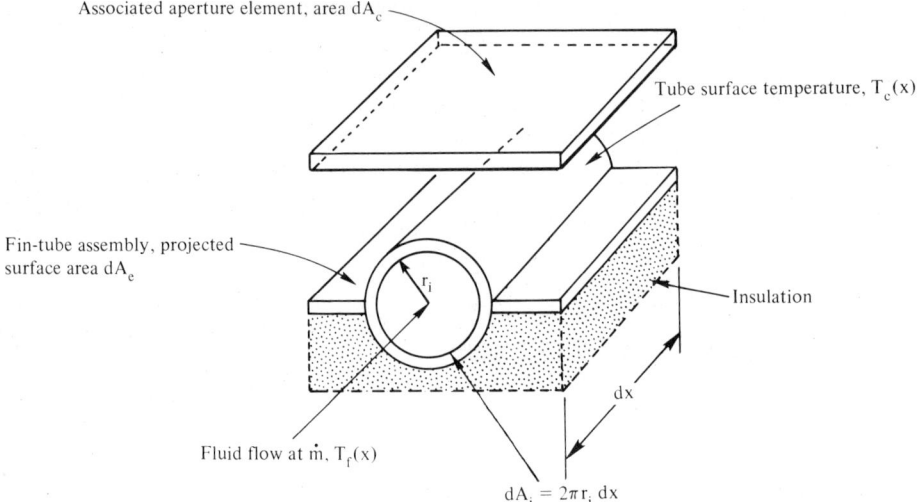

Figure 5-11 Elemental portion of solar collector.

the fin-tube absorber assembly surface at an average (over the transverse direction) temperature $\bar{T}_e(x)$ and ambient air at temperature T_a, and $(\tau\alpha)_{\text{eff}} I q_s dA_c$ is the net radiant energy gain of the assembly from a given area of collector aperture, dA_c. The dT_f is the change in fluid temperature across dx. For a flat-plate collector, $I\, dA_c = dA_e$. The I is the intercept factor defined in Sec. 5-2. Both q_s and $\bar{U}$ are assumed constant across the collector, and the tube temperature is assumed to be circumferentially invariant [12].

To eliminate $T_e(x)$ from Eq. (5-16), the energy transfer between the collector fluid and the fin-tube assembly is written

$$q_u(x)\, dA_e = \bar{U}_i[T_e(x) - T_f(x)]\, dA_i \tag{5-17}$$

where $\bar{U}_i$ is the overall thermal loss coefficient between the outer tube surface and the fluid, and may include the effects of fin-tube contact resistance, conduction through the tube wall, and convective transfer from the tube wall to the fluid. Note that $\bar{U}_i$ is based on the inner tube surface area $dA_i = 2\pi r_i dx$.

Substituting Eq. (5-17) into (5-16) to eliminate $T_e(x)$ gives

$$q_u(x)\, dA_e = \dot{m}c\, dT_f = \mu\{(\tau\alpha)_{\text{eff}} I q_s\, dA_c - \bar{U}[T_f(x) - T_a]\, dA_e\} \tag{5-18}$$

where

$$\mu = \frac{1}{1/\eta_{\text{fin}} + \bar{U}\, dA_e/\bar{U}_i\, dA_i} = \frac{1}{1/\eta_{\text{fin}} + \bar{U} A_e/\bar{U}_i A_i}$$

The ratio A_e/A_i is the absorber area divided by the total internal flow channel area for heat transfer. It is assumed that the element represented in Fig. 5-11 is typical of the entire absorber, that the tubes are equally spaced, and that the heat transfer in the headers and at the panel ends can be approximated by this representation. To determine the local fluid temperature $T_f(x)$, Eq. (5-18) is integrated from the inlet, where $T_f(x) = T_{f,\text{in}}$, to some position x:

$$\frac{\mu}{\dot{m}c} \int_0^{A_e(x)} dA_e = \int_{T_f = T_{f,\text{in}}}^{T_f = T_f(x)} \frac{dT_f}{(\tau\alpha)_{\text{eff}} I q_s (dA_c/dA_e) - \bar{U}[T_f(x) - T_a]}$$

which results in

$$\frac{-\gamma A_e(x)}{A_e} = \ln\left\{\frac{1 - \left[\dfrac{T_f(x) - T_a}{T_{f,\text{in}} - T_a}\right]\beta}{1 - \beta}\right\} \tag{5-19a}$$

or

$$T_f(x) = T_a + \frac{T_{f,\text{in}} - T_a}{\beta}\left\{1 - (1-\beta)\exp\left[-\frac{\gamma A_e(x)}{A_e}\right]\right\} \tag{5-19b}$$

Here,

$$\gamma = \frac{\bar{U}\mu}{Gc} \qquad \beta = \frac{\bar{U}[T_{f,\text{in}} - T_a]}{(\tau\alpha)_{\text{eff}} q_s C_R} \qquad C_R = I\frac{A_c}{A_e}$$

and $G = \dot{m}/A_e$ and $dA_c/dA_e = A_c/A_e$, which is constant by our assumptions.

It is most useful if the local and overall energy gains of the collector are found in terms of the inlet fluid temperature. This can be done by defining a function $F(x)$ such that the temperature difference between the fluid and ambient temperatures at any x is given by

$$\mu\{(\tau\alpha)_{\text{eff}} q_s C_R - \bar{U}[T_f(x) - T_a]\} = F(x)\{(\tau\alpha)_{\text{eff}} q_s C_R - \bar{U}[T_{f,\text{in}} - T_a]\} \quad (5\text{-}20)$$

Thus, the variation in $T_f(x)$ is accounted for on the right-hand side by $F(x)$, which must be determined. Equation (5-18) now becomes

$$q_u(x) \, dA_e = \dot{m}c \, dT_f = F(x)\{(\tau\alpha)_{\text{eff}} q_s C_R - \bar{U}[T_{f,\text{in}} - T_a]\} \, dA_e \quad (5\text{-}21)$$

and the total energy gained by the collector is

$$Q_u = \int_0^{A_e} q_u(x) \, dA_e = \int_0^{A_e} F(x)\{(\tau\alpha)_{\text{eff}} q_s C_R - \bar{U}[T_{f,\text{in}} - T_a]\} \, dA_e$$

$$\equiv F_R\{(\tau\alpha)_{\text{eff}} q_s C_R - \bar{U}[T_{f,\text{in}} - T_a]\} A_e$$

$$= F_R(\tau\alpha)_{\text{eff}} q_s A_e [1 - \beta] C_R \quad (5\text{-}22)$$

where
$$F_R = \frac{1}{A_e} \int_0^{A_e} F(x) \, dA_e \quad (5\text{-}23)$$

Now, $F(x)$ is found by substituting Eq. (5-20) into Eq. (5-19b), giving

$$F(x) = \mu e^{-\gamma A_e(x)/A_e} \quad (5\text{-}24)$$

and, from Eq. (5-23),

$$F_R = \mu(1 - e^{-\gamma})/\gamma \quad (5\text{-}25)$$

Substituting Eq. (5-25) into Eq. (5-22) gives

$$\frac{Q_u}{\mu(\tau\alpha)_{\text{eff}} q_s A_e (1 - \beta) C_R} = \frac{F_R}{\mu} = \frac{1 - e^{-\gamma}}{\gamma} \quad (5\text{-}26)$$

and the left-hand side is the ratio of useful energy collected to the useful energy collected if the entire collector were at the fluid inlet temperature. Note that in the limit as $\gamma \to 0$, L'Hospital's rule can be applied and Eq. (5-26) reduces to

$$\lim_{\gamma \to 0} \left[\frac{Q_u}{\mu(\tau\alpha)_{\text{eff}} q_s A_e (1 - \beta) C_R} \right] = \lim_{\gamma \to 0} \left(\frac{1 - e^{-\gamma}}{\gamma} \right) = 1$$

or, in physical quantities

$$\frac{Q_u(\gamma \to 0)}{(\tau\alpha)_{\text{eff}} q_s A_c C_R} = \mu(1 - \beta) = \mu \left\{ 1 - \frac{\bar{U}[T_{f,\text{in}} - T_a]}{(\tau\alpha)_{\text{eff}} q_s C_R} \right\} \quad (5\text{-}27)$$

This is the same result as for Eq. (5-22) or Eq. (5-26) with $F_R = 1$. Thus, Eq. (5-27) gives the useful energy gain over the absorbed solar energy for the case of the fluid temperature being uniform across the collector and equal to the inlet value.

The factor γ is the ratio of overall thermal loss coefficient between the fluid and ambient, $\mu \bar{U}$, to Gc, the thermal capacity rate of the fluid per unit of absorber area of the collector. Thus, the smaller γ is, the more efficient the collector should be.

Evaluation of Mean Temperatures

Generally, $\bar{U}$ is a function of absorber temperature. As most collectors are operated with a small temperature difference between inlet and outlet, it is reasonable to choose an average $\bar{U}$ value and assume it is constant across the collector. However, this value will depend on the mean temperature, $\bar{T}_e$. Integrating Eq. (5-17) results in

$$\bar{T}_e = \frac{1}{A_e} \int_0^{A_e} T_c(x)\, dA_e = \frac{1}{A_e} \int_0^{A_e} T_f(x)\, dA_e + \frac{1}{A_e} \int_0^{A_e} \frac{q_u(x)}{\bar{U}_i\, dA_i}\, dA_e$$

$$= \bar{T}_f + \frac{Q_u/A_e}{\bar{U}_i(dA_i/dA_e)} \tag{5-28}$$

where $dQ_u(x) = q_u(x)\, dA_e$. Substituting Eq. (5-26) for Q_u and Eq. (5-19b) for $T_f(x)$ when evaluating $\bar{T}_f$ gives

$$\frac{\bar{T}_e - T_a}{T_{f,\text{in}} - T_a} = \frac{1}{\beta} + \frac{\mu(\beta - 1)}{\beta \eta_{\text{fin}}}(1 - e^{-\gamma}) = \frac{1}{\beta} + \frac{\beta - 1}{\beta \eta_{\text{fin}}} F_R \tag{5-29}$$

Equating Eqs. (5-29) and (5-26) to eliminate F_R results, after substituting the definition of β, in

$$Q_u = \eta_{\text{fin}}[(\tau\alpha)_{\text{eff}}\, q_s\, C_R - \bar{U}(\bar{T}_c - T_a)] A_e \tag{5-30}$$

Discussion of the HWWB Collector Equation

Rewriting Eq. (5-22) as

$$\eta = \frac{Q_u}{q_s\left(\dfrac{A_e C_R}{I}\right)} = \frac{Q_u}{q_s A_c} = F_R\left[I(\tau\alpha)_{\text{eff}} - \frac{\bar{U}(T_{f,\text{in}} - T_a)}{q_s(C_R/I)}\right] \tag{5-31}$$

shows that, if collector efficiency is plotted against $[(T_{f,\text{in}} - T_a)/q_s]$, the efficiency should plot as a straight line with a y intercept of $IF_R(\tau\alpha)_{\text{eff}}$ and a slope of $(-I\bar{U}F_R/C_R) = (-\bar{U}F_R/C_g)$. For a concentrating or other collector operating at high temperature, radiative losses will cause the efficiency curve plotted in this way to be nonlinear. It is obviously desirable to keep F_R near its maximum value of unity, so that η will have a large value. Figure 5-12 shows how F_R/μ varies with γ, indicating that γ should be kept small in any good collector design. This can be done by increasing G, the mass flow per unit of collector area (which

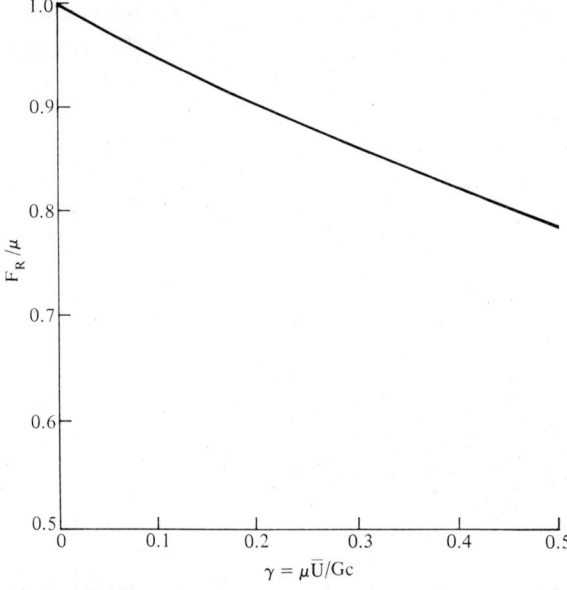

Figure 5-12 Collector temperature variation function F_R/μ versus flow parameter γ.

usually costs pumping power); by using a fluid with large c; by designing for a small value of overall thermal loss coefficient $\bar{U}$; and by minimizing the dimensionless thermal resistance μ between the absorber surface and coolant. The latter three factors usually involve design tradeoffs of cost and materials selection.

The Overall Thermal Loss Coefficient

Klein [13] has shown that $\bar{U}_{\text{top}}$, the overall thermal loss coefficient from the absorber surface to the environment through a glass cover assembly, can be predicted with a single equation for flat-plate collectors operating with average absorber temperatures in the range $40 < \bar{T}_e < 130°C$. Then, for $\bar{U}_{\text{top}}$ in W/(m² · °C), Duffie and Beckman [2] have presented a modified correlation for the daily average.

$$\bar{U}_{\text{top}} = \left[\frac{n}{(344/\bar{T}_e)\left[(\bar{T}_e - \bar{T}_a)/(n+f)\right]^{0.31}} + \frac{1}{h}\right]^{-1}$$
$$+ \frac{\sigma(\bar{T}_e + \bar{T}_a)(\bar{T}_e^2 + \bar{T}_a^2)}{[\epsilon + 0.0425n(1-\epsilon)]^{-1} + [(2n+f-1)/\epsilon_n] - n} \quad (5\text{-}32)$$

where n = number of glass cover plates
$f = (1.0 - 0.04h + 5.0 \times 10^{-4}h^2)(1.0 + 0.058n)$
ϵ = emittance of absorber plate
ϵ_n = emittance of glass plates
$\bar{T}_a$ = average ambient temperature, K
$\bar{T}_e$ = mean absorber plate temperature, Eq. (5-29), K
h = convective heat transfer coefficient, W/(m² · °C)

Example 5-2 Compute the overall loss coefficient $\bar{U}$ for the double-glazed collector shown in Fig. 5-9. For this collector,

$$n = 2$$
$$\epsilon = 0.5$$
$$\epsilon_n = 0.88$$
$$\bar{T}_a = 7 \text{ and } 35°C$$
$$h \text{ for 16 km/h wind}$$
$$\bar{T}_e = 100°C$$

First, h is found using the correlation in Appendix C as

$$h = 5.7 + 3.8\left[\frac{(16 \times 10^3) \text{ m}}{\text{h}} \frac{1 \text{ h}}{3600 \text{ s}}\right] = 22.6 \text{ W/(m}^2 \cdot °\text{C)}$$

Then

$$f = [1.0 - 0.04(22.6) + (5 \times 10^{-4})(22.6)^2][1 + 0.058(2)] = 0.392$$

Now, from Eq. (5-32), and $\bar{T}_a = 7°C$,

$$\bar{U}_{\text{top}} = \left[\frac{2}{(344/373)[93/2.392]^{0.31}} + \frac{1}{22.6}\right]^{-1}$$
$$+ \frac{(5.67 \times 10^{-8})(373 + 280)[(373)^2 + (280)^2]}{[0.5 + 0.0425(2)(1 - 0.5)]^{-1} + \{[2(2) + 0.392 - 1]/0.88\} - 2}$$
$$= 1.35 + 2.18 = 3.53 \text{ W/(m}^2 \cdot °\text{C)}$$

To get the overall loss coefficient $\bar{U}$, we must include the back-loss coefficient (assuming the h on the collector back and front are the same):

$$\bar{U}_{\text{back}} = \frac{1}{D_i/k_i + 1/h} = \frac{1}{1/0.7 + 1/22.6} = 0.7 \text{ W/(m}^2 \cdot °\text{C)}$$

for a $\bar{U}$ of

$$\bar{U} = \bar{U}_{\text{top}} + \bar{U}_{\text{back}} = 3.5 + 0.7 = 4.2 \text{ W/(m}^2 \cdot °\text{C)}$$

Note that $\bar{U}_{\text{top}}$ and $\bar{U}_{\text{back}}$ are additive because they are parallel conductances for the same area. The slope of the line on Fig. 5-9 is

$$\frac{\Delta \eta}{\Delta\left(\dfrac{\bar{T}_e - \bar{T}_a}{q_s}\right)} = \frac{\Delta\left[(\tau\alpha)_{\text{eff}} - \dfrac{\bar{U}(\bar{T}_e - \bar{T}_a)}{q_s}\right]}{\Delta\left(\dfrac{\bar{T}_e - \bar{T}_a}{q_s}\right)} = -\bar{U}$$

and the slope is then

$$-\bar{U} = \left[\frac{0.13 - 0.55}{0.15 - 0.10}\right] = -(0.42/0.1) = -4.2 \text{ W/(m}^2 \cdot {}^\circ\text{C)}$$

or $\quad \bar{U} = 4.2 \text{ W/(m}^2 \cdot {}^\circ\text{C)}$

so that the prediction by Klein's equation is in agreement with the more detailed computer solution.

5-5 APPLICATION OF THE ANALYSIS TO THE SOLSIM PROGRAM

SOLSIM uses Eq. (5-31) in the form

$$\frac{Q_u}{q_s A_c} = \eta = F_R \left\{ (\tau\alpha)_{\text{eff}} - \frac{\bar{U}(T_{f,\text{in}} - T_a)}{C_R q_s(t)} \right\}$$

$$= F_R \left[(\tau\alpha)_{\text{eff}} - \frac{b(\Theta_{f,\text{in}} - \Theta_a)}{\psi(t)} \right] \tag{5-33}$$

where $\quad b \equiv \dfrac{\bar{T}_a \bar{U}}{q_{s,\text{ref}} C_R} \quad \psi(t) = \dfrac{q_s(t)}{q_{s,\text{ref}}} \quad \Theta = T/\bar{T}_a$

In this relation as used in SOLSIM, the dimensionless fluid inlet temperature to the collector $\Theta_{f,\text{in}}$, the dimensionless ambient temperature Θ_a, and the insolation $q_s(t)$ all vary with time. The other factors in Eq. (5-33), including F_R, are all assumed to be constant.

Examination of Eq. (5-33) will show, however, that F_R, which depends on $q_s(t)$, $\Theta_{f,\text{in}}$, Θ_a, and γ, must also be a varying quantity. For the range of parameters typical of liquid systems, F_R tends to vary over a narrow range, while for air systems the value may vary considerably. For program simplicity, however, F_R is assumed constant.

5-6 COLLECTOR TESTING AND PERFORMANCE

As noted in Chap. 4, the quality of a solar collector must be viewed in terms of

- Its thermal performance
- Its durability (or useful life)
- Its initial cost
- Its effect on system design (pump sizing, tracking, etc.)

The ultimate criterion, of course, is the amortized cost of the useful energy produced over the collector lifetime in dollars per unit energy, and the entire system must be considered in this cost evaluation. (The economics of solar energy systems is dealt with in Chap. 9.) The durability of collectors and systems is frequently overlooked. Unfortunately, with a relatively new technology such as solar, insufficient data are presently available on the variety of components and systems available to make judicious selections in many cases. However, as the industry develops and more data on installations are collected, the quality systems and components will be recognized. Currently, accepted test procedures do provide some valid information on the durability and degradation of solar collectors. One important characteristic, thermal performance, can be measured, and is the main subject of this section.

Collector Efficiency

As discussed in Chap. 4, two efficiencies are used for collector evaluation: the *all-day* (or long-term) efficiency and the *instantaneous* efficiency. The all-day efficiency is the ratio of the useful energy collected during an entire day to the available insolation over the entire day. While this is a useful performance quantity and frequently is desired by the user, it is dependent on the system configuration (e.g., storage-to-collector size, thermal loss from piping), available insolation, load, collector transients, and other factors. All-day efficiency cannot therefore be calculated for a given collector unless the complete system is specified.

The instantaneous efficiency, which is most often quoted, is the ratio of the rate of useful energy collected to the rate of available insolation on the collector aperture when the collector is operating under steady-state conditions. This definition holds for both flat-plate and concentrating collectors, with the exception that q_s is generally defined as the incident global insolation (beam plus diffuse) for flat-plate collectors and as the beam insolation alone for concentrating collectors.

Hereafter, the instantaneous efficiency will be referred to simply as the efficiency. This quantity is independent of load and of system components other than the collector, and thus is a measure of the performance of the collector itself. As was discussed previously [Eq. (5-31)], the efficiency of a collector

$$\eta = F_R \left[I(\tau\alpha)_{\text{eff}} - I\bar{U}\left(\frac{T_{f,\text{in}} - T_a}{C_R q_s}\right) \right]$$

is a function of the collector design [$(\tau\alpha)_{\text{eff}}$, $\bar{U}$, F_R, C_R, I] and the quantity $(T_{f,\text{in}} - T_a)/q_s$. Although $T_{f,\text{in}}$, T_a, and q_s each individually affect efficiency, the parameter $(T_{f,\text{in}} - T_a)/q_s$ represents the single most important factor influencing the performance of a given collector.

Measurement of Efficiency

Solar collector efficiency can be determined experimentally by measuring

1. $\dot{Q}_u$ the rate at which energy is absorbed by the fluid in passing through the collector,

$$\dot{Q}_u = \dot{m}c(T_{f,\text{out}} - T_{f,\text{in}})$$

   during steady-state operation.

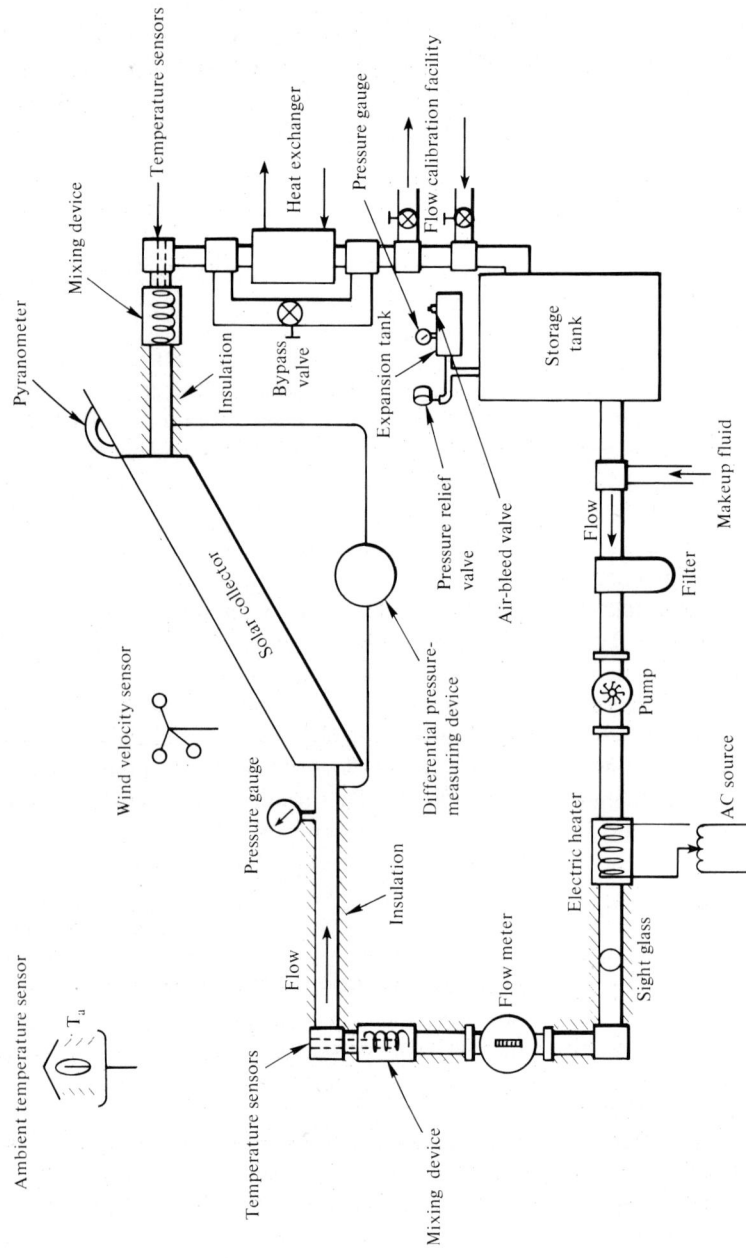

Figure 5-13 Closed-loop testing configuration for the solar collector when the transfer fluid is a liquid [14].

2. The rate of energy incident on the collector

$$\dot{Q}_s = q_s A_c \quad \text{(or } q_b A_c\text{)}$$

where q_s (or q_b) is the global (or beam) insolation per unit of aperture area and per unit time and A_c is the collector aperture area.

The flow rate $\dot{m}$ would normally be measured by a calibrated device like a rotameter or turbine-type flowmeter or possibly by measuring the time to collect a given mass of fluid. The temperatures could be measured by thermometers, resistance thermometers (RTDs), thermocouples, or thermistors. Measurement of the global solar flux q_s would be made with a pyranometer and of the beam component q_b with a pyrheliometer. A sketch of a fairly comprehensive test configuration for solar collector testing is shown in Fig. 5-13.

Solar Collector Test Standards

A standard method for evaluating the thermal performance of a solar collector has been more or less accepted in the industry. The actual testing procedure is described in a Standard (Ref. 14) published by the American Society of Heating, Refrigeration, and Air Conditioning Engineers (ASHRAE). The reader is referred to that Standard itself. In summary, however, the testing methods place strict requirements on the accuracy of the measuring instruments, duration of the tests, steady-state requirements, range of $(T_{f,\,in} - T_a)/q_s$ over which data is taken, and repeatability of data. In some cases of qualifying collectors it has also been required that tests be performed before and after a 30-day stagnation period to give an indication of the collector durability.

It should be pointed out that the standards are for the method of testing. That is, it is not necessary that a collector perform to some minimum level, but rather that the tests be performed according to a specified standard such that the performance results are reliable.

The Realities of Testing

The actual testing of a collector is time-consuming and therefore expensive. It is also apparent, [e.g., from the round-robin collector test program sponsored by the National Bureau of Standards (NBS), which resulted in a surprisingly wide range of performance ratings for the same collector (see Ref. 15).] that the testing procedures are either inadequately defined or not adhered to.

The testing of solar collectors takes a long time because testing must take place over a wide temperature range and under steady-state conditions. At each fluid inlet temperature the system must achieve thermal equilibrium before the performance can be determined. Test facilities adjust the inlet temperature by various methods: drawing fluid from a large tank with temperature controlled automatically by both heating and cooling units; using an automatically controlled (electric) heater in the flow line immediately before the inlet to the collector; mixing steam and cold water; and producing a known steady condition at some point in the fluid loop [16]. The various techniques suffer in varying degrees from the long system transients (in collector as well as test loop) and the fact that heat is being added by the collector and the transient environment (particularly insolation). Accurate fluid flow rate and temperature measurement are essential since fluid temperature rises across the collector are typically as low as only a few degrees. A test facility must therefore be carefully designed to achieve accurate measurement and steady-state operation.

The transient effects due to environment are very important. Except on very clear days, clouds interrupt the insolation. Even on clear days insolation is constantly varying throughout the day. Wind velocity and direction change and ambient temperature also vary. The outdoor environment cannot be controlled; however, testing should preferably be conducted during relatively clear days. An often overlooked source of error is the variation in the physical properties of the working fluid. Many organic additives used for freeze protection have properties that change with time, and are not accurately known even initially. In air systems, accurate humidity measurement is necessary because water vapor contributes significantly to the heat capacity of humid air. All of these factors contribute to uncertainties in the test results.

Simplified Testing

For a given flat-plate collector, it is possible to find values of $(\tau\alpha)_{\text{eff}}$, $\bar{U}$, and F_R in a relatively easy way if very accurate efficiency curves are not necessary. This is done as follows:

Effective transmittance-absorptance test Because the insolation q_s is either absorbed by the collector or reflected from it for any incidence angle of direct insolation, we can write (Fig. 5-14):

$$q_s(\theta) = \rho_c(\theta) q_s(\theta) + \alpha_c \, q_s(\theta)$$
$$= q_{s,r} + \alpha_c \, q_s(\theta)$$

or
$$\alpha_c = 1 - \frac{q_{s,r}}{q_s(\theta)} \tag{5-34}$$

where α_c is the fraction of the insolation absorbed by all parts of the collector. If it is assumed that energy absorbed in the covers finally ends up as useful energy, or that negligible energy is absorbed in the covers (i.e., in thin plastic films), then $(\tau\alpha)_{\text{eff}} \approx \alpha_c$. This assumption could lead to significant errors in evaluating $(\tau\alpha)_{\text{eff}}$.

The ratio of $q_{s,r}$ to $q_s(\theta)$ can be measured by placing a pyranometer with its axis normal to the collector face and measuring $q_s(\theta)$. Then the pyranometer is turned so that it faces the collector, and $q_{s,r}$ is measured. The value of $(\tau\alpha)_{\text{eff}}$ is thus determined.

No heat loss test If $T_{f,\text{in}}$ is adjusted to make it equal to the ambient temperature,

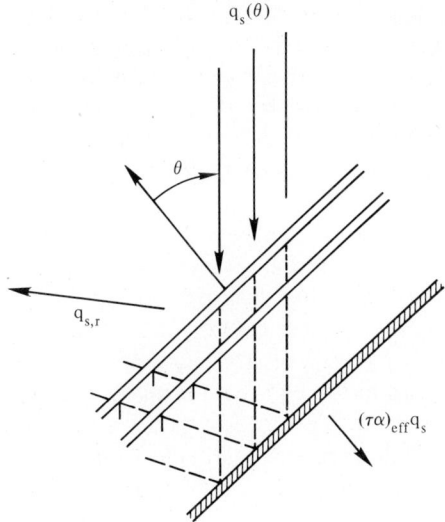

Figure 5-14 Overall collector radiant energy relations.

then Eq. (5-31) gives, with $I = 1$,

$$\frac{Q_u}{A_c} = F_R(\tau\alpha)_{\text{eff}}\, q_s = \dot{m}c[T_{f,\text{out}} - T_{f,\text{in}}(= T_a)] \tag{5-35}$$

and the value of F_R is calculated from the measured or known values of q_s, $\dot{m}$, c, $(\tau\alpha)_{\text{eff}}$, and the collector outlet temperature.

No insolation test If the collector is covered or operated at night, then $q_s = 0$ and Eq. (5-31) becomes

$$\frac{Q_u}{A_c} = -F_R \bar{U}[T_{f,\text{in}} - T_a] = \dot{m}c[T_{f,\text{out}} - T_{f,\text{in}}]$$

Thus, the product $F_R \bar{U}$ can be determined by measurement of $T_{f,\text{in}}$, $T_{f,\text{out}}$, T_a, $\dot{m}$, and c. Equation (5-35) can be solved for F_R, and $\bar{U}$ can then be determined. All the required characteristics for Eq. (5-31), $(\tau\alpha)_{\text{eff}}$, $\bar{U}$, and F_R are thus known from the three simple tests outlined. A further test can be carried out by allowing no flow so that the collector comes to its stagnation temperature T_{stag}. If the collector is isothermal at $\bar{T}_{\text{stag}}$, then $F_R = 1$ [see the analysis following Eq. (5-26)]. Thus Eq. (5-22) becomes

$$(\tau\alpha)_{\text{eff}}\, q_s = \bar{U}(\bar{T}_{\text{stag}} - T_a)$$

However, this $\bar{U}$ value will be based on $\bar{T}_{\text{stag}}$ and if $\bar{U}$ is dependent on the average plate temperature $\bar{T}_e$, it will not be the correct value for use in Eq. (5-31) because $\bar{T}_e$ during collector operation will be much lower.

Table 5-3 Comparison of calculated and experimental collector values of collector performance parameters†

Parameter	Calculated value	Experimental value
$(\tau\alpha)_{\text{eff}}$	0.729 (for $\alpha = 0.95$)	0.704
	0.693 (for $\alpha = 0.90$)	
$\bar{U}$‡	3.7	3.6
F_R	0.882	0.886

Note:
Conditions: Wind speed, 5 m/s; $T_{f,\text{in}} = 65°C$; flow, 0.775 L/s; collector tilt = 45°; $C_R = 1$ (flat plate).

Collector characteristics: 5 × 15 m (16 × 48 ft), double covers of double-strength window glass, aluminum absorber plate with internal tubes, water/glycol mixture as heat removal fluid, insulation behind absorber.

† Ref. 15.
‡ Measured in W/(m² · C).

Smith and Weiss [17] have applied the first three tests to the collectors on the Colorado State University Solar House I, and compared the experimental values with the values calculated from the known characteristics of the collectors. The results of their study, in terms of the parameters used here, are shown in Table 5-3.

PROBLEMS

5-1 A flat-plate solar collector of size 1 × 2 m is exposed to a solar flux of 800 W/m² and uses water as the coolant entering at 50°C and a rate of 60 kg/(h · m²). If the collector efficiency at this condition is 0.60, what are
 (a) The rate of useful energy collected?
 (b) The water exit temperature?
 Answer: (a) $\dot{Q}_u = 0.96$ kW. (b) $T_{f,\text{out}} = 57°C$.

5-2 A flat-plate solar collector of size 4 × 8 ft is exposed to a normal solar flux of 280 Btu/(h · ft²) with an ambient temperature of 85°F. The coolant fluid (50/50 water-ethylene glycol) inlet temperature is 140°F (data are in App. C). The quantities $F_R(\tau\alpha)_{\text{eff}}$ and $F_R \bar{U}$ for the collector are given in the collector test report as 0.72 and 6.4 W/(m² · °C) respectively. What are:
 (a) The collector efficiency?
 (b) The fluid outlet temperature for a flow rate of 6.0 lb$_m$/min?
 (c) The collector stagnation temperature?
 Answer: (a) $\eta = 0.50$. (b) $T_{f,\text{out}} = 156°F$. (c) $T_{\text{st}} = 284°F$.

5-3 A flat-plate solar collector of outside dimension 3 × 6 ft and aperture dimension 34.0 × 70.0 in has been tested with pressurized water as the test fluid to determine its efficiency over a range of operating conditions. The data collected for these 12 tests are included in the accompanying table. Construct an experimental efficiency versus $(T_{f,\text{in}} - T_a)/q_s$ curve by calculating η and $(T_{f,\text{in}} - T_a)/q_s$ for each of the tests and plot it on linear graph paper, where the outside collector dimensions are

Run	q_s, Btu/(h · ft²)	T_a, °F	$T_{f,in}$, °F	$T_{f,out}$, °F	$\dot{m}$, lb$_m$/min
1	119	76	153.2	155.3	4.1
2	181	81	128.9	135.8	3.9
3	235	86	113.1	124.6	3.8
4	273	91	109.0	122.9	3.8
5	292	93	149.1	160.8	4.1
6	299	97	149.3	161.6	4.1
7	301	98	187.0	197.5	4.3
8	295	101	205.2	214.4	4.4
9	271	102	207.1	214.5	4.4
10	231	102	208.0	213.8	4.5
11	199	101	207.8	211.8	4.4
12	146	99	204.1	205.6	4.4

used as the collector area. Determine (a) $F_R(\tau\alpha)_{eff}$ and (b) $F_R \bar{U}$ for the collector. (c) How would the measured efficiency be affected if it were based on the collector aperture rather than the outside dimension?

Answer: (a) $F_R(\tau\alpha)_{eff} \simeq 0.7$. (b) $F_R \bar{U} \simeq 0.74$ Btu/(h · ft² · °F).

5-4 A flat-plate collector was tested at only two test conditions using the following instruments and calibration data:

- *Temperatures*: Copper constantan thermocouple wire with a 0°C ice bath reference. Output in millivolts is measured with a potentiometer. (Obtain Cu-Const thermocouple data from "Reference Tables for Thermocouples," National Bureau of Standards Circular No. 561, April 1955, or later issues.)
- *Flow rate*: Water flow rate using a rotometer with calibration of 2.37 L/min at 100 percent scale reading.
- *Insolation*: A pyranometer with a sensitivity of 11.2 × 10^{-6} V per W/m², and a millivolt potentiometer for measurement.

The two tests provided the following data:

Run	Insolation, mv	Flow rate, % of scale	T_a, mv	$T_{f,in}$, mv	$T_{f,out}$, mv
1	8.33	78	1.194	1.821	2.095
2	7.06	90	1.539	4.161	4.263

If the collector dimensions are 0.92 × 1.84 m determine (a) $F_R(\tau\alpha)_{eff}$ and (b) $F_R\bar{U}$.

Answer: (a) $F_R(\tau\alpha)_{eff} = 0.776$. (b) $F_R \bar{U} = 5.62$ W/(m² · °C).

5-5 A flat-plate solar collector is designed with the characteristics shown in the accompanying table. Using the Hottel-Whillier-Woertz-Bliss method, predict the collector efficiency equation constants F_R and b for the equation

$$\eta = F_R \left[(\tau\alpha)_{eff} - \frac{b(\Theta_{f,in} - \Theta_a)}{\psi(t)} \right]$$

Collector characteristics

$\bar{U} = 6$ W/(m² · °C)	$G = 0.3$ kg/(min · m²)
$\eta_{\text{fin}} = 0.88$	$c = 4700$ J/(kg · °C)
$\bar{U}_i = 1200$ W/(m² · °C)	$q_{s,\text{ref}} = 700$ W/m²
$(\tau\alpha)_{\text{eff}} = 0.78$	$T_{f,\text{in}} = 85$°C
$q_s(t) = 850$ W/m²	$T_a = 25$°C

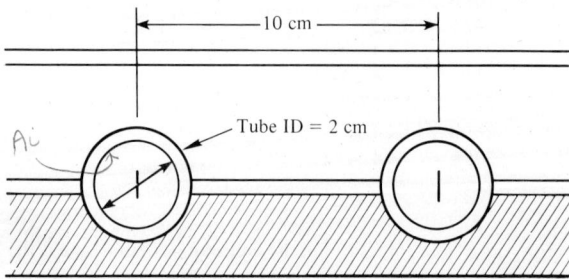

Figure P5-5

5-6 A flat-plate collector of standard design has the characteristics shown in column A of the accompanying table. A honeycomb material is placed between the absorber plate and the cover plate, resulting in the collector characteristics given in column B.

	Collector	
	A	B
F_R	0.9	0.9
$(\tau\alpha)_{\text{eff}}$	0.88	0.85
$\bar{U}$ Btu/(h · ft² · °F)	1.35	1.20

The values are for the HWWB modified equation

$$\eta = F_R\left[(\tau\alpha)_{\text{eff}} - \frac{\bar{U}(T_{f,\text{in}} - T_a)}{q_s(t)}\right]$$

(a) Explain with physical reasoning why you think the values of F_R, $(\tau\alpha)_{\text{eff}}$, and $\bar{U}$ might change in the manner shown in the table.

(b) Over what range of $(T_{f,\text{in}} - T_a)/q_s(t)$ is the collector efficiency increased by adding the honeycomb material?

(c) Why is the efficiency increased in some ranges of $(T_{f,\text{in}} - T_a)/q_s(t)$ and decreased in others by adding the honeycomb?

(d) If $T_{f,\text{in}} = 140$°F, $T_a = 80$°F, and $q_s(t) = 300$ Btu/(h · ft²), which of the two collectors would you purchase for these average conditions if the cost per unit area is the same? Why?

5-7 In testing the collector described in Prob. 5, a collector efficiency run in a National Bureau of Standards standard test apparatus in Madison, WI, gave different results than similar tests run in Melbourne, FL. In both tests, the collector was tilted at an angle equal to the local latitude and faced due south.

The following reasons were advanced for the differences. Discuss which of them you think might be valid and how you would set up an experiment to find out.

(a) Because of the different latitudes and tilts, the sun track relative to the collectors was different and the cover glass reflectivity variation with incident solar angle caused the difference.

(b) The free convection patterns in the cavity between the collector cover and absorber plates were different because of the differing tilt angles, leading to different loss coefficients.

(c) The atmospheric scattering conditions in the two locations are different, leading to different fractions of diffuse and direct insolation at each station.

(d) The pyranometer used for measuring the global insolation was mounted in the plane of the collector (with different tilt for the two locations), and thus did not read accurately.

(e) Everyone knows Florida is the "Sunshine State."

5-8 A moderately concentrating collector has a geometric concentration ratio of 3.0 and an intercept factor of 0.92. All other characteristics are the same as those of the collector discussed in Prob. 5-5.

(a) What are the collector constants F_R and b for the moderately concentrating collector?

(b) Plot the efficiency curves for the moderately concentrating and the flat-plate collectors on the same graph.

(c) What are the stagnation temperatures of the two collectors?

(d) At what temperature do the collectors each provide 30 percent efficiency?

5-9 An improved equation for calculating the surface convection heat-transfer coefficient is [18].

$$h = (k/L)(0.931)\, Re^{1/2}\, Pr^{1/3} \qquad Re < 4 \times 10^5$$

Calculate $\bar{U}$ for the collector described in Example 5-2 with the value of h evaluated with the above equation.

5-10 The figure below shows the normal incidence global spectral transmittance of 4-mil Tedlar™ PVF film (from "Glazings," DuPont Bulletin TD-31, March 1975) in the near infrared. Estimate the total infrared transmittance in this wavelength range. (*Note:* For glass, $\bar{\tau}_{ir} = 0$.)

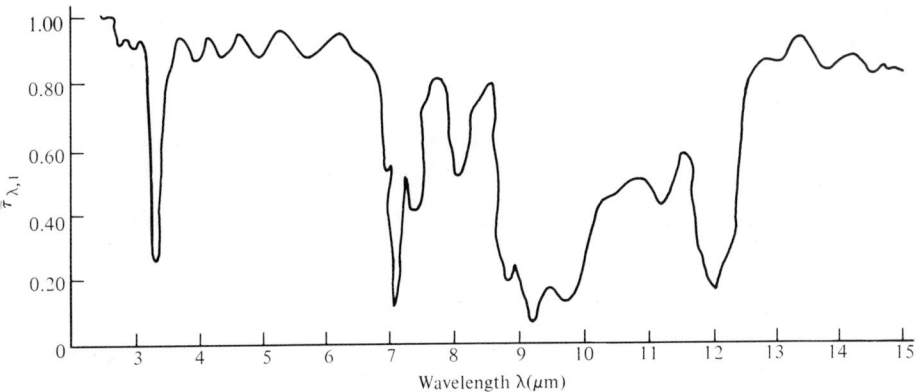

Figure P5-10

5-11 Repeat Example 5-1 for the collector described when a second glass cover is added. Compare $\bar{\tau}_2$ with Fig. 5-2.

5-12 A single-glazed flat-plate collector has the following characteristics:

- Cover plate: $\frac{1}{8}$-in glass, refractive index $n = 1.5$; absorption coefficient × thickness, $aL = 0.0524$; ir emittance of plate, 0.9
- Absorber plate: ir emissivity = solar absorptivity = 0.95; back insulation, $k_i = 0.05$ W/(m · °C), thickness = 4 cm
- Conditions: convective heat transfer coefficient to ambient on collector glazing and back, $h = 10$ W/(m² · °C); ambient temperature 20°C; collector average plate temperature 80°C

Find (a) the overall thermal loss coefficient $\bar{U}$ and, (b) the value of $(\tau\alpha)_{\text{eff}}$ for normal incidence.

5-13 Suppose you had some thin-strip glass mirrors, each 3 in wide. Design a 50× concentration ratio fixed-mirror tracking receiver collector (see Figs. 4-13 and 4-14). For design purposes, assume the incident beam is normal to the aperture. The design should include the specifications of the mirror placement and the description of the receiver. For your design construct a collector efficiency curve (for normal incidence).

5-14 Comment on the effect of off-axis (both east or west and north or south) insolation on the performance of the collector from Prob. 5-13. Assume the collector is aligned east-west.

5-15 Using insolation data from your location (or calculated from Chap. 3), compare the daily performance (energy delivered) of an evacuated tubular collector, a single-glass-cover selective-absorber flat-plate collector, and a parabolic trough collector (Figs. 4-10 and 4-23 may be useful). Do this for a summer day operating at a fluid inlet temperature of 100°C and for a winter day operating at 70°C. Assume that the nonconcentrating collectors are fixed and oriented to optimize summer performance (for air conditioning). The parabolic trough has two-axis tracking.

5-16 Suppose one of the collectors described in Prob. 5-15 is to be purchased. The evacuated tubular collector costs $25/ft², the flat-plate collector costs $15/ft² and the parabolic trough costs $20/ft². Which collector would you recommend? Assume all solar energy collected in the winter can be used, even though the collector array is probably oversized for winter use.

REFERENCES

1. D. K. Edwards, "Solar Absorption by Each Element in an Absorber-Coverglass Array," *Solar Energy*, vol. 19, no. 4, pp. 401–402, 1977.
2. J. A. Duffie, and W. A. Beckman, *Solar Energy Thermal Processes*, John Wiley & Sons, New York, 1974.
3. Robert Seigel and John R. Howell, *Thermal Radiation Heat Transfer*, 2d ed., McGraw-Hill Book Co., New York, 1981.
4. William A. Shurcliff, "Transmittance and Reflection Loss of Multi-Planar Window of a Solar-Collector: Formulas and Tabulations of Results for $n = 1.5$," *Solar Energy*, vol. 16, no. 3/4, pp. 149–154, 1974.
5. H. C. Hottel and B. B. Woertz, "The Performance of Flat-Plate Solar Heat Collectors," *Trans. ASME*, vol. 64, pp. 91–104, Feb. 1942.
6. H. C. Hottel and A. Whillier, "Evaluation of Flat-Plate Solar Collector Performance," *Trans. Conf. on the Use of Solar Energy*, vol. II, U. of Arizona, Tempe, 1958, pp. 74–104.
7. Raymond W. Bliss, Jr., "The Derivations of Several Plate-Efficiency Factors Useful in the Design of Flat-Plate Solar Heat Collectors," *Solar Energy*, vol. 3, pp. 55–64, 1959.
8. S. A. Klein, J. A. Duffie, and W. A. Beckman, "Transient Considerations of Flat-Plate Solar Collectors," ASME Paper 73-WA/SOL-1, Nov. 1973.
9. Peter J. Lunde, "New Heat Transfer Factors for Flat Plate Solar Collectors," *Solar Energy*, vol. 27, no. 2, pp. 109–113, 1981.
10. John R. Howell, "Flat-Plate Solar Collector Design and Sizing," *AIAA J. Energy*, vol. 3, no. 6, pp. 379–382, November–December, 1979.
11. Warren F. Phillips, "The Effects of Axial Conduction on Collector Heat Removal Factor," *Solar Energy*, vol. 23, no. 3, pp. 187–191, 1979.
12. E. M. Sparrow and R. J. Kroweck, "Circumferential Variations of Bore Heat Flux and Outside Surface Temperature for a Solar Collector Tube," *J. Heat Transfer*, vol. 99, pp. 360–366, Aug. 1977.
13. S. A. Klein, "Calculation of Flat-Plate Collector Loss Coefficients," *Solar Energy*, vol. 17, no. 2, pp. 79–80, 1975.
14. "Method of Testing Solar Collectors Based on Thermal Performance," ASHRAE Standard 93-77, ASHRAE, New York, January 1977.

15. E. R. Streed, et al., "Results and Analysis of a Round Robin Test Program for Liquid-Heating Flat-Plate Solar Collectors," *Solar Energy*, vol. 22, no. 3, pp. 235–249, 1979.
16. S. J. Kleis, R-T. Chen, and R. B. Bannerot, "Instantaneous Collector Thermal Efficiencies in Less Time," *Solar Energy*, vol. 24, no. 1, pp. 111–112, 1980.
17. Charles C. Smith and Thomas A. Weiss, "Design Application of the Hottel-Whillier-Bliss Equation," *Solar Energy*, vol. 19, no. 2, pp. 109–113, 1977.
18. E. M. Sparrow and K. K. Tien, "Forced Convection Heat Transfer at an Inclined and Yawed Square Plate-Application to Solar Collector," *J. Heat Transfer*, vol. 99, pp. 507–512, Nov. 1977.

CHAPTER
SIX

STORAGE OF THERMAL ENERGY

Energy storage is necessary for many solar heating, cooling, and water-heating installations and for most utility and industrial uses because the period of collection and time of demand are not coincident.

6-1 LIST OF SYMBOLS

A	area
Bi	Biot number, hD/k
c	specific heat
D	tube diameter
F_0	frozen fraction in radial element at entrance of latent storage unit
$\bar{F}$	average frozen fraction across latent storage unit
F	control index for stratified storage on collector flow, values 0 or 1
G	control index for stratified storage on load flow, values 0 or 1
h	convective heat-transfer coefficient
k	thermal conductivity
L	length of storage unit
l	summation index
M	total mass in storage
m	layer index in storage
$\dot{m}$	mass flow rate
NTU	number of transfer units
n	number of layers in storage
$\dot{Q}$	energy rate
T	temperature

t	time
U	overall thermal loss coefficient
x	coordinate
ϵ	void fraction in storage bed or heat exchanger effectiveness in latent storage
λ	heat of fusion
ρ	density
Φ	dimensionless group for rock bed storage, Eqs. (6-6) and (6-8)
Ψ	auxiliary function, Fig. 6-18
Ω	auxiliary function, Fig. 6-18

Subscripts

1, 2, ...	at layer 1, 2, ...
a	ambient
b	storage bed material
c	collector
f	fluid
L	load
sat	saturation (melting) value
st	storage
t	per individual tube
v	value based on volume
in, out	at inlet or outlet

6-2 METHODS OF STORAGE

Thermal energy may be stored in the forms of latent heat, sensible heat, or a combination of latent and sensible heats as well as by nonthermal means such as reversible chemical reactions. Only thermal storage will be addressed here.

Sensible Heat Storage

Sensible heat is stored when the temperature of a storage medium increases. Water and pebbles are the most common materials used for low-temperature energy storage because they are low in cost and readily available. (Rock and mineral oil have been proposed for high-temperature energy storage.) Any structurally and chemically stable solid or liquid, preferably having high specific heat and high density, may be used if the costs are justifiable.

The amount of energy required to raise the temperature of one unit of a material one degree is its specific heat (heat capacity). Table 6-1 lists the specific heats on both a mass and volume basis for a few common materials.

Note that water has three times the heat capacity of rock for the same volume. This means that a pebble-bed storage unit would be three times larger than a water tank storing the same amount of sensible heat.

Table 6-1 Specific heats of common materials

Material	Mass basis		Volume basis	
	kJ/(kg · K)	Btu/(lb$_m$ · °R)	kJ/(m^3 · K)	Btu/(ft^3 · °R)
Water	4.19	1.0	4190	62
Steel	0.46	0.11	2680†	40†
Rock [2–4 cm (0.75–1-in) diam]	0.84	0.2	1340†	20†
50/50 by volume water/glycol solution (15°C)	2.7	0.66	2450	36

† Assumes 70 percent packing density of solids in air.

Latent Heat Storage

Latent heat can be stored by melting a solid, such as wax. Table 6-2 lists several available latent heat storage materials, giving their melting points and heats of fusion both on a mass and volume basis. If a suitable material is used, solar energy at temperatures provided by collectors can be used to melt a solid and store the energy during the day, and then can be made to release the energy when

Table 6-2 Latent heat storage materials

	Melting point		Heat of fusion			
			Mass basis		Volume basis	
Material	K	°F	kJ/kg	Btu/lb$_m$	kJ/m^3	Btu/ft^3
Water	273	32	335	144	301 × 10^3	8,087
Hydrated inorganic salts						
Ortho phosphoric acid	303	84.8	144	61.6	261 × 10^3	7,010
Calcium chloride	304	86.4	170	73.2	282 × 10^3	7,570
Glauber salt	306	90.4	238	102	347 × 10^3	9,320
Disodium phosphate	308	94.2	282	121	430 × 10^3	11,550
Calcium nitrate	316	108.6	142	61.1	257 × 10^3	6,900
Sodium thiosulphate	322	119	95	40.7	176 × 10^3	4,710
Sulfur trioxide	336	144	319	137	624 × 10^3	16,750
Anhydrous inorganic salts						
Meta phosphoric acid	316	109	108	46.2		
Phosphoric acid	343	158	157	67.4		
Waxes and organic solids						
Anthracine	369	205	105	45.2	130 × 10^3	3,480
Naphthalene	353	176	151	64.9	172 × 10^3	4,620
Naphthol	368	203	163	70.0	197 × 10^3	5,280
Beeswax	335	143	177	76.2	168 × 10^3	4,500
Stearic acid (tallow)	349	169	199	85.4	168 × 10^3	4,500
Amorphous paraffin wax	348	166	231	99.0	183 × 10^3	4,900

the liquid cools and reverts back to solid form. In ideal latent heat storage systems a very small temperature difference is sufficient to change the phase from solid to liquid or liquid to solid.

One advantage of latent heat storage is that very large quantities of energy can be stored and released per pound of material, resulting in less volume to store the energy required by the system unless a large temperature swing is allowed in the sensible storage system. A second advantage is that the temperature remains nearly constant during the phase change, which is advantageous for both the load and the collector.

One disadvantage of latent heat storage is that many known latent heat materials will last only a few years or less and then must be replaced, and most are expensive compared to water or pebble-bed storage. More research is needed for development of practical latent heat storage units.

Table 6-3 presents a comparison of some sensible and latent heat storage materials, including advantages and disadvantages. The fourth column compares the storage volume required for 10^6 kJ of capacity over a 25-K temperature difference and includes the latent plus sensible heat storage associated with a 25-K temperature rise. Figure 6-1 presents a graphical comparison for the three most common materials: water, rocks, and Glauber's salt assuming an approxi-

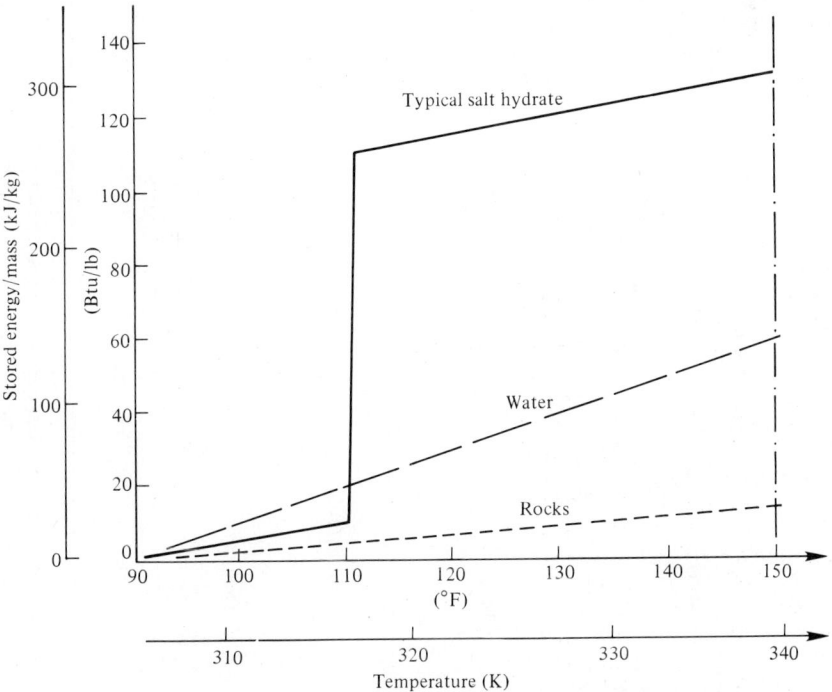

Figure 6-1 Relative thermal storage capacities of rocks, water, and hydrated salts.

Table 6-3 Storage Materials Characteristics

Material	Bulk density kg/m³	Specific heat kJ/(kg · K)	Latent heat, kJ/kg	Storage volume,† m³ (10⁶ kJ, 25-K diff.)	Comments
Rocks	1670‡	0.84	...	6.1	Cheap, noncorrosive, good for heating systems
Water	1000	4.2	...	2.0	Cheap, mildly corrosive, general utility, very wide use, high specific heat, good heat-transfer characteristics
Phase change					
Paraffin	660‡	2	230	1.4	Moderate cost, noncorrosive, poor heat transfer, little used
Salt hydrates§	1200‡	2	225	0.8	Fairly cheap, mildly corrosive, hysteresis effects, little used
Sodium metal	960	1.4	114	1.9	Moderately expensive, corrosive, dangerous, good conductivity, not for low-temperature storage

† For 10⁶ kJ storage and 25-K temperature swing.
‡ Assuming approximately 70 percent of solid density for transfer medium flow passages.
§ Typical average values for several possible hydrates.

mate 35-K (60°F) temperature swing. Note the relative storage capacities per unit mass of material.

6-3 THERMAL STORAGE REQUIREMENTS

Duration

Both demand (load) and insolation have periodic daily and seasonal variations as well as intermittent characteristics due usually to weather variations. While there is some interest in seasonal storage (between heating and cooling seasons), the following discussion does not consider that possibility, which is presently only in the developmental/experimental stage. Water heating is a fairly regular daily requirement (except for possible specific variations during the week such as washday in residences and weekend versus week days in commercial buildings) and requires a storage period of 1 day (plus). Most systems are sized on the basis of approximately $1\frac{1}{2}$ days' storage.

Because space heating represents a more intermittent load which tends not to be coincident with insolation, the recommended storage capacity for space-heating applications is generally larger—possibly two days. However, optimization analyses on storage capacity show that the overall system economics are weakly dependent on storage capacity near the optimum, and thus storage capacity varies widely, typically from one to several days. Figure 6-2 presents a qualitative sketch of the solar system unit cost (installed system cost per unit of annual energy delivered) as a function of the storage volume (or mass) per unit of collector area. If the storage is too small, the potential energy collection cannot be stored (energy wasted), while, if it is too large, the incremental cost of additional storage increases faster than the incremental fuel savings. Also, as storage volume increases, storage losses tend to become smaller due to lower time-average storage temperatures, but tend to become larger due to increased surface

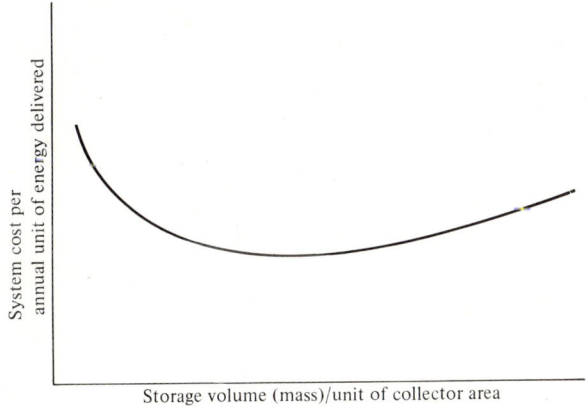

Figure 6-2 Effect of storage on system economics.

area. The combined effect of these trends can push losses higher or lower depending on storage capacity and geometry.

The most practical sizes to consider for a heating system are approximately 35–70 kg of water per square meter of collector area (1–2 gal per square foot) for a water system and approximately 0.3 m^3 of rock storage (pebbles) per square meter of collector (1 ft^3 per square foot) for an air system.

Temperature Requirements

For residential hot water, storage temperatures of 325–330 K (130–140°F) are typical, while for commercial dishwashing applications, a temperature of 355 K (180°F) is recommended. In solar heating applications, storage temperatures of approximately 330 K (140°F) are typical. While the energy may be delivered to the air (liquid-air exchanger) at a substantially lower temperature of 305 K ($\simeq 90°$F), the higher storage temperature reduces storage volume. Heat-driven cooling cycles (i.e., absorption) require temperatures of approximately 350 K (170°F) minimum, and thus storage between 360 and 380 K (190 and 220°F) would be required.

Solar energy storage units may be sited above or below grade and either inside the building or outdoors. In most locations the storage unit should be placed above ground and within the building enclosure whenever possible and close to other solar equipment. An indoor storage location is protected from moisture and cold, and the energy loss from the storage unit will assist in heating the building in winter. The disadvantages are that energy loss to the building adds to the cooling load during the summer, and a finite amount of relatively expensive space must be provided. Thus, in climates where cooling loads dominate, a location outside conditioned space may be preferable.

6-4 WATER ENERGY STORAGE

Water is recommended for energy storage if a liquid is used as the energy transport medium in the solar collectors. If an antifreeze solution is required in the collector loop, a heat exchanger between the collector fluid loop and the storage fluid may be necessary because using antifreeze as the storage medium is prohibitively expensive. If freezing is not a concern, liquid from storage may be circulated directly through the collectors (without a heat exchanger) and the storage water pumped directly to the heating coils or the absorption cooling unit.

Four conditions must be avoided when using liquid for energy storage:

1. Freezing, in cold climates
2. Boiling, with resultant buildup of pressure in the system
3. Corrosion of the storage tanks and pipes
4. Leakage

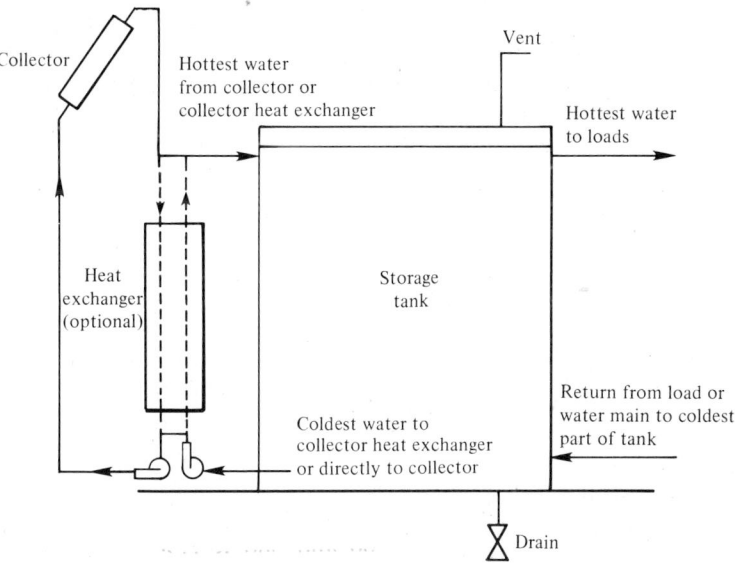

Figure 6-3 Operation of a water heat storage tank.

Freezing can be prevented by locating the vessel inside, or if it must be outside, then by burying it below the frost line. While boiling is uncommon for well-designed systems during the heating season, it may be expected during the summer if the load diminishes. The steam that is produced can easily be vented outside the building to prevent pressure buildup in the tank and to "dump" the heat from the system. A pipe from the top of the tank to the outdoors with a low-pressure relief valve is sufficient. For a nonpressurized system, a pressure relief valve is not required, but a float-controlled valve to provide makeup water in the storage tank should be provided. Frequent boiling of storage water will cause buildup of mineral deposits in the collector and storage.

Corrosion often occurs at the pipe connections to the tank. If dissimilar metals are used, galvanic corrosion will result. Therefore, neoprene rubber hoses or dielectric unions should be used to connect copper pipes to steel tank fittings.

Water leakage must be prevented because it can damage tank insulation and other materials near the tank. Because it is difficult to locate a leak after insulation is applied, the tank should be leak-tested with all fittings in place before insulation is applied.

The fluid at the top of the storage tank will generally be hotter than that at the bottom because the lower density of the hot fluid causes it to migrate upward through the more dense cold fluid. This effect can be magnified by manifolding the inlet pipes to the storage tank. The magnitude of the temperature difference is a function of tank height and cross section as well as the location of plumbing ports and the circulation rate. Any temperature stratification achieved can be useful if the piping is connected appropriately, because the lowest temperature possible can then be pumped to the collector, which improves the performance of the collectors, and the highest temperature available can be delivered to the load heating coils. This arrangement is illustrated in Fig. 6-3. Stratification can also be enhanced by the use of multiple tanks and by various types of baffles.

Modeling the Thermal Behavior of Liquid Storage—Simple Systems

In the simple model discussed in Chap. 2, the storage is assumed to be a well-mixed tank of liquid. The tank thus comes to a new uniform temperature whenever energy is added to or withdrawn from the tank by replacing tank water with water at a different temperature. Such a modeling scheme ignores the presence of temperature gradients present in real storage tanks. The following section deals with stratified storage systems.

Modeling the Thermal Behavior of Liquid Storage—Stratified Systems

To model a stratified storage system, the liquid storage tank of height L is assumed to be made up of n incremental elements of height Δx. If conduction between fluid elements and in the wall are negligible, and if the storage volume is large and the velocities within the tank are small, the energy exchange between differential elements within the tank can be modeled as follows: Hot fluid from the collector enters the storage tank at the top. If the temperature $T_{f,\text{out}}$ of this stream is greater than the temperature of the top layer of the tank ($m = 1$), the top layer internal energy is increased by an amount $\dot{m}_c c(T_{f,\text{out}} - T_1)\, dt$. If, however, the inlet temperature is below T_1, the inlet mass flow is assumed to simply sink through layer 1 and succeeding layers until it reaches layer m, where $T_{f,\text{out}} > T_m$. Conditions in higher layers are thus unaffected. The energy of the first affected layer is increased by $\dot{m}_c c(T_{f,\text{out}} - T_m)\, dt$. Once the inlet mass flow reaches the layer where $T_{m-1} > T_{f,\text{out}} > T_m$, all layers *below* layer m must also change in energy, since it is assumed that the mass from the collector entering layer m will displace an equal mass of lower temperature fluid at T_m, pushing it into layer $m + 1$ and increasing the internal energy of that layer by $\dot{m}_c c(T_m - T_{m+1})\, dt$. This cascading process will continue until all elements *below* m have increased their energy.

In a similar manner, cold return flow from the load loop enters the storage tank at $m = n$ at temperature $T_{L,\text{out}}$. If the load return fluid is colder than T_n, the internal energy of element n is changed by $\dot{m}_L c(T_{L,\text{out}} - T_n)\, dt$. If $T_{L,\text{out}} > T_n$, the return flow is assumed to rise through each layer without disturbance until it reaches layer m where $T_{m+1} < T_{L,\text{out}} < T_m$. The energy of layer m is then decreased by $\dot{m}_L c(T_{L,\text{out}} - T_m)\, dt$. All layers *above* m now have mass displaced by upward flow from layers below, and their energies will decrease by $\dot{m}_c c(T_m - T_{m+1})\, dt$. The process is illustrated in Figs. 6-4 and 6-5. The F and G are control functions having values of 0 or 1, and are set to inject the collector and load return flows into the element having the correct temperature

An energy balance on element m (Fig. 6-5) then can be written at time t as

$$\begin{pmatrix} \text{Rate of} \\ \text{change of} \\ \text{energy in} \\ \text{element } m \end{pmatrix} = \begin{pmatrix} \text{rate of change of} \\ \text{enthalpy due to} \\ \text{possible injection} \\ \text{of collector liquid} \end{pmatrix} + \begin{pmatrix} \text{rate of change of} \\ \text{enthalpy due to pos-} \\ \text{sible injection from} \\ \text{element above} \end{pmatrix}$$

$$- \begin{pmatrix} \text{rate of change of} \\ \text{enthalpy due to} \\ \text{possible flow to} \\ \text{element below} \end{pmatrix} + \begin{pmatrix} \text{rate of change of} \\ \text{enthalpy due to pos-} \\ \text{sible injection of} \\ \text{load return} \end{pmatrix}$$

$$+ \begin{pmatrix} \text{rate of change of} \\ \text{enthalpy due to} \\ \text{possible flow from} \\ \text{element below} \end{pmatrix} - \begin{pmatrix} \text{rate of change of} \\ \text{enthalpy due to} \\ \text{possible flow to} \\ \text{element above} \end{pmatrix} - \begin{pmatrix} \text{rate of} \\ \text{energy} \\ \text{loss to} \\ \text{surroundings} \end{pmatrix}$$

From Fig. 6-4, this can be written as

$$(M_c)_{st} \frac{dT_m}{dt} = F_m \dot{m}_c c T_{f,\text{out}} + \dot{m}_c c T_{m-1} \sum_{l=1}^{m-1} F_l - \dot{m}_c c T_m \sum_{l=1}^{m} F_l$$

$$+ G_m \dot{m}_L c T_{L,\text{out}} + \dot{m}_L c T_{m+1} \sum_{l=m+1}^{n} G_l - \dot{m}_l c T_m \sum_{l=m}^{n} G_l$$

$$- \bar{U}_{st} A_m (T_m - T_a) \qquad (6\text{-}1)$$

In this equation, M_m is the mass of liquid in storage volume element m, A_m is the surface area of storage associated with element m, $\bar{U}_{st}$ is the overall loss coefficient for exposed storage surface, and the control functions F and G are found from the relations

$$F_m = 1 \quad \text{if } T_{m-1} > T_{f,\text{out}} > T_m$$
$$F_m = 0 \quad \text{otherwise}$$
$$G_m = 1 \quad \text{if } T_{m+1} < T_{L,\text{out}} < T_m$$
$$G_m = 0 \quad \text{otherwise}$$

Equation (6-1) can be rewritten as

$$T_m(t + \Delta t) = T_m(t) + \frac{\dot{m}_c \Delta t}{M_m} \Bigg\{ F_m[T_{f,\text{out}}(t) - T_m(t)] + [T_{m-1}(t) - T_m(t)] \sum_{l=1}^{m-1} F_l$$

$$+ \frac{\dot{m}_L}{\dot{m}_c} \Bigg[G_m[T_{L,\text{out}}(t) - T_m(t)] + [T_{m+1}(t) - T_m(t)] \sum_{l=1}^{n} G_l \Bigg]$$

$$- \frac{\bar{U}_{st} A_m M_m}{\dot{m}_c \Delta t} [T_m(t) - T_a(t)] \Bigg\} \qquad (6\text{-}2)$$

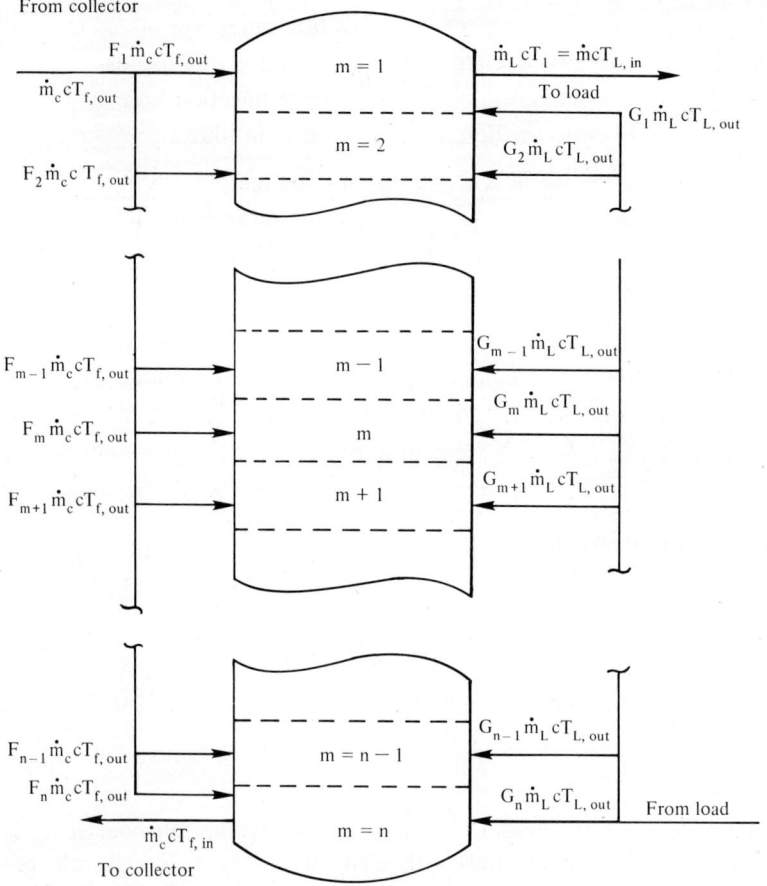

Figure 6-4 Model of a stratified liquid storage tank.

From the known initial temperature distribution in the storage tank, $T_m(0)$, the temperature distribution can be found at a later time for any inlet temperature variations $T_{f,\,out}(t)$ and $T_L(t)$.

In order to assure numerical stability of this method, we must make certain the time increment Δt is small enough that the energy added to a given volume element m in Δt will be less than the storage capacity of the element.

In terms of the symbols used above, for energy added from the collector,

$$\dot{m}_c c[T_{m+1}(t) - T_m(t)] \Delta t \leq (M_m c)_{st}[T_m(t + \Delta t) - T_m(t)] \qquad (6\text{-}3)$$

If the temperature change of element m in time Δt is held to be no more than the temperature difference between element m and the adjacent element $m + 1$, then the temperature difference terms on both sides of Eq. (6-3) cancel. This can

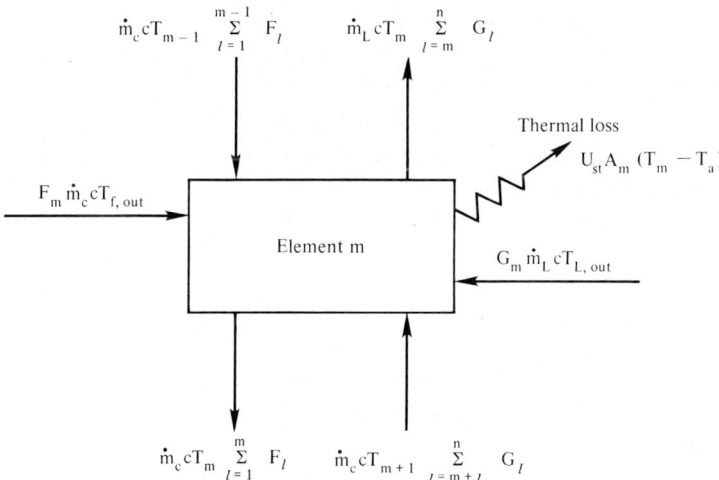

Figure 6-5 Energy balance on liquid storage element *m*.

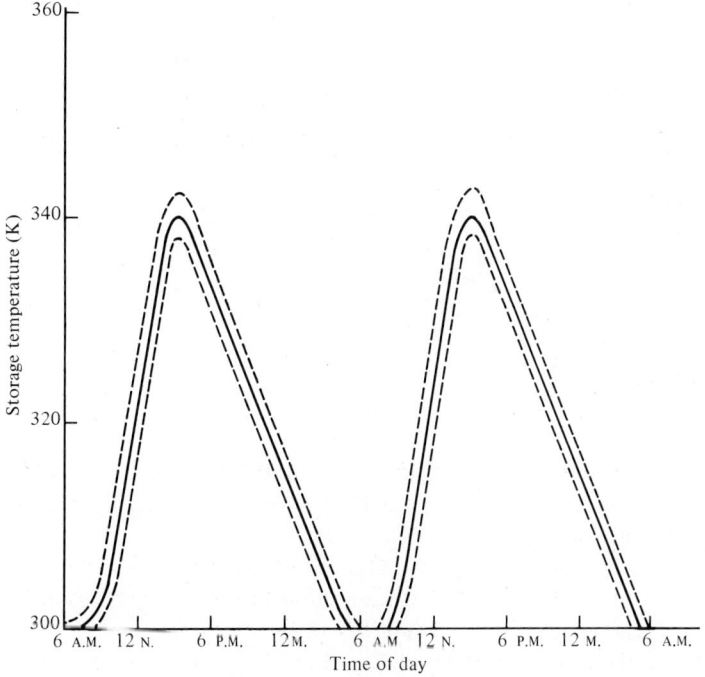

Figure 6-6 Comparison of mixed and stratified storage temperatures for the base case. ——— Mixed storage (supplied 85.1 percent of the load); – – – – stratified storage, top and bottom (supplied 89.6 percent of the load).

only occur if, from Eq. (6-3), (for all M_m equal),

$$\Delta t \le \frac{(M_m c)_{st}}{\dot{m}_c c} = \frac{(Mc)_{st}}{\dot{m}_c cn}$$

Using this criterion, the temperature of a given element cannot rise to a value above the temperature of the adjacent element in time Δt.

A similar criterion exists for the energy decrease of m by the fluid returning from the load. In this case

$$\Delta t \le \frac{(M_m c)_{st}}{\dot{m}_L c} = \frac{(Mc)_{st}}{\dot{m}_L cn}$$

To assure stability in general, then, if the same fluid is used in both loops and in the storage tank,

$$\Delta t \le \frac{M}{\dot{m}n} \qquad (6\text{-}4)$$

where $\dot{m}$ is the smaller of $\dot{m}_c$ and $\dot{m}_L$.

Using this model in the computer simulation, the effect of thermal stratification on storage temperature can be determined. Figure 6-6 shows, for the base case of Table 2-2, the top and bottom element temperatures in a 10-division stratified storage tank and the mixed temperature for isothermal storage. Note that the bottom element temperature is generally below the mixed-storage prediction of tank temperature. Thus the inlet liquid to the collector in the stratified

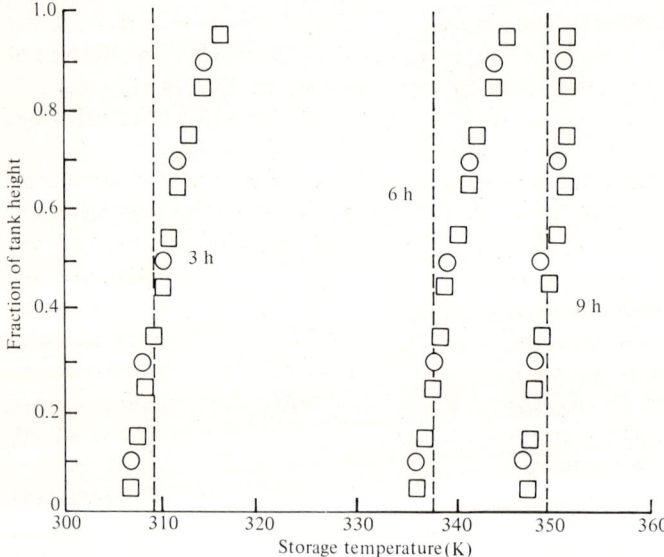

Figure 6-7 Effect of stratification on tank temperature profile—base case. — — — Base case, mixed; ○ stratified, five divisions; □ stratified, 10 divisions.

system will be lower, leading to higher collector efficiencies than for the system with mixed storage. The collected solar energy and, therefore, system collection efficiency are increased over the mixed-storage results. In addition, the temperature from the tank top delivered to the load is higher in the stratified case; thus, the load heat exchanger will transfer energy at a greater rate. Profiles at 3, 6, and 9 h past sunrise are shown in Fig. 6-7.

The model used here predicts the stratification for an ideal system in that no conductive or turbulent convective energy transfer between layers is assumed to occur. The actual stratification and system performance should lie between those predicted by this analysis and those for mixed storage.

6-5 PEBBLE-BED STORAGE

General Considerations

In an air system, solar heated air is passed directly through the pebble bed from top to bottom. As the air passes through the pebbles, the rock temperature rises as energy is transferred from the air to the rocks. The cooled air which leaves the bottom of the pebble bed is returned to the collectors to be reheated. Pebble-bed storage will maintain thermal stratification (top warmer than bottom) better than a liquid system because the effective conduction through the bed is small and there is no convective mixing as is possible in liquid systems due to liquid injection.

The stored heat is delivered to the building by circulating room air through the pebble bed in the opposite direction. As cooled air flows upward through the bed it is heated and the warmed air is recirculated to the rooms. The bottom of the pebble bed is at the lowest (near room) temperature. During the next collection period, air leaving the bed at the bottom will be cool and the collectors will operate at maximum efficiency.

A maximum depth of about 2 m (6 ft) of pebbles is recommended for acceptable floor loading and air pressure drop. The pressure drop also depends upon size and uniformity of the pebbles. At a typical air velocity of about 0.1 m/s (0.3 ft/s) through 1.5 m (5 ft) of gravel 1.8 to 3.6 cm (0.75 to 1.5 in) in diameter the pressure drop will be about 75 Pa (0.25 in H_2O or 0.01 lb_f/in^2).

As shown in Fig. 6-8, the pebbles are supported on a wire screen, which in turn is supported on blocks for maximum area for air flow in the lower plenum. Coverage of the bottom by the supporting blocks should be about 50 percent for light-weight screen support. If a heavy-mesh woven or welded wire screen is used, the block spacing can be greater.

Horizontal flow has occasionally been used in pebble beds; however, heat exchange effectiveness is lower than in vertical flow beds, because of the tendency for warm air to flow through the upper part and cool air through the bottom. If a horizontal arrangement cannot be avoided, vertical baffles should be provided to enhance the proper flow through the bed.

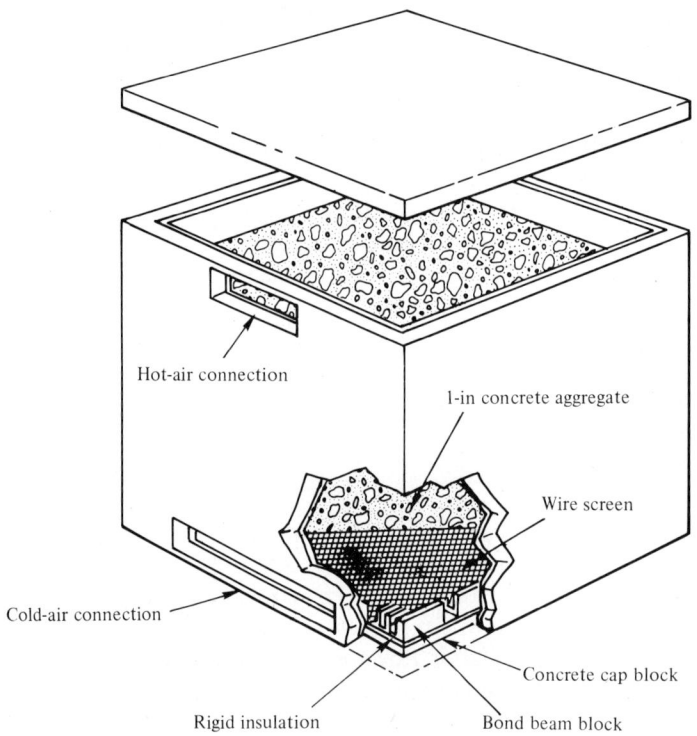

Figure 6-8 Pebble-bed heat storage unit.

Analysis

Mumma and Marvin [1] have documented a straightforward simulation method for the behavior of rock-bed storage systems. The simulation is based on a one-dimensional transient analysis of energy exchange between the air stream and the pebbles, using a finite-difference representation. Sowell and Curry [2] present a discussion of the method of Mumma and Marvin, along with more exact methods, and note that the results given in Ref. 1 are generally accurate. The following analysis is based on Ref. 1.

The bed is of length L in the flow direction and of cross-sectional area A. It is divided into n differential elements of length Δx. An energy balance on the fluid over the length $\Delta x = L/n$ is (Figs. 6-9 and 6-10),

$$\dot{m}_f c_f \, dT_f \simeq \dot{m}_f c_f (T_{f,x} - T_{f,x+\Delta x}) = h_v A \, \Delta x (T_{f,x} - T_{b,x}) \tag{6-5}$$

where h_v is the volumetric heat transfer coefficient between the air and the pebbles. Löf and Hawley [3] show the value of h_v to be about 7 kW/(m³ · °C) [400 Btu/(h · ft³ · °F)] for conditions typical of pebble-bed storage. Equation

STORAGE OF THERMAL ENERGY **175**

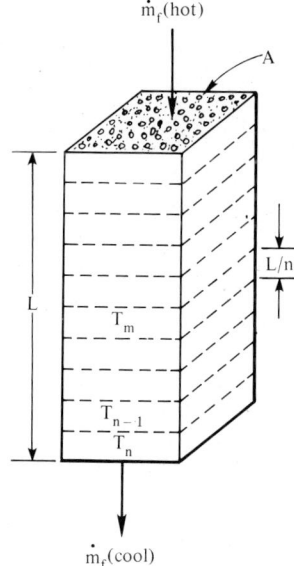

Figure 6-9 Pebble bed.

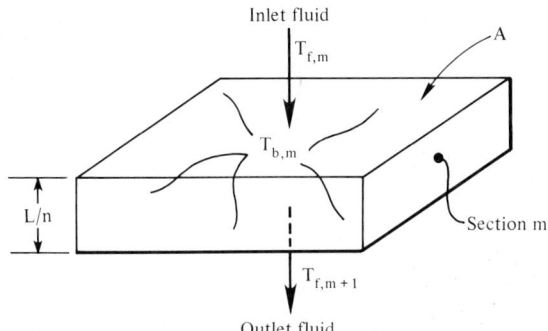

Figure 6-10 Section m of the bed.

(6-5) can be integrated to find $T_{f, m+1}$ by noting

$$\int_0^{\Delta x} \frac{dT_f}{(T_f - T_b)} = \int_0^{\Delta x} \frac{h_v A\, dx}{\dot{m}_f c_f} = \frac{h_v A L}{n \dot{m}_f c_f} = \Phi_1$$

or
$$T_{f, m+1} = T_{b, m} + (T_{f, m} - T_{b, m}) \exp(-\Phi_1) \tag{6-6}$$

An energy balance on the pebbles in element m gives, for time increment dt (Fig. 6-11),

Rate of change in internal energy of pebbles
= rate of energy gain from air
− rate of energy loss to surroundings

or

$$(\rho A\, \Delta x)(1 - \epsilon) c_b \frac{dT_{b,m}}{dt} = \dot{m}_f c_f (T_{f,m} - T_{f, m+1}) - U_{\text{st}} A_{\text{st}, m}(T_{b, m} - T_a) \tag{6-7}$$

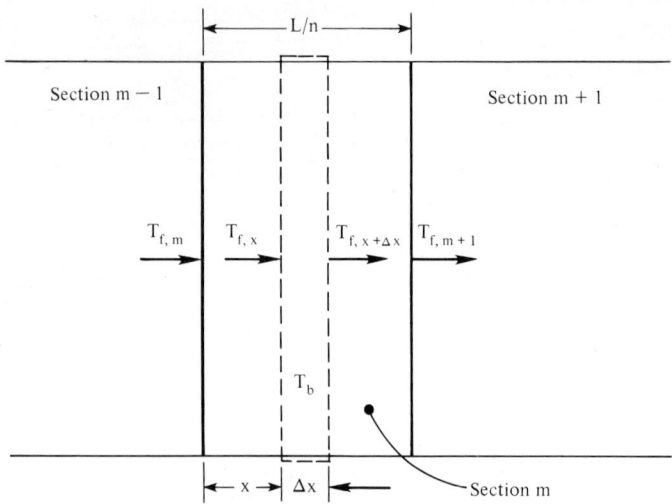

Figure 6-11 Control volume for section m energy balance.

Here, ϵ is the void fraction of the pebble bed, ρ is the rock density, c_b is the heat capacity of the rock alone, U_{st} is the overall loss coefficient between the bed and the surroundings, and $A_{st,m}$ is the external area for thermal loss of element m of the bed.

In finite-difference form, Eq. (6-7) becomes

$$T_{b,m}(t + \Delta t) = T_{b,m}(t) + [\Phi_2(T_{f,m} - T_{f,m+1}) - \Phi_3(T_{b,m} - T_a)] \Delta t$$

where
$$\Phi_2 = \frac{(\dot{m}c)_f n}{\rho A L (1 - \epsilon) c_b} \qquad \Phi_3 = \frac{U_{st} A_{st}}{(\dot{m}c)_f} \Phi_2 \qquad (6\text{-}8)$$

Equations (6-6) and (6-8) are all that are necessary to calculate the system behavior once Φ_1, Φ_2, and Φ_3 are specified along with initial conditions of bed temperature.

In Ref. 1 it is stated that this numerical analysis provides stable predictions of bed temperatures that compare well with analytical solutions if the time increment and bed element size are related by

$$\Delta t \leq \{\Phi_2[1 - \exp(-\Phi_1)]\}^{-1}$$

However, Sowell and Curry [2] note that this criterion may not be adequate for general variations in inlet air temperature. It was derived for a step input; however, it is probably adequate for practical inlet variations.

Some predictions of bed temperature profiles calculated from this method are presented in Figs. 6-12 and 6-13, taken from Ref. 1.

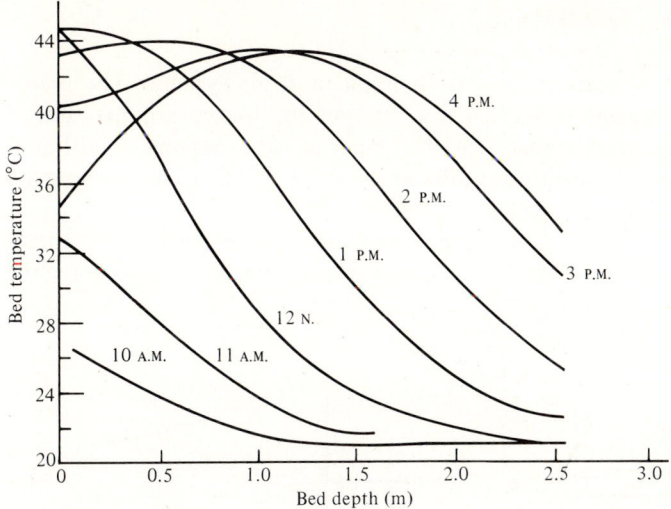

Figure 6-12 Bed heating simulation. $A = 65$ m^2; $\dot{m}_f = 1.4$ kg/s; $A_{st} = 6.8$ m^2; $L = 2.5$ m; $D = 0.025$ m; $\rho = 2400$ kg/m^3; $c_f = 1.012$ kJ/(kg · °C); $c_b = 0.837$ kJ/(kg · °C); $\epsilon = 0.4$; n = 15; $\Delta t = 5$ s; Biot number $= h_v AL/k = 0$

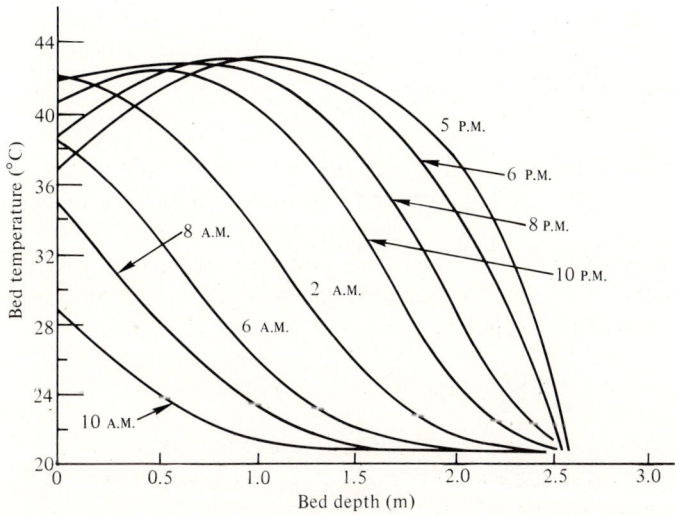

Figure 6-13 Bed cooling simulation. Values for various parameters the same as those given for Fig. 6-12.

6-6 LATENT HEAT STORAGE

Latent heat storage has been used in both liquid and air systems. The major advantage of latent storage is the great reduction in storage volume that is possible. However, the heat transfer in the presence of a moving solid-liquid boundary is quite complex, leading to difficult modeling if all of the phenomena are completely accounted for.

Shamsundar has shown [4–6] that accurate approximation to three-dimensional modeling of a shell-and-tube latent storage device is possible for the

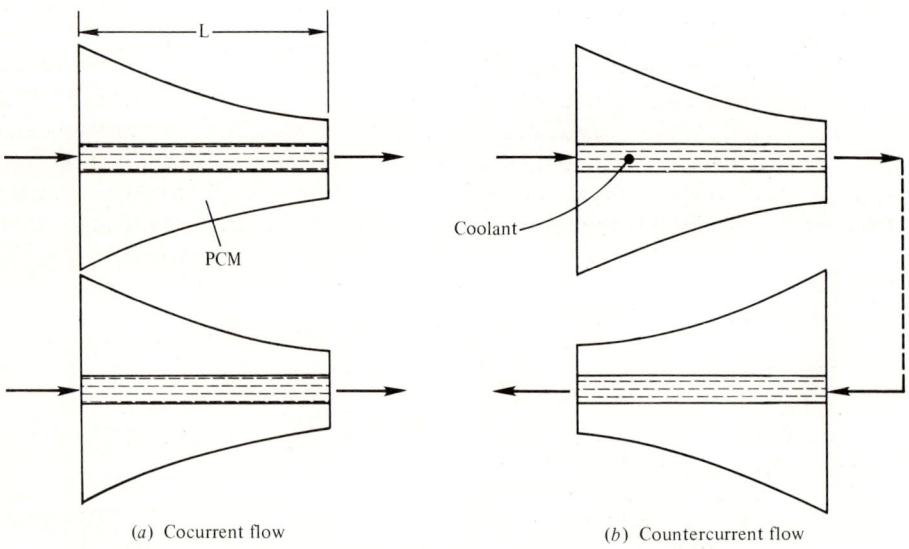

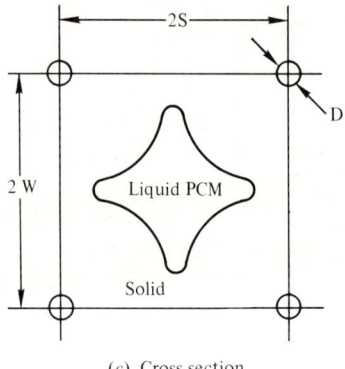

Figure 6-14 Schematic of shell-and-tube PCM heat exchanger showing three-dimensional nature of solidification.

case of constant heat input or constant load. In the shell-and-tube design, the phase-change material (PCM) is contained in the shell, and the working fluid in the collector or load loop passes through the tubes.

In the analysis, it is assumed that all energy transport within the PCM is by conduction, (neglecting axial conduction); that sensible heat storage within the PCM is negligible in comparison with the latent storage (although this restriction can be relaxed); and that the coolant inlet bulk temperature is constant with time while the heat-transfer coefficient between the fluid and tube wall is constant along the tube.

Figure 6-14 shows the layout of a typical cross section in the storage vessel, along with schematics of the solid-liquid interface configuration when adjacent channels are in co- and countercurrent flow.

Using the assumptions noted, Shamsundar and Srinivasan [5] computed the equivalent effectiveness ϵ of latent heat storage units as a function of the number of transfer units (NTUs) for various fractions of the PCM being frozen. The ϵ and NTU are the standard shell and tube heat-exchanger design factors. Figures 6-15 through 6-17 present these results for various values of the Biot number (Bi = hD/k) where h is the fluid-to-tube wall heat-transfer coefficient, D is the tube

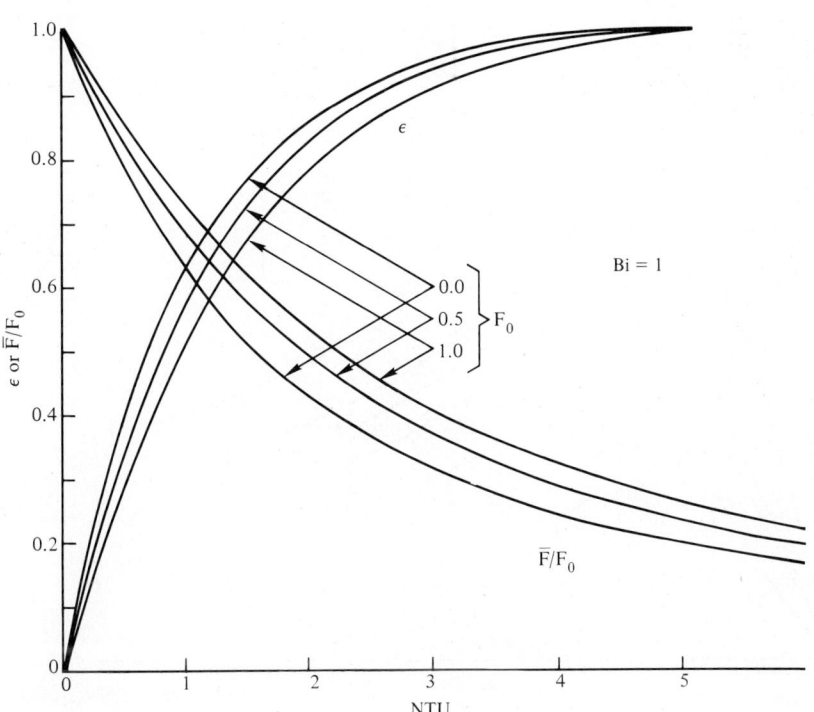

Figure 6-15 Effectiveness and mean frozen fraction/frozen fraction at inlet versus NTU for cocurrent flow; Biot number = 1.

diameter and k is the thermal conductivity of the *solid* PCM. All results are for cocurrent (parallel) flow.

The effectiveness ϵ is defined for latent storage as the energy actually transferred to or from the latent heat storage device divided by the energy that could be transferred if the coolant left the end of the storage device at the solidification temperature of the PCM, T_{sat}. Thus

$$\epsilon = \frac{\dot{m}c_f[T_f(L) - T_f(0)]}{\dot{m}c_f[T_{\text{sat}} - T_f(0)]} = 1 - \frac{T_{\text{sat}} - T_f(L)}{T_{\text{sat}} - T_f(0)} \qquad (6\text{-}9)$$

where $T_f(0)$ and $T_f(L)$ are the fluid inlet and outlet temperatures.

The number of transfer units (NTUs) is defined as $\pi D h L/(\dot{m}c)_f = A_t h/(\dot{m}c)_f$. Here, h and D are defined as above, $\dot{m}c$ is the mass-flow-rate–heat-capacity product for the fluid (per tube) and A_t is the heat-transfer area per tube, πDL.

Figures 6-15 to 6-17 show the parameter F_0 on the effectiveness curves. The F_0 is the fraction of PCM that is frozen in a radial plane element at the inlet to the storage device. Plotted on the figures are a set of curves showing the ratio $\bar{F}/F_0$ versus NTU, with F_0 again as a parameter. $\bar{F}$ is the fraction of PCM in the entire storage device that is frozen.

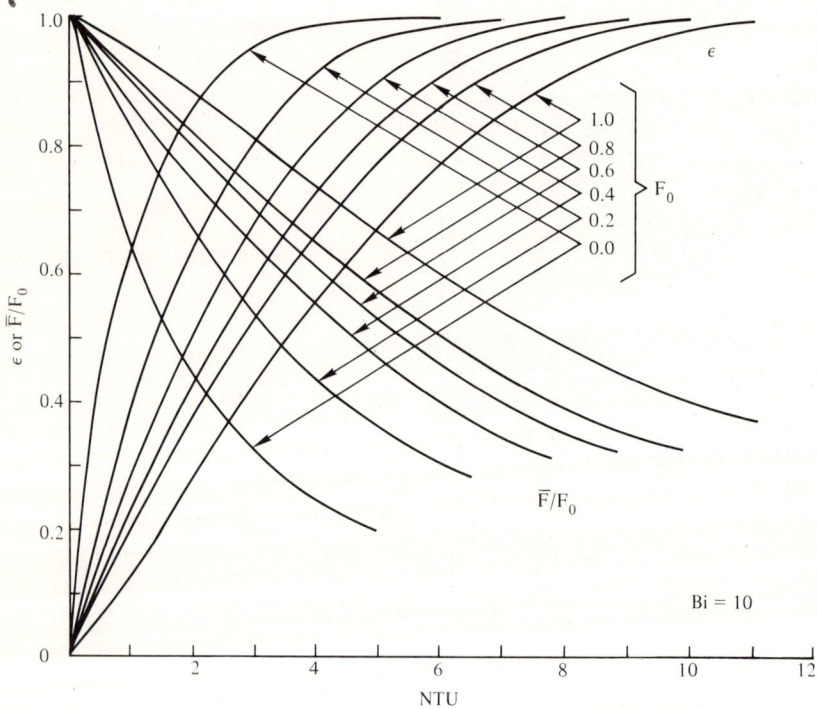

Figure 6-16 Effectiveness and mean frozen fraction/frozen fraction at inlet versus NTU for cocurrent flow; Biot number = 10.

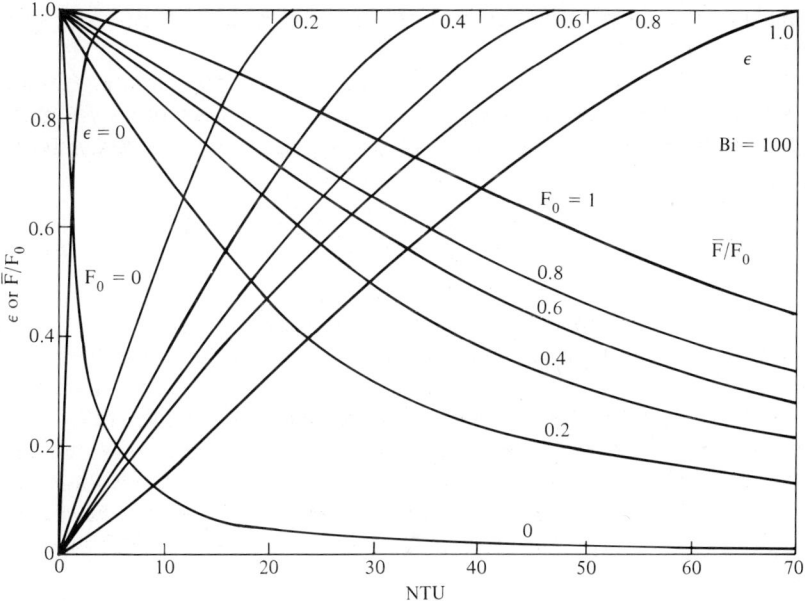

Figure 6-17 Effectiveness and mean frozen fraction/frozen fraction at inlet versus NTU for cocurrent flow; Biot number = 100.

To use the graphical information, the following relations are necessary to relate the time of flow to the degree of melting or freezing of the PCM, and to compute the NTU-ϵ relation for the Biot numbers other than those presented in Figs. 6-15 to 6-17:

$$t_0 = B\left[\frac{F_0}{Bi} + \psi(F_0)\right] \tag{6-10}$$

$$\text{NTU} = \ln\frac{1}{1-\epsilon} + \text{Bi}\left\{\Omega(F_0) - \Omega[(1-\epsilon)F_0]\right\} \tag{6-11}$$

$$\bar{F} = \{F_0\epsilon + \text{Bi}\left[\Psi(F_0) - \Psi[(1-\epsilon)F_0]\right]\}/\text{NTU} \tag{6-12}$$

and the functions $\Psi(F_0)$ and $\Omega(F)$ are plotted in Fig. 6-18. The dimensionless time t_0 is defined by

$$t_0 = \int_0^t \frac{k[T_{\text{sat}} - T_f(0)]}{\rho\lambda D^2}\,dt \tag{6-13}$$

where λ is the heat of fusion, and the geometric parameter B is

$$B = [(4SW/D^2) - \pi/4]/\pi$$

and D, S, and W are defined on Fig. 6-14.

Some examples of the use of the storage relations are now presented, based on material in Ref. 5.

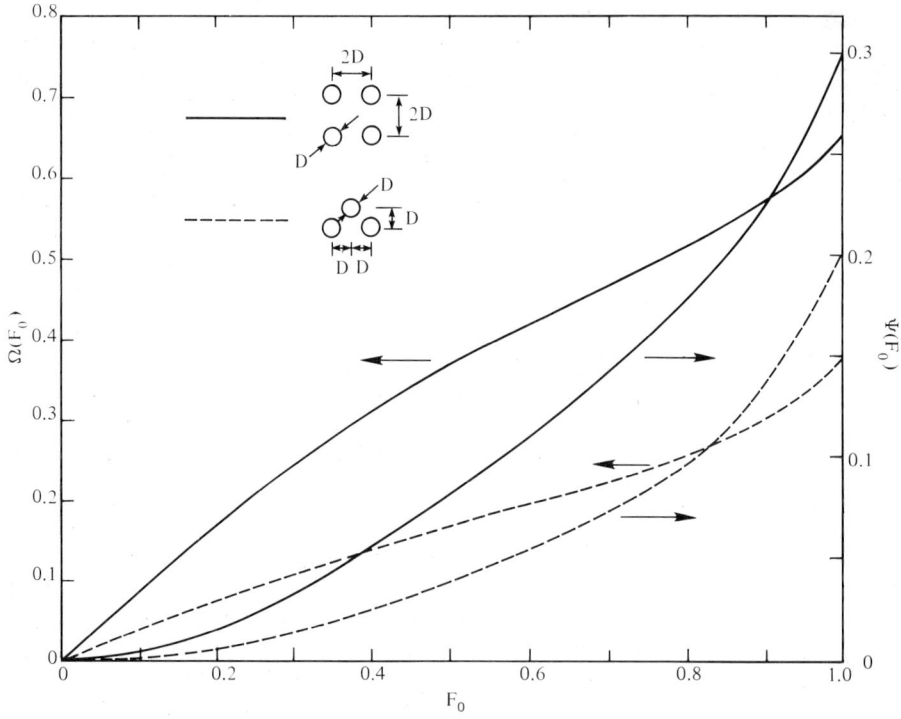

Figure 6-18 Auxiliary functions for phase change storage calculations.

Example 6-1 A parallel flow heat exchanger is built up of 2-cm OD, 1.8-cm ID, 5-m-long tubes arranged in a square pattern with a pitch of 4 cm. The PCM used has a melting point of 220°C. The coolant is compressed water at 2 MPa, entering the tubes at 200°C, with a flow rate of 20 kg/h per tube. Both $\dot{m}$ and $T_f(0)$ are constant in time. Calculate the heat recovery rate and frozen fraction for 1 h past the commencement of heat recovery if the Biot number for the PCM-coolant system is 10 and the PCM properties are $k = 0.49$ W/(m · °C), $\rho = 2180$ kg/m³, and $\lambda = 137$ kJ/kg.

SOLUTION Using the definition of the dimensionless time, Eq. (6-13),

$$t_0 = \int_0^t \frac{k[T_{\text{sat}} - T_f(0)]}{\rho \lambda D^2}$$

and noting that $T_f(0)$ is constant,

$$t_0 = \frac{k}{\rho \lambda D^2} [T_{\text{sat}} - T_f(0)]t$$

$$= \frac{[0.49 \text{ W}/(m \cdot °C)](220 - 200)(°C)(1 \text{ h})}{(2180 \text{ kg/m}^3)(137 \text{ kJ/kg})(0.02 \text{ m})^2} \frac{(3600 \text{ s})(1 \text{ kW})}{(1 \text{ h})(1000 \text{ W})}$$

$$= 0.30$$

For this case, NTU = $A_t h/\dot{m}c_f$. The value of h is $h = (\text{Bi} \cdot k)/D$ so that

$$\text{NTU} = \frac{\pi D_i L \text{ Bi } k}{\dot{m}c_f D}$$

$$= \frac{\pi \ (5 \text{ m})(0.018 \text{ m})(10)[0.49 \text{ W}/(\text{m} \cdot {}°\text{C})](3600 \text{ s})}{(20 \text{ kg/h})[4187 \text{ J}/(\text{kg} \cdot {}°\text{C})](1 \text{ h})(0.02 \text{ m})}$$

$$= 3.0$$

Using Eq. (6-10) and Fig. 6-18 for $\psi(F_0)$, the value of F_0 can be found by trial and error. For the geometry of this problem, the solid curve is appropriate. From Eq. (6-10), noting that $B = (4/\pi - \frac{1}{4})$

$$t_0 = 1.023\left[\frac{F_0}{10} + \psi(F_0)\right]$$

Now, values of F_0 are chosen, and t_0 is calculated as shown in the accompanying table.

F_0	$\psi(F_0)$	t_0
0.8	0.185	0.265
0.9	0.225	0.315

Interpolation for F_0 at $t_0 = 0.30$ gives $F_0 = 0.87$. Now given Bi = 10, NTU = 3.0, and $F_0 = 0.87$, Fig. 6-16 is used to give $\epsilon = 0.47$. From Eq. (6-9)

$$\epsilon = \frac{\dot{Q}}{\dot{m}c_f[T_{\text{sat}} - T_f(0)]}$$

or $\quad \dot{Q} = (0.47)(20 \text{ kg/h})[4187 \text{ J}/(\text{kg} \cdot {}°\text{C})](1 \text{ h}/3600 \text{ s})(220 - 200)({}°\text{C})$

$$= 219 \text{ W/tube}$$

and, also from Fig. 6-16, the frozen fraction $\bar{F}/F_0$ is about 0.75.

Example 6-2 The same problem exists as for Example 6-1, except that the PCM is the eutectic mixture of $NaNO_3$ and KNO_3. The Biot number is not known. Find the heat recovery rate and frozen fraction.

SOLUTION. In order to calculate the Biot number, it is necessary to calculate the convective coefficient on the inside of the tubes. Using the Sieder-Tate equation (App. C, Sec. C-2) for this purpose, we get

Re = 3000   Pr = 0.93   Nu = 13.5   Bi = 20.1   NTU = 6.1

By again using Eq. (6-13) we calculate $t_0 = 0.30$. Next, we use Eq. (6-10) and Fig 6-18 to obtain F_0 at this value of t_0 by trial and error as shown in the first accompanying table. For the geometric arrangement given, B =

($4/\pi - \frac{1}{4}$). By interpolation, we get $F_0 = 0.93$. (More accuracy is achieved by plotting t_0 against F_0, and reading the resulting curve at $t_0 = 0.30$.)

F_0	$\psi(F_0)$	t_0
0.90	0.225	0.276
0.95	0.258	0.312

Next, we use Eq. (6-11) and Fig. (6-18) to calculate the effectiveness as follows. We assume a few values of ϵ and calculate NTU for them, as shown in the second accompanying table below. Note that $\Omega(F_0) = 0.590$ for $F_0 = 0.93$. By interpolation with NTU = 6.1 we get $\epsilon = 0.54$. From the effectiveness, we get the total instantaneous heat flux as

$$\dot{Q} = \epsilon(\dot{m}c)_f[T_{sat} - T_f(0)] = 250 \text{ W/tube}$$

ϵ	$(1-\epsilon)F_0$	$\Omega[(1-\epsilon)F_0]$	NTU
0.57	0.40	0.308	6.5
0.52	0.45	0.338	5.8

To obtain the frozen fraction, we use Eq. (6-12). From Fig. (6-18), we find that

$$\psi[(1-\epsilon)F_0] = 0.062$$

and

$$\psi(F_0) = 0.242$$

Therefore,

$$\bar{F} = [(0.54)(0.93) + (20.1)(0.242 - 0.063)]/6.1 = 0.67$$

It is interesting to note that the values predicted by a two-dimensional calculation are $\dot{Q} = 213$ W and $\bar{F} = F_0 = 0.93$ [4].

Example 6-3 For the heat exchanger of Example 6-2, find the time required to complete heat recovery.

SOLUTION We are required to find t such that $\bar{F} = 1$. First, we find t_1, which corresponds to $F_0 = 1$, by using Eq. (6-10), as

$$t_1 = (4/\pi - \tfrac{1}{4})(1/20.1 + 0.300) = 0.36$$

The time for complete heat removal is t_1 plus the time to freeze the remaining liquid, or

$$t_0 = t_1 + [\pi BkL/(\dot{m}c)_f] = 0.67.$$

The physical time is obtained from Eq. (6-9) as

$$t = 2.2 \text{ h}$$

Again, the prediction of the two-dimensional calculation may be noted. It is $t = 1.2$ h [4].

In Ref. 5, extensions to the cases of time-varying $T_f(0)$ and $\dot{m}_f$ are given, as well as material on handling counterflow storage devices.

PROBLEMS

6-1 For the base case, compare the percent of load supplied over a 3-day period for the cases of mixed and stratified storage with five tank divisions. How do the present values (allowable capital costs) of the systems compare with and without stratification?

6-2 The storage system described in Prob. 2-5 is to be replaced with a latent heat storage system. The phase change material is to be naphthalene (MP = 176°F, heat of fusion = 64 Btu/lb$_m$). Assuming ideal heat removal is possible (that is, no loss of energy occurs due to conduction heat-transfer effects and no sensible heat storage is considered), what mass of naphthalene is required for the 2-day storage to provide a constant load of 30,000 Btu/h?

6-3 For a load of 500,000 J/min (constant), size the collector area and storage capacity necessary to provide at least 70 percent of the load. Assume the SOLSIM base case for the collector efficiency parameters and insolation. Rough out the answer by hand calculation, then check it out using the computer program.

6-4 Repeat Prob. 6-3, but use an evacuated tabular collector.

6-5 Repeat Prob. 6-3, but assume that the load requirement rate is twice as high and occurs only from 9 A.M. to 3 P.M. Note that this problem requires input to SOLSIM of a daily load profile.

6-6 A solar tower system is designed to operate with a 100-MW (input) low-pressure turbine that accepts minimum inlet conditions of saturated steam at 500°F. It is proposed to design a storage system that will provide energy to the low-pressure turbine from storage for a 4-h period after the storage is fully charged to a temperature of 750°F. Estimate the required storage volume if

(*a*) Pressurized liquid water is used as the storage medium.

(*b*) Sodium hydroxide [melting point, 604°F; heat of fusion, 72 Btu/lb$_m$; heat capacity (solid and liquid), 0.3 Btu/(lb$_m$ · °F); density (solid and liquid), 120 lb$_m$/ft^3] is used as the storage medium. Note any assumptions that you make.

REFERENCES

1. S.A. Mumma and W.C. Marvin, "A Method of Simulating the Performance of a Pebble-Bed Thermal Energy Storage and Recovery System," ASME Paper 76-HT-73, ASME/AIChE National Heat Transfer Conf., St. Louis, August 1976.
2. E.F. Sowell and R.L. Curry, "A Convolution Model of Rock Bed Thermal Storage Units," *Solar Energy*, vol. 24, no. 5, pp. 441–449, 1980.
3. G.O.G. Löf and R.W. Hawley, *Ind. Eng. Chem.*, vol. 40, 1061–1070, 1948.
4. N. Shamsundar, "Heat Transfer in Thermal Storage Systems," Department of Mechanical Engineering Final Report No. ERDA-77-C-04-3974/EFT-5, University of Houston, May 1978.
5. N. Shamsundar and R. Srinivasan, "Effectiveness—NTU Charts for Heat Recovery from Latent Heat Storage Units," *J. Solar Engineering*, vol. 102, pp. 263–271, Nov. 1980.
6. N. Shamsundar and R. Srinivasan, "A New Similarity Method for Analysis of Multi-Dimensional Solidification," *J. Heat Transfer*, vol. 101. pp. 585–591, Nov. 1979.

CHAPTER
SEVEN
LOAD MODELING

The energy load for any solar energy system is the quantity that determines the characteristics and allowable cost of the solar energy system. However, good data on load characteristics are in some cases difficult for the designer to obtain. In this chapter, the loads typical of those to be supplied by solar-thermal conversion systems of various types are examined. In particular, heating and cooling loads for residential and commercial buildings, domestic hot-water loads, swimming pool loads, industrial loads, and electrical generation loads are considered. Some other loads for solar use are also mentioned.

Because of the large effect of climate, building type, degree of building use, etc., on heating and cooling, hot water, swimming pool, and electrical loads, it is very difficult to give simple predictions of any of these time-varying loads. Thus load estimation relies to a great extent on the experience of the estimator coupled with careful data collection and interpretation.

7-1 LIST OF SYMBOLS

A	area
DD	degree-days
$\dot{Q}$	energy rate
T	temperature
U	overall thermal loss coefficient

Subscripts

d	design value
in	indoor

max, min	maximum or minimum value
m	monthly total
o	outdoor
r	required

7-2 HEATING AND COOLING LOADS

Obviously, the major factor in determining residential heating and cooling loads is local climate. If the local climate had ideal conditions for human comfort the year round, no heating or cooling systems would be required. The more extreme the local climatic conditions become, on the other hand, the more energy must be expended to mitigate those extremes.

To design carefully for heating and cooling systems, the engineer must fully understand all the factors that determine human comfort. The most important of those for the active solar system designer are air temperatures and humidity, but the effects of air movement, human exposure to direct insolation, radiation to and from cold or hot walls, or windows, and a host of other conditions can affect human comfort in particular situations. In addition, body comfort is strongly affected by the type and degree of activity being carried out by an individual. If individuals can accept wider ranges of conditions than are normally considered comfortable (for example, by the use of more appropriate clothing), then the required heating and cooling loads can be reduced.

In residential design, the major loads are those caused by energy exchanges with the environment through the building shell. This is in contrast with large, commercial buildings, where the loads are usually dominated by so-called internal loads.

Internal loads include "metabolic" loads (energy released by persons within the building), energy from lighting systems, and energy from devices within the building (computer systems, electrical equipment, restaurant stoves, advertising signs, etc.). Because of the diversity of possible internal loads, design for such buildings must be on a case-by-case basis. The presence of internal loads in many buildings reduces winter heating requirements. Many large buildings are actually air-conditioned year-round.

Estimation of Residential Heating Loads

To accurately calculate residential loads requires a complex computer program which will handle the detailed calculations of thermal losses through a completely specified structure (glass area, window type, insulation and structural details, slab thickness, wall and roof area and orientation, etc.) and detailed weather data (hourly temperature, insolation, wind speed and direction, and humidity).

Simpler although less accurate approaches are also possible, and those are probably adequate for the designer of active systems. In passive designs, much more care in load calculations may be required. Loads are generally calculated in

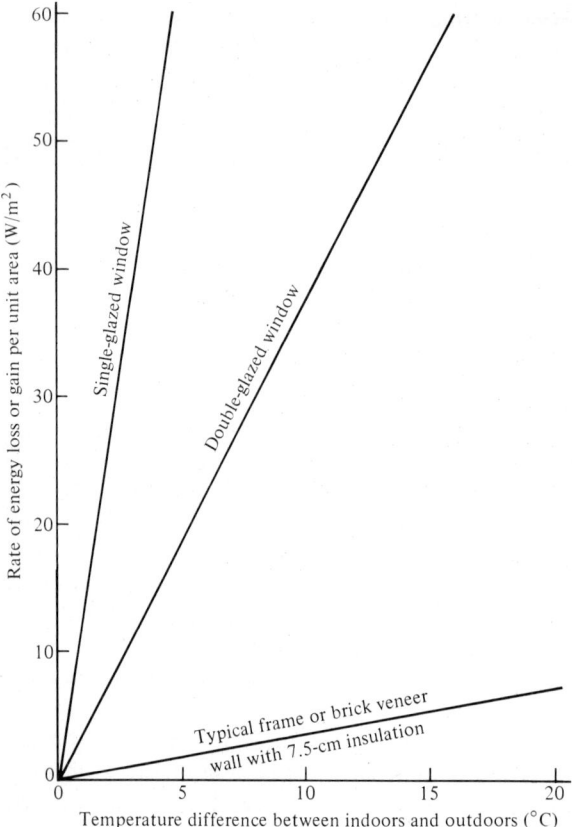

Figure 7-1 Heat losses through building skin components.

units of watts or Btu's per hour. The Btu is still the most common measure of thermal energy used by U.S. architects and heating and cooling engineers, so the common English system units will be given in parentheses following SI units throughout the chapter.

Buildings can gain energy from various sources. Large buildings, when there is a lot of lighting and many people, generally require little or no additional heating because of these sources. Residential buildings, because of their relatively higher losses through the structure's skin, usually require heating to augment these internal sources. Each person adds energy to the structure at a rate that depends on his or her physical activity. This body thermal energy production is shown in Table 7-1.

Thermal losses through walls and windows are in direct proportion to the difference in temperature between the air in the house and the air outside. For example, Fig. 7-1 shows predicted thermal losses per unit area through single- and double-glazed windows and through common structural walls.

Table 7-1 Metabolic loads—average person

Activity	Load W	Btu/h
Walking, 3.5 km/h (2 mi/h)	230	800
Running	1400	4750
Typewriting	180	600
Handsawing	530	1800
Sleeping	90	300
Sitting/resting	120	400

Because thermal losses through the skin of a building are proportional to the indoor-to-outdoor temperature difference, losses can be roughly estimated by the "degree-day" method. Suppose we wish to hold the temperature inside our house to 18°C (65°F). On a given day, the temperature outside might average 7°C (45°F). The difference is 11°C (20°F) for that day. If we sum up the daily average differences over a month, the total degrees difference is called the *degree-days* for that month. The number of degree-days is proportional to the required energy for heating for that month, since heating and degree-days both are directly proportional to the indoor-outdoor temperature difference.

For selected cities, measured degree-days for heating are shown in Table 7-2. More complete data are given in App. B.

Thermal losses from well-designed single-story residences of about 170 m² floor area are in the range 8–13 kWh/°C-day (15000–25000 Btu/°F-day). Rough estimates can also be obtained from the thermal loss coefficients given in Table 7-3.

Table 7-2 Degree-days heating based on 18.3°C (65°F) indoor temperature†

City	July	Aug.	Sept.	Oct.	Nov.	Dec.	Jan.	Feb.	Mar.	Apr.	May	June	Total
Fairbanks, AK	95 (171)	184 (332)	357 (642)	668 (1203)	1018 (1833)	1252 (2254)	1411 (2539)	1056 (1901)	966 (1739)	593 (1068)	314 (555)	123 (222)	8037 (14459)
Chicago, IL	0 (0)	0 (0)	37 (66)	155 (279)	392 (705)	584 (1051)	639 (1150)	556 (1000)	482 (868)	272 (489)	126 (226)	27 (48)	3270 (5882)
San Francisco, CA	107 (192)	97 (174)	57 (102)	66 (118)	128 (231)	215 (388)	246 (443)	186 (336)	177 (319)	155 (279)	133 (239)	100 (180)	1667 (3001)
Houston, TX	0 (0)	0 (0)	0 (0)	3 (6)	102 (183)	171 (307)	213 (384)	160 (288)	107 (192)	20 (36)	0 (0)	0 (0)	776 (1396)
Key West, FL	0 (0)	0 (0)	0 (0)	0 (0)	0 (0)	16 (28)	22 (40)	17 (31)	5 (9)	0 (0)	0 (0)	0 (0)	60 (108)

Source: ASHRAE Handbook and Product Directory; 1980 Systems, American Society of Heating, Refrigerating and Air Conditioning Engineers, Inc., New York, 1980.

† Values are °C-days, with °F-days shown in parentheses below.

Table 7-3 Estimated thermal loss coefficients for various types of residential construction

Construction	Net loss coefficient U (based on floor area)	
	kWh/(m² · °C-day)	Btu/(ft² · °F-day)
A. Brick veneer, four-bedroom house, asphalt roof, storm windows, no insulation, 24-km/h (15-mi/h) wind	0.086	15.3
B. Same as A, but with R-11 insulation in walls and attic	0.053	9.3
C. Same as B, but with R-19 insulation in attic	0.049	8.7
D. Same as B, but with R-38 insulation in attic	0.048	8.4
E. Stucco over frame, four-bedroom house, shake roof, R-19 insulation in attic only	0.080	14.1
F. Frame, three-bedroom, heated basement, R-11 insulation in walls, R-14 batts in ceiling	0.066	11.7

Example 7-1 In Houston, the average winter temperature in January is taken as 16°C (61°F), with a design (minimum) temperature of 0°C (32°F). Find the design daily heat load, average heat load, and required energy from the heating system for a well-insulated house of 200 m² floor area.

SOLUTION The maximum daily thermal load for design purposes is, assuming a 23.9°C (75°F) comfort temperature

$$\dot{Q}_{d,\,max} = UA(T_{in} - T_{min}) = 0.048 \text{ kWh/(m}^2 \cdot \text{°C-day)}$$
$$\times 200 \text{ m}^2 \, [23.9 - (0)]\text{°C} = 229 \text{ kWh/day} \qquad (7.8 \times 10^5 \text{ Btu/day})$$

where the value of U is taken from Table 7-3. The average daily winter design loss for a very-well-insulated house in Houston, with a floor area of 200 m², is

$$\dot{Q}_d = UA(T_{in} - T_d) = (0.048)(200)(18.3 - 16)$$
$$= 22 \text{ kWh/day} \qquad (74.8 \times 10^3 \text{ Btu/day})$$

To calculate the monthly energy required, $\dot{Q}_r$, from the heating system, use

$$\dot{Q}_r = \frac{\dot{Q}_d(DD)_m}{T_{in} - T_o} \qquad (7\text{-}1)$$

where $\dot{Q}_d$ = design heat loss, energy/day
$(DD)_m$ = monthly degree-days from Table 7-2
T_{in} = indoor temperature desired
T_o = outdoor design temperature

For the Houston example for January,

$$\dot{Q}_r = \frac{22}{(18.3 - 16)} 213 = 2040 \text{ kWh/month} \quad (7 \times 10^6 \text{ Btu/month})$$

The efficiency of the heating system may greatly increase the amount of energy needed to meet this load. Gas furnaces are typically 60 percent efficient (the remainder is lost in the flue gases or through incomplete combustion or other losses), so that 3400 kWh/month (11.7 × 10⁶ Btu/month) of energy in the form of natural gas would have to be purchased.

Knowing the monthly heating load requirements from Eq. (7-1) for a residential structure, we can proceed to consider the design of the solar energy system. (Note that the above methods of calculation do not apply to large structures and are only estimates for residential housing.) Further, since the size and cost of the solar system are proportional to the heating load, it pays to design the residence carefully for good insulation, sealing, passive solar collection, and other factors that will help minimize the solar heating requirements.

Residential Cooling Loads

For cooling, the same type of load analysis can be done as for heating. However, degree-days for cooling are based on days when the average temperature is above, rather than below, the reference temperature of 18°C (65°F). For example, for Houston, TX, degree-days of cooling by month are as shown in Table 7-4.

Table 7-4 Degree-days cooling based on 18.3°C (65°F) indoor temperature for Houston, TX

January	9	(16)
February	12	(22)
March	33	(59)
April	86	(155)
May	186	(335)
June	268	(483)
July	315	(567)
August	317	(570)
September	237	(426)
October	115	(207)
November	21	(38)
December	6	(11)
Yearly total	1,605	(2,889)

To calculate the cooling load, again estimate the daily load and then find the monthly cooling requirement from Eq. (7-1). This calculation neglects the latent heat load necessary to reduce the water content (humidity) of the makeup air by condensation, which is sizable in humid climates such as Houston's.

For air-conditioning systems, sizing is conventionally done in terms of *tons*, where a ton is the rate at which heat must be removed from a ton of water initially at the freezing temperature to produce a ton of ice per day. The amount is 84.4 kWh/day-ton (288,000 Btu/day-ton). Once the daily heat load $\dot{Q}_d$ is determined, the tons of air conditioning required are determined by

$$\dot{Q}_d(\text{tons}) = \frac{\dot{Q}_d(\text{Btu/day})}{288,000}$$

$$\dot{Q}_d(\text{tons}) = \frac{\dot{Q}_d(\text{kWh/day})}{84.4}$$

(7-2)

7-3 DOMESTIC HOT-WATER SYSTEMS

Most domestic hot-water (DHW) systems are sized to provide 75–110 L/day (20–30 gal/day) per person of water at 60°C (140°F). The daily load profile for the use of this water does not correlate well with the availability of insolation. Usage in most residences peaks in the evening hours for cooking, dishwashing, and

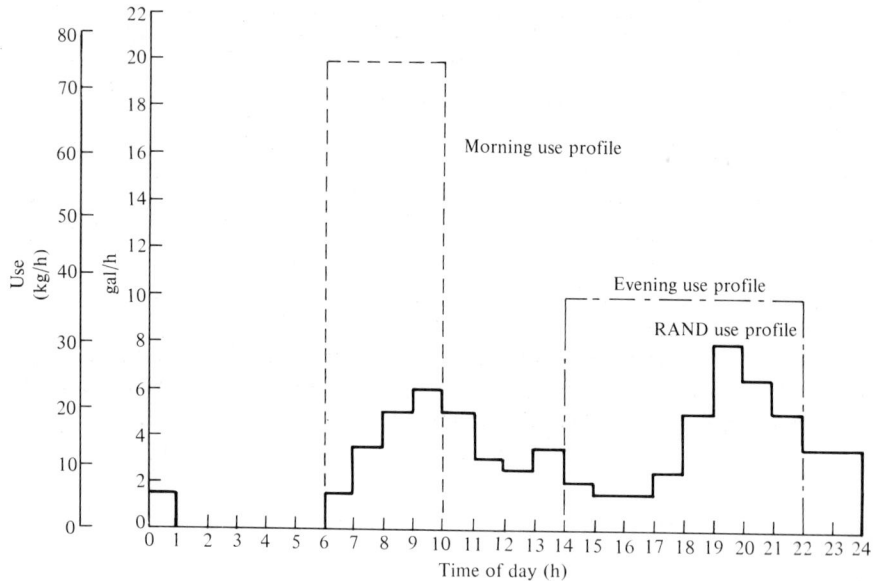

Figure 7-2 Assumed domestic hot-water demand profiles, described in terms of use versus time of day [1].

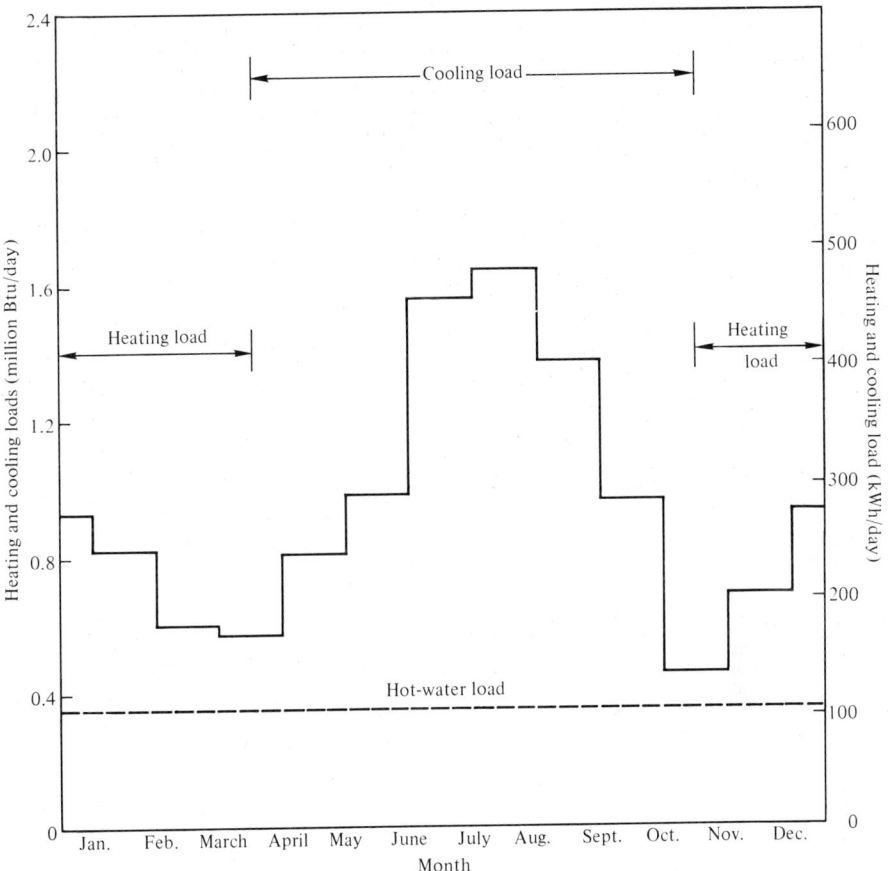

Figure 7-3 Heating and cooling loads for a 12-unit apartment house in Austin, TX.

bathing and again in the early morning for showers and cooking. Household habits may dictate additional high-use periods for laundry or other purposes. Storage is thus necessary for most DHW systems. A positive factor in DHW system design is the relatively constant monthly average load required.

Some assumed domestic load profiles are shown in Fig. 7-2, indicating loads with high morning, high evening, and distributed values [1].

An example of a seasonal load profile for apartments is indicated in Fig. 7-3. The profile is for a 12-unit apartment located in Austin, TX. The location accounts for the relatively large cooling requirement.

7-4 SWIMMING POOL HEATERS

Solar pool heaters can be viewed as solar collector systems with both the collector bank and the pool itself acting as solar collectors, the pool acting as the storage medium, and the energy loss from the pool being the load.

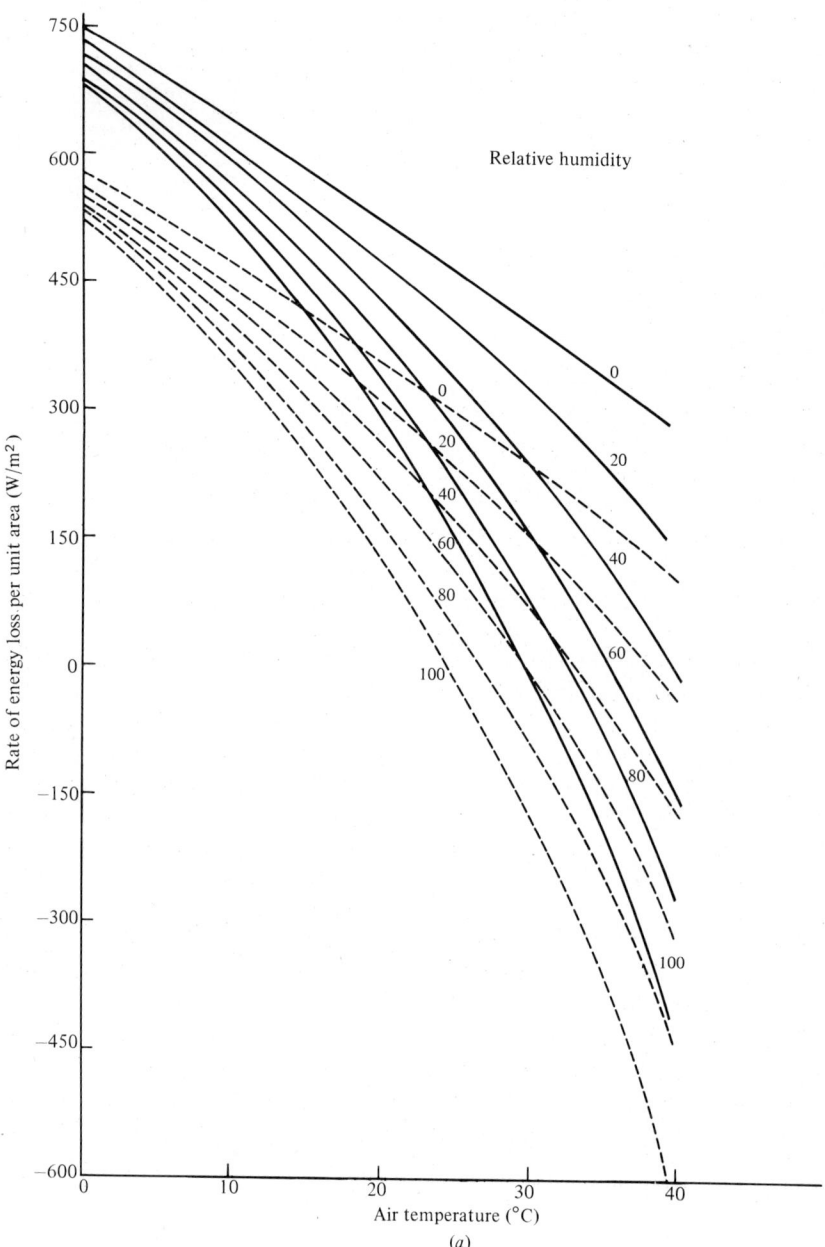

Figure 7-4 Net energy loss per unit area from pools of water. (a) Wind velocity = 10 km/h; (b) wind velocity = 20 km/h. Pool temperatures for both (a) and (b): ——— 30°C; – – – – 25°C.

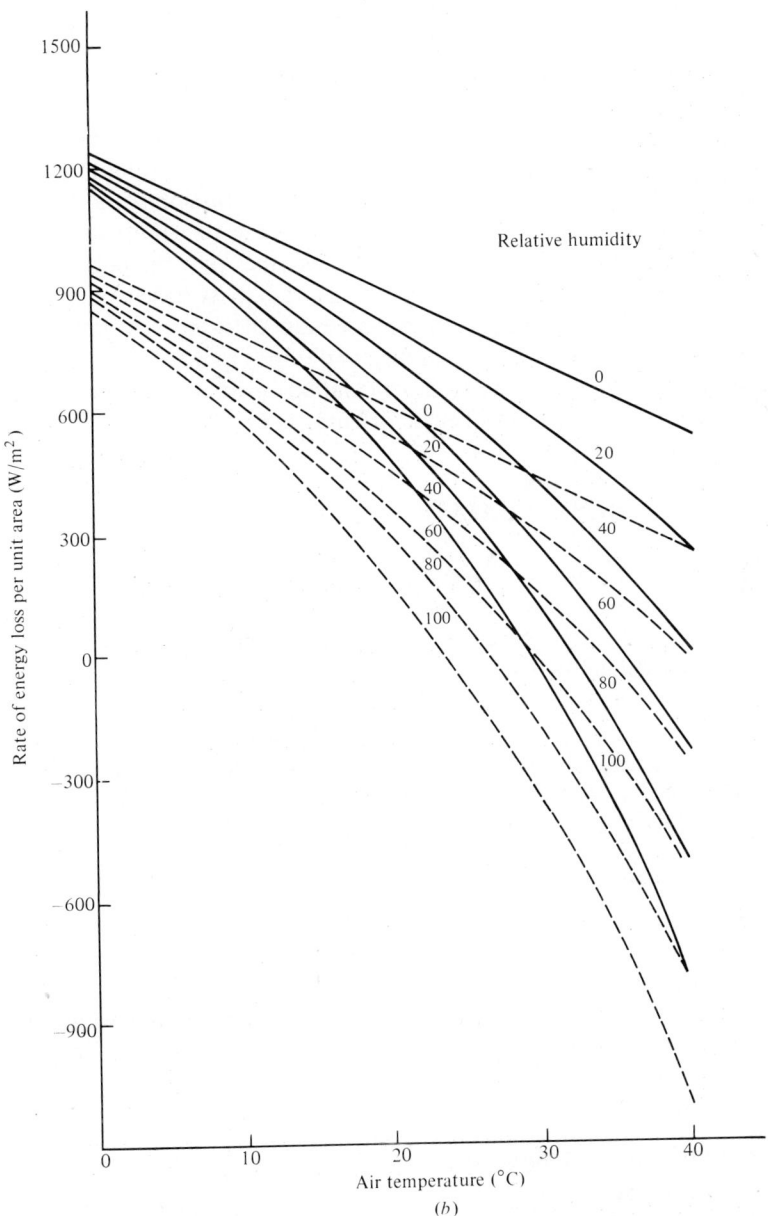

Figure 7-4 (*Continued*)

Thermal Load Calculations

Energy is lost from pools in four ways—by radiation to the sky, by convection to the air, by evaporation, and by conduction through the pool walls and bottom. For pools in the ground, the last is negligible compared with the others.

Figure 7-4 shows the total energy loss from water held at a given temperature or, in another interpretation, the energy that must be added to the pool to maintain it at 30°C (86°F) or 25°C (77°F) under various values of air temperature, relative humidity, and wind speed. Evaporation is the chief contributor to the energy loss under many conditions. In addition, radiation emission from the sky, infrared emission from the pool, and convective losses due to wind can be important. Figure 7-4 includes all of these effects, and assumes that the pool has an infrared emissivity of 0.97 and receives emitted sky radiation based on the air temperature. Pool equilibrium temperatures based on these curves have been found to agree with measured values. Since the conditions of air temperature and humidity constantly change, the energy loss calculations must be made hourly to accurately determine how a solar pool heater will function.

7-5 INDUSTRIAL ENERGY LOADS

Industrial energy requirements are extremely varied. Figure 7-5 shows the fraction of energy required below various temperatures for a group of industries that use some 48 percent of the total energy consumed by U.S. industry [2]. Note that less than 5 percent is used below 100°C (212°F).

If the energy consumed is assumed to be used for heating of a process stream from an initial temperature of 16°C (60°F) to its final use temperature, then a considerably larger fraction of energy is actually used at lower temperatures. For example, some 30 percent of total consumption occurs below 100°C (212°F) if preheating is included. (The preheat curve is also shown on Fig. 7-5.) The potential for solar applications, even with flat-plate collectors, in preheating is thus very large.

For a better estimate of industrial solar application potential, Table 7-5 shows the energy used at various temperatures in selected industries.

Load profiles for many industrial processes are constant with respect to time. They are thus simpler to model than the time-varying residential, hot water and commercial loads. Modeling of one industrial system will be outlined in Chap. 11.

Electrical Power Plant Load Requirements

Electrical utility loads are strongly dependent on the mix of customers and on the climatic characteristics of the utility service area. If the utility load has a large fraction of industrial users, then daily and seasonal loads may be relatively constant. However, if the utility serves chiefly residential customers, the seasonal and daily loads may vary widely. In the South, for example, summer peaks occur because of the large air-conditioning loads, and without industrial customers the winter loads are quite small. An example of the peak demand profile for Austin, TX is shown in Fig. 7-6. Figure 7-7 shows daily variations in the instantaneous load in the summer for the same utility. Some northern utilities with many customers using electric heat have peak loads in the winter.

Southern utility load profiles, both seasonal and daily, fit particularly well with the availability of solar energy. Thus, solar air conditioning or solar steam generation should be a great help to these utilities when the economics of these solar systems become attractive.

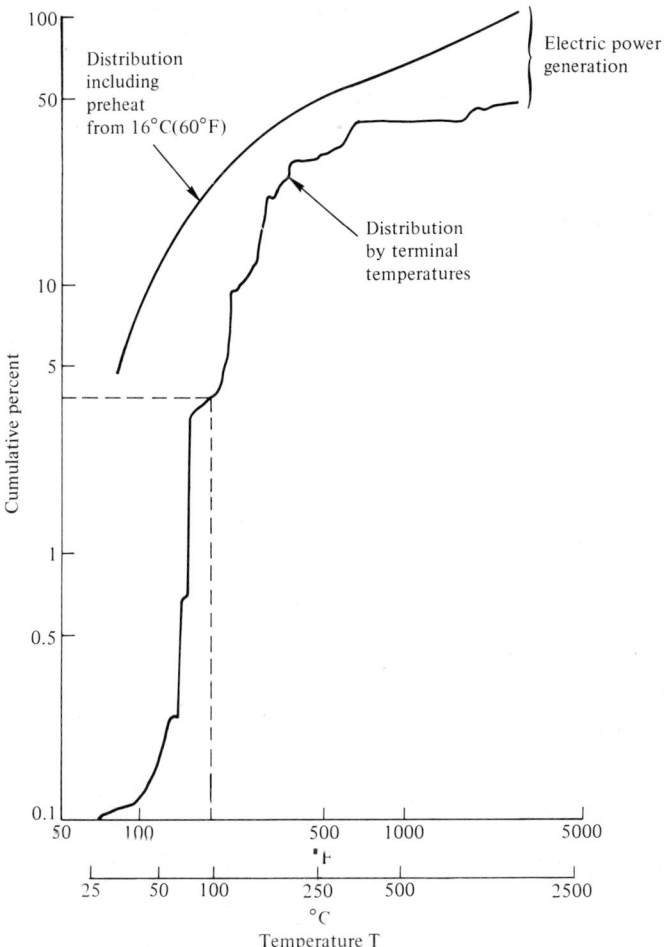

Figure 7-5 Percent industrial heat required at T or less for 67 U.S. industry groups accounting for approximately 48 percent of total annual industrial energy consumption.

Table 7-5 Summary of industrial process heat requirements, 10^9 kWh/year (10^{12} Btu/year) [2]

Industry/segment	Hot water <100°C (<212°F)	Steam 100–177°C (212–350°F)	Steam >177°C (>350°F)	Direct heat/hot air <100°C (<212°F)	Direct heat/hot air 100–177°C (212–350°F)	Direct heat/hot air >177°C (>350°F)	Totals
Aluminum/primary			11.3 (38.4)			18.5 (63.1)	29.8 (101.5)
Automobiles and trucks	3.8 (13.0)	0.42 (1.4)		6.24 (21.3)	2.93 (10.0)	0.3 (0.9)	13.7 (46.6)
Portland cement kilns						144 (490)	144 (490)
Concrete block and brick		4.2 (14.3)					4.2 (14.3)
Ceramics/glass	2.3 (8.0)	0.88 (3.0)			4.83 (16.5)	90.8 (310)	98.8 (338)
Ceramics/gypsum					3.28 (11.2)	9.35 (31.9)	12.7 (43.1)
Inorganic chemicals	7.0 (24.0)	28.1 (96.0)		0.26 (0.9)		41.6 (142)	77.0 (263)
Copper, primary and secondary	4.2 (14.2)			0.70 (2.4)		16.2 (55.4)	21.1 (72.0)
Lumber	1.0 (3.5)	3.84 (13.1)	0.73 (2.5)	20.0 (67.5)	0.76 (2.6)	13.1 (44.7)	39.4 (134)
Mining/Frasch sulfur	12.8 (43.7)						12.8 (43.7)
Pulp and paper†		40.0 (137)					40.0 (137)
Petroleum refining‡						760 (2600)	760 (2600)
Selected plastics		2.2 (7.6)	2.8 (9.5)				5.0 (17.1)
Rubber/SBR manufacture		1.6 (5.4)					1.6 (5.4)
Textiles	2.6 (9.0)	58.3 (199)	1.3 (4.3)		8.58 (29.3)	3.96 (13.5)	74.7 (255)
Totals	33.7 (115)	140 (477)	16.1 (54.7)	27.2 (92.1)	20.4 (69.6)	1098 (3752)	1335 (4560)

† After credit for by product steam generation.
‡ Includes fired heaters only.

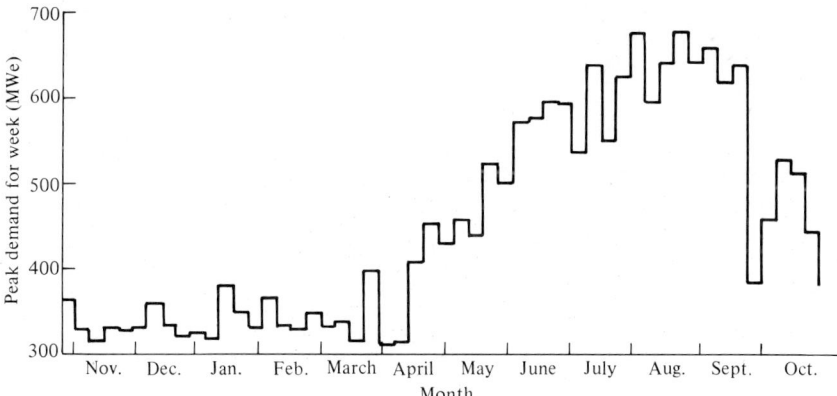

Figure 7-6 Peak electrical demand of week versus week of year, November 1974 through October 1975, for Austin, TX.

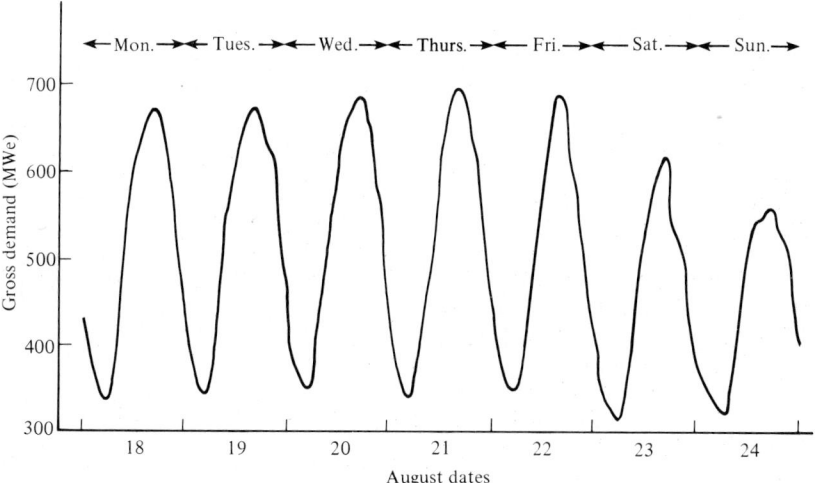

Figure 7-7 Gross instantaneous system electrical load versus time, August 18–24, 1975, for Austin, TX.

PROBLEMS

7-1 Estimate the use of hot water in your house (apartment, etc.) and compare it to the rule-of-thumb value.

7-2 Chart the use versus time of day for hot-water use in your dwelling and plot a demand curve analogous to that shown in Fig. 7-2.

7-3 Use the SOLSIM base case with your hot-water demand profile to size a collector and storage system to supply 80 percent of your hot water year round. (Use three representative periods, one each in winter, spring, and summer.)

7-4 Using the performance data of an evacuated tubular collector, size a collector and storage system to meet 60 percent of the summer cooling load depicted in Fig. 7-3. Assume the solar derived heat is used to drive an absorption cooler with a thermal COP of 0.5 and a 190°F activation temperature.

7-5 Appendix Table B-5 lists heating design data for selected cities in the United States and Canada. (The $97\frac{1}{2}$ percent winter design temperature is the temperature above which the actual temperature remains for $97\frac{1}{2}$ percent of the time, and can be used as the "design" temperature.) Find the design heating load, the average heat load, and the required energy for the heating season for a well-insulated 200-m^2 floor area house in January in (a) Flagstaff, AZ, (b) Minneapolis, MN, (c) Seattle, WA.

7-6 Estimate the net loss coefficient (based on floor area) for the building in which you live. Estimate the energy requirements to heat your building by month for a year. Try to obtain information on the actual energy used (electric bills, gas bills, etc.) in your building. Compare these with your calculations.

7-7 Size an array of unglazed, nonselective absorber collectors (see Fig. 4-10) so that the pool losses shown in Fig. 7-4 can be regained by a combination of insolation to the pool directly and to the collector array; the date is November, the location is San Diego, CA.

7-8 Suppose you are to design a solar system to supply energy to a process requiring energy at 150°C. Since this temperature is above the practical limit of a flat-plate collector, either an evacuated tubular collector or a concentrating collector is required. Assume that an evacuated tubular collector is available at \$25/ft^2 and that it has the performance illustrated in Fig. 4-10. A cheaper system might be possible if a flat-plate collector is used to preheat. Assume that flat-plate collector D in Fig. 4-10 is available at \$10/ft^2, that the ambient temperature is 25°C, and that water is available at 25°C. Using only the collector efficiency curves and cost, determine the most cost-effective combination of collectors to supply 1 MW of thermal energy at 150°C at an insolation of 900 W/m^2.

7-9 Modify the SOLSIM program so that the system described in Prob. 7-8 can be analyzed. Compare the long-term economics of the design you developed in Prob. 7-8 with one utilizing *only* evacuated tube collectors.

7-10 A three-bedroom frame house in Austin, TX, with heated basement, 1500-ft^2 floor area, R-11 wall insulation, and R-14 ceiling insulation is being purchased by a buyer from Schenectady. He finds that electricity in Austin costs 5.2¢/kWh. Electric resistance heating is used in the house. Estimate

(a) The peak load required from the furnace in order to heat the house under design conditions.

(b) The expected utility bill for January for heating.

7-11 Calculate the equilibrium temperature of a swimming pool in July in your city. Use local weather bureau data for average wind speed, air temperature, and relative humidity. Use weather bureau data for insolation if available; otherwise, use one of the methods from Chap. 3 for predicting local insolation.

REFERENCES

1. Rob. B. Farrington, L. M. Murphy, and Darryl L. Noreen, "A Comparison of Six Generic Solar Domestic Hot Water Systems," Solar Energy Research Institute Report No. ERI/RR-351-413, Golden, CO, April 1980.
2. Malcolm D. Fraser, "Survey of the Applications of Solar Energy to Industrial Process Heat," *Proceedings of the 1976 Annual Meeting of the Am. Sec. of ISES, Sharing the Sun*, vol. 5, Winnipeg, 1976, pp. 46–57.

CHAPTER
EIGHT

CONTROL OF SOLAR THERMAL SYSTEMS

Devices for control of solar-thermal energy systems should have three major attributes: They should be able to maximize overall system efficiency, should be reliable, and should be economical. System control may be achieved either manually or automatically. Our later discussions will include manual control but will emphasize automatic control, which is usually implied when we refer to system control. The control logic and level of complexity is application-specific, ranging from the entirely passive control in thermosiphon water heating, to the relatively simple control of active water-heating systems, to the progressively more sophisticated control logic of heating, cooling, and power systems. These will be addressed herein. However, the control logic is also function-specific, e.g., collector loop circulation control and freeze prevention, which may be common to several applications. The following discussion addresses the control system by function, followed by discussion of controls for specific applications.

8-1 COLLECTOR LOOP CIRCULATION

Active control of collector loop fluid circulation is common to essentially all solar applications involving solar collectors. One exception is a thermosiphon hot-water system, illustrated in Fig. 8-1. In this application fluid circulation occurs naturally with sufficient insolation, since the heated fluid in the collector and return leg (left) is less dense than fluid in the supply leg (right). However, under conditions of no or low insolation, the temperatures of the fluid in the collector and supply legs are similar and circulation automatically stops, leaving the warm and less dense fluid in the tank above the cooler and denser fluid in the collector and plumbing lines. The thermosiphon is thus well suited to solar applications in

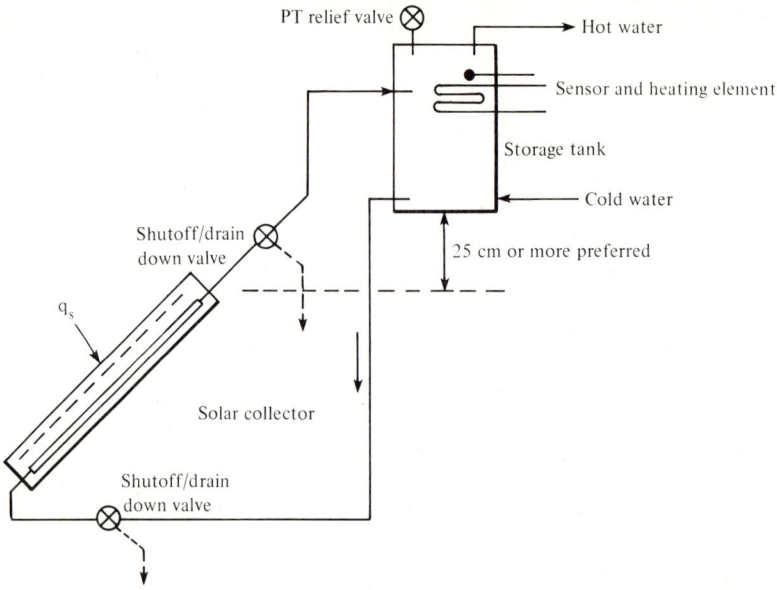

Figure 8-1 Basic thermosiphon configuration (single tank and direct heating).

that circulation is automatically controlled, and in fact, the circulation rate is modulated according to the insolation intensity.

Frequently, for aesthetic, architectural, security, or structural reasons, it is desirable to mount the collector above the storage tank. In such situations a circulation pump is required to circulate the warmer fluid downward from the collector to the storage and the cooler fluid from the bottom of the storage up to the collector. To discuss the collector loop fluid circulation control for these "active" systems, the domestic solar water heating system, Fig. 8-2, is used. A differential thermostat controller in such a system senses the difference between the temperature of the collector fluid near the outlet fitting and the temperature near the bottom of the storage tank. When, under exposure to insolation, the collector outlet temperature exceeds the storage temperature by a preset amount ΔT_{Hi}, the controller turns on the pump. The pump remains on provided this temperature difference remains above a preset lower value ΔT_{Lo}. Usually during startup in the morning there is some on-off cycling of the pump. This occurs because, after circulation is initiated, the fluid may cool the collector such that the difference drops below ΔT_{Lo}; however, as the insolation increases, the cycling stops. Late in the day, when the insolation decreases to the point that the temperature difference drops below ΔT_{Lo}, the pump is turned off. Again, some pump cycling may occur at this time for the same reason as given above in regard to startup.

A common design error in this sort of on-off control is setting ΔT_{Hi} and ΔT_{Lo} too close together. This will result in excessive cycling, which causes wear on both the controller and pump. Another error is to set both ΔT_{Hi} and ΔT_{Lo} too large, which delays turning on the pump and turns it off too soon, reducing the

CONTROL OF SOLAR THERMAL SYSTEMS **203**

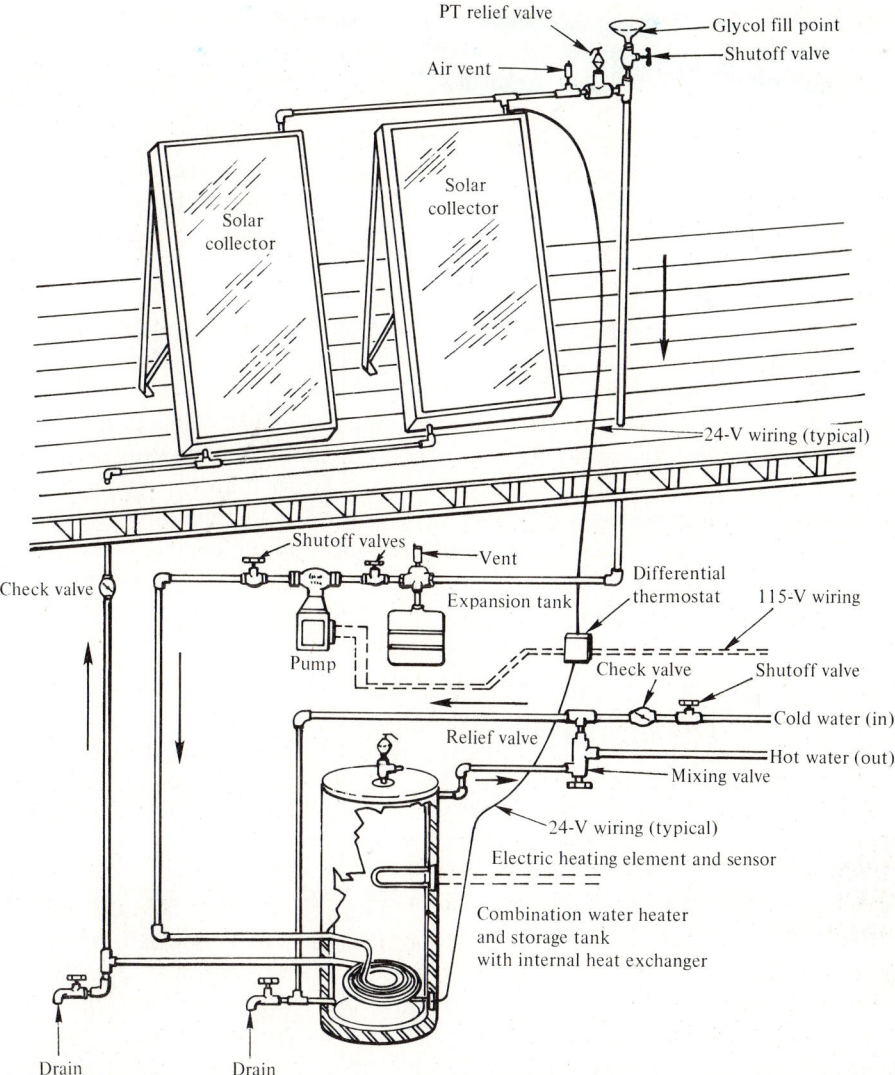

Figure 8-2 Domestic solar water-heating system illustrating combination water heater and storage tank with internal heat exchanger—all piping insulated. (*Courtesy Revere Copper and Brass, Inc.*)

potential for collecting solar energy. Placement of the collector temperature sensor on the collector outlet fitting or even in the upper manifold (rather than on the absorber panel itself) reduces the cycling problem; however, if the collector sensor is placed too far from the collector exit, the collector performance will deteriorate, because the collector pump will be delayed in coming on.

Rules of thumb for liquid systems are to set ΔT_{Hi} and ΔT_{Lo} such that $\Delta T_{Hi}/\Delta T_{Lo} \cong 5$ with ΔT_{Hi} between 5 and 8°C. The value selected for ΔT_{Hi} depends on the fluid circulation rates the collectors are designed for. With relatively large circulation rates per unit of collector aperture area [~ 1.5 L/(min · m²)], the lower value would be acceptable, while for smaller rates [~ 1.0 L/(min · m²)], the higher difference is advisable. For air systems in which the fluid temperature increase across the collector is allowed to be considerably higher to reduce flow rate and pumping power, ΔT_{Hi} would be 20 to 35°C referenced to the bottom storage temperature. It is important that the collector circulation loop be designed as a system and after installation that the circulation rate and temperature difference levels be adjusted to result in acceptable operation.

The on-off control system is most common and it is used in the SOLSIM program for collector loop control.

An alternative to simple on-off control is proportional control (i.e., varying pump speed to control mass flow through the collector to limit collector temperature rise and thus maintain high collector efficiency). This requires a proportional controller and a variable-speed pump or blower. The proportional control design still includes differential temperature limits for actuating the pump and turning it off, though they are set at lower levels than for simple on-off control. While an argument for proportional control is the enhanced collection efficiency, it has been found that a well designed on-off control system will achieve nearly the same efficiency. Also, the additional cost and larger parasitic power requirement for proportional control systems generally do not make them more cost-effective, except possibly in large-system applications.

8-2 FREEZE PREVENTION

A salient advantage of air systems is the absence of the freezing problem. However, for liquid (water) systems, damage due to freezing must be avoided and this may be a dominant design feature in moderate to severe freezing climates. To avoid damage from freezing of the fluid in the collector of liquid systems, several methods are possible:

- Antifreeze solutions   ⎫
- Expandable collector materials ⎬ Passive
- Drainback (automatic or manual) ⎭
- Draindown (automatic or manual) ⎫
- Storage fluid circulation ⎬ Active
- Collector heating elements ⎭

The first three methods are passive in that they involve no active control element for freeze protection. The last three of these methods may be classed as active, in that each requires a sensor on the collector absorber panel to detect a potential freezing condition and subsequently to activate valves, pump, or heater as appropriate. The location of the sensor and the temperature for activation are crucial. Water undergoes a density maximum at 4°C (39°F); thus, because of the complex natural circulation patterns that occur in the vicinity of the freezing point, the activation temperature is usually set around 3.5–4.5°C (38–40°F) to ensure that the freeze protection function is initiated before any location on the panel reaches 0°C (32°F). The sensor must definitely be mounted on the absorber panel in the collector enclosure, preferably near the lower middle of the panel. A freeze sensor located on the external piping will generally not be indicative of the panel temperature.

"Draindown" refers to draining the collector fluid out of the system, while "drainback" refers to draining the collector fluid back into the storage vessel. While either method can be used for unpressurized systems, drainback cannot be used in a pressurized application such as solar domestic water heating. Thus, for applications such as domestic water heating, which invariably involves pressurized storage, or many other liquid-heating and -cooling systems that use pressurized storage, drainback cannot be used. However, for solar swimming pool heating and heating (or cooling) systems which involve an open (atmospheric pressure) storage vessel, the drainback concept is satisfactory, and in fact,

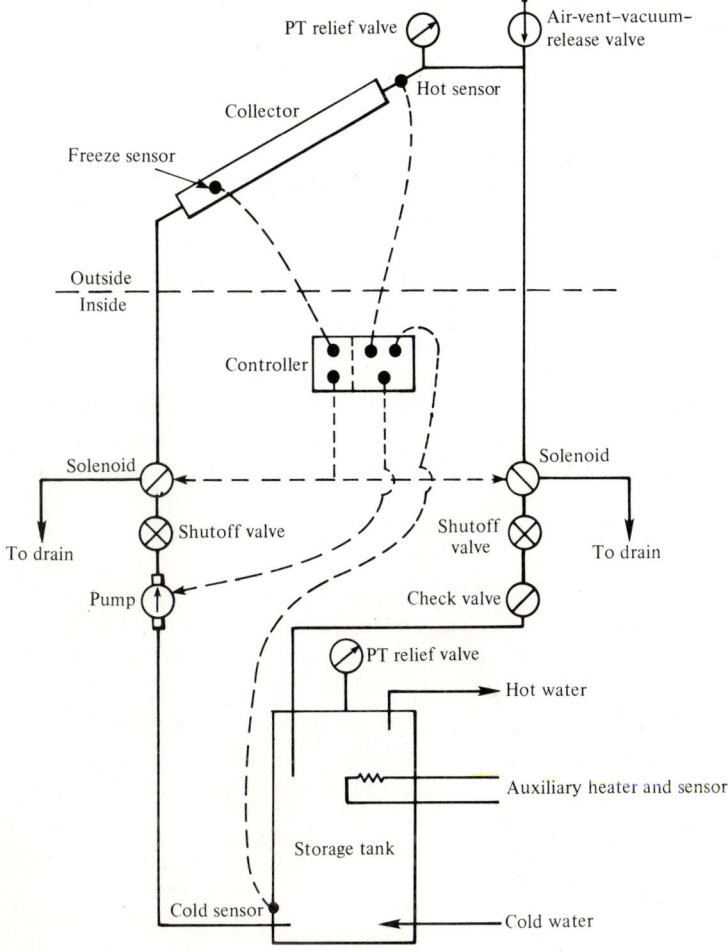

Figure 8-3 Draindown freeze prevention.

is generally preferred to draindown because the fluid (which may be conditioned) is not lost from the system.

Draindown systems (Fig. 8-3) use two three-way solenoid valves. When a potential freezing condition is sensed by the sensor on the absorber panel, both valves are activated to isolate the collector from the pressurized supply and open the collector supply and return lines to the drain. To ensure adequate drainage, the collector should be oriented so that all fluid passages drain easily, and a vacuum release valve should be located near the collector outlet. This vacuum release valve normally also serves the function of an air vent valve that allows air which comes out of the solution during normal operation (particularly for solar domestic water heating) to be vented from the collector. The solenoid valves should be located in the building enclosure to ensure that they and any undrained piping near them are not exposed to a freezing environment. Manual draindown control is possible, but if the owner or operator is not always present, automatic control is preferable.

Drainback systems (Fig. 8-4) use a somewhat simpler concept, in that only a one-function controller is needed to turn on the pump for fluid circulation when the hot sensor indicates useful energy can be collected. When the hot sensor registers lower than the cold sensor, the pump is off and with a vented tank below the collector the fluid in the collector automatically drains back into the storage vessel, provided a vacuum release valve is located near the collector exit to permit breaking the vacuum in the piping. In this concept the hot sensor needs to be located on the absorber panel or very near the collector exit to reduce the delay time for the pump to be activated. Also, the pump needs to be located below the fluid level in the tank and have sufficient head capability to lift the fluid to the collector exit at low flow. A higher head pump is needed with drainback than draindown because of the hydrostatic head.

In storage fluid circulation freeze prevention (Fig 8-5), the pump is activated when the sensor detects a near-freezing condition. Fluid from the bottom of storage is pumped to the collector until the collector is warmed to the point that the sensor deactivates the pump. The cycle is repeated whenever the collector cools toward freezing. Such a system works well where freezing conditions are relatively infrequent and mild.

Another active option is to use the freeze sensor to activate an electrical strip-heating element located near the bottom of the absorber plate. This option is only viable in very mild climates, but where freezing may occur on occasion. Since thermosiphon water heaters are used only in mild climates and have no other ready means for freeze prevention, this method is appropriate to their design. The element can be quite small in capacity, as the energy dissipated must only balance the top loss term $\bar{U} A_c (\bar{T}_e - T_a)$ from the collector absorber which is quite well insulated. For typical values of $\bar{U} = 4$ W/(m² · °C), $T_a = -10°$C, and $\bar{T}_e = +4°$C, a heat input of 52 W/m² is required, the power input of a small lightbulb per square meter of collector.

To avoid the freeze problem, an antifreeze solution can be used in the collector loop, with a heat exchanger between it and the storage fluid, as illustrated for the solar domestic hot-water system shown in Fig. 8-2. For domestic water

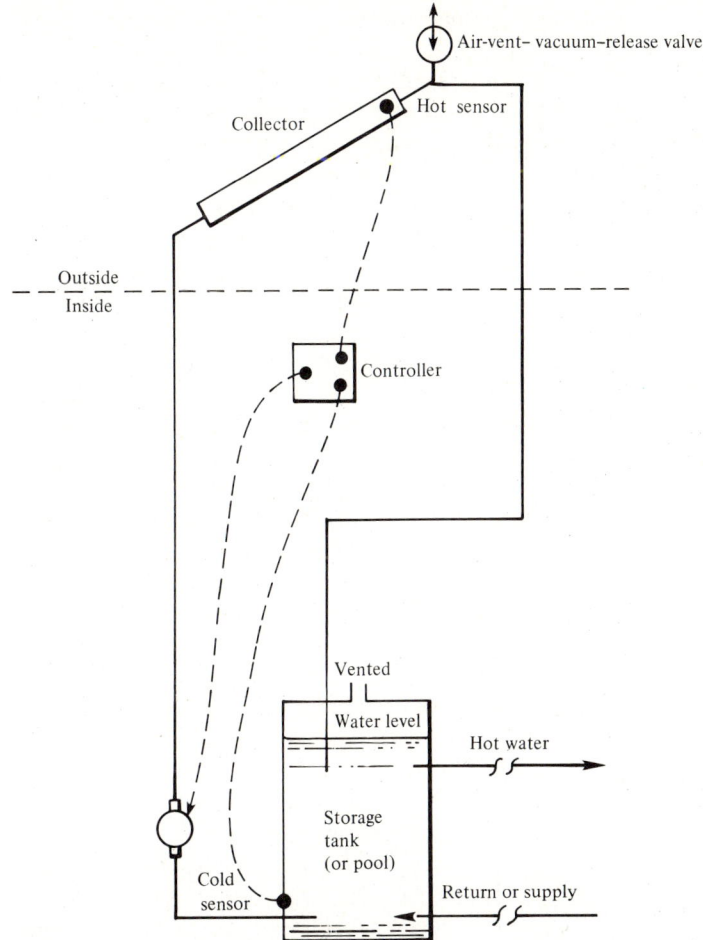

Figure 8-4 Drainback freeze prevention.

heating some building codes require a double-wall heat exchanger to ensure that there will be no contamination of the potable water. Even in heating systems a heat exchanger is required because it is expensive to use antifreeze in the entire storage-collector system. For example, for a 50/50 solution of ethylene glycol/water, a system with a 4000-L capacity uses 2000 L of glycol. At a cost of about $1/L, this represents a significant investment. Thus the alternative is to put antifreeze in the collector loop and use a heat exchanger to transfer heat to water storage. The heat exchanger itself is a moderate-cost item and its presence imposes a performance penalty of typically a 3–5°C increase in collection temperature to get the same temperature in storage. However, antifreeze protection is a good choice in more severe climates.

Finally, research has been done on designing collector absorber panels that are not susceptible to freeze damage. Some plastics have sufficient elastic strain

208 SOLAR-THERMAL ENERGY SYSTEMS

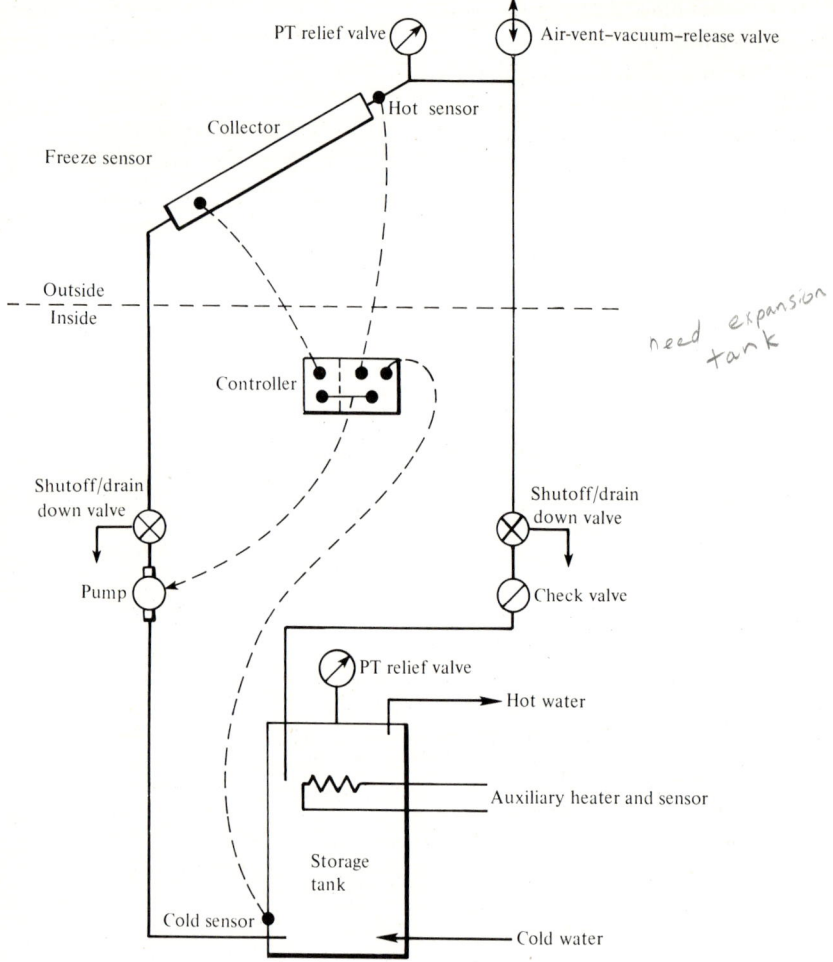

Figure 8-5 Fluid circulation freeze prevention.

to accommodate the expansion of water to ice. So far, these have found only modest use, largely because they are new and because plastics have a poor thermal conductivity, which may jeopardize collector performance.

8-3 OTHER COLLECTOR LOOP PASSIVE CONTROL COMPONENTS

As mentioned above, an air-vent–vacuum-release valve is normally required on liquid collectors to vent air that may come out of solution and to serve as a vacuum release (vent to atmosphere) when the collector is drained down or back, either automatically or manually. Second, a pressure temperature relief valve should be mounted near the collector exit if shutoff valves are located in the supply and return lines. This valve protects the collector against excessive pressure and temperature should the shutoff valves be unintentionally closed during a clear day, and also permits venting of steam and therefore limits the absorber panel temperature should power fail (no circulation) on a clear day. Finally, a check valve should be installed in the return line to prevent reverse thermosiphon action during cold nights, i.e., circulation of hot fluid up the return line to the collector where it cools and flows back to the storage via the supply line. Otherwise, a significant thermal loss occurs at night, resulting in a degradation of performance. This is not required in thermosiphon or drainback systems.

8-4 TRACKING SYSTEMS

Control of tracking systems used with concentrating collectors can be done in many ways. Tracking is commonly done by feedback control, computer control, or by "passive" control.

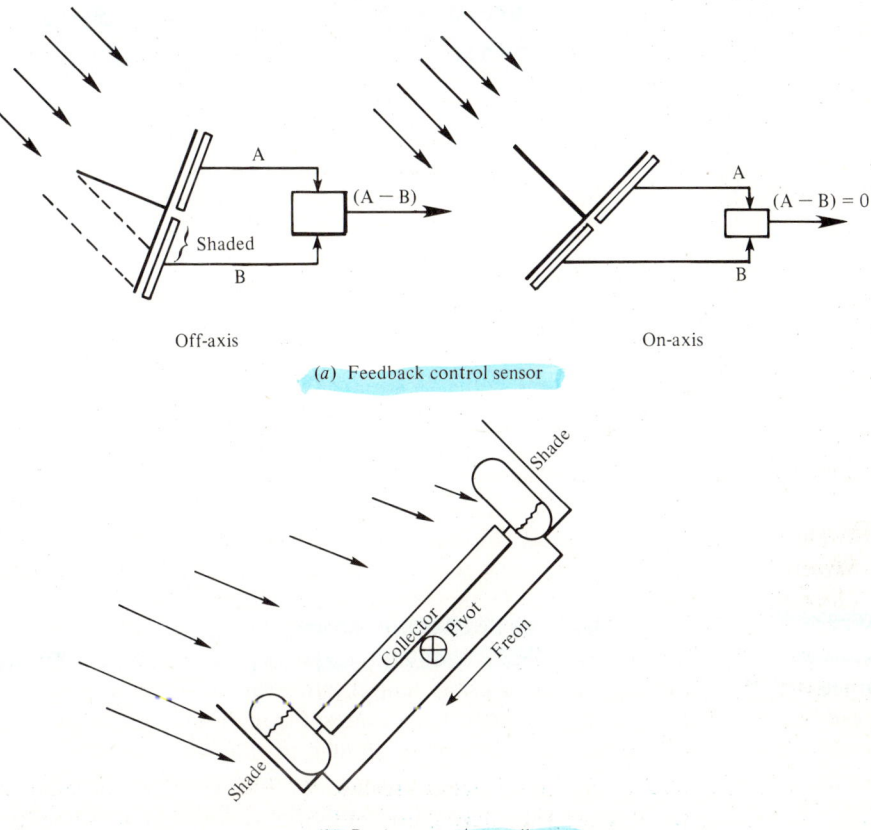

Figure 8-6 Control methods for tracking. In (b), Freon boils in unshaded bottle and condenses in shaded bottle. Unbalanced weight causes the collector to track.

Feedback control uses an element to sense whether focus is achieved, and if it is not, activates the tracking system in an appropriate direction until it is. For single-axis tracking, commonly two photocell or temperature (thermocouple or thermistor) detectors are separated by a shading device and located on a surface parallel to the concentrator aperture (Fig. 8-6a). When the concentrator aperture is not normal to the solar beam radiation, the detector senses a difference in the signal from the two detectors, and the controller rotates the collector until the difference is nulled. Two-axis tracking can use two sets of detectors. The actual tracking of the collectors is achieved through a drive operating on a mechanical linkage or pulley-cable system.

Computer control uses stored information about the sun's position to control the position of individual concentrators or banks of concentrators in large-scale systems such as those used in power production, where thousands or tens of thousands of collectors must be oriented continuously and where individual control systems for each collector would be expensive and present a difficult maintenance problem.

Passive control uses the sun's energy to run the tracking system directly. Various systems have used a low-boiling-point material such as Freon which moves from one chamber to another in response to the presence of insolation. The resulting changing weight distribution or pressure, depending on the design, reorients the collector (Fig. 8-6b). Other designs have proposed the use of bimetallic elements to open and shut slats or louvers over a surface in order to control the insolation available to the surface, or in some cases plastics with large coefficients of thermal expansion have been used to orient solar devices.

8-5 CONTROL OF SPECIFIC SYSTEMS

The controls associated with the collector loop have been described above, and are generally common to all solar energy systems. Other control aspects become application-specific, however, and several typical applications are addressed here, primarily in terms of the use of auxiliary energy.

Swimming Pool Heaters

When solar heaters are incorporated into swimming pool heating systems, the same pump is used for pool water circulation and for circulation through the collectors. The control consists of a differential temperature sensor (collector to pool) which activates a valve and diverts water through the collectors when useful energy can be collected in a manner similar to that used for solar water heating. In addition, to prevent overheating of the pool in summer, a temperature-limiting control function will override the differential control if the pool temperature is too high, thus bypassing the collectors. The drainback concept is used to prevent freezing.

Domestic water heating

Solar domestic water-heating systems are either the single- or dual-tank design (Fig. 8-7). In the single-tank concept (Figs. 8-1 to 8-5 and 8-7a), the auxiliary element and sensor are located approximately at a distance one-third of the tank

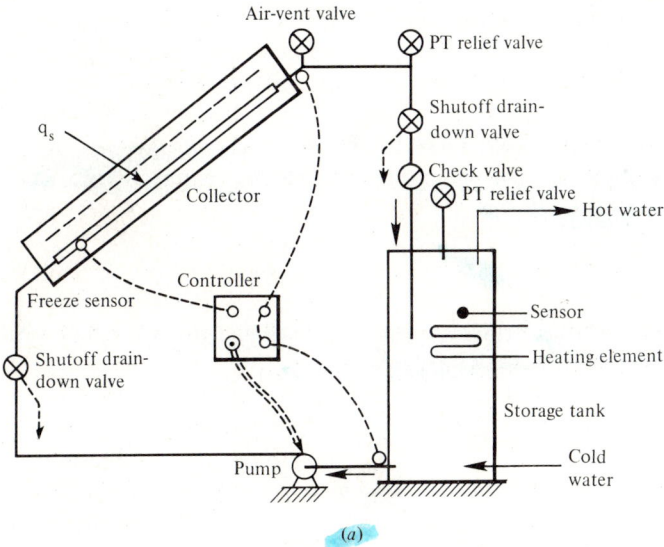

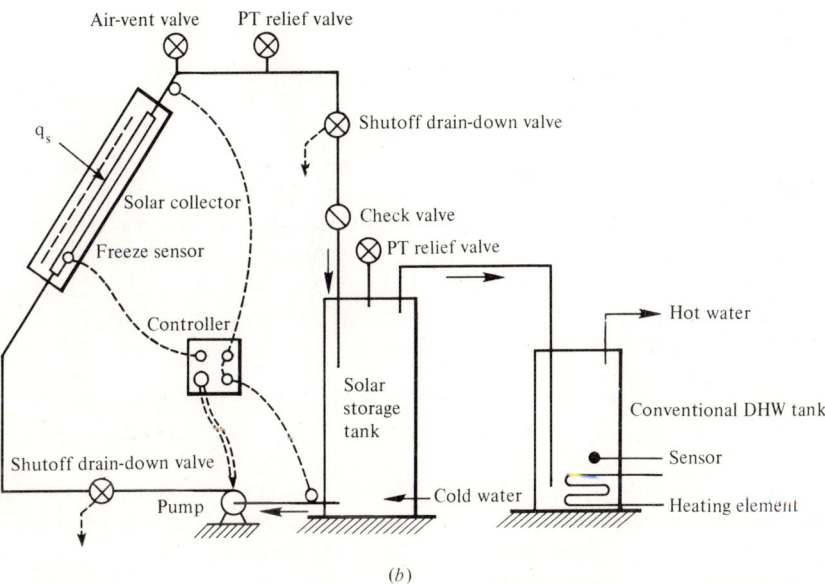

Figure 8-7 Basic pumped circulation solar water heater. (*a*) Single tank and direct heating; (*b*) two-tank configuration with direct heating.

height from the tank top. Thus, the auxiliary element maintains the upper third of the tank at the thermostat setting and the lower two-thirds is available for solar storage. In the two-tank concept, about two-thirds of the total storage volume is allocated to the solar tank and one-third to the conventional tank with the heater (gas or electric) at the bottom of it. When hot water is drawn from the conventional tank, solar preheated water from the top of the solar tank enters the bottom of the conventional tank and cold water enters the bottom of the solar tank. In essence, the two concepts operate in much the same way. In the single-tank concept an electric element is feasible, but gas is not. Auxiliary energy is demanded automatically when water at the auxiliary element level is below the thermostat setting. One feature of the solar domestic water-heating system is that even when the solar preheated water is below the thermostat setting the solar heating still provides a fraction

$$f = \frac{T_{st} - T_{sup}}{T_{th} - T_{sup}}$$

of the water heating, where T_{st}, T_{th}, and T_{sup} refer to the storage water, thermostat, and supply water temperatures, respectively.

Residential Solar Heating and Cooling Systems

A reasonably comprehensive plumbing system is shown in Fig. 8-8 for a solar water- and space-heating and -cooling system. In contrast to domestic water heating, in a space-heating or -cooling system there is normally some minimum temperature required for the system to function (meet the load). In the heating mode this is the inlet temperature to the water-to-air heat exchanger, typically 32°C (90°F) minimum. In the cooling mode typically this may be 77°C (170°F). Thus, if the storage temperature is above the minimum, water from the storage tank is pumped directly to the "heating coil" in the heating mode or to the "generator of the absorption chiller" in the cooling mode. If below the minimum, the hot storage tank is bypassed and the necessary heating is provided by the auxiliary boiler (heater). If the hot storage liquid is slightly below the minimum, there is the possibility of circulating it to the auxiliary boiler to provide some solar contribution, i.e., to use the boiler in series. However, this is generally not warranted, since using the boiler in parallel permits the hot storage to build up to the level where it meets the minimum. Thus, on-off (parallel) operation is usually preferred to the series configuration.

Power Production Systems

In solar electric or other high-temperature systems, the large concentration ratios used make possible extremely high stagnation temperatures. Thus, an additional safety problem is present, since if flow through the absorber element is lost the absorber can reach its melting point and be destroyed in a short time.

The absorber elements are thus monitored for temperature, and if a preset upper limit is exceeded the tracking system is deactivated to defocus the col-

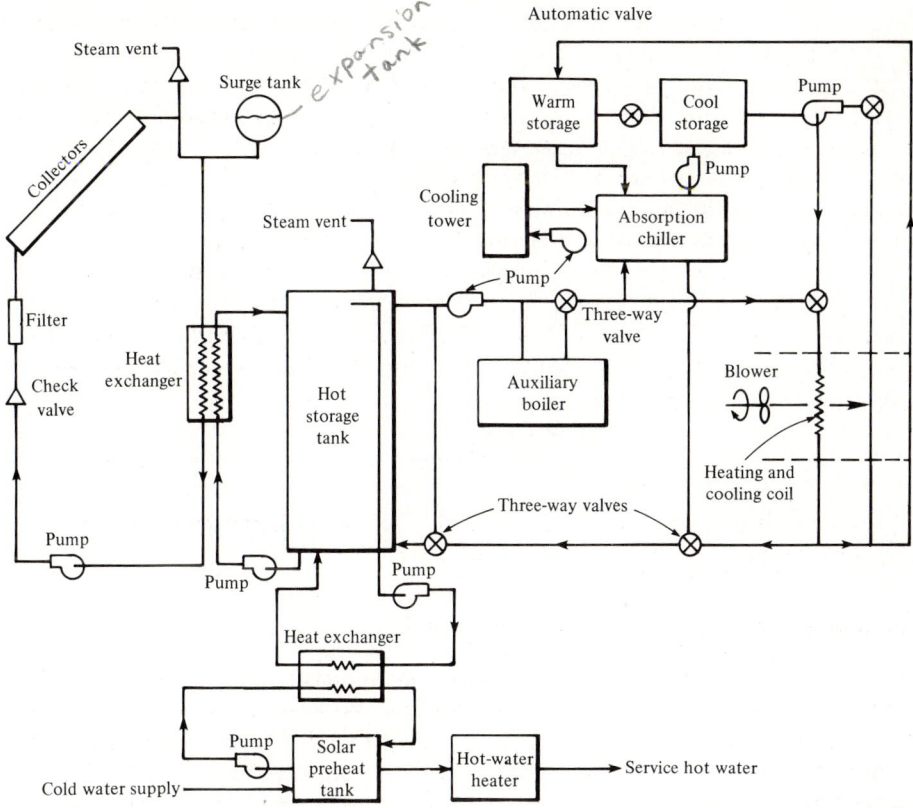

Figure 8-8 Solar heating and cooling system.

lectors. Additionally, automatic control is usually designed so that reflecting surfaces can be inverted or oriented to avoid damage in the event of high winds, hail, or sandstorms.

Other Systems

Industrial heat and other solar systems use much the same control philosophies as outlined above, but the variety of uses for the thermal energy produced makes detailed comments impossible here.

PROBLEMS†

Use the base case in SOLSIM for the following problems, but modify the control functions as described below. For each suggested modification run a 3-day simulation. Let the 3-day run be after startup from an equilibrium storage temperature to avoid startup transients. Compare your results with the original base case and discuss the significance of the design change.

† Problems 8-1 to 8-6 require modification of SOLSIM.

214 SOLAR-THERMAL ENERGY SYSTEMS

8-1 In the base case $(T_{stag} - T_{f,in}) \geq 6°C$ before $\delta_c = 1$. Try other values for this threshold temperature (e.g., 10°C, 25°C, etc.).

8-2 In the base case $(T_{f,out} - T_{f,in}) \geq 2°C$ before $\delta_c = 1$. Try other values (e.g., 1°C, 5°C, 7°C, etc.). Reduce the collector flow rate if necessary so that the collector can operate at larger values of $T_{f,out} - T_{f,in}$.

8-3 Reduce the maximum storage temperature $T_{st,max}$ and run the base case. With the reduced $T_{st,max}$, increase storage capacity and rerun the base case.

8-4 When the properties of the heat-transfer fluid change, the control setting may also have to be changed. In App. Table C-2 are listed the thermal properties of selected liquids. Discuss the changes in the control temperatures and other system parameters (flow rates, storage capacities, etc.) required if the following fluids were utilized in the indicated temperature ranges: (a) Pressurized water to 260°C, (b) A 50/50 solution of ethylene glycol in water up to 60°C, and (c) Dow Therm A in the range 300–350°C.

8-5 Suppose you were going to use a 50/50 solution of ethylene glycol in a domestic water-heating system (to prevent freezing). Assume that the entire storage volume is also the 50/50 solution (not likely due to cost). Modify the base case collector parameters in SOLSIM and the other fluid properties as necessary. Run a 3-day simulation (for the base case except as noted above). Compare this result with a 3-day simulation with water only. Change the control variables as you see fit to improve the performance of the water/glycol system and discuss the results.

8-6 Incorporate a heat exchanger and provisions for different collector and storage fluids into SOLSIM. Repeat Prob. 8-5 with 50/50 ethylene glycol and water in the collector loop and water in the storage. Assume an appropriate heat exchange effectiveness (~ 0.85).

8-7 An electric heater with a turn-on temperature of 3°C is to be used in a collector to prevent freezing. Use the SOLSIM base case to determine the daily useful solar collection with a daily average ambient temperature of 5°C and temperature swing of $\pm 15°C$, maximum insolation of 700 W/m², and day length of 10 h. Determine the relative solar energy collection to electric heater dissipation. (The electric energy use must be calculated by hand. Assume the collector is at ambient temperature unless the heater and/or pump are activated.)

The remaining problem does not require the use of SOLSIM.

8-8 A solar collector has efficiency of

$$\eta = 0.8 - 2 \frac{T_{f,in} - T_a}{q_s(t)}$$

and on a particular day, the insolation is given by

$$q_s(t) = 300 \sin\left[\left(\frac{t - 420}{600}\right)\pi\right] \quad 420 < t < 1020$$

where t is in minutes, $q_s(t)$ in Btu/(h · ft²), and the ambient temperature is 80°F.

The designer of an active solar hot-water system wants the system pump to turn on at 2 h past sunrise and not to cycle on and off after that time until near sunset. What temperature differences above the storage tank bottom temperature should be set for:

(a) Turning on the collector loop pump based on collector stagnation temperature?
(b) Turning off the collector loop pump based on fluid outlet temperature from the collector?

Assume that the storage has a constant temperature of 100°F, and the flow per unit of collector area is 0.03 gal/min of water.

CHAPTER
NINE

SIZING SOLAR ENERGY SYSTEMS

Several methods of sizing solar energy systems are developed in this chapter to varying degrees of sophistication. The most comprehensive methods involve computer simulation. From these analyses as well as from experience gained with the operation of solar installations, some "rules of thumb" for sizing have been proposed which are quite good guidelines, provided the user recognizes their limitations. Also, some "fraction from solar" or "f" methods are described developed from computer simulations qualified by actual data. Most methods are based purely on performance and predict the fraction of the load met by solar. Some include economic considerations. We will now proceed to discuss the various methods of sizing solar heating and cooling components and systems.

9-1 LIST OF SYMBOLS

A_c	collector aperture
c	specific heat
C_c, C_s, C_h, C_a	heat capacity rates, see Fig. 9-3
C_m	lesser of C_c and C_s
DD	number of degree-days per month
DTD	design temperature difference
D_1, D_2	parameters in the f-chart method, Figs. 9-4 and 9-5
f	monthly solar fraction
f_a	annual solar fraction
F_R	collector heat removal factor
F'_R	see Fig. 9-6
$F_{\tau\alpha}$	normalized $(\tau\alpha)_{\text{eff}}$
H	insolation on a horizontal surface
$\bar{H}$	monthly average daily insolation on a horizontal surface
$\bar{H}_0$	monthly average daily extraterrestrial insolation on a horizontal surface

$\bar{K}_T$	$\bar{H}/\bar{H}_0$
K_1, K_2, K_3, K_4	parameters in the f-chart method [Eqs. (9-3) to (9-5); Fig. 9-7]
L	load for a particular month
$(Mc)_s$	thermal capacity of storage
ND	number of days in a month
N_H	number of hours in a month
$\dot{Q}_{d,SH}$	average daily space heating load
$\dot{Q}_d$	design load
$\dot{Q}_{d,DHW}$	average daily hot water demand
$\dot{Q}_u$	collected insolation
q_s	insolation on a tilted surface
$\bar{q}_s$	monthly average daily insolation on a tilted surface
q'_s	insolation absorbed by collector $= q_s \times F_{\tau\alpha}$
q_u	collected insolation per unit of aperture area
R	$\bar{q}_s/\bar{H}$
r	fraction of the daily insolation available during a given hour
S_t	January insolation on a tilted surface
S	monthly insolation on a tilted surface $= \bar{q}_s \times$ ND
T	temperature
T_a	ambient temperature
T_0	monthly average daytime temperature
$T_{f,in}$	collector fluid inlet temperature
T_{load}	hot water demand temperature
T_{ref}	reference temperature (Eq. 9-5)
T_{sup}	hot water supply temperature
U	collector thermal loss coefficient
V	volume
α	solar absorptance
α_s	solar altitude
δ	declination
ϵ_{cs}	heat exchanger effectiveness
η	collector efficiency
τ	transmittance
$(\tau\alpha)_{eff}$	fraction of insolation absorbed by collector
θ	angle between surface (collector) normal and sun
ρ	density
Φ	latitude

9-2 RULES OF THUMB

Table 9-1 lists, for some common solar applications, the recommended relative sizes of collector and storage as well as collector tilt and fluid circulation rates.

In general, for heating systems the recommended collector slope (tilt toward

Table 9-1 Rules of thumb for sizing[†]

Design element	SI values	English system values
Solar air-heating systems		
Collector slope	Latitude + (15 to 20°)	Latitude + (15 to 20°)
Collector air flow rate	0.15–0.2 m^3/min per 1 m^2 of collector	1.5–2 ft^3/min per 1 ft^2 of collector
Pebble-bed storage size	0.15–0.3 m^3 of rock per 1 m^2 of collector	0.5–1 ft^3 of rock per 1 ft^2 of collector
Rock depth	1–2 m in air flow direction	3–6 ft in air flow direction
Pebble size	2–3-cm concrete aggregate	0.75–1-in concrete aggregate
Solar space heating system		
Collector slope	Latitude + (15 to 20°)	Latitude + (15 to 20°)
Collector flow rate	1.0–1.6 L/min per 1 m^2 of collector	0.025–0.04 gal/min per 1 ft^2 of collector
Water storage size	0.06–0.15 m^3 per 1 m^2 of collector	1.5–2.5 gal per 1 ft^2 of collector
Solar domestic water heating systems		
Collector slope	Latitude + 10°	Latitude + 10°
Collector size	3–4.5-m^2 double- or single-glazed selective for family of 4	35–50-ft^2 double- or single-glazed selective for family of 4
Collector flow rate	1.0–1.6 L/min per 1 m^2 of collector	0.025–0.04 gal/min per 1 ft^2 of collector
Water storage size	50–70 L/m^2 of collector	1.2–1.8 gal/ft^2 of collector
Solar swimming pool heating system		
Collector size/ Collector slope (tilt = latitude)	50–65% of pool area	50–65% of pool area
Collector slope (horizontal)	60–75% of pool area	60–75% of pool area

† Partially abstracted from Ref. 1.

the equator) is latitude plus 15–20°, while for domestic water heating a tilt of latitude plus 10° is near optimum. The reason for the difference is that domestic hot water use is year round while space heating is a seasonal (winter) demand, and thus the collectors are tilted at a higher angle for the heating application to maximize collection during the heating season. In space-heating applications, domestic hot water is usually also incorporated, but since the collector array in this case is much oversized relative to the domestic hot-water needs in the nonspace-heating seasons, orienting the collector at a higher tilt than optimum for domestic hot water only still results in more energy collection during the summer than needed.

In air space-heating systems, the storage-to-collector area should be approximately 0.15–0.3 m^3 of rock per 1 m^2 of collector, while, for water space-heating systems, the storage should be about 0.06–0.15 m^3 of water per 1 m^2 of collector. The reasons for these values are that if storage volume is too small the collectors will not be used effectively, while if storage volume is too large the expense cannot be justified and storage losses increase. The listed pebble size for rock storage permits adequate heat transfer while not resulting in too high a pressure drop.

Temperature increases across collectors are typically 30–40°C and 3–10°C for air and liquid collectors, respectively. The fluid flow rates are proportioned to collector area and the recommended flow rates (per unit of collector area) are listed in Table 9-1. For air systems, the temperature rise across the collector is allowed to be higher to reduce pressure drop by reducing flow rate; a second reason for this is that larger temperature gradients can be more readily sustained in rock storage than in water storage.

The required collector area for space heating will depend on the location, size, and construction of the dwelling. For domestic water heating, 3–4.5 m^2 of collector are reasonable for a family of four. For swimming pools, collectors are normally unglazed because of the low collector temperature and the considerably lower cost of unglazed collectors. To extend the swimming season by 2–3 months longer than usual, the collector areas listed in Table 9-1 under "solar swimming pool heating system" are recommended.

In general, in most solar installations the energy fraction to be met by solar should be 60–75 percent. Attempting to meet much more than this will result in poor utilization of the additional collector area, while meeting much less than this will result in excessive fixed costs that increase the cost per unit of energy delivered. This will be shown in later discussions.

9-3 EARLY OPTIMIZATION STUDIES

One of the earliest studies of the economics and optimization of residential solar systems was that of Löf and Tybout [2, 3], who about a decade ago examined solar heating, solar cooling, and combined heating and cooling for several cities in the United States. The study considered typical residential applications in

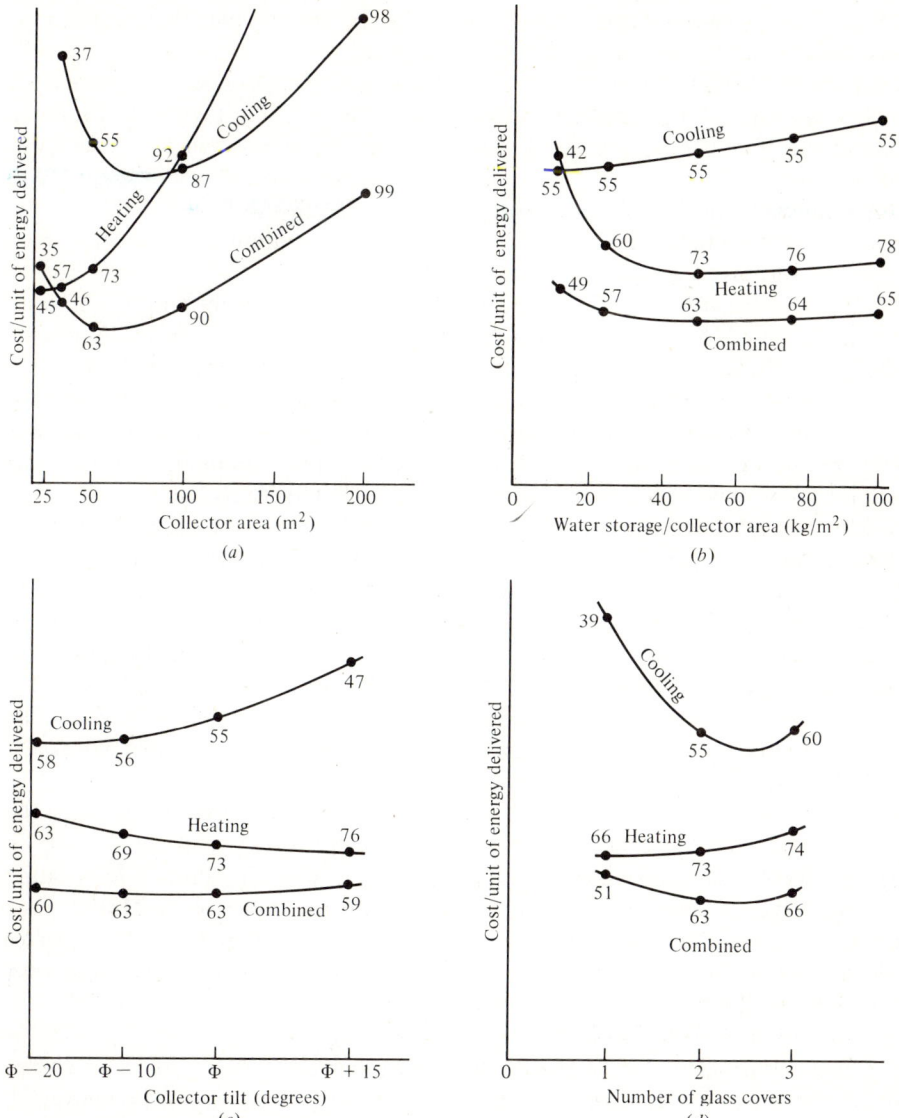

Figure 9-1 Optimization of collector area, storage size, glazings, and collector tilt for Albuquerque, NM. (*Abridged from Ref. 2.*)

these locations and used an optimization of the total cost of energy (solar plus conventional) per unit of energy used. The primary variables considered were collector area, storage mass per unit of collector area, collector tilt, and the number of glazings. While the costs assumed for backup fuel and most of the components in the system are unrealistic in view of today's prices, particularly for the collectors, several trends in the resulting data provide valuable insight into the performance as affected by component size.

Figure 9-1a–d taken from Ref. 2 for Albuquerque show the cost per unit of energy used as a function of the above four variables, each for the three applications: heating, cooling, and both. The following comments will be restricted to heating only or both, since cooling only is not a reasonable choice. The numbers on the curves indicate the percent of annual load met by solar. The scale for the cost per unit of energy delivered has intentionally been omitted from the figures since the objective here is to point out trends.

Figure 9-1a indicates a fairly strong minimum in cost (optimum) for each application, with the percent optimum from solar varying from 50 to 70 percent and the optimum collector area varying from 30 to 60 m^2 (350 to 700 ft^2). The optimum for the combined application shifts to larger collector areas and to a lower unit cost because of the higher use factor for the equipment. The unit energy cost increases with collector area beyond the optimum, because each additional increment in collector area is utilized less effectively, i.e., the condition of 100 percent from solar is approached asymptotically. As area is reduced below the minimum, the fixed costs result in an increase in the unit energy cost. Although a minimum cost is exhibited at a finite collector area, it should be noted that a lower cost possibly occurs for conventional energy only.

The effect of storage size on unit energy cost presented in Fig. 9-1b, in terms of storage mass per unit of collector area, is seen to exhibit a very weak optimum near 50 kg/m^2 (1.2 gal/ft^2), which results in a 60–75 percent solar fraction. The broad optimum occurs because, as storage is increased, additional energy is collected but both storage losses and cost increase. If storage is too small relative to collector area, the potential solar energy collection cannot be stored; thus the fraction from solar decreases and unit energy cost increases.

Figure 9-1c indicates an optimum tilt of about latitude plus 15° for heating and slightly less than latitude for combined heating and cooling. However, in either case the range of values near optimum is quite broad. Figure 9-1d indicates two glazings to be near optimum. (However, if a selective surface were to be used, a single glazing might be equivalent in some cases to double glazing with a nonselective absorber, which was used in the study.)

In general, except for very undersized systems, it is seen that the unit energy cost is reduced for the combined case compared to heating only. The reason for this is the better annual utilization of the collectors. It should be emphasized, however, that in these studies the assumed costs for providing cooling are the least realistic and current economics generally favor heating only over combined heating and cooling.

9-4 MANUAL DESIGN CALCULATIONS

Whereas conventional heating and cooling equipment is sized on the basis of near-worst-case (design) conditions, solar energy equipment is sized on the basis of meeting some fraction of the demand over a period of time, typically a month.

Load Calculations

The space-heating load calculations can be made on the basis of the average daily load by month using the effective UA (overall heat-transfer coefficient × area) value for the building and the DD (number of degree-days per month).† The effective UA value for the building is the sum of the products of the overall heat-transfer coefficients (U_i) and building areas (A_i) for the building's various exterior surfaces: slab, walls, windows, and roof. The number of degree-days is the total over the month of all the negative differences between 18.3°C and each average daily temperature. The details and data for these calculations can be obtained from Ref. 4, and are outlined in Chap. 7. Thus the average daily space-heating load, $Q_{d,\,SH}$, for a month of ND days would be, from Eq. (7-1),

$$\dot{Q}_{d,\,SH} = 24(DD)\frac{UA}{ND} \qquad (9\text{-}1)$$

An alternative method of determining the average daily load is from the design load (q_d) and design temperature difference (DTD). The design load (in kilowatts) is the heating load on the building, occurring when the outside design temperature prevails. The DTD is the difference between the inside and outside design temperatures. If the design load is known (either from calculations or possibly from the nameplate capacity on an installed furnace), then the average daily space-heating load by month is

$$\dot{Q}_{d,\,SH} = \frac{24 q_d}{DTD}\frac{DD}{ND} \qquad (9\text{-}2)$$

The space-cooling load can also be determined from the number of cooling degree-days, but less accurately because of the complication of the influence of internal loads, humidity, and direct solar gain.

The domestic hot-water (DHW) load can be calculated from the average daily hot-water demand and the temperature increase from the supply water temperature to the demand temperature:

$$\dot{Q}_{d,\,DHW} = 0.00116(NL)(T_{load} - T_{sup})$$

where NL is the daily demand in liters and T_{load} and T_{sup} are the demand and supply temperatures, respectively. $\dot{Q}_{d,\,DHW}$ is in units of kilowatthours per day.

---

† Note that in most U. S. sources used for load calculations, UA has units of Btu/(h · °F) and DD has units of °F-days/month; however, in this text the units to be used are W/°C and °C-days/month and the average daily heating load is Wh/day. See Chap. 7 for further discussion of degree-days.

Collection

An efficiency curve $\eta = f_n[(T_{f,\text{in}} - T_a)/q_s]$ should be obtained for the collector to be used. Several such curves for specific collectors are included on Fig. 4-10. The curve to be used in the following sample calculation is a hypothetical curve with an intercept of $F_R(\tau\alpha)_{\text{eff}} = 0.74$ and slope of $F_R \bar{U} = 4.07$ W/(m² · K).

Table 9-2 presents the sequence of calculations to be done to obtain the average daily solar energy collection for 1 month. These involve hourly calculations over the day, which are summed to obtain average daily collection. The particular example is for July, for San Antonio, TX, assuming a collection temperature of 96°C (205°F) (cooling application) and a collector tilt of 18° (11° less than latitude). For San Antonio, the average daily global horizontal solar radiation for July is 7.44 kWh/(m² · day) [2357 Btu/(ft² · day)].

The hourly ambient daytime temperature T_a is given in column 2 and the hourly insolation on a horizontal surface H in column 4. Hourly solar data are frequently not readily available but can be approximated knowing the day length and global insolation. Figure 3-16, taken from a paper by Liu and Jordan [5], indicates the proportioning factor to be used to obtain hourly data from global daily data. Local monthly average global daily insolation can be obtained from data in App. B or from any other reliable source. The hourly average insolation values are tabulated in column 4 of Table 9-2, a result of determining the factors r for each hour (column 3) from Fig. 3-16 and then the product of r and 7.44 kWh/(m² · day).

For the chosen collector tilt, compute hourly both the cosine of the angle of the sun to the Horizon (column 5) and the cosine of the angle of sun to the collector surface normal (column 6). The values can be obtained from equations defining the solar angles or from Fig. 3-8 (taken from Hottel and Woertz [6]) which gives the cosine of the incidence angles as functions of declination, latitude, tilt, and time. The insolation q_s can therefore be approximated by $H \cos\theta/\sin\alpha_s$ (column 8). [Note that this does not discriminate between the beam and diffuse components as described in Eq. (3-21); i.e., it is assumed that $R = R_b$.] Since the collector efficiency data given in Fig. 4-10 is based on an insolation which is incident normal to the collector, a further correction for increased loss due to the angle of incidence with the collector (column 9) may be warranted.† The transmittance of a typical double glazing as a function of angle and the absorptance of a flat black paint as a function of angle are as indicated in Fig. 9-2a. If the $(\tau\alpha)_{\text{eff}}$ product is normalized to 1.0 for normal incidence, one obtains the correction factor $F_{\tau\alpha}$ shown in Fig. 9-2b. These hourly correction factors are listed in column 9. The corrected solar flux per unit area of collector (corrections for angle as well as for absorptance-transmittance loss) is therefore $q'_s = H \cos\theta/\sin\alpha_s \cdot F_{\tau\alpha}$ and the hourly values are included in column 10.

---

† Note that since this correction differs significantly from 1 only for angles 60° or more from the normal, this correction may not be considered necessary since only a small fraction of the total available energy is beyond 60° from normal incidence.

Table 9-2 Sample collector calculations for July† for San Antonio, TX

(1) Solar time	(2) T_a, °C (°F)	(3) r	(4) H, kW/m² [Btu/(h·ft²)]	(5) $\sin \alpha_s$	(6) $\cos \theta$	(7) $\cos \theta / \sin \alpha_s$	(8) q_s, kW/m² [Btu/(h·ft²)]	(9) $F_{\tau\alpha}$	(10) q'_s, kW/m² [Btu/(h·ft²)]	(11) $\Delta T/q_s$ (m²·°C)/kW [(ft²·h·°F)/Btu]	(12) η	(13) q_u kW/m² [Btu/(h·ft²)]
5-6	24 (75)	0.006	0.04 (13)	0	0	0	0	0	0	0	0	0
6-7	24 (75)	0.024	0.18 (57)	~0.28	~0.15	~0.54	0.097 (31)	0.1	0.01 (3)	0	0	0
7-8	25 (77)	0.05	0.37 (117)	0.49	0.44	0.90	0.333 (106)	0.6	0.200 (63)	213 (1.21)	0	0
8-9	27 (81)	0.08	0.59 (187)	0.67	0.64	0.96	0.566 (179)	0.85	0.481 (152)	122 (0.69)	0.16	0.091 (28.8)
9-10	29 (84)	0.10	0.74 (235)	0.82	0.80	0.98	0.725 (230)	0.96	0.696 (221)	92.4 (0.53)	0.35	0.254 (80.5)
10-11	31 (88)	0.115	0.86 (273)	0.94	0.92	0.98	0.843 (267)	1.0	0.843 (267)	77.1 (0.438)	0.43	0.362 (115)
11-12	33 (91)	0.126	0.94 (298)	0.99	0.98	0.99	0.930 (295)	1.0	0.930 (295)	67.7 (0.384)	0.46	0.438 (139)
12-1	34 (93)	0.126	0.94 (298)	0.99	0.98	0.99	0.930 (295)	1.0	0.930 (295)	66.7 (0.379)	0.47	0.437 (139)
1-2	34 (93)	0.115	0.86 (273)	0.94	0.92	0.98	0.843 (267)	1.0	0.843 (267)	73.5 (0.417)	0.44	0.371 (118)
2-3	35 (95)	0.10	0.74 (235)	0.82	0.80	0.98	0.725 (230)	0.96	0.696 (221)	84.1 (0.48)	0.38	0.276 (87.5)
3-4	36 (97)	0.08	0.59 (187)	0.67	0.64	0.96	0.566 (179)	0.85	0.481 (152)	106 (0.60)	0.13	0.074 (23.5)
4-5	35 (95)	0.05	0.37 (117)	0.49	0.44	0.90	0.333 (106)	0.60	0.200 (63)	183 (1.04)	0	0
5-6	34 (93)	0.024	0.18 (57)	~0.28	~0.15	~0.54	0.097 (31)	0.1	0.01 (3)	0	0	0
6-7	33 (91)	0.006	0.04 (13)	0	0	0	0	0	0	0	0	0

Average daily July insolation is 7.44 kWh/(m²·day) [2357 Btu/(ft²·day)]; 6.988 kW/m² day (2215 Btu/ft² day)

∴ Average Daily Efficiency = $\frac{2.303}{6.988}$ = 33%

2.303 kWh/(m²·day) (731 Btu/ft²·day)

† Average daily July insolation is 7.44 kWh/(m²·day) [2357 Btu/(ft²·day)]; $\delta = +22$; latitude = 30°N; tilt = 18°; $T_{f,\text{in}}$ = 96°C; day length = 13.6 h.

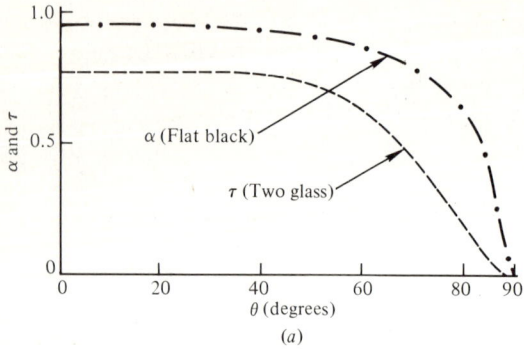

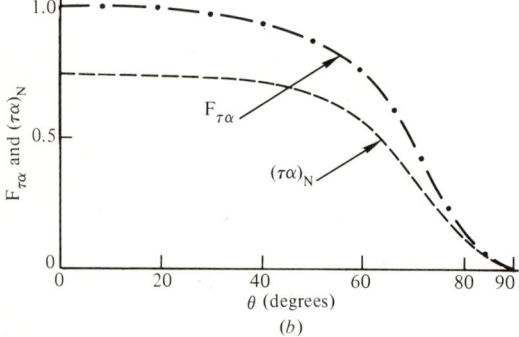

Figure 9-2 (*a*) Typical variations of transmittance (double glass) and absorptance (flat back) with angle of incidence. (*b*) Transmittance-absorptance product and normalized correction factor $F_{\tau\alpha}$.

The average temperature over which the energy is to be collected needs to be considered. For the winter (heating season) 35–60°C is reasonable. For the summer (cooling season) select the necessary temperature at which energy is required to power the cooling unit. In the present example, a value of 96°C can be chosen for a water-lithium bromide absorption cooling unit. The collector efficiency values (column 12) are determined from Fig. 4-10, where the hourly solar flux q_s and ambient temperature T_a were used to compute $(T_{f,\,in} - T_a)/q_s$ with $T_{f,\,in}$ assumed to be 96°C. The useful energy collected (column 13) is thus the product of effective solar flux and the efficiency: $q_u = q'_s \eta$. The overall daily collector efficiency is the ratio of the total daily useful energy collected (total of column 13) and the available insolation (total of column 8). Whereas the collector efficiency near noon is approximately 47 percent, the overall daily collection efficiency is approximately 33 percent.

There are several things to be noted from these calculations, the most essential one being that most (over 92 percent) of the useful collection occurs during the middle 6 h of the day even though the day length is 13.6 h. In fact, for an operating solar facility it is likely that the control system would allow little or no collection outside of this 6-h period because of the larger relative parasitic power requirements. The reasons for the narrow collection band are greater atmospheric attenuation at shallow angles, the cosine effect, the poor optical efficiency of the collector at shallow angles, and the decrease in collector efficiency for low insolation.

Table 9-3 Summary of annual collection, load, and deficiency for an apartment in Austin, TX

(1) Month	(2) $\bar{H}$, kWh/(m² · day) [Btu/(ft² · day)]	(3) q_u kWh/(m² · day) [Btu/(ft² · day)]	(4) $\dot{Q}_L$,† kWh/day (10⁶ Btu/day)	(5) Available,‡ kWh/day (10⁶ Btu/day)	(6) Deficiency,§ kWh/day¶ (10⁶ Btu/day)¶
January	3.36 (1065)	1.59 (504)	267 (0.91)	444 (1.51)	
February	4.05 (1284)	1.99 (631)	255 (0.87)	555 (1.89)	
March	4.74 (1501)	2.37 (751)	223 (0.76)	661 (2.25)	
April	5.37 (1703)	2.68 (850)	176 (0.60)	748 (2.55)	
May	6.22 (1972)	1.78 (565)	721 (2.46)	497 (1.70)	224 (0.76)
June	7.01 (2223)	2.06 (652)	970 (3.31)	575 (1.96)	395 (1.35)
July	7.44 (2357)	2.30 (730)	1123 (3.83)	642 (2.19)	481 (1.64)
August	6.72 (2130)	2.03 (642)	868 (2.96)	566 (1.93)	302 (1.03)
September	5.74 (1821)	1.67 (530)	542 (1.85)	466 (1.59)	76 (0.26)
October	4.62 (1465)	1.26 (400)	170 (0.58)	351 (1.20)	
November	3.62 (1147)	1.77 (560)	88 (0.30)	494 (1.68)	
December	3.15 (998)	1.48 (470)	255 (0.87)	413 (1.41)	

† Load determined for an apartment building in Austin, TX with an absorption cooling unit COP of 0.6 for May to October.

‡ For 279 m² (3000 ft²) of collectors.

§ A noncoincidence factor (greater than one) may be justified to account for a noncoincidence of collection and demand. This would more likely be the case if there were deficiencies in the winter months when cold days are frequently associated with overcast or low insolation. In the above examples when the deficiencies occur only in the cooling period a noncoincidence factor is not used.

¶ Totals for year for column 6: 44,340 kWh/year (151.2 × 10⁶ Btu/year).

If similar calculations are done for an average day for each month of the year, the results would be as shown in Table 9-3, where for the months of May through October the load is cooling and energy is collected at 96°C, while in the heating season of November through April energy is collected at an assumed average temperature of 60°C. Column 4 indicates the monthly average daily load determined for a two story, 14 unit apartment building in Austin, TX (including space heating and cooling and water heating). The climate in Austin is comparable to that in San Antonio, so San Antonio climate and insolation data were used for the Austin design. The cooling load for the Austin apartment building accounts for an absorption cooling unit COP of 0.6 for the months of May through October. Column 5 indicates the monthly average daily energy collected for 279 m² (3000 ft²) of collector. Column 6 indicates the deficiency of collection relative to load by month. The total of column 6 times 30 days/month is the annual deficiency.

For this system, about 74 percent of the building energy requirements would be supplied by solar.

Storage

For the present application it is assumed that the cooling unit draws on a hot-water storage vessel to produce cooling as needed and only a small chilled storage vessel is required to buffer the cooling unit from the load. The hot storage vessel is used for thermal storage during the heating season.

It is considered practical to provide thermal storage for only 1 or possibly 2 days of collection during the heating season. Thus, using the data given in Table 9-3, the most severe heating loads occur for January where the collection capability is 444 kWh/day (1.5 × 10⁶ Btu/day). Assuming a storage temperature swing between 60 and 38°C (140 and 100°F), the approximate volume V for 1 day of storage is

$$V = \frac{\dot{Q}_u}{\rho c \, \Delta T} = \frac{(444 \text{ kWh})(3600 \text{ s/h})}{(10^3 \text{ kg/m}^3)[4.186 \text{ kJ/(kg} \cdot \text{°C)}](22°\text{C})} = 17.4 \text{ m}^3$$

$$= 614 \text{ ft}^3 = 4600 \text{ gal}$$

In the cooling season (July) the collection is 642 kWh/day and if it is assumed that a swing of 20°C is available in hot storage† then this volume would store

$$\frac{(17.2 \text{ m}^3 \times 10^3 \text{ kg/m}^3)[4.186 \text{ kJ/(kg} \cdot \text{°C)}](20°\text{C})}{3600 \text{ s/h}} = 400 \text{ kWh}$$

or approximately 62 percent of the available daily collection. More than likely this will be satisfactory because there is a significant deficiency of collection in the

---

† It is assumed here that if energy is collected at 96°C and a 20°C drop occurs in the cooling unit regenerator, the useful temperature swing for storage is limited to 20°C because the cooling unit does not perform efficiently below about 75°C inlet temperature.

cooling months (Table 9-3, column 6) and because a substantial portion of the cooling load occurs during the period of collection when the collected energy can be used immediately.

Cooling Unit Capacity

The capacity of the cooling unit should be approximately equal to the hourly peak cooling load during the cooling season.

General Comments

It is to be emphasized that the optimum system will vary considerably from one application to the other, and as fuel costs increase the optimum would shift to a larger capacity being carried by solar. Furthermore, the above calculations and discussion are guidelines to sizing the system components and do not represent an economically optimum system. To obtain an optimum design, several collector areas would have to be considered and each system amortized over its life, including fuel and maintenance costs. Finally, this method of calculation does not account for the dynamics of the system and the intermittent variations of load and insolation. This requires a more detailed dynamic computer modeling of the system or some other method that accounts for these factors.

9-5 COMPUTER MODELING

A number of computer modeling programs have been developed to simulate the performance of solar energy systems. Of these, most predict the performance of a system, given the system configuration (components and the time-dependent load and solar energy available). Generally they do not optimize the system either on a performance or on an economic basis. Probably the best known of these and also the most sophisticated is TRNSYS, developed by the University of Wisconsin. The SOLCOST program, developed by International Business Services Solar Group, is less sophisticated in predicting performance, but does analyze the system economics and compute the payback for the system. These programs can be run only on a large computer and are thus not readily available to most individual users.

The SOLSIM program developed in the text does provide a dynamic simulation, though it does not provide the sophistication of a program such as TRNSYS.

9-6 f-CHART METHOD

Realizing the need for a sizing procedure which is relatively accurate but which can be done without the aid of a computer, personnel at the University of Wisconsin have developed a "monthly solar fraction calculational procedure,"

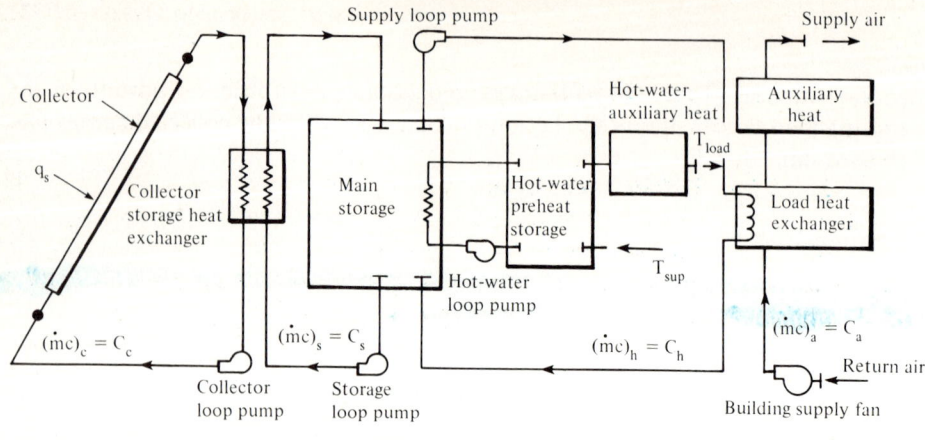

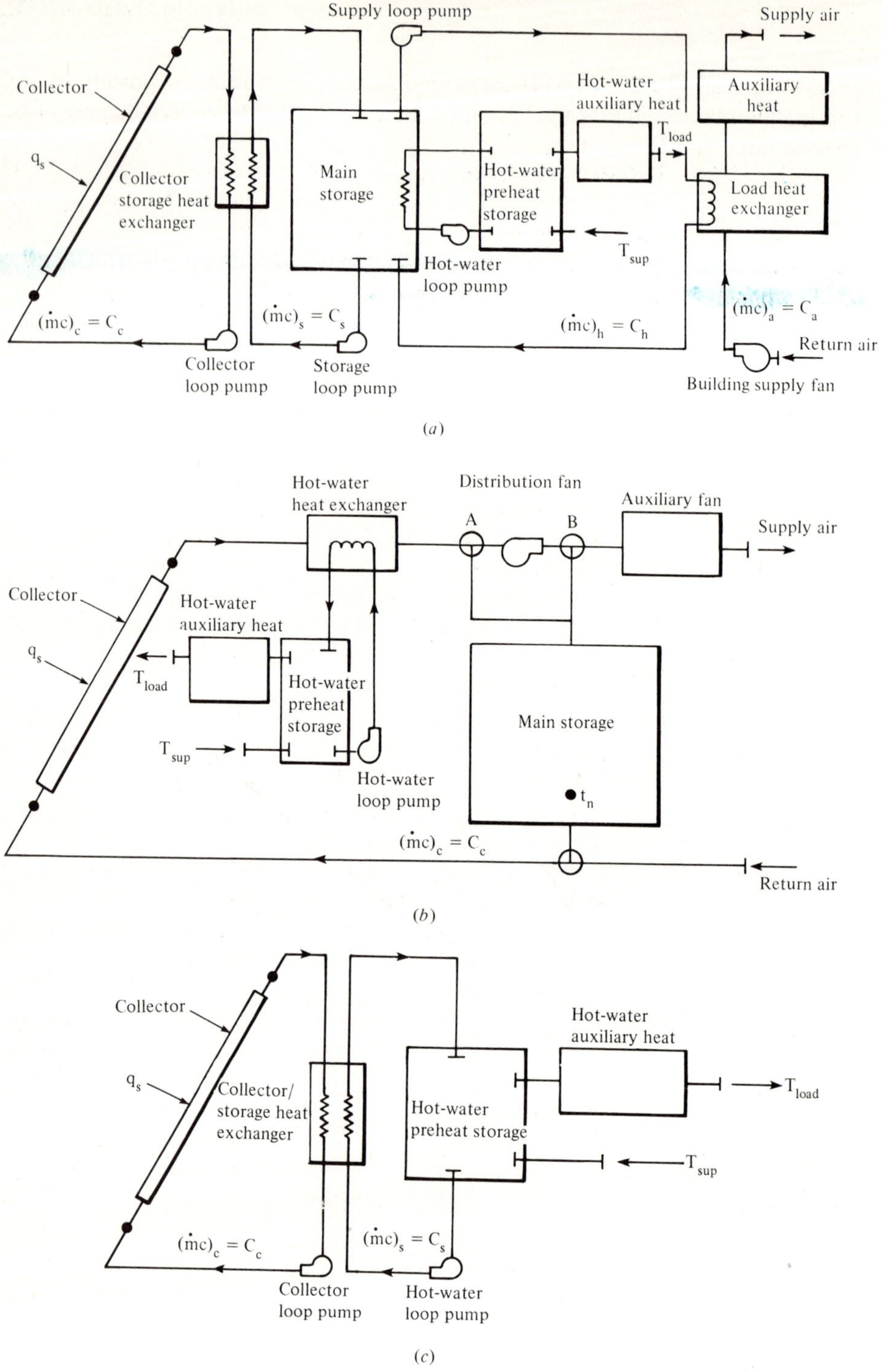

Figure 9-3 Schematics for (a) a liquid transfer system for space heating and domestic hot water, (b) an air transfer system for space heating and domestic hot water, and (c) a liquid transfer system for domestic hot water only.

hereafter called *f-chart*. The *f*-chart procedure is detailed in both Refs. 7 and 8, to which the reader is referred. The *f*-chart calculational procedure is summarized below.

Applications include water or air transfer systems meeting the demands for domestic hot water only or space and water heating combined. Figure 9-3a, b, and c shows schematics of the combined heating applications (water and air) and domestic hot water only. From extensive computer runs using TRNSYS, it was found that the monthly solar fraction, f, for solar space and domestic water-heating applications could be correlated by two parameters D_1 and D_2. Figures 9-4 and 9-5 present the graphical correlations of $f = f_n(D_1, D_2)$ for water and air transfer systems, respectively. The parameters are individually defined and interpreted as

$$D_1 = \frac{A_C F_R(\tau\alpha)_{\text{eff}} S}{L} \left(\frac{F'_R}{F_R}\right) K_4 \equiv \left(\frac{\text{Energy absorbed by collector}}{\text{Total heating load}}\right) \quad (9\text{-}3)$$

$$D_2 = \frac{A_C F_R \bar{U}(T_{\text{ref}} - T_0) N_H}{L} \left(\frac{F'_R}{F_R}\right)(K_1 K_2 K_3)$$

$$\equiv \left(\frac{\text{Reference collector thermal loss}}{\text{Total heating load}}\right) \quad (9\text{-}4)$$

where
- A_c = collector aperture area in m² (ft²). This must be consistent with the collector manufacturer's performance data and in most instances equals A_a used in the collector thermal performance test, ASHRAE Standard 93-77.
- S = global insolation on collector aperture for an average month in kWh/(m² · month) [Btu/ft² · month)].
- L = total heating and hot water load for the particular month in kWh/month (Btu/month).
- $F_R(\tau\alpha)_{\text{eff}}$, $F_R\bar{U}$ = collector performance characteristics that are obtained from the experimentally determined efficiency plot for the collector. The $F_R(\tau\alpha)_{\text{eff}}$ product is dimensionless and the $F_R\bar{U}$ product has units of kW/(m² · °C) [Btu/(h · °F · ft²)].
- T_{ref} = 100°C (212°F), reference temperature (arbitrarily chosen).
- T_0 = monthly average daytime temperature in °C (°F). Appendix Table B-2 lists this value for many locations.
- N_H = total number of hours for the particular month.
- K_1 = air collector flow capacitance rate factor, dimensionless (K_1 = 1 for liquid collectors).
- K_2 = storage mass capacitance factor, for liquid and air heating and hot-water systems, dimensionless.
- K_3 = hot-water factor for liquid hot-water only systems, dimensionless.
- K_4 = load heat exchanger factor for liquid heating systems, dimensionless. (K_4 = 1 for air systems.)

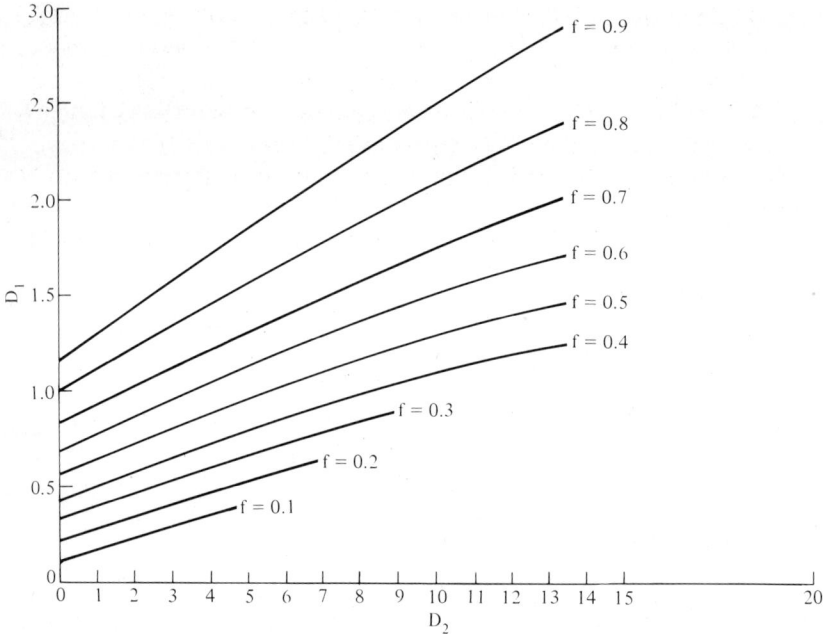

Figure 9-4 An *f*-chart for solar liquid-heating systems. (*Replotted from Ref. 8.*)

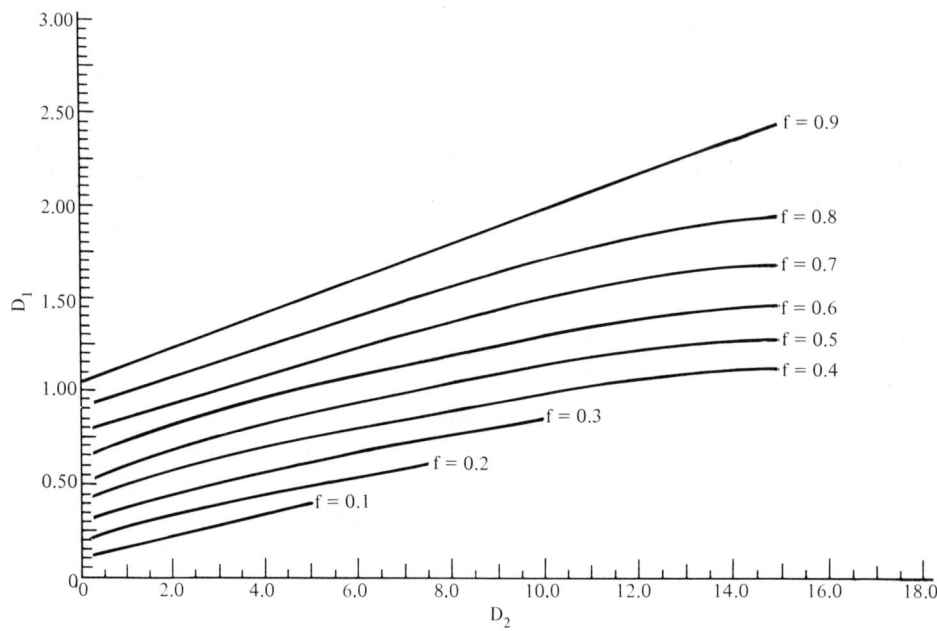

Figure 9-5 An *f*-chart for solar air-heating systems. (*Replotted from Ref. 8.*)

The ratio F'_R/F_R can be obtained as a function of $A_c(F_R \bar{U})/C_c$ and $C_c/\epsilon_{cs} C_{min}$ from Fig. 9-6. The quantity C_c is the collector loop flow thermal capacitance, $C_c = \dot{m}_c c$ in kW/°C [Btu/(h · °F)] and C_{min} is the minimum of the collector loop or storage loop flow thermal capacitance through the collector loop heat exchanger. ϵ_{cs} is the collector loop heat exchanger effectiveness.

The hot-water factor K_3 is influenced by the cold-water supply and hot-water supply temperature as depicted in Fig. 9-7.

The factors K_1, K_2, and K_4 can be computed from the following equations:

Collector capacitance factor K_1

$$\left. \begin{array}{ll} \text{Air:} & K_1 = 0.51 \, (C_c/A_c)^{0.27} \\ \text{Water:} & K_1 = 1.0 \end{array} \right\} \quad (9\text{-}5)$$

where C_c and A_c were defined previously [C_c/A_c in kW/(m² · °C)].

Storage capacitance factor K_2

$$\left. \begin{array}{ll} \text{Air:} & K_2 = 5.51[(Mc)_{st}/A_c]^{-0.3} \\ \text{Water:} & K_2 = 4.19[(Mc)_{st}/A_c]^{-0.25} \end{array} \right\} \quad (9\text{-}6)$$

where $(Mc)_{st}$ is the product of storage mass and specific heat in kJ/°C.

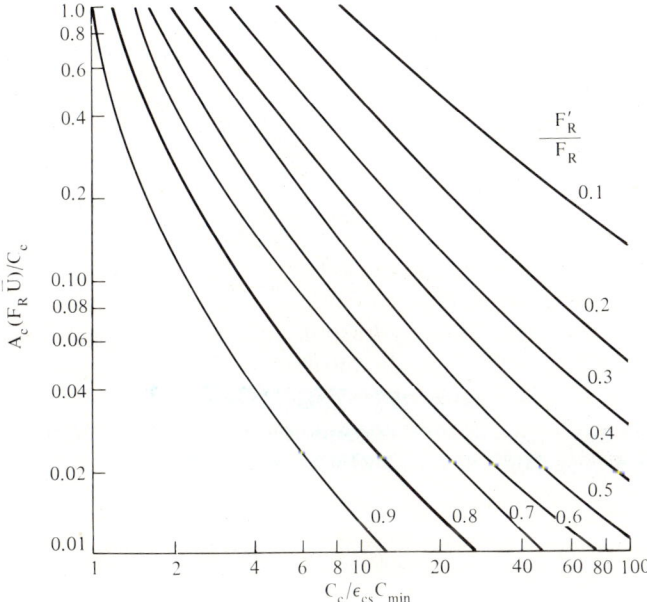

Figure 9-6 Relationship used to determine ratio F'_R/F_R.

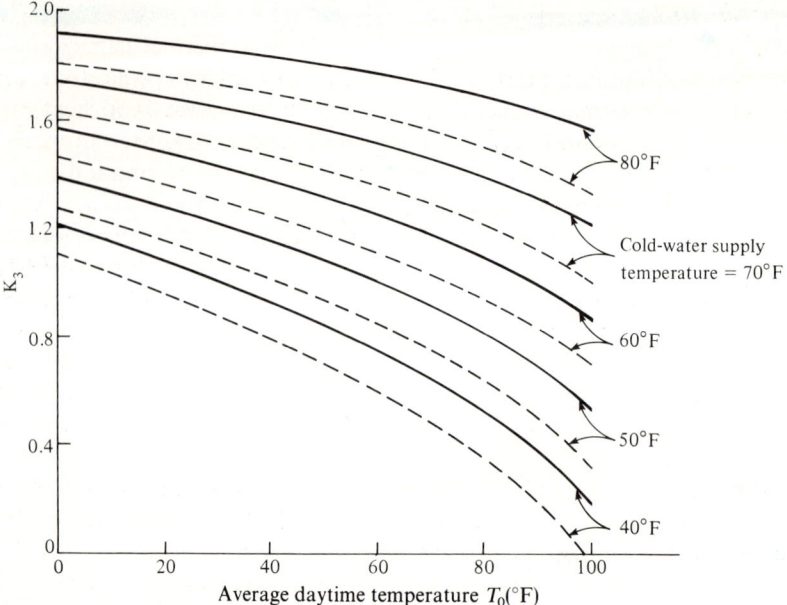

Figure 9-7 Hot-water factor K_3. —— Hot-water supply temperature = 140°F; --- hot-water supply temperature = 120°F.

Load heat exchanger factor K_4

$$\text{Air:} \quad K_4 = 1.0$$
$$\text{Water:} \quad K_4 = 1.05 - 0.088[\epsilon_L C_{min}/(UA)_{bldg}]^{-0.78} \quad (9\text{-}7)$$

where ϵ_L is the load heat exchanger effectiveness, C_{min} is the minimum flow thermal capacitance ($\dot{m}_L c$) for the load heat exchanger in kW/°C, and $(UA)_{bldg}$ is the building thermal loss rate per degree, also in kW/°C.

For each month of the year the necessary quantities must be determined to calculate D_1 and D_2, from which the monthly fraction f can be determined from Fig. 9-4 or 9-5. From these results the annual contribution and thus the annual solar fraction can be determined. An example is on pages 259–260.

For DHW systems without space heating, the monthly fraction is found from the heating plus DHW charts by modifying the value of D_2 to

$$D_{2,\text{DHW only}} = D_2 \frac{(11.6 + 1.18 T_{load} + 3.86 T_{sup} - 2.32 T_0)}{(100 - T_0)}$$

where all temperatures are Celsius.

9-7 ANNUAL f-CHART METHOD

To facilitate the prediction of the annual performance of solar systems, attempts have been made to base the annual fraction on only one month of the year, usually the peak heating month of January. One such method has been proposed by Huck and Winn [9].

This method is based on several hundred applications of the f-chart program to typical solar water- and/or space-heating systems. The heating and hot-water applications involved a design house heating requirement of 7.9 kWh/DD (15,000 Btu/DD) and a daily hot-water use of 303 L/day (80 gal/day), with the collector slope equal to the local latitude. Simulations were also run to determine the reliability of the curves for design conditions which deviate from those assumed. The house design heating requirement was varied from 5.3 to 15.8 kWh/DD (10,000 to 30,000 Btu/DD), and the collector slope was varied over local latitude $\pm 15°$. Little difference in performance was observed for these variations in parameters. Therefore, the curves may be considered to be useful guides for design conditions between 5.3 and 15.8 kWh/DD (10,000 and 30,000 Btu/DD) and for collector slopes in the range of local latitude $\pm 15°$. Thirty cities throughout the continental United States were used in developing the design curves. Collectors representing a wide range of performances were included in the original study, but only four are included here.

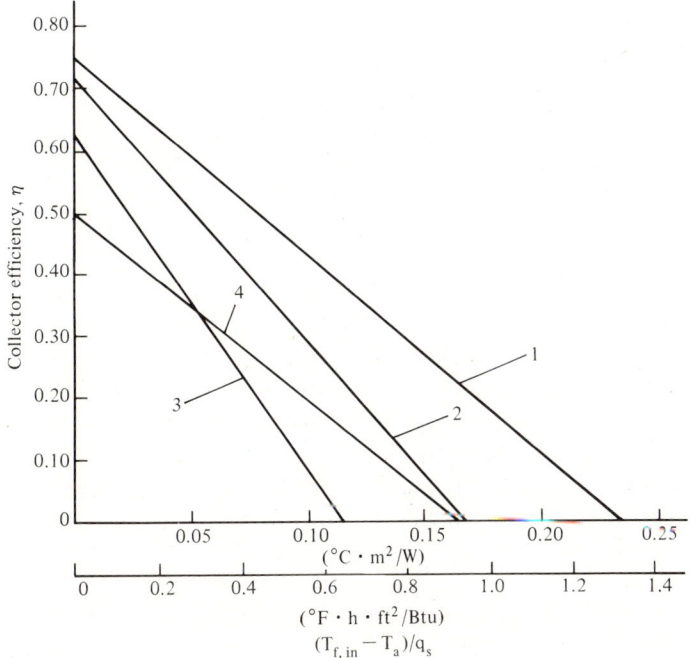

Figure 9-8 Solar collector performance curves. (See Table 9-4.)

Table 9-4 Intercepts and slopes of selected collectors (for Fig. 9-8)

Collector number	Type	$F_R(\tau\alpha)_{eff}$	$F_R \bar{U}$ W/(m² · °C)	$F_R \bar{U}$ Btu/(h · ft² · °F)
1	Water	0.74	3.12	0.55
2	Water	0.71	4.20	0.74
3	Water	0.62	5.34	0.94
4	Air	0.49	2.95	0.52

The approach involved running the *f*-chart program using incremental increases in collector area for each collector type, application, and location, and computing the annual fraction met by solar for each. Results are presented only for the solar fraction f_a in the range 0.1–0.9, since for very low values of f_a, fixed costs of the system degrade the economics and above 0.9 the incremental collector area becomes ineffective. It was found that the parameter $S_t A_c/L$ best correlated the annual solar fraction f_a. In this parameter, S_t is the January insolation on the tilted collector surface, A_c is the collector area, and L is the January total heating load.

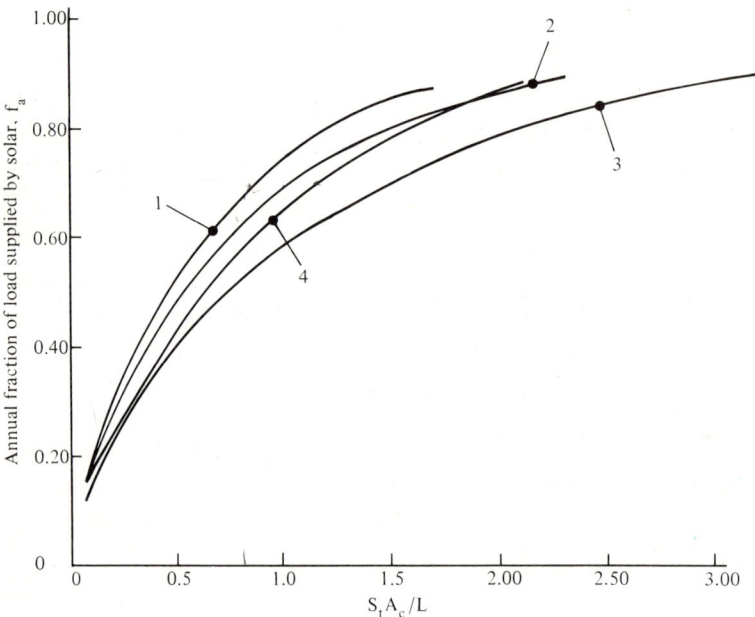

Figure 9-9 Annual solar fraction f_a for combination space and water heating. A_c, solar collector area; S_t, January solar insolation on a tilted surface; L, total January heating load. For any location. (*From Ref. 9.*)

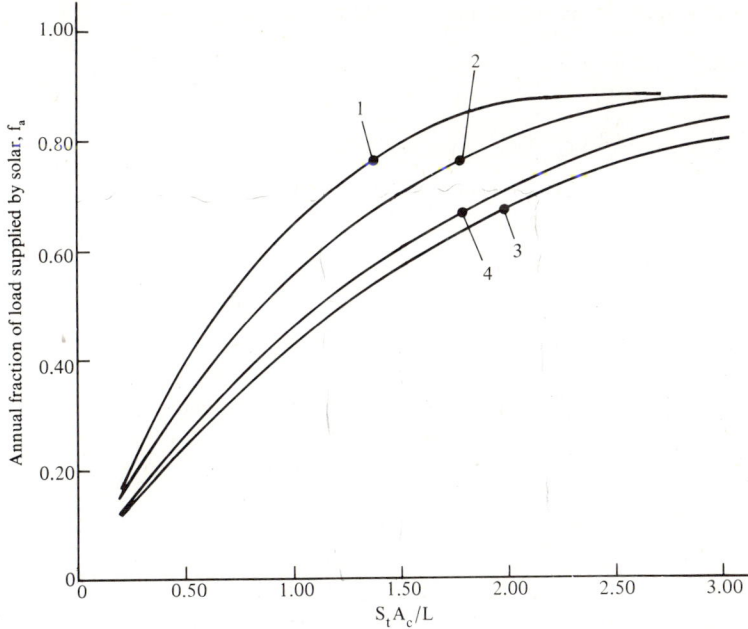

Figure 9-10 Annual solar fraction f_a for domestic water heating only. Collector tilt equal to latitude; for any location. (*From Ref. 9.*)

The collector types considered are included in Fig. 9-8, where the performances are presented in the standard form of efficiency versus $(T_{f,\text{in}} - T_a)/q_s$. Table 9-4 includes the values of the intercepts $F_R(\tau\alpha)_{\text{eff}}$ and slopes $F_R \bar{U}$ for the four collectors considered, three liquid and one air. The user should pick the collector efficiency curve which best agrees with that of the collector to be used and base the sizing on it.

Figures 9-9 and 9-10 present the results in the form of the annual solar fraction f_a as a function of the parameter $S_t A_c/L$. The curves presented represent best fits through the data obtained for various locations. These curves can be used either to estimate the annual solar fraction for a given collector, area, location and application or to estimate the required area for a specific collector, given the desired annual fraction, location, and application. The following example illustrates the use of the annual solar fraction method.

Example 9-1 Combination space and water heating for San Antonio, TX. Determine the annual solar fraction f_a for a San Antonio residence with the following characteristics: (*a*) 27.9 m² (300 ft²) of solar collector similar to No. 2 in Fig. 9-8, and tilted at 45°; (*b*) design heating load of 10.3 kW and 24°C indoor, −1°C outdoor (35,000 Btu/h, 75°F indoor, 30°F outdoor); (*c*) four occupants.

SOLUTION

Step 1 January total load.
    Water heating [heated to 54°C from 18°C (130°F to 64°F)]:

$$\dot{Q}_{d,\,\text{DHW}} = (4 \text{ people})\,(76 \text{ L/person-day})\,(31 \text{ days})$$
$$\times (1 \text{ kg/L})\,(1/3600 \text{ h/s})\,[4.186 \text{ kJ/(kg} \cdot {}^\circ\text{C})]$$
$$\times [(54 - 18){}^\circ\text{C}] = 400 \text{ kWh}$$
$$= 1.3 \text{ million Btu}$$

Space heating:

$$\dot{Q}_{d,\,\text{SH}} = (10.3 \text{ kW})\,(24 \text{ h/day}) \left(\frac{1}{[24 - (-1)]} \right) (238 \text{ degree-days})$$
$$= 2350 \text{ kWh}$$
$$= 8.0 \times 10^6 \text{ Btu}$$

January total load $= 390 + 2350 = 2740 \text{ kWh}$
$$= 9.3 \times 10^6 \text{ Btu}$$

Step 2 January insolation on tilted surface. The solar radiation on the collectors tilted at 45° will be determined for the month of January for San Antonio, TX. From App. Table B-2: latitude $= 29.5°$, $\bar{H} = 1045$ Btu/(ft^2 · day) [$=3.29$ kWh/(m^2 · day)], and $\bar{K}_T = 0.541$. From App. Table B-3, interpolating between the charts for $\bar{K}_T = 0.50$ and 0.60 for the month of January at a latitude of $29.5° \cong 30°$ and a latitude minus tilt of $29.5° - 45° = -15.5° \cong -15°$, one obtains $\bar{R} = 1.53$. Therefore, the January global insolation on the collectors tilted at 45° in San Antonio is [3.29 kWh/(m^2 · day)] (31 days)(1.53) = 156 kWh/(m^2 · month).

Step 3 Parameter $S_t A_c/L$. Substituting the January total heating load and the insolation information in the parametric group $S_t A_c/L$, we have:

$$S_t A_c/L = \frac{156 \text{ (kWh/m}^2)(27.9 \text{ m})}{2740 \text{ kWh}} = 1.58$$

Step 4 Annual solar fraction. From Fig. 9-9, for heating and water heating using a collector similar to No. 2 with $S_t A_c/L = 1.58$, the resulting annual solar fraction is approximately 0.81. This is a somewhat high value, so a smaller collector area could be considered.

It is important to note that the results using this method are approximate and should be used only as an initial estimate which should be subsequently refined using a more detailed analysis including economics.

9-8 FLAT-PLATE COLLECTOR ORIENTATION AND SHADING

Three important considerations in mounting solar collectors are their tilt, azimuth, and shading.

Collector Tilt

The tilt of a collector is the angle between the collector and the local horizontal. In most solar heating applications, a tilt of approximately latitude plus 10° to 15° is near optimum because it is necessary to favor the winter season, when the collector operates under adverse conditions (due to lower ambient temperature) and load is greatest (higher space-heating requirements and lower domestic water supply temperatures). Figure 9-11 taken from Ref. 10 presents collector effectiveness as a function of tilt, which indicates the fraction of energy delivered at a particular tilt compared to that at optimum tilt (approximately latitude plus 10° to 15°). It is seen that while latitude plus 10° to 15° is approximately optimal, the tilt can deviate from this by up to 15° with only 5 percent degradation in annual performance.

Collector Azimuth

The azimuth is the angle the collector faces relative to south. In the northern hemisphere, due south facing is preferred. Figure 9-12, also from Ref. 10, presents collector effectiveness as a function of the azimuth, indicating the fraction of energy delivered at a particular azimuth compared to that for due south. It can be seen that the orientation may deviate as much as 30° from south with only a 5 percent degradation in annual performance.

Collector Shading

While the problem of collector shading is very real, it is frequently overestimated in that early morning and late afternoon shading will have little effect on overall

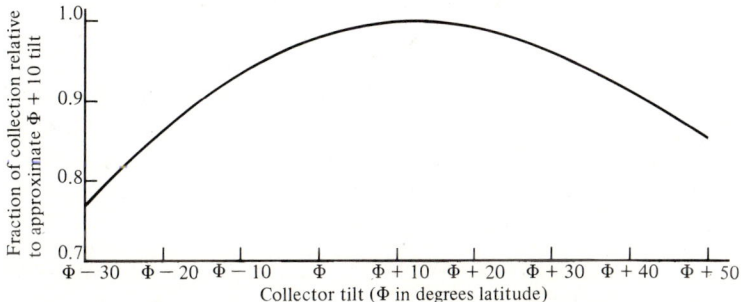

Figure 9-11 Effect of solar collector tilt on annual heating performance. (*From Ref. 10.*)

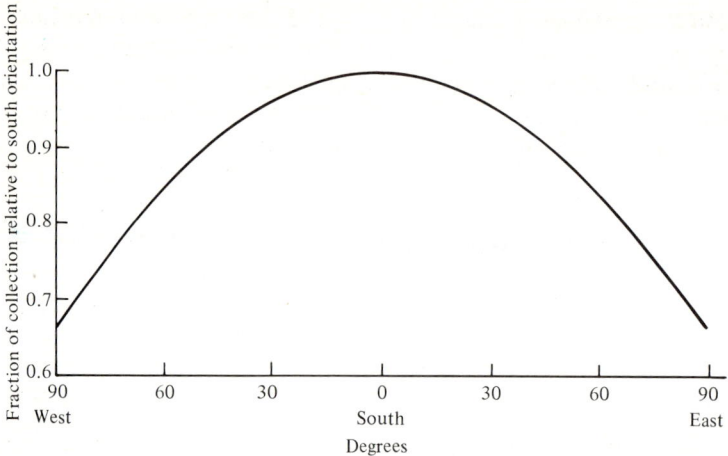

Figure 9-12 Effect of solar collector orientation (azimuth) on annual heating performance. (*From Ref. 10.*)

daily performance. In a typical installation, 85–95 percent of collection occurs in the middle 6 h of the day. Figure 9-13 shows the effect of shading on a collector relative to a completely unshaded collector for a particular installation in Austin, TX. This figure shows that approximately 50 percent of the potential collection occurs before and 50 percent after 12:30 P.M. solar time. The reason for the curve not being symmetrical about solar noon is that collection is slightly better in the solar P.M. due to higher ambient temperatures.

The curve can be further interpreted as follows: Assuming the collector is completely shaded up to 10 A.M., then approximately 28 percent of the potential

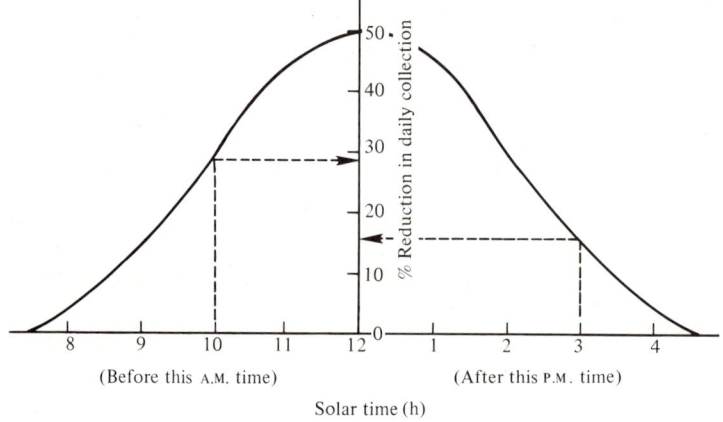

Figure 9-13 Effect of shading on daily collection. Double-glazed flat-plate tilt, 18° (latitude −12°); collection at 96°C; location, Austin, TX, in July.

collection is lost, and if completely shaded only after 3 P.M., then approximately 16 percent of the potential collection is lost. If completely shaded during both of these periods then $28 + 16 = 44$ percent of the potential collection is lost, or during the period between 10 A.M. and 3 P.M. the collection is approximately 56 percent of the potential for a completely unshaded collector.

PROBLEMS

9-1 Check out the rules of thumb for a solar domestic hot-water system for your location for a family of four. Use insolation expected at your location for 3-day periods in the winter, fall, and summer. Discuss your results. What do you think about the rules of thumb? How would you change those "rules" for your location? Assume the SOLSIM base case with DHW ($T_{sup} = 18°C$) except as noted above:

9-2 For the system described in Table 9-2, change the collector to an evacuated tubular collector (Fig. 4-10) and rework the example. Assume $F_{\tau\alpha}$ is the same as for a flat plate.

9-3 Use the rule-of-thumb design method and the annual f-chart method to design a domestic hot-water system for a family of four in San Antonio and also for Minneapolis. For both methods, indicate the required collector area and storage volume for a solar load fraction of 70 percent.

9-4 Use the monthly f-chart method to size the system described in Prob. 9-3.

9-5 The UA value for a building in Madison, Wisconsin is 440 Btu/(h · °F). What is the average daily heating load for the building in January and in November?

9-6 The nameplate capacity of a furnace located in a house in Lincoln, Nebraska shows 40,000 Btu/h. Assuming the furnace was selected to meet the design heating load, determine the average load for the months of January and November.

9-7 A building in Boston, Massachusetts has a UA value of 400 Btu/(h · °F), a hot water demand of 100 gal/day (at 140°F) and is to use the collector No. 2 of Table 9-4 as the heart of a solar space heating and hot-water system that is to provide approximately 60 percent of the annual load (from solar). (*a*) Use the annual f-chart method to size the collectors for a 60 percent solar fraction. (*b*) Use the monthly f-chart method to determine the monthly fractions and annual fractions from solar. (*c*) Using SOLSIM, determine the approximate annual solar fractions by selecting three or four periods of the year to simulate. Use the rules of thumb to select the sizes of other components after selecting the collector area from (*a*). Discuss the comparative results.

REFERENCES

1. *Active Solar Energy Systems—A Design Installation Manual*, Texas Solar Energy Society, Austin, TX, 1980.
2. G.O.G. Löf, and R.A. Tybout, "The Design and Cost of Optimized Systems for Residential Heating and Cooling by Solar Energy" (unpublished paper).
3. G.O.G. Löf, and R.A. Tybout, "Cost of House Heating with Solar Energy," *Solar Energy*, vol. 14, pp. 253–278, 1973.
4. *ASHRAE Handbook of Fundamentals*, ASHRAE, New York, 1974.
5. B.Y.H. Liu, and R.C. Jordan, "The Long-Term Average Performance of Flat-Plate Solar Energy Collectors," *Solar Energy*, vol. 7, p. 53, 1963.
6. H.C. Hottel and B.B. Woertz, "Performance of Flat-Plate Solar-Heat Collectors," *Trans. ASME*, vol. 64, p. 91, 1942.
7. W.A. Beckman, S.A. Klein, and J.A. Duffie, "Solar Heating Design—By the *F*-Chart Method," John Wiley & Sons, New York, 1977.

8. *HUD Intermediate Minimum Property Standards Supplement, 1977, Solar Heating and Domestic Hot Water*, Appendix A: "Calculation Procedures for Determining the Thermal Performance of Active Solar Space Heating and Domestic Hot Water Systems," Dept. of Housing and Urban Development, Washington, DC, 1977.
9. Steve E. Huck and Byron C. Winn, "Design Charts for Residential Solar Heating Systems," Master's thesis, Colorado State University, Boulder, 1976.
10. *Solar Heating and Cooling of Residential Buildings, Vol. I, Design of Systems* (SN 003-011-00084-4), and *Vol. II, Sizing, Installation and Operating of Systems* (SN 003-011-00085-2), U. S. GPO, Washington, DC, 1977. Available from the U.S. Government Printing Office.

CHAPTER
TEN

ECONOMICS OF SOLAR-THERMAL ENERGY CONVERSION SYSTEMS

10-1 GENERAL ECONOMIC CHARACTERISTICS OF SOLAR-THERMAL CONVERSION

As we have seen, solar conversion systems have characteristics that make it very difficult to compare the cost of solar energy with more conventional forms. These characteristics include the large capital costs and low operating (fuel) costs as compared with the relatively low capital costs and high operating costs of competing energy.

How are the relative costs of solar and other systems to be fairly compared? First, let us agree to base our comparison on the direct monetary outlay of the user, ignoring external costs of the systems such as social and pollution costs and other noneconomic factors such as status value, novelty value, freedom from the utility grid, etc. Even though these factors are quite important to many potential buyers of solar systems, they are difficult to quantify in a way that is acceptable to proponents of both solar and competing systems. Standard economic methods *do* allow quantification and comparison of monetary outlays, however. Thus, if the economic comparison shows costs to be close, the choice can then be made by examining the noneconomic factors.

We should account for the special economic benefits available to solar users. These may include federal, state, and local income, sales, and other tax credits and other possible rebates and incentives.

To compare the costs of solar and conventional systems, one approach is to determine *all* costs over the lifetime of the system. These would include capital costs, interest on the capital, fuel costs, operating and maintenance costs, replacement cost of components, etc., over the useful life of the system. This total is termed the *life-cycle cost* of the system; it offers a fair method of comparing solar and other systems, and is probably the most widely used method of comparison.

10-2 LIST OF SYMBOLS

A	end of period uniform payment, dollars
B	cost of building modification, dollars
B_N	net benefit over N years, dollars
C	total tax credit, dollars
C_s, C_c	present value of solar and conventional systems, dollars
D_j	value of depreciation taken in year j, dollars
e	annual escalation rate of fuel cost
F_c, F_s	present value of fuel costs over system life for conventional and solar systems, dollars
F_0	present annual cost of fuel, dollars
F	future sum of money, dollars
G_j	assessed value of system in year j
i	interest rate
I	initial cost, dollars
L_j	additional outstanding mortgage loan in year j, dollars
m	mortgage interest rate
M	annual maintenance cost, dollars
N	number of interest periods, or expected system lifetime
$P, (PV)_0$	present value, dollars
Q	cost of space, dollars
R_j	replacement and repair cost in year j, dollars
$\bar{S}_j$	salvage value, dollars
t	personal income tax rate in highest bracket paid
t	property tax rate
Y	yearly annual income, dollars

10-3 LIFE-CYCLE COSTS OF SOLAR HEATING AND AIR-CONDITIONING SYSTEMS

A major part of the life-cycle costs of solar systems is the capital costs; and the major capital cost items are listed in Table 10-1. The listing is of major subsystems, and each of these should be understood to include the pumps, fans, blowers, ducts, piping, fittings, wiring, etc., that go with it.

Other costs to be included in life-cycle costing are listed in Table 10-2.

Table 10-1 Capital cost items for solar subsystems

Collector
Storage
Domestic hot-water system
Air-conditioning components
Auxiliary energy system
Distribution system
Controls and electrical

Table 10-2 Other costs considered in the life-cycle

Cost of acquisition (search, comparison)
Repair and replacement
Operation (energy for operation and backup
  systems, other operation costs)
Maintenance
Insurance
Taxes and tax credits
Salvage value
Structure modification (roof strengthening, etc.)
Siting
Credit for roofing replacement
Use of space (both additional land and
  building space)

Note that certain parts of the systems being compared may be identical. For example, the duct system for hot-air distribution may be the same whether solar heat or gas heat is used. In a cost *comparison* of the two systems, then, the cost of the ducts may be omitted from both.

One problem in comparing systems arises in comparing cash flows; that is, *when* is the money spent? If, for example, the money is paid out in increments for fuel over the life of the system, then the money could collect interest during the

Table 10-3 Discount formulas

Standard nomenclature	Use when†	Standard notation†	Algebraic form†
Single compound amount formula	Given P; to find F	(SCA, i, N)	$F = P(1+i)^N$
Single present worth formula	Given F; to find P	(SPW, i, N)	$P = F \dfrac{1}{(1+i)^N}$
Uniform compound amount formula	Given A; to find F	(UCA, i, N)	$F = A \dfrac{(1+i)^N - 1}{i}$
Uniform sinking fund formula	Given F; to find A	(USF, i, N)	$A = F \dfrac{i}{(1+i)^N - 1}$
Uniform capital recovery formula	Given P; to find A	(UCR, i, N)	$A = P \dfrac{i(1+i)^N}{(1+i)^N - 1}$
Uniform present worth formula	Given A; to find P	(UPW, i, N)	$P = A \dfrac{(1+i)^N - 1}{i(1+i)^N}$

Source: Gerald W. Smith, *Engineering Economy: Analysis of Capital Expenditure*, 3d ed., Iowa State Univ. Press, Ames, 1979, p. 47.

† *Note*: P = a present sum of money
      F = a future sum of money, equivalent to P at the end of N periods of time at an interest of i
      i = an interest rate, decimal
      N = number of interest periods
      A = an end-of-period payment (or receipt) in a uniform series of payments (or receipts) over N periods at i interest rate, usually annually

period before it is paid, and we should credit the system for this. If the money is spent for the initial capital cost of the system, then the credit for earned interest is lost.

To account for these differences in cash flow, all the expenditures can be brought to a common time base, usually the present value. This is known as *discounting*. The discount rate is generally taken as the rate of return that could be earned on the next most lucrative investment.

Table 10-3 shows the commonly used discounting formulas.

Residential Systems

For the homeowner, the present value $(PV)_0$ of an energy system can be expressed, using the discounting formulas shown in Table 10-3, as

$$(PV)_0 = I - C + B + Q + \sum_{j=1}^{N} \frac{R_j - \bar{S}_j}{(1+i)^j} + M\left[\frac{(1+i)^N - 1}{i(1+i)^N}\right]$$

$$+ F_0 \sum_{j=1}^{N} \left(\frac{1+e}{1+i}\right)^j + (1-\bar{t}) \sum_{j=1}^{N} \frac{tG_j}{(1+i)^j}$$

$$- \sum_{j=1}^{N} \frac{\bar{t}(L_j m)}{(1+i)^j} \tag{10-1}$$

where I = initial system acquisition cost, including capital cost, installation, search costs.
C = federal and state tax credits available to homeowners installing solar.
B = cost of building modifications necessary to support the system.
Q = cost of space occupied by the system components.

$\sum_{j=1}^{N} \frac{R_j - \bar{S}_j}{(i+1)^j}$ = present value of replacement and repair costs R_j in year j less the salvage value $\bar{S}_j$ of those parts replaced in year j, for all N years.

$M\left[\dfrac{(1+i)^N - 1}{i(1+i)^N}\right]$ = present value of annual maintenance cost M at present prices. No inflation rate is included. If it were, the term would be modified to the same form as the fuel cost term.

$F_0 \sum_{j=1}^{N} \left(\dfrac{1+e}{1+i}\right)^j$ = present value of fuel used over the system lifetime, based on F_0, the cost of fuel used in the present year, and e, the annual escalation rate in fuel price.

$(1-\bar{t}) \sum_{j=1}^{N} \dfrac{tG_j}{(1+i)^j}$ = present value of the net property taxes paid on the energy system, including the effect of reduced income taxes due to property tax credit. The factors include the personal income tax rate $\bar{t}$ (in the highest bracket paid); the property tax rate t; and G_j, the assessed value of the system in year j.

$$\sum_{j=1}^{N} \frac{\bar{t}(L_j m)}{(1+i)^j} = \text{present value of the income tax reduction due to extra mortgage interest paid on loan to cover the solar energy system,}$$

where L_j is the additional outstanding mortgage loan because of the energy system in year j, based on the mortgage interest rate m.

In all of these terms, N is the expected life of the system, which can be taken as the building life. However, shorter lifetimes are often taken.

Commercial Systems

The chief difference between the economics of residential and commercial systems is that additional tax deductions for investment credit and depreciation are available for commercial capital investments. Additional aftertax revenue such as rental income may be credited to the presence of the system.

The present value of depreciation taken in year j, D_j, for all the N years of the system lifetime, is

$$(PV)_D = \sum_{j=1}^{N} \frac{(D_j \bar{t})}{(1+i)^j}$$

and the present value of additional yearly income Y is

$$(PV)_Y = (1 - \bar{t}) Y \left[\frac{(1+i)^N - 1}{i(1+i)^N} \right]$$

These two terms are subtracted from Eq. (10-1) in order to find the present value of a commercial solar energy system.

10-4 COST-BENEFIT COMPARISON OF ENERGY SYSTEMS

The possible benefits of a solar energy system compared to a conventional system over their life-cycle can be cast in the form of the savings of fuel due to the presence of the solar system less the additional cost of capital for the solar system. Thus the net benefit of owning the solar system over N years, B_N, is

$$B_N = (F_c - F_s) - (C_s - C_c) \tag{10-2}$$

where $F_c - F_s$ is the difference in the present value of fuel costs for the conventional and solar systems, and $C_s - C_c$ is the present value of the extra investment in the solar over the conventional system. If B_N is positive in value, it shows a net benefit for the solar over the conventional system.

Table 10-4 Assumed additional capital (including installation) and maintenance costs for a solar system

	Capital costs	Maintenance costs
Collector	$16,000/20 years (80 m² @ $200/m²)‡	$25/5 years†
Thermal storage tank	$600/20 years (3630 kg of fluid)	$25/year
Pipes and fittings	$200/20 years	
Motors and pumps	$200/10 years	
Heat exchanger	$200/20 years	
System control	$150/20 years	
Building modifications:		
Roof	$100/20 years	
Insulation	$75/20 years	
Basement space	$500 (2.5 m² @ $200/m²)	
Auxiliary heating unit	(Same as conventional system)	

Totals from above:

$18,025	First cost for the solar system
$18,189	Present value of additional capital cost including parts replacements over 20 years (pump, discounted over 10 years)
$ 470	Present value of additional maintenance cost (i.e., $25 every 5 years plus $25 every year, for 20 years)
$18,659	Present value of capital and maintenance cost for the solar system

† The notation $/year indicates the cost and frequency of occurrence, e.g., $25/5 years indicates an expenditure of $25 which will have to be duplicated every 5 years.

‡ This estimate consists of $150/m² for materials and $50/m² for installation.

Example 10-1† A residence has an annual heating load of 89×10^6 kJ/year. A solar system has been designed that will carry 70 percent of this load, with a fuel oil heating system that acts as a backup carrying 30 percent of the load. The efficiency of the backup is 55 percent. The system life is expected to be 20 years. A real discount rate of 2 percent (that is, 2 percent above inflation) is assumed. Fuel costs are expected to escalate at 5 percent/year in real terms. The solar system costs are shown in Table 10-4. Assume the fuel oil has a heating value of 40,000 kJ/L (140,000 Btu/gal). Find the net benefit (if any) of installing the solar heating system.

SOLUTION The energy costs of driving the motors and pumps on the solar system are taken as $20/year, and the cost for fuel for the backup system (at $0.27/L or $1.00/gal) is

† This example is adapted from Ref. 1.

$$(0.30)(89)(10^6 \text{ kJ heating/year}) \left(\frac{1 \text{ J input}}{0.55 \text{ J heating}}\right)$$

$$\times \left(\frac{1 \text{ L}}{40{,}000 \text{ kJ}}\right)(\$0.27/\text{L}) = \$327/\text{year}$$

for a total energy cost for the present year of $347.

For the conventional system, assuming it uses fuel oil at the same price but has a higher efficiency (60 percent) because of less intermittent operation, the fuel cost for the present year is

$$\$10 \text{ (pumps, blowers)} + 89 \times 10^6 \text{ kJ/year}(1/0.6)\left(\frac{1 \text{ L}}{40{,}000 \text{ kJ}}\right)(\$0.27/\text{L})$$

$$= \$10 + \$1000 = \$1010/\text{year}$$

Thus the first-year fuel saving for the solar system is $1010 − $347 = $663.

Now, the net benefit of the solar system is

$$B_N = (F_c - F_s) - (C_s - C_c)$$

$$= \sum_{j=1}^{20} \left[663\left(\frac{1+0.05}{1+0.02}\right)^j\right] - \$18{,}659$$

$$= \$20{,}241 - \$18{,}659 = \$1582$$

showing a net benefit for the solar system of over $1500 over the 20-year life of the system.

Note that Eq. (10-2) can be cast in the form that, for positive net benefits, reads

$$(C_s - C_c) < (F_c - F_s)$$

or the present value of capital cost for the solar system (over the conventional system) must be less than the present value of the fuel savings over the system life. This relation is used in SOLSIM to determine the allowable system cost based on fuel savings calculated from the simulation.

10-5 INFLUENCE OF ASSUMPTIONS

The outcome of Example 10-1 can be easily changed by assuming different fuel escalation rates, discount rates, system lifetimes, or solar system costs. Generally, high discount rates decrease the net benefit, higher fuel escalation rates increase net benefits, and cost reductions in solar equipment aid net benefits.

No credit on the capital cost was taken for the federal income tax credit in the example. As tax laws change, the credit may vary, but it will always tend to increase the net benefit of the solar system.

10-6 ECONOMIC ANALYSES OF SOLAR SYSTEMS

In Ref. 2, an analysis was carried out to determine the length of time required for the compounded net fuel savings to pay off the remaining mortgage on the solar equipment. This is called the *payback period*. The analysis assumed state-of-the-art solar technology for domestic hot-water and space heating; $215/m^2 ($20/ft^2) installed collector costs; complete conventional backup; annual maintenance of 2 percent of installed initial system cost (escalated at 6 percent annually); collector sized to minimize life-cycle costs; load for a new, detached, 140-m^2 (1500-ft^2) four-person residence of good energy-conscious design; a 30-year, 8.5 percent mortgage; a homeowner in the 30 percent incremental tax bracket; no property tax on the solar equipment; 6 percent annual inflation; and a 10 percent (4 percent real) annual fuel cost escalation. No income tax credit was given. The study was carried out for various cities, with comparison against various conventional systems. The payback periods calculated are shown in Table 10-5. Whenever the system life exceeds the payback period, positive benefits will accrue to the solar system.

An analysis was also carried out to determine the time to "positive savings," the time at which the annual fuel bill for the conventional system would exceed the annual payments plus fuel costs on the solar system. These times are also shown in Table 10-5.

Table 10-5 Economics of solar hot-water and heat, cost reduction scenario for 1980,† $108/m^2 ($10/ft^2) installed system cost

	Years to positive savings				Years to payback			
	Electricity	H.P.‡	Oil	Gas	Electricity	H.P.‡	Oil	Gas
Atlanta	1	2	1	5	9	13	11	16
Bismarck	1	1	1	§	8	11	11	§
Boston	1	2	3	4	9	13	14	14
Charleston	1	1	1	6	6	12	11	17
Columbia	1	3	2	§	8	13	13	§
Dallas/Ft. Worth	1	3	2	3	8	14	12	13
Grand Junction	1	1	1	5	7	10	10	16
Los Angeles	1	1	1	4	5	10	10	15
Madison	1	1	2	§	8	11	13	§
Miami	1	1	1	4	4	9	9	15
New York City	1	1	3	2	7	12	14	13
Seattle	§	§	3	§	§	§	13	§
Washington, DC	1	3	3	5	9	14	13	15

† This analysis uses fuel costs escalated at 10 percent/year from 1976 to 1980.
‡ Electrically driven heat pump.
§ Insufficient solar load, i.e., less than 40 percent hot water and heat or less than 50 percent hot water.

10-7 OTHER ECONOMIC FACTORS

A number of factors that militate against solar systems exist in the tax and rate structures. These must be recognized as real barriers to the implementation of solar energy in certain cases.

Many utilities charge energy rates that are in a "declining block" structure. The first energy sold by the utility to a consumer is at a high rate, and the more energy purchased, the lower is the unit cost of the energy. When a solar system is installed to reduce energy use, it replaces only the lowest-cost energy sold by the utility. Thus, the economics of the solar system are in a sense penalized by this type of rate structure, which is prevalent for commercial and industrial customers.

Tax laws for business and industry allow corporations to deduct fuel costs as a business expense. Capital investment, however, is given a much smaller tax credit. Thus capital investment by industry often goes to more attractive investments than for fuel-saving equipment.

Utilities are allowed to pass fuel cost increases directly along to consumers in so-called "pass-through" charges. The utility does not then have to request a rate change from the regulatory agencies each time fuel costs change. The incentive to invest in solar energy or in other fuel-conserving measures by the utilities is therefore small. Utilities do gain a return based on their capital investments, but capital for energy-efficient equipment enjoys no advantage over other opportunities.

Incentives for solar energy exist in the form of significant federal income tax credits and, in many states, state income or other tax credits. The consumer must decide whether investment of significant capital in a solar energy system is a worthwhile use of his or her money.

PROBLEMS

10-1 Solar collectors for a domestic hot-water system can be obtained for $150/m^2, storage costs $6/kg water, other components combined cost about 10 percent of collector costs, installation is 25 percent of the total system cost, and the system replaces electricity at $0.052/kWh. Use your best guess for interest rate, system life, fuel escalation rate, etc., to find the present value of the system you designed in Probs. 9-3 and 9-4.

10-2 Repeat Prob. 10-1 with the solar DHW system replacing natural gas as the thermal source. Natural gas costs $4.50/1000 ft^3, and a gas water heater has 60 percent combustion efficiency when used alone or 50 percent efficiency when used as an auxiliary to solar (due to intermittent operation).

REFERENCES

1. Rosalie T. Ruegg, "Solar Heating and Cooling in Buildings: Methods of Economic Evaluation," National Bureau of Standards Center for Building Technology Report No. NBSIR 75-712, July 1976.

2. "An Economic Analysis of Solar Water and Space Heating," ERDA Division of Solar Energy Report No. DSE-2322-1, November 1976.
3. Jeanne W. Powell, "An Economic Model for Passive Solar Designs in Commercial Environments," National Bureau of Standards, Center for Building Technology, Building Sci. Ser. No. 125, June 1980.
4. Rosalie T. Ruegg and G. Thomas Sav, "Microeconomics of Solar Energy," National Bureau of Standards, Washington, DC, 1980 (in press).

CHAPTER
ELEVEN
CASE STUDIES

In this chapter, four case studies are presented. They are of a typical solar domestic hot-water system, a solar heated commercial building, an industrial solar-thermal system, and a solar-driven power generation system. Each is examined in some detail, and the performance of each system is computed. Design methods and details are given, referring to the methods and data of previous chapters. Nomenclature here is the same as for Chaps. 2, 3 and 9.

11-1 DOMESTIC HOT-WATER SYSTEM DESIGN

The U.S. National Bureau of Standards (NBS) has carried out a series of tests on six generic systems for solar domestic hot water [1, 2]. In the case study presented in this section, a double-tank domestic hot-water (DHW) system using direct heating (i.e., no glycol loop through the collector) is studied.

The particular system studied here has the characteristics given in Table 11-1, and its behavior will be predicted in two ways. First, the annual fraction of the load supplied by the system will be predicted by the annual f-chart method described in Chap. 9. Next, the system behavior for a typical 10-day period in January and in July will be simulated using SOLSIM. The predicted behavior from these two methods will then be compared with the measured system behavior given by the NBS tests as reported in Ref. 3.

Table 11-1 Characteristics of direct-heating DHW system

Location: Washington, DC

Collector characteristics:
  Type: Flat plate, double-glazed, with low-iron, glass black-
    chrome selective surface
  Area = 5.02 m² (54 ft²)
  $(\tau\alpha)_{\text{eff}} = 0.850$
  $\bar{U} = 4.21$ W/(m² · °C) [0.741 Btu/(h · ft²°F)]
  $F_R = 0.854$
  $\dot{m}_c = 4.94$ kg/min (10.9 lb$_m$/min)

Storage:
  Tank 1 (Solar storage):
    $(Mc)_{\text{st}} = 1294$ kJ/K (681 Btu/°F, 82 gal)
    $U_{\text{st}} = 0.931$ W/(m² · °C) [0.164 Btu/(h · ft² · °F)]

Load:
  Daily RAND use profile, Fig. 7-2
  Water main supply temperature = 289 K (60°F)
  Required load temperature = 322 K (120°F)

Annual f-Chart Method

For this method, we must calculate three factors: the collector area A_c, the January load on the system L, and the average daily insolation on the collector. From these factors and the collector efficiency data, the annual fraction of the load that is supplied can be calculated.

Calculation of load: The daily load is assumed here to be that calculated from the RAND load profile shown in Fig. 7-2. We take the city main temperature as 60°F, and the temperature at which hot water is supplied, T_{load}, as 120°F. Summing the hourly hot-water requirements from Fig. 7-2 gives

$$L = \sum_{i=1}^{24} \dot{m}_i c(T_{\text{load}} - T_{\text{sup}}) = 35{,}500 \text{ Btu/day}$$

Average daily January insolation: From Table B-2 in App. B, the value of the monthly average daily horizontal global insolation is found for Washington, DC for January to be $\bar{H} = 632.4$ Btu/(ft² · day) and $\bar{K}_T = 0.445$. For this $\bar{K}_T$, Table B-3 gives (for a collector tilted to the latitude and after interpolation) $\bar{R} = 1.65$. Thus

$$S_t = (1.65)(632.4) = 1044 \text{ Btu/(h · ft}^2)$$

Therefore, the dimensionless group needed for the annual f-charts is

$$\frac{S_t A_c}{L} = \frac{[1044 \text{ Btu/(ft}^2 \cdot \text{day})](54 \text{ ft}^2)}{35{,}500 \text{ Btu/day}} = 1.59$$

Collector characteristics: Comparing the collector characteristics listed in Table 11-1 with the collector efficiency curves shown in Fig. 9-8, we find that the collector for this problem lies between the curves for collectors 1 and 2. For $S_t A_c/L = 1.59$, Fig. 9-10 gives an annual solar fraction of between 0.72 and 0.8.

SOLSIM simulation: In order to run a simulation of the DHW system, insolation curves must be available for the location of the collectors. Using the monthly average daily global horizontal values found as noted under the annual *f*-chart method, the hourly values of horizontal global insolation H for an average day of the month can be found by using Fig. 3-16. Day lengths for Washington, DC, are found for the "average" days in January and July (Table 3-2) by using Eqs. (3-1) and (3-5).

From these hourly values of horizontal global insolation, the values of $q_s(t)$ on the collectors (assumed to be tilted at the latitude) are found from $q_s(t) = HR$. The values of R are found as outlined in method 1 of Example 3-8 (assuming that the hourly K_T values can be taken as equal to $\bar{K}_T$ from App. Table B-3).

The results of these calculations are shown in Fig. 11-1, where sine curves have been fitted to the calculated data points. The best fit occurs when the day length for the collectors, Eq. (3-13), is used for t_d in the sine curve equations. Note that if (for July) the actual day length is used, then there is a poor fit. The reason is that early and late in the solar day during the summer months, the sun shines on the back of the tilted collectors.

The hourly RAND load profile shown in Fig. 7-2 was used in SOLSIM. Runs were also made with a constant load, and with and without stratified storage. The results (for the RAND profile case with stratified storage) are compared in Table 11-2 with test results from NBS and the annual *f*-chart predictions.

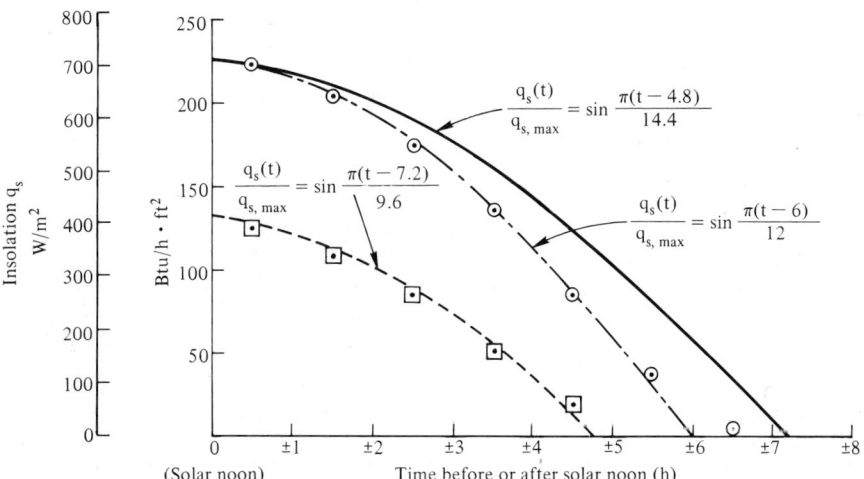

Figure 11-1 Insolation curves for SOLSIM simulation. Calculated values (see Table 11-1): ⊙ July; ☐ January; — · — SOLSIM July profile, — — — SOLSIM January profile; ——— sine profiles using actual July day length.

Table 11-2 Comparison of results for two-tank direct DHW system

	Fraction of load supplied by solar, %	Average daily collector efficiencies, %
NBS test results	Annual—52 (6-month data)	January—47 July—39; 41 (6-month average)
By f-chart prediction	Annual—72–80	
SOLSIM:		
January	45.3	43
July	99.9	47

It was found in the SOLSIM simulation that the system performed poorly under January conditions unless the mass flow to the collector was reduced from the design value shown in Table 11-1. This reduction was necessary in order to allow the control system to keep the collector pump on through much of the day when insolation was available. At full design flow, which worked well under July conditions, the controller cycled the pump on and off through the January typical day.

Table 11-3 Comparison of SOLSIM results—various assumptions

Month	Load	Storage	Fraction of load supplied, %	Daily average collector efficiency, %
January	Constant	Mixed	34.2	33.2
		Stratified	39.0	36.2
	RAND Profile	Mixed	39.8	39.4
		Stratified	45.3	43.2
July	Constant	Mixed	99.0	47.3
		Stratified	99.7	48.3
	RAND Profile	Mixed	99.6	46.1
		Stratified	99.9	47.2

Table 11-2 indicates that the NBS tests show the system to provide less of the load than is predicted by the annual f-chart method and SOLSIM. The test results, however, are based on only 6 months of data and the SOLSIM results are based on only two periods. The average predicted daily collector efficiencies agree somewhat more closely.

In Table 11-3, the effects of different assumptions on system behavior are shown. It is seen that stratification improves system efficiency in all cases, while hourly variations in load as assumed in the RAND profile also help system efficiency. The latter result is found because the RAND profile predicts that most energy use is during the daylight hours, so that the storage is at relatively low temperatures overnight (reducing heat losses) and at morning startup (increasing collector efficiency).

The SOLSIM results could be improved by assuming that the collector is tilted to an angle higher than the latitude. This would not only increase the January fraction of load supplied, but would slightly decrease the July fraction, thus improving comparison with the NBS test results.

11-2 CONCEPTUAL DESIGN OF A COMMERCIAL OFFICE HEATING SYSTEM

A small, one-story office has a floor area of 80 m^2. Collectors have been chosen with the characteristics shown in Fig. 11-2. Each collector has an aperture area of 3.22 m^2. The peak heating load for the office has been calculated as 12.8 kW, and a typical hourly load profile is shown in Fig. 11-3. It is desired to size the collector area and storage volume, determine the fraction of annual load supplied by the solar energy system, and simulate the system behavior. The office will be located in Tokyo. The system will be designed using guidelines discussed in Chap. 9, and the system performance will then be assessed using the monthly f-chart method and SOLSIM.

Design

Tokyo is at 36°N lat. From Table 3-3, the monthly average daily extraterrestrial insolation H_0 on a horizontal surface at that latitude is 17.6 MJ/(m^2 · day). From the data presented in Table 11-4, the monthly average daily insolation on a collector tilted at 50° in Tokyo is 12.8 MJ/(m^2 · day) in January. To provide the peak heating load of 12.8 kW for 8 h/day requires a total collector bank area of 28.8 m^2 (assuming an average collector efficiency of 37 percent based on Fig. 11-2). Since the individual collectors have areas of 3.22 m^2 each, a convenient collector array of 10 units is chosen to give a total collector area of 32.2 m^2.

The collector loop flow rate using the rule of thumb of 1.0 L/(min · m^2) is then $\dot{m}_c = [1.0 \text{ L/(min · m}^2)] (32.2 \text{ m}^2)(1 \text{ kg/L}) = 32 \text{ kg/min}$.

Storage volume can be sized to provide storage of one clear day's collection of energy, which will provide about 1 day's load. Assuming a storage circulation

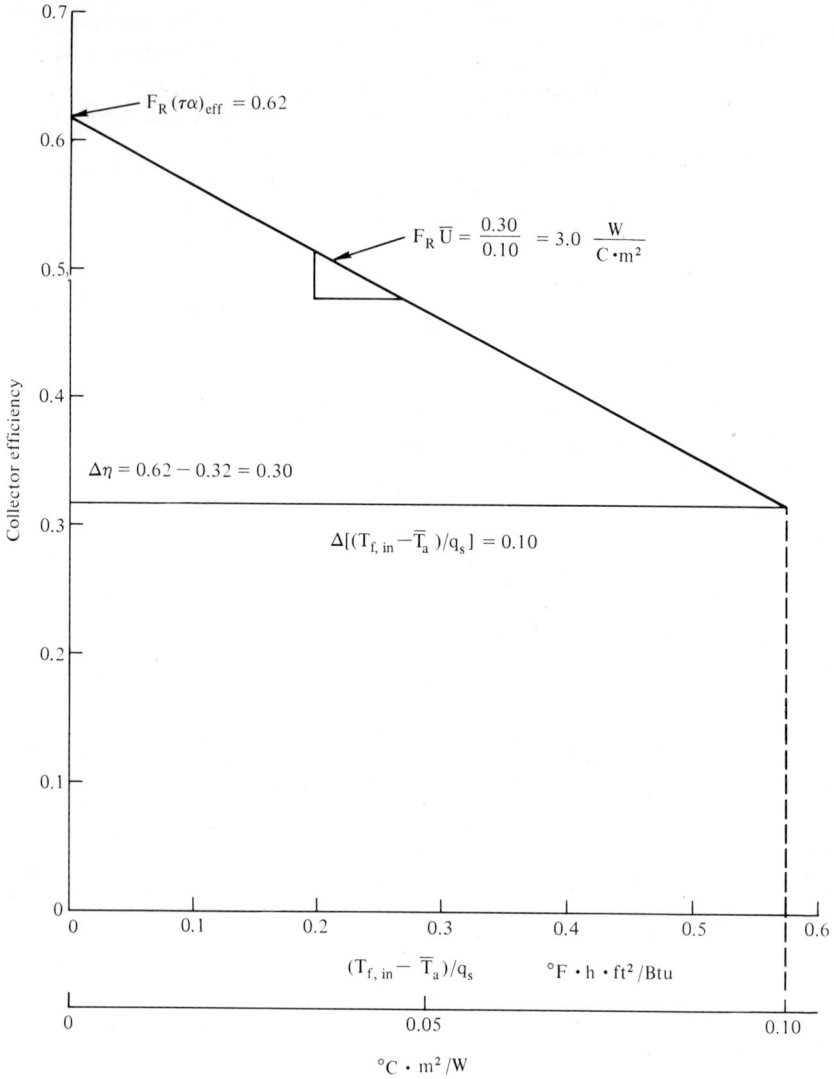

Figure 11-2 Efficiency curve for commercial office solar collector.

system to avoid freezing, water storage will be used. The storage capacity, then, assuming a 37 percent daily average collector efficiency and a 15-K swing in temperature from charge to discharge, is

$$(Mc)_{st} = \frac{\bar{\eta}\bar{q}_s A_c}{(T_{max} - T_{min})_{st}} = \frac{(0.37)[12.8 \text{ MJ}/(m^2 \cdot \text{day})](32.2 \text{ m}^2)(1 \text{ day})}{15 \text{ K}}$$

$$= 10.3 \text{ MJ/K}$$

and the storage mass is

$$M_{st} = \frac{10.3 \text{ MJ/K}}{4.19 \text{ kJ/kg}} = 2460 \text{ kg}$$

Required storage volume is then

$$V_{st} = M_{st}/\rho = (2460 \text{ kg})\left(\frac{1}{1000 \text{ kg/m}^3}\right) = 2.5 \text{ m}^3$$

The initial design of the system is now complete and is summarized in Table 11.5.

Behavior Prediction

Now the behavior of the system can be predicted. First, the monthly and annual f-chart method will be used to predict annual heating load supplied.

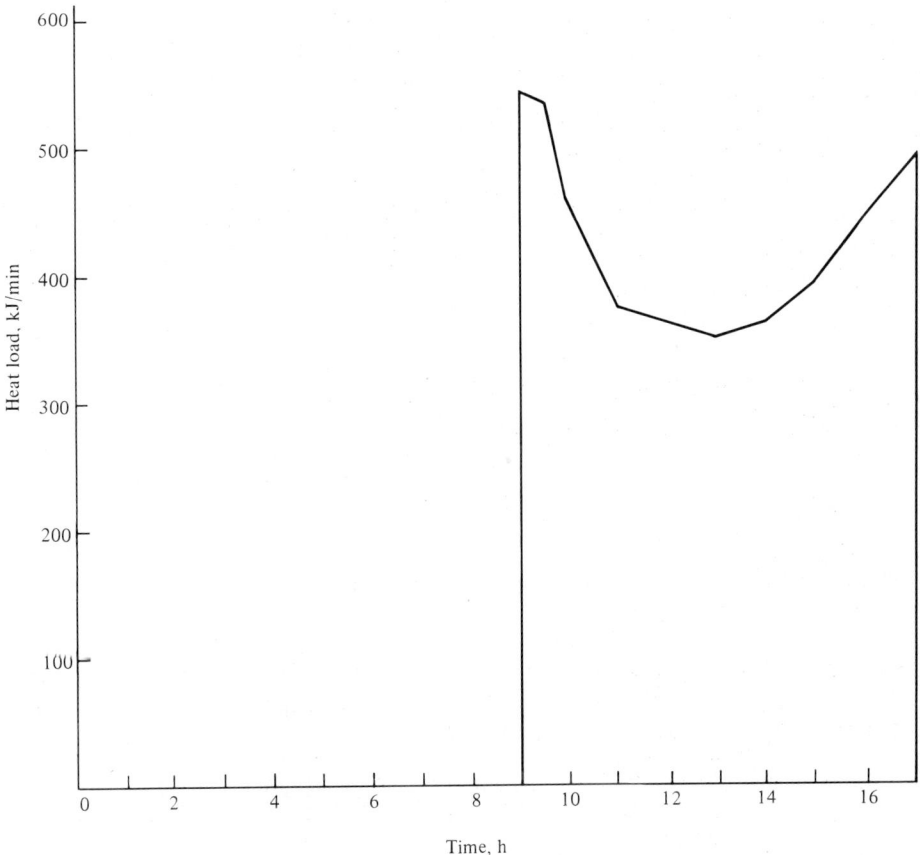

Figure 11-3 Typical January daily load characteristics for small commercial office.

Table 11-4 Insolation data for tilted surface† (Tokyo)

	Month												
	Jan.	Feb.	Mar.	Apr.	May	June	July	Aug.	Sept.	Oct.	Nov.	Dec.	Ref.
$\bar{H}_0$, MJ/(m² · day)	17.6	22.5	28.8	35.3	39.4	41.1	40.3	36.8	30.9	24.3	18.7	16.1	Table 3-3
$\bar{H}$, MJ/(m² · day)	8.0	9.7	11.5	13.1	14.4	12.7	14.1	14.2	10.6	8.5	7.7	7.1	‡
$\bar{K}_T = \bar{H}/\bar{H}_0$	0.454	0.431	0.399	0.371	0.365	0.309	0.350	0.386	0.343	0.350	0.412	0.441	
$\bar{R}$	1.65	1.42	1.10	0.91	0.81	0.79	0.79	0.86	1.00	1.20	1.52	1.71	Table B-3
$\bar{q}_s = \bar{R}\bar{H}$, MJ/(m² · day)	12.8	13.8	12.7	11.9	11.7	10.0	11.1	12.2	10.6	10.2	11.7	12.1	

† Conditions: Collector tilt angle = 50°.
‡ B. deJong, *Net Radiation Received by a Horizontal Surface at the Earth*, Delft University Press, Delft, Netherlands, 1973.

Table 11-5 Design of conceptual solar heating system

Collector:
$F_R(\tau\alpha)_{eff} = 0.62$
$F_R \bar{U} = 3.0 \text{ W}/(°C \cdot m^2)$
$A_c = 32.2 \text{ m}^2$
$\dot{m}_c = 32 \text{ kg/min}$

Storage:
Volume $= 2.5 \text{ m}^3$
$(Mc)_{st} = 10.3 \text{ MJ/K}$
$\dot{Q}_L$ from Fig. 11-3

Monthly f-chart calculation For this system, no heat exchanger is incorporated in the collector-storage flow loop. The required factors are

$$K_1 = 1.0$$

$$K_2 = 4.19[(10.3 \times 10^3 \text{ J/K})/(32.2 \text{ m}^2)]^{-0.3} = 0.742$$

$$K_3 = 1 \quad \text{(assumed for no hot water supplied)}$$

$$K_4 = 1.05 - 0.088 \left\{ \frac{1[(32 \times 4.190)/(60)] \text{ kW/°C}}{(UA)_{bldng}} \right\}^{-0.78}$$

For Tokyo, the winter design temperature is $-1°C$ (30°F) for an indoor temperature of 18.7°C. The maximum heat load on the building was given as 12.8 kW, so that

$$(UA)_{bldng} = \frac{\dot{Q}_{max}}{(\Delta T)_{max}} = \frac{12.8 \text{ kW}}{[18.7 - (-1)]°C} = 0.65 \text{ kW/°C}$$

Therefore,

$$K_4 = 1.05 - 0.088 \left(\frac{2.23}{0.65} \right)^{-0.78} = 1.02$$

Finally, F'_R/F_R is found using Fig. 9-6. The values for that figure are

$$\frac{C_c}{\epsilon_{cs} C_{min}} = 1$$

(since for no heat exchanger $C_c = C_{min}$ and $\epsilon_{cs} = 1$) and

$$\frac{A_c F_R \bar{U}}{C_c} = \frac{(32.2 \text{ m}^2)[3.0 \text{ W}/(°C \cdot m^2)](1 \text{ kJ} \cdot s/1000 \text{ W})}{(4.19 \text{ kJ/kg})(32 \text{ kg/min})(1 \text{ min}/60 \text{ s})} = 0.043$$

From Fig. 9-6, $F'_R/F_R = 1.0$.

Now, the f-chart fractions can be found from Fig. 9-4. The monthly required loads are backcalculated from the January peak load using the degree-day method. It is assumed that heating is only required 8 h/day. The annual amount

Table 11-6 Tokyo system performance by f-chart method

Month	$\bar{T}_a$, °C†	Degree-days,† °C	L, MJ/month	D_1	D_2	f	Load provided by solar MJ/month
January	4	330	6,200	1.26	4.56	0.69	4,280
February	4	357	6,670	1.14	3.84	0.68	4,540
March	9	311	5,830	1.30	4.59	0.87	5,070
April	15	153	2,870	2.46	8.46	0.95	2,730
May	19	56	1,030	6.9	>20	1	1,030
June	22	11	200	>10	>20	1	200
July	26	0	0	...	...	1	
August	26	1	0	...	...	1	
September	24	10	170	>10	>20	1	170
October	19	83	1,570	4.0	15.2	1	1,570
November	13	174	3,270	2.12	7.6	0.89	2,910
December	6	336	6,300	1.17	4.4	0.65	4,100
Annual		1822	34,070			0.78	26,580

† Calculated from *Facility Design and Planning Engineering Weather Data*, Dept. of the Air Force Manual No. AFM-88-29, July 1, 1978. Computed using total degree-day data for Tokyo and monthly adjusted degree-day data for Itazuke Auxiliary AFB, Japan.

of heating energy provided is found to be over 26,000 MJ, or 78 percent of the total required load as presented in Table 11-6.

The annual f-chart method can be used for a similar determination. The collector used here falls between collector types 2 and 3 in Table 9-4. The parameter $S_t A_c/L$ has the value

$$S_t A_c/L = \frac{[12.8 \text{ MJ}/(\text{m}^2 \cdot \text{day})](31 \text{ days/month})(32.2 \text{ m}^2)}{(6200 \text{ MJ/month})} = 2.06$$

which, using Fig. 9-9, predicts an annual fraction of about 81 percent, in good agreement with the monthly f-chart method.

SOLSIM simulation Data have been taken on a system similar to the one designed here, and are reported in Ref. 4. The system characteristics noted in Table 11-5 were used in SOLSIM to predict system behavior on a winter day. The insolation curve measured for the simulation day is shown in Fig. 11-4, and the sine curve used in SOLSIM is superimposed.

Figure 11-4 also shows the comparison of the measured collector efficiency and the efficiency as predicted by SOLSIM. The measured collector efficiency is somewhat below the predicted values early in the solar day, but the match is quite good over the period of high insolation. The daily average collector efficiency measured for the system was 0.391, and the predicted value from SOLSIM was 0.418 over this particular day.

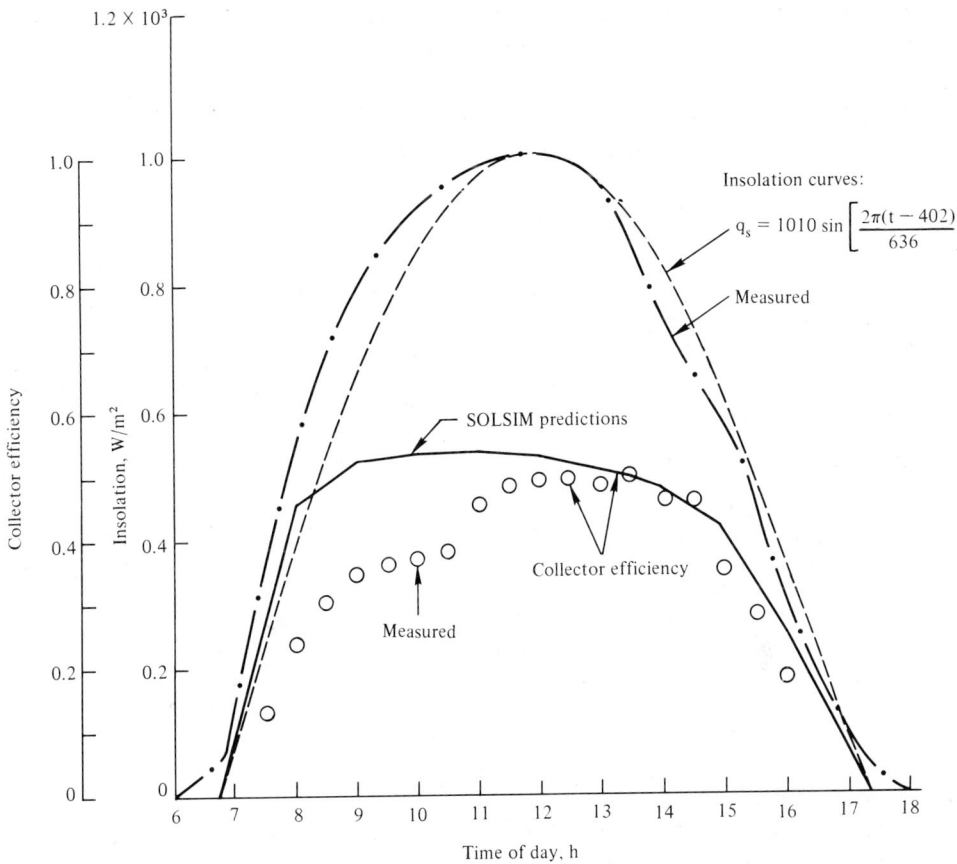

Figure 11-4 Measured and predicted characteristics of commercial office heating system.

Figure 11-5 shows measured and predicted collector outlet and bottom storage tank (collector inlet) temperatures. The SOLSIM simulation used a five-division stratified storage model and the conventional simulation with time-dependent load as given in Fig. 11-3. The actual system uses a mixing valve to hold the collector inlet temperature at 317 K, and the effect of this valve on the plotted temperatures is obvious from the measured results. In the early hours, heated water from the top of the storage tank is recycled to the collector inlet, causing the rapid early temperature rise. The valve then begins to mix the colder water from the bottom of the storage tank in order to hold the collector inlet temperature constant at 317 K. This variation in collector inlet temperature causes low collector efficiencies during the early hours of operation, but increases the collector efficiency during peak collection hours over the efficiencies predicted by SOLSIM. This is reflected in the efficiency curves shown in Fig. 11-4.

The performance prediction of the initial design is now complete. Further improvements and economic evaluation would be pursued in a complete study.

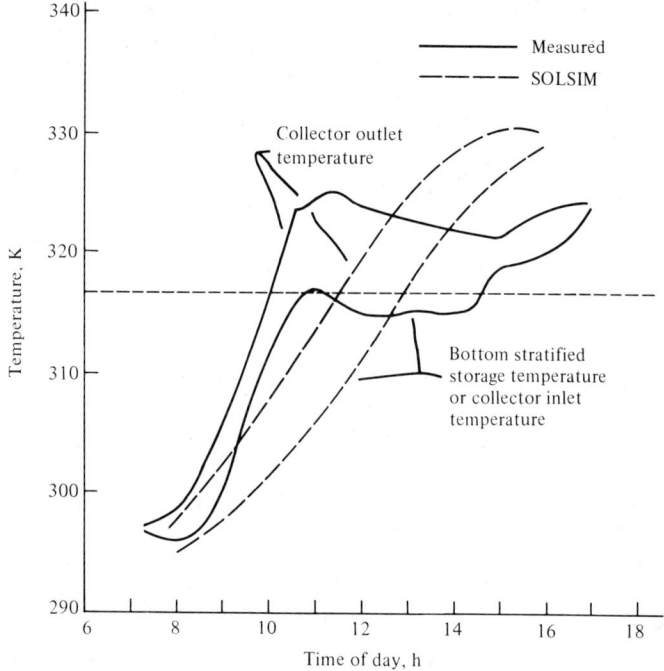

Figure 11-5 Collector outlet and bottom storage tank temperature for commercial office heating system.

11-3 SOLAR INDUSTRIAL PROCESS HEAT

The potential application of solar process heat in industry is quite large. Recent surveys (Refs. 5–7) show that up to 24 percent of all industrial heat is directly used in processes at temperatures below 177°C. Of the other 76 percent, a significant amount of energy is used to preheat up to 177°C. If this energy is added to that used for processes below 177°C, then some 40 percent of all process energy is found to be needed in the temperature range 16–177°C (see Fig. 7-5).

In certain industries, much larger percentages of energy are used at these relatively low temperatures. In Ref. 5, it is noted that 100 percent of all process heat (steam) is required in the range 16–177°C for the pulp and paper industry, and at temperatures at or below 100°C for the Frasch process for sulfur (see Table 7-5 in this book also).

It is relatively easy to provide solar energy in the form of low-grade (<100°C) heat. However, as the required supply temperature increases, the cost of a unit of solar energy usually increases. This increase in cost occurs because the increasing heat losses in solar collectors operating at high temperatures reduce their efficiency. Reduction in collector heat losses requires collector design modifications such as the addition of selective surfaces, evacuated chambers, or concentrating elements such as reflectors or lenses along with requirements for one- or two-axis tracking. Such modifications generally increase the cost per unit area of the collector. Because the cost of a unit of energy from fossil fuels is essentially independent of the temperature at which it is used (up to some fraction of the adiabatic flame temperature, at least), solar energy competes less and less favorably at higher process temperatures. As fossil fuel costs increase, the comparative cost of solar energy will become favorable at progressively higher temperatures. It is quite feasible to provide solar energy from highly concentrating systems at temperatures in excess of 1000°C.

General Limitations on Solar Process Heat

The operational limitations on solar energy conversion are caused by the intermittent nature of the availability of insolation because of diurnal and weather variations, and by the diffuse nature of insolation that requires large areas for collection of the energy needed to supply major process heat requirements.

To overcome the intermittency problems, storage of energy must often be incorporated into the solar system. Reference 8 has shown that storage volumes for systems using sensible heat storage can become extremely large, even for modest storage times Δt_s of about $\frac{1}{2}$ h. Yet for major continuous industrial processes that require on the order of kilowatts of thermal energy, some storage is probably necessary, if only to allow bringing auxiliary energy sources on line in the event of unexpected cloudiness. Under some special plant situations and for batch processes, it is possible to eliminate storage requirements.

The cost of providing storage is quite high because of the need to provide extra collector area to charge the storage system during times of available insolation in addition to the area needed to provide process heat during that period.

Process systems can be lumped under two categories: Direct systems (Fig. 11-6), in which a process fluid is directly heated; and indirect systems (Fig. 11-7),

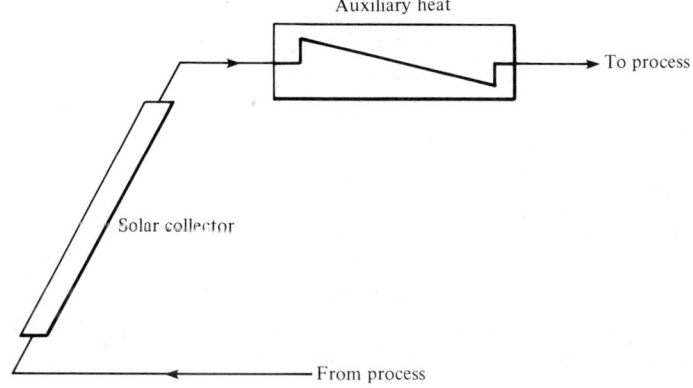

Figure 11-6 Direct solar system.

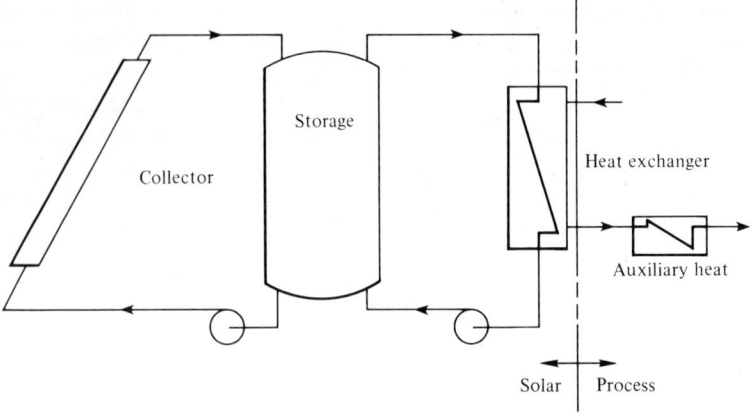

Figure 11-7 Indirect solar heating.

in which a secondary fluid is heated and transfers its heat to the process stream or load through a heat exchanger. The type of system to be used depends, of course, on the process being provided with heat.

The range of applications for process heat is large, and collectors that provide the higher process temperatures needed in some cases are more expensive. The choice of collector can be made using the approach outlined in Ref. 8, which led to the comparison shown in Fig. 4-24 between the efficiencies of different collectors and their relative costs through the cost effectiveness index, or CEI. The collector with the highest CEI (which is simply the collector efficiency divided by the cost per square foot) is the most appropriate one for use at a given temperature.

Case Study

An indirect system with minimal storage is analyzed here for use with a constant-load refinery unit. This is an instructive design case, and it shows many of the problems that can arise in large-scale implementation of solar energy.

The process requiring heat is taken as a reboiler requiring 40×10^6 kJ/h to vaporize a hydrocarbon at 66°C. This is typical of conditions in some units in the chemical process industries, this particular unit being representative of a deisobutanizer. It is assumed that all liquid throughput not vaporized in the heat exchanger (reboiler) is passed on to a second unit where auxiliary heat completes the vaporization. The system used in the model is illustrated in Fig. 11-8 and consists of an array of flat-plate collectors, a thermally stratified storage tank, pumps, piping, valves, and controls. Heat is transported directly from the collectors to the storage tank through water-to-water mixing in the tank. Heat is transferred to the process from storage through a shell-and-tube exchanger. Any auxiliary heat required to supplement that supplied by the solar system is added on the process side.

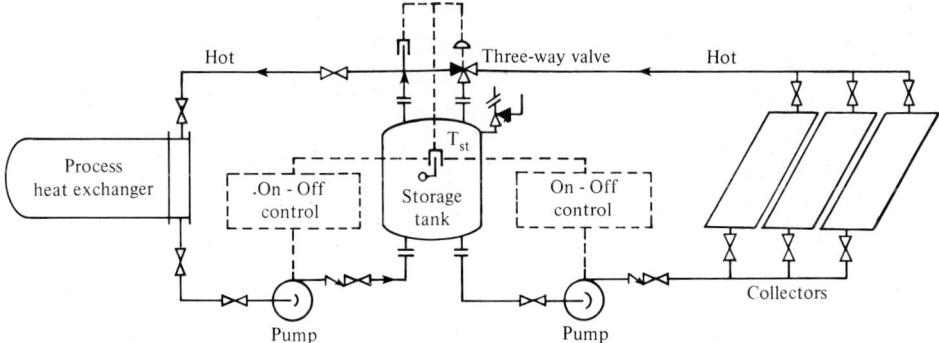

Figure 11-8 Indirect solar system model.

Thermal Analysis

The analytical method is to model the system with the control system and component models used in SOLSIM. Stratified storage is used, and collectors are assumed to have a linear efficiency relation.

Once the system is sized and the useful energy saved is calculated, the computer program determines the total cost of the solar system and compares it to the cost of fuel saved. Knowing that a system saves money is not necessarily justification for installing it. Earlier in this case study it was argued that a solar system can only supplement a process system. Therefore, it is also important to know what percentage of the total annual heat load the solar system provides, because the saving of valuable fossil fuels is one major objective in using solar. The computer economic program provides all the necessary data.

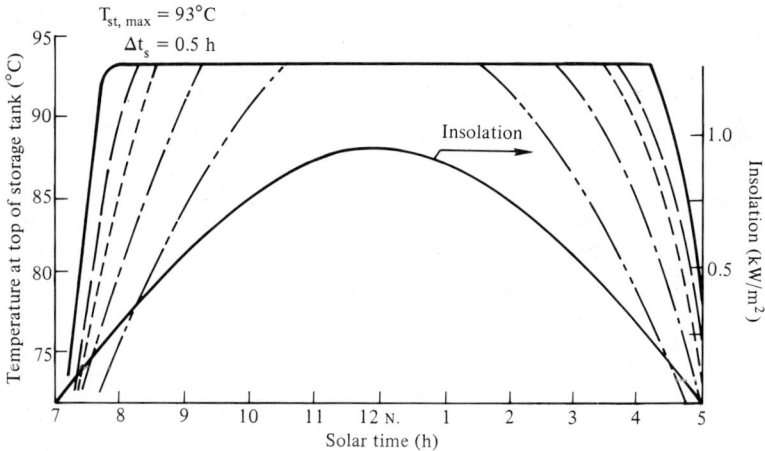

Figure 11-9 Daily variation of tank temperature and insolation for a four-layer tank with a 5×10^4-L capacity. Collector area: ——— 2.0×10^5 m^2; ——— 1.3×10^5 m^2; ----- 1.0×10^5 m^2; ——-— 0.75×10^5 m^2; ----- 0.5×10^5 m^2.

The process is modeled as having a constant energy requirement of 40×10^6 kJ/h to be supplied at or above a temperature of 66°C. The solar system is modeled as follows: The maximum temperature allowable in the storage tank and the length of time the storage is required to provide the process with 100 percent solar heat are specified, and the storage mass and volume are computed from these. The collector area is calculated by assuming that the collectors must supply 100 percent of the process heat at solar noon, when operating at 35 percent efficiency. This sizing allows the collectors to provide heat whenever insolation is available.

Insolation is modeled as a sine function peaking at solar noon with a symmetrical shape going to zero insolation at 7 A.M. and 5 P.M., solar time. Flow to the collector bank is sized to provide a 5.6°C temperature rise across the bank at solar noon, and flow rate to the process is sized to provide the required process heat at the maximum storage temperature. The control system is assumed to provide flow to the collectors as long as the outlet temperature of the collector bank is sufficient to provide energy to the process or to storage. Flow is provided to the process as long as at least a 6°C approach temperature is provided at the reboiler by the stream from the storage tank. Initial storage conditions are taken to be 66°C.

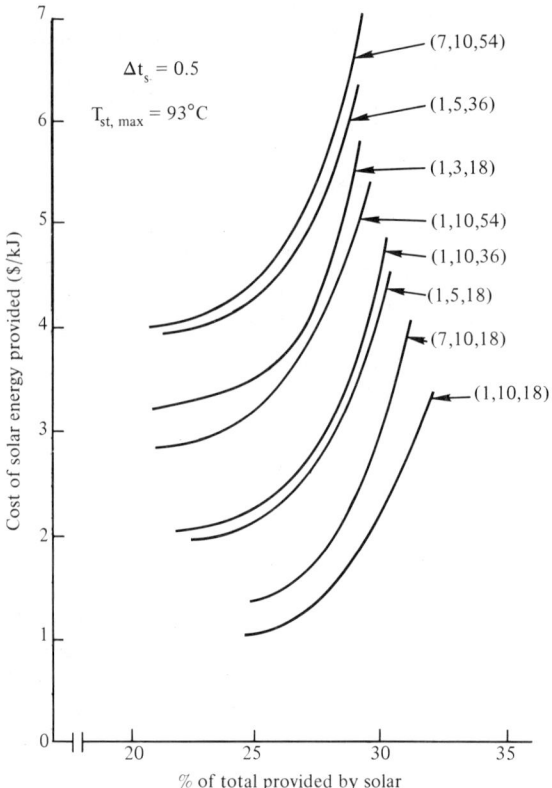

Figure 11-10 Cost of solar for various economic conditions. Parenthetical values refer to interest rate (percent), amortization time (years), and collector cost ($/m²).

Some results typical of the analysis of this model are illustrated in Fig. 11-9, which shows the top temperature of the stratified storage tank (four layers are assumed) and the solar energy available for a typical day. The simulated collector had a stagnation temperature at solar noon of 177°C with $b = 1.47, (\tau\alpha)_{eff} = 0.85$. The effects of various collector areas are demonstrated. Also available from the model are the temperature distribution in the tank, collector efficiency, collector outlet temperature, and other parameters describing the system operation.

Economic Evaluation

Before examining the economics of the proposed system, it is worthwhile to examine certain cost factors that may impact on the solar design. First, because of the large areas required for collectors, the question of land cost for plant expansion may arise. Even for extremely low collector costs of $2.00/ft² (about $20/m²) and extremely high land costs of $10,000/acre ($2.50/m²), it is obvious that collector costs are governing. However, land availability may be a consideration over and above cost. Second, the cost analysis shows that the collectors are by far the overriding cost factor in comparison with pumps, valves, piping, storage, etc. This is more evident for large-scale process heat systems that for residential solar heating since in the latter relatively large storage requirements, and therefore costs, can be a significant fraction of the system cost.

Many parameters determine the economics of the solar system, and only a few of the results of the economic analysis can be presented here. In Fig. 11-10 is

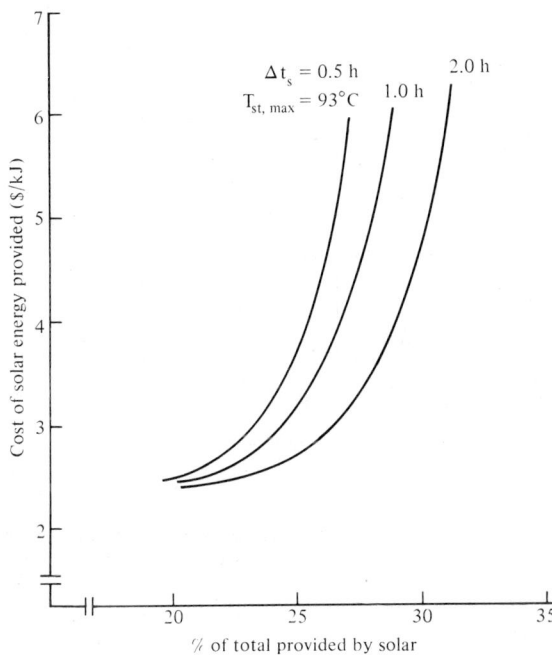

Figure 11-11 Effect of storage size on solar energy available to process. Interest rate = 1 percent; amortization = 10 years; collector cost = $36/m².

shown the cost per 10^6 kJ of provided solar energy as a function of collector cost, amortization period, and interest rate for various fractions of total energy provided to the process by the solar system. The effective interest rate shown is the difference between the actual interest rate and the fuel escalation rate. For our example system, natural gas was used as the alternate energy source at a cost of about $2.60/$10^6$ kJ. This is quite low in today's market. In Fig. 11-11 the effects of storage time on the cost and percentage of solar energy are demonstrated.

Under certain conditions of very low collector costs and subsidized interest rates or high escalation rates, solar energy can provide up to about 30 percent of process heat at costs competitive with other fuels.

11-4 HIGH-TEMPERATURE APPLICATIONS— SOLAR-DRIVEN HEAT ENGINES

In Chap. 2, SOLSIM was used to analyze the behavior of an idealized solar concentrator system for providing electric power. That system provides an interesting case study for examining the behavior of an energy-producing system formed by linking solar-derived thermal energy to a heat engine for converting that thermal energy into mechanical work.

Here we shall examine some of the limiting thermodynamics of such systems, and then see how the results of the SOLSIM simulation can be modified using this fundamental information.

Basic Thermodynamics

The efficiency of collection of solar energy depends on a number of factors, including the design of the collection equipment (Chaps. 4 and 5).

However, for a given collector design and conditions, we have seen that the efficiency tends to decrease with increasing temperature of the working fluid in the collector. This effect is a result of the increase in heat losses at the higher temperatures. For high-temperature collectors, the relationship may become nonlinear, and Eq. (2-2) must be applied without dropping the nonlinear term.

A simplified expression for the collector energy gain is thus given by Eqs. (2-1) and (2-2). This expression applies for a collector providing $\dot{Q}_u$ useful energy at a mean collector temperature $\bar{T}_e$. The collector has an aperture area A_c, and concentrates the insolation $q_s(t)$ onto an absorbing surface area A_e. The environment is at temperature T_a. For the meaning of other symbols, see Sec. 2-1.

Note that, for a concentrating system, A_c is interpreted as the projected interception area or aperture area of the entire array of collectors. For a flat-plate collector, A_c is simply the area of the collector and is taken as equal to the projected absorber area A_e. Letting $q_s(t) = q_{s,\text{ref}}$ at a given instant, Eq. (2-2) becomes

$$\eta = (\tau\alpha)_{\text{eff}} - a(\bar{\theta}_e^4 - \theta_a^4) - b(\bar{\theta}_e - \theta_a) \tag{11-1}$$

Physically, η is the ratio of useful energy provided by the collector to the insolation and a and b are parameters giving the relative importance of radiation and convection losses.

Suppose now that the useful energy $\dot{Q}_u$ at $\bar{T}_e$ is used to drive a heat engine. Because the efficiency of any heat engine increases with increasing heat-addition temperature, while solar collector efficiency decreases with increasing collection temperature, some optimum temperature must provide the most work output from the collector-heat engine system.

For a given thermodynamic cycle, the fraction of useful energy used in the cycle can be found by multiplying Eq. (11-1) by the thermodynamic efficiency of the cycle, η_t. Assume that η_t is some function of $\bar{\theta}_e$. Then defining Φ as the fraction of insolation that is converted to useful work, we obtain

$$\Phi = \eta_t \eta = \eta_t[(\tau\alpha)_{\text{eff}} - a(\bar{\theta}_e^4 - \theta_a^4) - b(\bar{\theta}_e - \theta_a)] \tag{11-2}$$

To maximize Φ for a given collector, the derivative of Φ can be taken with respect to temperature ratio $\bar{\theta}_e$ as

$$\frac{d\Phi}{d\bar{\theta}_e} = \frac{\partial \Phi}{\partial \bar{\theta}_e}\bigg|_{a,\,b\text{ constant}} = \frac{\partial \eta_t}{\partial \bar{\theta}_e}[(\tau\alpha)_{\text{eff}} - a(\bar{\theta}_e^4 - \theta_a^4)$$
$$- b(\bar{\theta}_e - \theta_a)] - \eta_t[4a\bar{\theta}_e^3 + b] = 0 \tag{11-3}$$

In this step, it is assumed that a and b are not functions of $\bar{\theta}_e$. This implies that the physical properties and convective heat-transfer coefficients of the solar collector are taken to be independent of temperature, a fair but not strictly valid assumption.

As an example of the use of Eq. (11-3) for the Carnot cycle,

$$\eta_t = 1 - \frac{T_a}{T_e} = 1 - \frac{1}{\bar{\theta}_e} \tag{11-4}$$

where T_a is the reservoir temperature for heat rejection, taken here as equal to the ambient temperature. Substituting into Eq. (11-3), we find the relation between $\bar{\theta}_e$, a, and b that is required for maximum work output from a collector driving a Carnot engine to be

$$\bar{\theta}_e^5 - \tfrac{3}{4}\theta_a\bar{\theta}_e^4 + \frac{b}{4a}\bar{\theta}_e^2 - \theta_a\left[\frac{(\tau\alpha)_{\text{eff}} + a + b}{4a}\right] = 0 \tag{11-5}$$

Values of $\bar{\theta}_e$ for given a and b values are plotted in Fig. 11-12.

The values of a and b for the given collector are first calculated. Figure 11-12 is then used to determine the temperature at which the collector should be run to attain the maximum work output for the prescribed cycle. When this temperature is found, it can be used in Eqs. (11-1), (11-2), and (11-4) to find η, Φ, and η_t, respectively.

In calculating values for a and b, the values should be based on the approximate optimum temperature, since a and b are both functions of temperature.

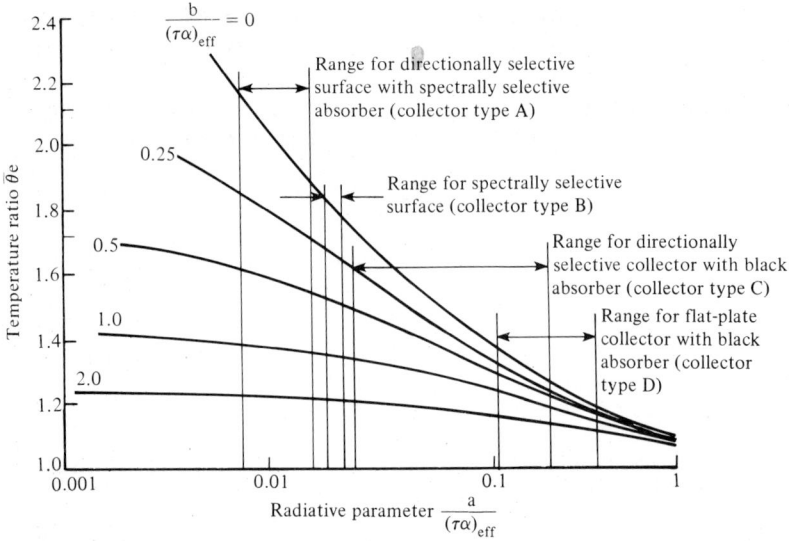

Figure 11-12 Collector temperatures for maximum work output from a Carnot engine.

Comparison of Various Collector Types

In Fig. 11-12 are plotted behavior ranges for various collector classes. All comparisons assume an average daily environment temperature $\bar{T}_a = 21°C$ and an insolation value of 1 kW/m².

For the nonselective flat-plate collector, $A_e = A_c$ and α/ϵ ranges from unity up to about 3 to allow for cover glass selectivity. This gives $0.13 < a/(\tau\alpha)_{eff} < 0.34$ (collector type D).

For nontracking concentrators (concentration ratio ≤ 3) using nonselective absorbers, the absorber area A_e for the best designs is about one-third the collector area, $A_c/3$, and again the cover-glass selectivity and internal reflection losses on the concentrators allow α/ϵ to range from a minimum of about 0.6 to a maximum of about 3. Thus, the parameter $a/(\tau\alpha)_{eff}$ lies in the range 0.04–0.21 (collector type C) for these surfaces. The worst of these designs overlap with the best flat-plate designs.

For spectrally selective flat-plate collectors, $A_e = A_c$ and α/ϵ may range from 10 to perhaps 15 for presently available coatings. This gives $a/(\tau\alpha)_{eff}$ in the range of 0.026–0.035 (type B). Combining both spectral selectivity and directional selectivity (moderate concentration) gives $a/(\tau\alpha)_{eff}$ lying the range 0.009–0.022 (collector type A). Sophisticated tracking concentrators achieve an A_c/A_e ratio on the order of 1000 or higher and yield very low values for $a/(\tau\alpha)_{eff}$, that is, <0.001.

For a "conventional" design that gives no special attention to convection suppression, the value of the convective loss coefficient is approximately A_e/A_c. Hence, for collector types A and C, $b/(\tau\alpha)_{eff}$ is approximately 1. For collector types B and D, $b/(\tau\alpha)_{eff}$ ranges from 0.3 to 1. The approximate operating ranges with respect to $\alpha/(\tau\alpha)_{eff}$ are indicated in Fig. 11-12.

Of course the Carnot cycle is only an idealized representation of the efficiency of a real heat engine. The method outlined above can be applied easily to any cycle with an efficiency that can be expressed as a function of the heat-addition temperature. In Ref. 9 this was done for idealized Sterling, Ericsson, and Brayton cycles as well as absorption and other refrigeration cycles.

Application to the "Power Tower"

For the large-scale power generation system analyzed in Chap. 2, using SOLSIM, the values of the collector parameters were taken to be

$$a = 6.2 \times 10^{-4}$$
$$b = 4.04 \times 10^{-3}$$
$$(\tau\alpha)_{\text{eff}} = 0.9$$

For these values, Eq. (11-5) can be solved to find the collector outlet temperature (and heat engine inlet temperature) as $\bar{\theta}_e = 3.4$, or, for an ambient temperature of 285 K, $\bar{T}_e = 970$ K. Equations (11-1), (11-2), and (11-4) then give the optimum values of

- Collector efficiency, $\eta = 0.808$
- Fraction of insolation converted to useful work, $\Phi = 0.570$
- Thermodynamic efficiency of heat engine, $\eta_t = 0.705$

The SOLSIM simulation shown in Fig. 2-11 indicates a peak collector efficiency of about 0.88 after 3 days of operation, and a heat engine efficiency of 0.6, giving a fraction of insolation converted to useful work of 0.53. The difference of these values from the optimum values calculated here are because of the storage included in the SOLSIM simulation. The maximum storage temperature was limited to 750 K in the simulation, a limit placed by materials problems at higher temperatures. With this storage temperature, the collector will operate at a higher efficiency than is optimum for the best system (because it operates at a lower temperature) and the heat engine at a lower efficiency because it will have a lower energy input temperature. The overall system efficiency of the SOLSIM model is, as expected, lower than the optimum that is found from thermodynamic considerations. Of course, a real system will have even more constraints on its performance than does the system in the SOLSIM model. The Carnot efficiency itself is, of course, much higher than can be obtained in any practical heat engine.

PROBLEMS

11-1 A home in Houston, TX, is to be designed with a solar heating system, and with provision for the later addition of solar air conditioning. The home is to have 2400 ft^2 on one floor and will be used as a private residence. Design a solar heating system for this house that specifies the following:
   (a) The collector area required, in ft^2.
   (b) The storage volume required, in ft^3; note type.

(c) Pump or fan capacity and horsepower.
(d) A diagram of the piping and/or ducting system. Include sizes of piping and/or ducting.
(e) A control system diagram.
(f) Calculation of energy saved by the solar system.

Assume about 10,000 Btu heat loss per degree day for this house, and that Houston has 1434 degree-days of heating per year based on a 65°F interior comfort temperature, with January having 416 °F-days. Please discuss in full how you made your choices, your assumptions, and your calculations.

11-2 When ambient air temperatures drop below about 50°F, auxiliary heat must be supplied to the evaporator of a heat pump in order to continue efficient operation. The auxiliary heat is usually supplied by an electrical resistance heater, but can just as easily (and perhaps more cheaply) be supplied from solar energy.

Model a residential or commercial-scale heat pump system, and size the solar-assist portion (collector, storage) in order to provide operation through cold-weather periods typical of the Columbus, OH, climate.

Results should include (1) A flow chart of the computer model used, (2) results of sizing study for collector and storage, and (3) economic study of solar versus electrical assist.

11-3 The computer model SOLSIM is not suitable for systems using air as a working fluid. Air systems generally use rock-bed storage, which behaves much differently from mixed or stratified liquid storage.

Modify the existing program to handle air systems, putting in the correct model for rock-bed storage (which is available in Chap. 6), and air collectors. Some examples should then be worked out.

Results should include (1) a report on the details of the model, (2) a printout of the modified program, and (3) results of an example system for residential heating.

11-4 In many utility service areas in the South, large air-conditioning loads cause the peak power demands to occur on hot afternoons in late summer. These peaks should correlate well with the availability of solar thermal energy.

Design a solar tower system to provide steam on an "as-available" basis to a plant presently burning natural gas at a cost of $4.00/$10^6$ Btu. The plant can produce 400 MWe, while operating at an overall efficiency of 40 percent. Use the load curve for the Austin Municipal Electric Utility for late summer conditions (Figs. 7-7 and 7-8).

Results should include (1) design of a solar power system with sufficient collector area and storage to provide 25 percent of the total load over the period from 12 N. to 6 P.M. and (2) calculation of the capital cost allowable for the system. How much must fuel costs increase before the solar power system is competitive if it is not competitive now?

11-5 Assume that the "Solar Car Wash (Wash your car in Sunshine)" outfit wants to heat its wash water 100 percent with solar energy. It is willing to stay open on sunny days only, so storage can be minimized.

Estimate the load requirements (check with a carwash) in terms of gallons of water per car, number of cars per hour, etc., and design a solar system to provide the needed hot water. Think about peak versus average loads, whether storage is necessary or desirable, and whether such a system can use any methods of waste heat recovery in a useful way.

Results should include (1) completed design of the solar system and (2) an economic analysis of whether the system is cheaper than a natural gas system. (Think about whether the advertising costs should be included in the cost analysis.)

REFERENCES

1. A.H. Fanney, *Experimental Validation of Computer Programs for Solar Domestic Hot Water Heating Systems*, National Bureau of Standards Center for Building Technology, Washington, DC, July 1978.

2. A.H. Fanney and S.T. Liu, "Experimental System Performance and Comparison with Computer Predictions for Six Solar Domestic Hot Water Systems," *Proc. 1979 Int. Congress of the Int. Solar Energy Soc.*, Atlanta, May 28 to June 1, 1979.
3. Rob B. Farrington, L.M. Murphy, and Darryl L. Noreen, "A Comparison of Six Generic Solar Domestic Hot Water Systems," Solar Energy Research Institute Rept. No. RR-351-412, Golden, CO, April 1980.
4. Nobuo Nakahara, Yasuyuki Miyakawa, and Mitsunobu Yamamoto, "Experimental Study on House Cooling and Heating with Solar Energy Using Flat Plate Collector," *Solar Energy*, vol. 19, no. 6, pp. 657–662, 1977.
5. Malcolm D. Fraser, "Survey of the Applications of Solar Thermal Energy to Industrial Process Heat," *Proceedings of the 1976 Annual Meeting of the Am. Sec. of ISES, Sharing the Sun*, vol. 5, Winnipeg, August 15–20, 1976, pp. 46–57.
6. Elton Hall, and John G. Rupert, "Survey of the Applications of Solar Thermal Energy Systems to Industrial Process Heat," *Proc. of Solar Industrial Process Heat Workshop*, College Park, MD, June 28–29, 1976, pp. 8–13.
7. D.P. Grimmer and K.C. Herr, "Solar Process Heat from Concentrating Flat-Plate Collectors," *Proceedings of the 1976 Annual Meeting of the Am. Sec. of ISES, Sharing the Sun*, vol. 2, Winnipeg, August 15–20, 1976, pp. 351–374.
8. Richard Reimels and John R. Howell, "Solar Energy for Process Heat," *Proceedings of the 1976 Annual Meeting of the Am. Sec. of ISES, Sharing the Sun*, vol. 5, Winnipeg, August 15–20, 1976, pp. 58–76.
9. John R. Howell and Richard B. Bannerot, "Optimum Solar Collector Operation for Maximizing Cycle Work Output," *Solar Energy*, vol. 19, no. 2, pp. 149–153, 1977.

APPENDIX
A

FUNDAMENTAL CONSTANTS AND CONVERSION FACTORS

Table A-1 Radiation constants

Symbol	Definition	Value
C_1	Constant in Planck's spectral energy distribution	0.18892×10^8 (Btu · μm^4)/(h · ft^2) 0.59544×10^8 (W · μm^4)/m^2 0.59544×10^{-16} W · m^2
C_2	Constant in Planck's spectral energy distribution	25,898 μm · °R 14,388 μm · K
C_3	Constant in Wien's displacement law	5216.0 μm · °R 2897.8 μm · K
σ	Stefan-Boltzmann constant	0.1712×10^{-8} Btu/(h · ft^2 · °R^4) 5.6696×10^{-8} W/(m^2 · K^4)
I_0	Solar constant	428 ± 7 Btu/(h · ft^2) 1353 ± 21 W/m^2†
T_{solar}	Effective surface radiating temperature of the sun	5780 K, 10,400°R

† Expected to be changed in 1982 to 1377 W/m^2.

Table A-2 Conversion factors

Length

	mi	km	m	ft	in
1 mi =	1	1.609	1609	5280	6.336×10^4
1 km =	0.6214	1	10^3	3.281×10^3	3.937×10^4
1 m =	6.214×10^{-4}	10^{-3}	1	3.281	39.37
1 ft =	1.894×10^{-4}	3.048×10^{-4}	0.3048	1	12
1 in =	1.578×10^{-5}	2.540×10^{-5}	2.540×10^{-2}	8.333×10^{-2}	1
1 cm =	6.214×10^{-6}	10^{-5}	10^{-2}	3.281×10^{-2}	0.3937
1 mm =	6.214×10^{-7}	10^{-6}	10^{-3}	3.281×10^{-3}	0.0394
1 μm =	6.214×10^{-10}	10^{-9}	10^{-6}	3.281×10^{-6}	3.937×10^{-5}
1 nm =	6.214×10^{-13}	10^{-12}	10^{-9}	3.281×10^{-9}	3.937×10^{-8}
1 Å =	6.214×10^{-14}	10^{-13}	10^{-10}	3.281×10^{-10}	3.937×10^{-9}

	cm	mm	μm	nm	Å
1 mi =	1.609×10^5	1.609×10^6	1.609×10^9	1.609×10^{12}	1.609×10^{13}
1 km =	10^5	10^6	10^9	10^{12}	10^{13}
1 m =	10^2	10^3	10^6	10^9	10^{10}
1 ft =	30.48	3.048×10^2	3.048×10^5	3.048×10^8	3.048×10^9
1 in =	2.540	25.40	2.540×10^4	2.540×10^7	2.540×10^8
1 cm =	1	10	10^4	10^7	10^8
1 mm =	10^{-1}	1	10^3	10^6	10^7
1 μm =	10^{-4}	10^{-3}	1	10^3	10^4
1 nm =	10^{-7}	10^{-6}	10^{-3}	1	10
1 Å =	10^{-8}	10^{-7}	10^{-4}	10^{-1}	1

Area

1 ft^2 = 0.0929030 m^2
1 in^2 = 6.4516×10^{-4} m^2
1 m^2 = 10.7639 ft^2

Mass

1 lb$_m$ = 0.453592 kg
1 kg = 2.20462 lb$_m$

Volume

1 ft^3 = 0.028317 m^3 = 7.48 gal
1 m^3 = 35.315 ft^3 = 264 gal
1 gal = 0.133 ft^3 = 3.79×10^{-3} m^3
1 L = 1000 cm^3 = 0.264 gal

Density

1 lb$_m$/ft^3 = 16.0186 kg/m^3
1 kg/m^3 = 0.062428 lb$_m$/ft^3

Energy

(1 kJ = 1 kW · s)
1 kJ = 0.94782 Btu† = 0.23885 kcal†
1 Btu = 1.0551 kJ = 0.25200 kcal
1 kcal = 4.1868 kJ = 3.9683 Btu
1 kWh = 3.60×10^6 J

Energy rate per unit area (heat flux, insolation)

1 W/m^2 = 0.31700 Btu†/(h · ft^2)
1 Btu/(h · ft^2) = 3.1546 W/m^2
1 Langley/s = 1 cal/(cm^2 · s) = 4.1868 W/cm^2
   = 1.327×10^4 Btu/(h · ft^2)

Table A-2 (*Continued*)

Energy rate	Heat-transfer coefficient
1 W = 3.4121 Btu†/h 1 Btu/h = 0.29307 W 1 horsepower = 2545 Btu/h	1 W/(m² · K) = 0.17611 Btu†/(h · ft² · °R) 1 Btu/(h · ft² · °R) = 5.6783 W/(m² · K)
Thermal conductivity	**Specific heat**

1 W/(m · K) = 0.57779 Btu/(h · ft · °R)
 	
1 Btu/(h · ft · °R) = 1.7307 W/(m · K)
	1 kJ/(kg · K) = 0.23885 Btu/(lb$_m$ · °R)
	    = 0.23885 kcal/(kg · K)
	1 Btu/(lb$_m$ · °R) = 4.1868 kJ/(kg · K)
	    = 1.0000 kcal/(kg · K)
	1 kcal/(kg · K) = 4.1868 kJ/(kg · K)
	    = 1.0000 Btu/(lb$_m$ · °R)

Energy per unit mass	**Temperature**
1 kJ/kg = 0.42992 Btu/lb$_m$ = 0.23885 kcal/kg 1 Btu/lb$_m$ = 2.3260 kJ/kg = 0.55556 kcal/kg 1 kcal/kg = 4.1868 kJ/kg = 1.8000 Btu/lb$_m$	K = 5/9°R = 5/9(°F + 459.67) = °C + 273.15 °R = 9/5 K = 9/5(°C + 273.15) = °F + 459.67 °F = 9/5°C + 32 °C = 5/9(°F − 32)

† Proc. Sixth International Conf. on the Properties of Steam, Issued by ASME, New York, October 1963.

Table A-3 Names and symbols for multipliers

Multiplier	Symbol	Prefix
10^{12}	T	tera
10^{9}	G	giga
10^{6}	M	mega
10^{3}	k	kilo
10^{2}	h	hecto
10^{1}	da	deka
10^{-1}	d	deci
10^{-2}	c	centi
10^{-3}	m	milli
10^{-6}	µ	micro
10^{-9}	n	nano
10^{-12}	p	pico

APPENDIX
B

INSOLATION AND WEATHER DATA

Table B-1 Average daily global horizontal insolation patterns†

	Pacific northwest (Tacoma)	Pacific southwest (L.A.)	North central (Bismarck)	Central (Dodge City)	Southwest (El Paso)	Gulf coast (Mobile)	Great Lakes (Madison)	Northeast (Boston)	Southeast (Charleston)
Jan.	1.5	3.5	1.9	3.1	3.8	3.0	2.0	2.0	2.9
Feb.	1.9	4.1	2.9	3.7	4.9	3.9	3.1	2.8	3.6
Mar.	3.5	6.0	4.1	4.8	6.2	4.3	4.1	4.0	4.3
Apr.	4.6	7.5	5.3	6.0	7.6	6.0	5.2	4.9	6.0
May	5.8	8.1	6.2	6.8	9.0	6.5	6.1	5.5	6.8
Jun.	6.8	8.5	7.1	7.3	8.5	6.5	7.0	6.0	6.2
Jul.	7.3	8.2	7.0	7.1	7.9	6.1	7.0	5.8	6.0
Aug.	5.8	7.0	6.2	6.8	7.4	5.7	6.1	5.5	5.8
Sept.	4.0	6.1	4.6	5.3	6.2	5.4	4.5	4.2	4.0
Oct.	2.5	5.0	3.2	4.3	5.2	4.4	3.0	3.0	4.2
Nov.	1.5	3.7	1.9	3.2	3.8	3.2	2.0	2.0	3.1
Dec.	0.9	3.2	1.5	2.5	3.5	2.8	1.6	1.6	2.8
Average	3.84	5.91	4.33	5.08	6.17	4.82	4.31	3.94	4.64
Mean deviation	1.88	1.69	1.74	1.48	1.66	1.22	1.68	1.39	1.27
% Mean deviation (% deviation from mean)	49.0	28.6	40.2	29.1	26.9	25.3	39.0	35.3	27.4

Source: E.C. Boes, I.J. Hall, R.R. Prairie, R.P. Stromberg, and H.E. Anderson, "Distribution of Direct and Total Solar Radiation Available for the USA," *Proceedings of the 1976 Annual Meeting of the Am. Sec. of ISES, Sharing the Sun*, Winnipeg, 1976, pp. 238–263.
† All values in $kWh/(m^2 \cdot day)$.

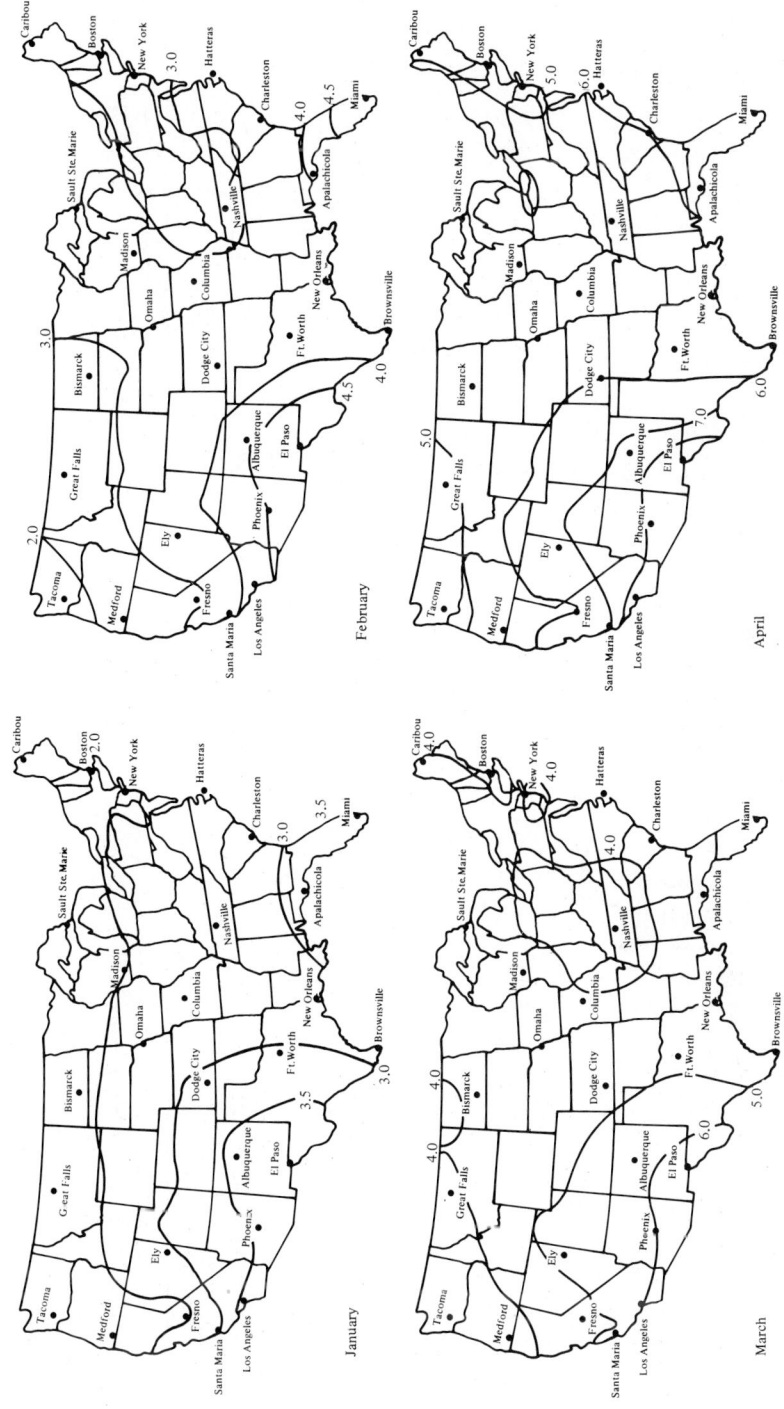

Figure B-1 Average daily global horizontal solar insolation kWh/(m² · day). (*Source*: E. C. Boes, I. J. Hall, R. R. Prairie, R. P. Stromberg, and H. E. Anderson, "Distribution of Direct and Total Solar Radiation Available for the USA," *Proceedings of the 1976 Annual Meeting of the Am. Sec. of ISES, Sharing the Sun, Winnipeg, 1976, pp. 238–263.*)

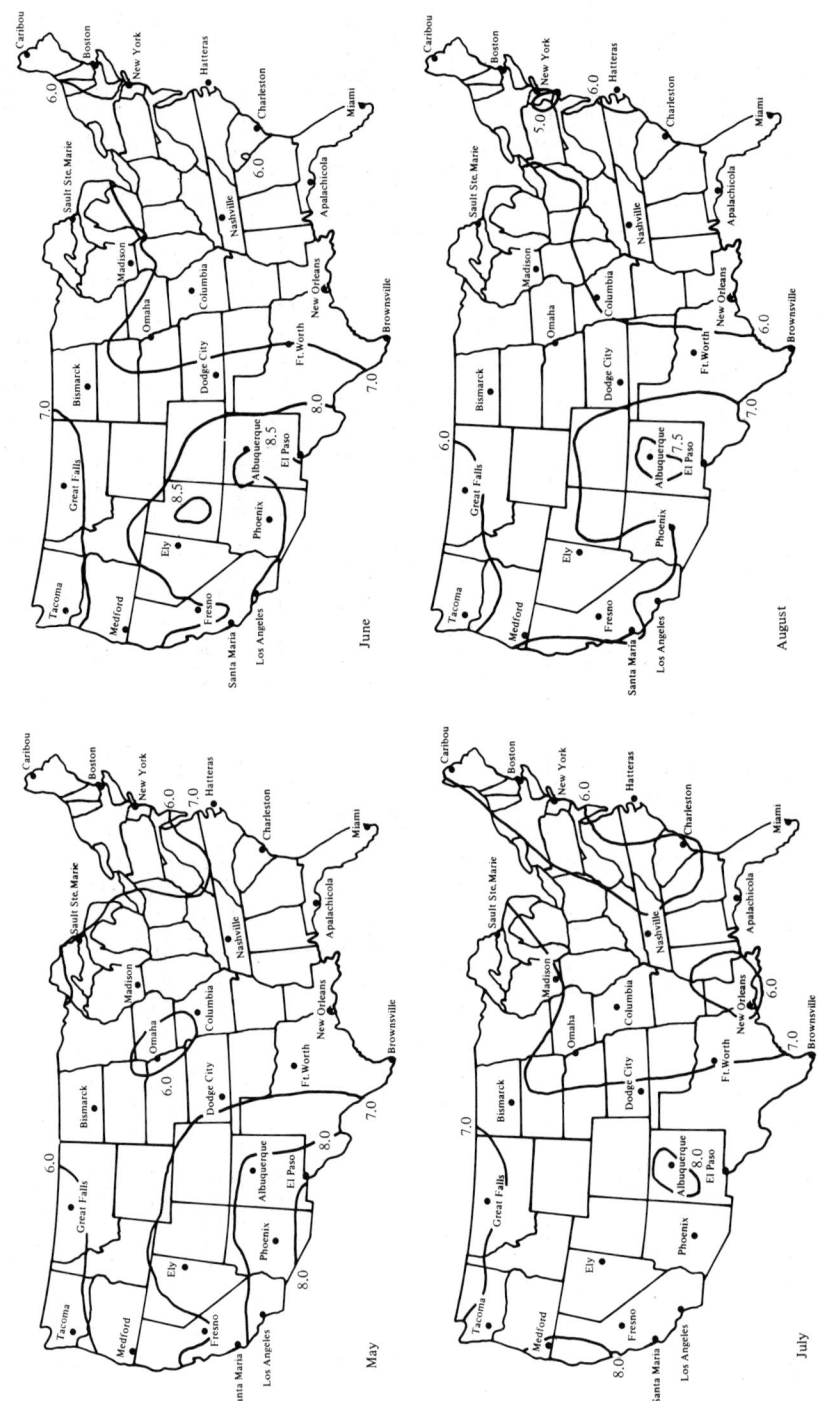

Figure B-1 (*Continued*)

Figure B-1 (*Continued*)

Table B-2 Insolation and Other Data for 80 Locations in the United States and Canada

$\bar{H}$ = Monthly average daily global insolation on a horizontal surface, Btu/day · ft^2; $\bar{K}_T$ = monthly average clearness index, $\bar{T}_a$ = average daytime ambient temperature, °F

		Jan.	Feb.	Mar.	Apr.	May	June	July	Aug.	Sept.	Oct.	Nov.	Dec.
ALASKA													
Annette Is	$\bar{H}$	236.2	428.4	883.4	1357.2	1634.7	1638.7	1632.1	1269.4	962	454.6	220.3	152
Lat. 55°02′N	$\bar{K}_T$	0.427	0.415	0.492	0.507	0.484	0.441	0.454	0.427	0.449	0.347	0.304	0.361
El. 110 ft	$\bar{T}_a$	35.8	37.5	39.7	44.4	51.0	56.2	58.6	59.8	54.8	48.2	41.9	37.4
Barrow	$\bar{H}$	13.3	143.2	713.3	1491.5	1883	2055.3	1602.2	953.5	428.4	152.4	22.9	
Lat. 71°20′N	$\bar{K}_T$	—	0.776	0.773	0.726	0.553	0.533	0.448	0.377	0.315	0.35	...	
El. 22 ft	$\bar{T}_a$	−13.2	−15.9	−12.7	2.1	20.5	35.4	41.6	40.0	31.7	18.6	2.6	−8.6
Bethel	$\bar{H}$	142.4	404.8	1052.4	1662.3	1711.8	1698.1	1401.8	938.7	755	430.6	164.9	83
Lat. 60°47′N	$\bar{K}_T$	0.536	0.557	0.704	0.675	0.519	0.458	0.398	0.336	0.406	0.432	0.399	0.459
El. 125 ft	$\bar{T}_a$	9.2	11.6	14.2	29.4	42.7	55.5	56.9	54.8	47.4	33.7	19.0	9.4
Fairbanks	$\bar{H}$	66	283.4	860.5	1481.2	1806.2	1970.8	1702.9	1247.6	699.6	323.6	104.1	20.3
Lat. 64°49′N	$\bar{K}_T$	0.639	0.556	0.674	0.647	0.546	0.529	0.485	0.463	0.419	0.416	0.47	0.458
El. 436 ft	$\bar{T}_a$	−7.0	0.3	13.0	32.2	50.5	62.4	63.8	58.3	47.1	29.6	5.5	−6.6
Matanuska	$\bar{H}$	119.2	345	...	1327.6	1628.4	1727.6	1526.9	1169	737.3	373.8	142.8	56.4
Lat. 61°30′N	$\bar{K}_T$	0.513	0.503	...	0.545	0.494	0.466	0.434	0.419	0.401	0.390	0.372	0.36
El. 180 ft.	$\bar{T}_a$	13.9	21.0	27.4	38.6	50.3	57.6	60.1	58.1	50.2	37.7	22.9	13.9
ALBERTA													
Edmonton	$\bar{H}$	331.7	652.4	1165.3	1541.7	1900.4	1914.4	1964.9	1528	1113.3	704.4	413.6	245
Lat. 53°35′N	$\bar{K}_T$	0.529	0.585	0.624	0.564	0.558	0.514	0.549	0.506	0.506	0.504	0.510	0.492
El. 2219 ft	$\bar{T}_a$	10.4	14	26.3	42.9	55.4	61.3	66.6	63.2	54.2	44.1	26.7	14.0
ARKANSAS													
Little Rock	$\bar{H}$	704.4	974.2	1335.8	1669.4	1960.1	2091.5	2081.2	1938.7	1640.6	1282.6	913.6	701.1
Lat. 34°44′N	$\bar{K}_T$	0.424	0.458	0.496	0.513	0.545	0.559	0.566	0.574	0.561	0.552	0.484	0.463
El. 265 ft	$\bar{T}_a$	44.6	48.5	56.0	65.8	73.1	76.7	85.1	84.6	78.3	67.9	54.7	46.7
ARIZONA													
Phoenix	$\bar{H}$	1126.6	1514.7	1967.1	2388.2	2709.6	2781.5	2450.5	2299.6	2131.3	1688.9	1290	1040.9
Lat. 33°26′N	$\bar{K}_T$	0.65	0.691	0.716	0.728	0.753	0.745	0.667	0.677	0.722	0.708	0.657	0.652
El. 1112 ft	$\bar{T}_a$	54.2	58.8	64.7	72.2	80.8	89.2	94.6	92.5	87.4	75.8	63.6	56.7

Location		Jan	Feb	Mar	Apr	May	Jun	Jul	Aug	Sep	Oct	Nov	Dec
Tucson	$\bar{H}$	1171.9	1453.8	...	2434.7	...	2601.4	2292.2	2179.7	2122.5	1640.9	1322.1	1132.1
Lat. 32°07'N	$\bar{K}_T$	0.648	0.646	...	0.738	...	0.698	0.625	0.640	0.710	0.672	0.650	0.679
El. 2556 ft	$\bar{T}_a$	53.7	57.3	62.3	69.7	78.0	87.0	90.1	87.4	84.0	73.9	62.5	56.1
CALIFORNIA													
Davis	$\bar{H}$	599.2	945	1504	1959	2368.6	2619.2	2565.6	2287.8	1856.8	1289.5	795.6	550.5
Lat. 38°33'N	$\bar{K}_T$	0.416	0.490	0.591	0.617	0.662	0.697	0.697	0.687	0.664	0.598	0.477	0.421
El. 51 ft	$\bar{T}_a$	47.6	52.1	56.8	63.1	69.6	75.7	81	79.4	76.7	67.8	57	48.7
Fresno	$\bar{H}$	712.9	1116.6	1652.8	2049.4	2409.2	2641.7	2512.2	2300.7	1897.8	1415.5	906.6	616.6
Lat. 36°46'N	$\bar{K}_T$	0.462	0.551	0.632	0.638	0.672	0.703	0.682	0.686	0.665	0.635	0.512	0.44
El. 331 ft	$\bar{T}_a$	47.3	53.9	59.1	65.6	73.5	80.7	87.5	84.9	78.6	68.7	57.3	48.9
Inyokern	$\bar{H}$	1148.7	1554.2	2136.9	2594.8	2925.4	3108.8	2908.8	2759.4	2409.2	1819.2	3170.1	1094.4
Lat. 35°39'N	$\bar{K}_T$	0.716	0.745	0.803	0.8	0.815	0.830	0.790	0.820	0.834	0.795	0.743	0.742
El. 2440 ft	$\bar{T}_a$	47.3	53.9	59.1	65.6	73.5	80.7	87.5	84.9	78.6	68.7	57.3	48.9
Los Angeles, (WBO)	$\bar{H}$	911.8	1223.6	1640.9	1866.8	2061.2	2259	2428.4	2198.9	1891.5	1362.3	1053.1	877.8
Lat. 34°03'N	$\bar{K}_T$	0.538	0.568	0.602	0.571	0.573	0.605	0.66	0.648	0.643	0.578	0.548	0.566
El. 99 ft	$\bar{T}_a$	57.9	59.2	61.8	64.3	67.6	70.7	75.8	76.1	74.2	69.6	65.4	60.2
Los Angeles, (WBAS)	$\bar{H}$	930.6	1284.1	1729.5	1948	2196.7	2272.3	2413.6	2155.3	1898.1	1372.7	1082.3	901.1
Lat. 33°56'N	$\bar{K}_T$	0.547	0.596	0.635	0.595	0.610	0.608	0.657	0.635	0.641	0.574	0.551	0.566
El. 99 ft	$\bar{T}_a$	56.2	56.9	59.2	61.4	64.2	66.7	69.6	70.2	69.1	66.1	62.6	58.7
Riverside	$\bar{H}$	999.6	1335	1750.5	1943.2	2282.3	2492.6	2443.5	2263.8	1955.3	1509.6	1169	979.7
Lat. 33°57'N	$\bar{K}_T$	0.589	0.617	0.643	0.594	0.635	0.667	0.665	0.668	0.665	0.639	0.606	0.626
El. 1020 ft	$\bar{T}_a$	55.3	57.0	60.6	65.0	69.4	74.0	81.0	81.0	78.5	71.0	63.1	57.2
Santa Maria	$\bar{H}$	983.8	1296.3	1805.9	2067.9	2375.6	2599.6	2540.6	2293.3	1965.7	1566.4	1169	943.9
Lat. 34°54'N	$\bar{K}_T$	0.595	0.613	0.671	0.636	0.661	0.695	0.690	0.678	0.674	0.676	0.624	0.627
El. 238 ft	$\bar{T}_a$	54.1	55.3	57.6	59.5	61.2	63.5	65.3	65.7	65.9	64.1	60.8	56.1
COLORADO													
Grand Junction	$\bar{H}$	848	1210.7	1622.9	2002.2	2300.3	2645.4	2517.7	2157.2	1957.5	1394.8	969.7	793.4
Lat. 39°07'N	$\bar{K}_T$	0.597	0.633	0.643	0.632	0.643	0.704	0.690	0.65	0.705	0.654	0.59	0.621
El. 4849 ft	$\bar{T}_a$	26.9	35.0	44.6	55.8	66.3	75.7	82.5	79.6	71.4	58.3	42.0	31.4
Grand Lake	$\bar{H}$	735	1135.4	1579.3	1876.7	1974.9	2369.7	2103.3	1708.5	1715.8	1212.2	775.6	660.5
Lat. 40°15'N	$\bar{K}_T$	0.541	0.615	0.637	0.597	0.553	0.63	0.572	0.516	0.626	0.583	0.494	0.542
El. 8389 ft	$\bar{T}_a$	18.5	23.1	28.5	39.1	48.7	56.6	62.8	61.5	55.5	45.2	30.3	22.6

Table B-2 (*Continued*)

		Jan.	Feb.	Mar.	Apr.	May	June	July	Aug.	Sept.	Oct.	Nov.	Dec.
DISTRICT OF COLUMBIA													
Washington (WBCO)	$\bar{H}$	632.4	901.5	1255	1600.4	1846.8	2080.8	1929.9	1712.2	1446.1	1083.4	763.5	594.1
Lat. 38°51′N	$\bar{K}_T$	0.445	0.470	0.496	0.504	0.516	0.553	0.524	0.516	0.520	0.506	0.464	0.460
El. 64 ft	$\bar{T}_a$	38.4	39.6	48.1	57.5	67.7	76.2	79.9	77.9	72.2	60.9	50.2	40.2
FLORIDA													
Apalachicola	$\bar{H}$	1107	1378.2	1654.2	2040.9	2268.6	2195.9	1978.6	1912.9	1703.3	1544.6	1243.2	982.3
Lat. 29°45′N	$\bar{K}_T$	0.577	0.584	0.576	0.612	0.630	0.594	0.542	0.558	0.559	0.608	0.574	0.543
El. 35 ft	$\bar{T}_a$	57.3	59.0	62.9	69.5	76.4	81.8	83.1	83.1	80.6	73.2	63.7	58.55
Gainesville	$\bar{H}$	1036.9	1324.7	1635	1956.4	1934.7	1960.9	1895.6	1873.8	1615.1	1312.2	1169.7	919.5
Lat. 29°39′N	$\bar{K}_T$	0.535	0.56	0.568	0.587	0.538	0.531	0.519	0.547	0.529	0.515	0.537	0.508
El. 165 ft	$\bar{T}_a$	62.1	63.1	67.5	72.8	79.4	83.4	83.8	84.1	82	75.7	67.2	62.4
Miami	$\bar{H}$	1292.2	1554.6	1828.8	2020.6	2068.6	1991.5	1992.6	1890.8	1646.8	1436.5	1321	1183.4
Lat. 25°47′N	$\bar{K}_T$	0.604	0.616	0.612	0.600	0.578	0.545	0.552	0.549	0.525	0.534	0.559	0.588
El. 9 ft	$\bar{T}_a$	71.6	72.0	73.8	77.0	79.9	82.9	84.1	84.5	83.3	80.2	75.6	72.6
Tampa	$\bar{H}$	1223.6	1461.2	1771.9	2016.2	2228	2146.5	1991.9	1845.4	1687.8	1493.3	1328.4	1119.5
Lat. 27°55′N	$\bar{K}_T$	0.605	0.600	0.606	0.602	0.620	0.583	0.548	0.537	0.546	0.572	0.590	0.589
El. 11 ft	$\bar{T}_a$	64.2	65.7	68.8	74.3	79.4	83.0	84.0	84.4	82.9	77.2	69.6	65.5
GEORGIA													
Atlanta	$\bar{H}$	848	1080.1	1426.9	1807	2618.1	2002.6	2002.9	1898.1	1519.2	1290.8	997.8	751.0
Lat. 33°39′N	$\bar{K}_T$	0.493	0.496	0.522	0.551	0.561	0.564	0.545	0.559	0.515	0.543	0.510	0.474
El. 976 ft	$\bar{T}_a$	47.2	49.6	55.9	65.0	73.2	80.9	82.4	81.6	77.4	66.5	54.8	47.7
Griffin	$\bar{H}$	889.6	1135.8	1450.9	1923.6	2163.1	2176	2064.9	1961.2	1605.9	1352.4	1073.8	781.5
Lat. 33°15′N	$\bar{K}_T$	0.513	0.517	0.528	0.586	0.601	0.583	0.562	0.578	0.543	0.565	0.545	0.487
El. 980 ft	$\bar{T}_a$	48.9	51.0	59.1	66.7	74.6	81.2	83.0	82.2	78.4	68	57.3	49.4
IDAHO													
Boise	$\bar{H}$	518.8	884.9	1280.4	1814.4	2189.3	2376.7	2500.3	2149.4	1717.7	1128.4	678.6	456.8
Lat. 43°34′N	$\bar{K}_T$	0.446	0.533	0.548	0.594	0.619	0.631	0.684	0.660	0.656	0.588	0.494	0.442
El. 2844 ft	$\bar{T}_a$	29.5	36.5	45.0	53.5	62.1	69.3	79.6	77.2	66.7	56.3	42.3	33.1

		Jan	Feb	Mar	Apr	May	Jun	Jul	Aug	Sep	Oct	Nov	Dec
ILLINOIS Lemont Lat. 41°40'N El. 595 ft	$\bar{H}$ $\bar{K}_T$ $\bar{T}_a$	(590) (0.464) 28.9	879 0.496 30.3	1255.7 0.520 39.5	1481.5 0.477 49.7	1866 0.525 59.2	2041.7 0.542 70.8	1990.8 0.542 75.6	1836.9 0.559 74.3	1469.4 0.547 67.2	1015.5 0.506 57.6	(639) (0.433) 43.0	(531) (0.467) 30.6
INDIANA Indianapolis Lat. 39°44'N El. 793 ft	$\bar{H}$ $\bar{K}_T$ $\bar{T}_a$	526.2 0.380 31.3	797.4 0.424 33.9	1184.1 0.472 43.0	1481.2 0.47 54.1	1828 0.511 64.9	2042 0.543 74.8	2039.5 0.554 79.6	1832.1 0.552 77.4	1513.3 0.549 70.6	1094.4 0.520 59.3	662.4 0.413 44.2	491.1 0.391 33.4
KANSAS Dodge City Lat. 37°46'N El. 2592 ft	$\bar{H}$ $\bar{K}_T$ $\bar{T}_a$	953.1 0.639 33.8	1186.3 0.598 38.7	1565.7 0.606 46.5	1975.6 0.618 57.7	2126.5 0.594 66.7	2459.8 0.655 77.2	2400.7 0.652 83.8	2210.7 0.663 82.4	1841.7 0.654 73.7	1421 0.650 61.7	1065.3 0.625 46.5	873.8 0.652 36.8
KENTUCKY Lexington Lat. 38°02'N El. 979 ft	$\bar{H}$ $\bar{K}_T$ $\bar{T}_a$	... ... 36.5	... ... 38.8	... ... 47.4	1834.7 0.575 57.8	2171.2 0.606 67.5	... ... 76.2	2246.5 0.610 79.8	2064.9 0.619 78.2	1775.6 0.631 72.8	1315.8 0.604 61.2	... ... 47.6	681.5 0.513 38.5
LOUISIANA Lake Charles Lat. 30°13'N El. 12 ft	$\bar{H}$ $\bar{K}_T$ $\bar{T}_a$	899.2 0.473 55.3	1145.7 0.492 58.7	1487.4 0.521 63.5	1801.8 0.542 70.9	2080.4 0.578 77.4	2213.3 0.597 83.4	1968.6 0.538 84.8	1910.3 0.558 85.0	1678.2 0.553 81.5	1505.5 0.597 73.8	1122.1 0.524 62.6	875.6 0.494 56.9
MAINE Caribou Lat. 46°52'N El. 628 ft	$\bar{H}$ $\bar{K}_T$ $\bar{T}_a$	497 0.504 11.5	861.6 0.579 12.8	1360.1 0.619 24.4	1495.9 0.507 37.3	1779.7 0.509 51.8	1779.7 0.473 61.6	1898.1 0.522 67.2	1675.6 0.527 65.0	1254.6 0.506 56.2	793 0.455 44.7	415.5 0.352 31.3	398.9 0.470 16.8
Portland Lat. 43°39'N El. 63 ft	$\bar{H}$ $\bar{K}_T$ $\bar{T}_a$	565.7 0.482 23.7	874.5 0.524 24.5	1329.5 0.569 34.4	1528.4 0.500 44.8	1923.2 0.544 55.4	2017.3 0.536 65.1	2095.6 0.572 71.1	1799.2 0.554 69.7	1428.8 0.546 61.9	1035 0.539 51.8	591.5 0.431 40.3	507.7 0.491 28.0
MANITOBA Winnipeg Lat. 49°54'N El. 786 ft	$\bar{H}$ $\bar{K}_T$ $\bar{T}_a$	488.2 0.601 3.2	835.4 0.636 7.1	1354.2 0.661 21.3	1641.3 0.574 40.9	1904.4 0.550 55.9	1962 0.524 65.3	2123.6 0.587 71.9	1761.2 0.567 69.4	1190.4 0.504 58.6	767.5 0.482 45.6	444.6 0.436 25.2	345 0.503 10.1
MASSACHUSETTS Blue Hill Lat. 42°13'N El. 629 ft	$\bar{H}$ $\bar{K}_T$ $\bar{T}_a$	555.3 0.445 28.3	797 0.458 28.3	1143.9 0.477 36.9	1438 0.464 46.9	1776.4 0.501 58.5	1943.9 0.516 67.2	1881.5 0.513 72.3	1622.1 0.495 70.6	1314 0.492 64.2	941 0.472 54.1	592.2 0.406 43.3	482.3 0.436 31.5

Table B-2 (*Continued*)

		Jan.	Feb.	Mar.	Apr.	May	June	July	Aug.	Sept.	Oct.	Nov.	Dec.
MASSACHUSETTS (*Cont'd.*)													
Boston Lat. 42°22'N El. 29 ft	$\bar{H}$ $\bar{K}_T$ $\bar{T}_a$	505.5 0.410 31.4	738 0.426 31.4	1067.1 0.445 39.9	1355 0.438 49.5	1769 0.499 60.4	1864 0.495 69.8	1860.5 0.507 74.5	1570.1 0.480 73.8	1267.5 0.477 66.8	896.7 0.453 57.4	535.8 0.372 46.6	442.8 0.400 34.9
East Wareham Lat. 41°46'N El. 18 ft	$\bar{H}$ $\bar{K}_T$ $\bar{T}_a$	504.4 0.398 32.2	762.4 0.431 31.6	1132.1 0.469 39.0	1392.6 0.449 48.3	1704.8 0.480 58.9	1958.3 0.520 67.5	1873.8 0.511 74.1	1607.4 0.489 72.8	1363.8 0.508 65.9	996.7 0.496 56	636.2 0.431 46	521 0.461 34.8
MICHIGAN													
East Lansing Lat. 42°44'N El. 856 ft	$\bar{H}$ $\bar{K}_T$ $\bar{T}_a$	425.8 0.35 26.0	739.1 0.431 26.4	1086 0.456 35.7	1249.8 0.406 48.4	1732.8 0.489 59.8	1914 0.508 70.3	1884.5 0.514 74.5	1627.7 0.498 72.4	1303.3 0.493 65.0	891.5 0.456 53.5	473.1 0.333 40.0	379.7 0.349 29.0
Sault Ste. Marie Lat. 46°28'N El. 724 ft	$\bar{H}$ $\bar{K}_T$ $\bar{T}_a$	488.6 0.490 16.3	843.9 0.560 16.2	1336.5 0.606 25.6	1559.4 0.526 39.5	1962.3 0.560 52.1	2064.2 0.549 61.6	2149.4 0.590 67.3	1767.9 0.554 66.0	1207 0.481 57.9	809.2 0.457 46.8	392.2 0.323 33.4	359.8 0.408 21.9
MINNESOTA													
St. Cloud Lat. 45°35'N El. 1034 ft	$\bar{H}$ $\bar{K}_T$ $\bar{T}_a$	632.8 0.595 13.6	976.7 0.629 16.9	1383 0.614 29.8	1598.1 0.534 46.2	1859.4 0.530 58.8	2003.3 0.533 68.5	2087.8 0.573 74.4	1828.4 0.570 71.9	1369.4 0.539 62.5	890.4 0.490 50.2	545.4 0.435 32.1	463.1 0.504 18.3
MISSOURI													
Columbia Lat. 38°58'N El. 785 ft	$\bar{H}$ $\bar{K}_T$ $\bar{T}_a$	651.3 0.458 32.5	941.3 0.492 36.5	1315.8 0.520 45.9	1631.3 0.514 57.7	1999.6 0.559 66.7	2129.1 0.566 75.9	2148.7 0.585 81.1	1953.1 0.588 79.4	1689.6 0.606 71.9	1202.6 0.562 61.4	839.5 0.510 46.1	590.4 0.457 35.8
MONTANA													
Glasgow Lat. 48°13'N El. 2277 ft	$\bar{H}$ $\bar{K}_T$ $\bar{T}_a$	572.7 0.621 13.3	965.7 0.678 17.3	1437.6 0.672 31.1	1741.3 0.597 47.8	2127.3 0.611 59.3	2261.6 0.602 67.3	2414.7 0.666 76	1984.5 0.630 73.2	1531 0.629 61.2	997 0.593 49.2	574.9 0.516 31.0	428.4 0.548 18.6
Great Falls Lat. 47°29'N El. 3664 ft	$\bar{H}$ $\bar{K}_T$ $\bar{T}_a$	524 0.552 25.4	869.4 0.596 27.6	1369.7 0.631 35.6	1621.4 0.551 47.7	1970.8 0.565 57.5	2179.3 0.580 64.3	2383 0.656 73.8	1986.3 0.627 71.3	1536.5 0.626 60.6	984.9 0.574 51.4	575.3 0.503 38.0	420.7 0.518 29.1

		Jan	Feb	Mar	Apr	May	Jun	Jul	Aug	Sep	Oct	Nov	Dec	Ann
NEBRASKA														
Lincoln Lat. 40°51'N El. 1189 ft	$\bar{H}$ $\bar{K}_T$ $\bar{T}_a$	712.5 0.542 27.8	955.7 0.528 32.1	1299.6 0.532 42.4	1587.8 0.507 55.8	1856.1 0.522 65.8	2040.6 0.542 76.0	2011.4 0.547 82.6	1902.6 0.577 80.2	1543.5 0.568 71.5	1215.8 0.596 59.9	773.4 0.508 43.2	643.2 0.545 31.8	
NEVADA														
Ely Lat. 39°17'N El. 6262 ft	$\bar{H}$ $\bar{K}_T$ $\bar{T}_a$	871.6 0.618 27.3	1255 0.660 32.1	1749.8 0.692 39.5	2103.3 0.664 48.3	2322.1 0.649 57.0	2649 0.704 65.4	2417 0.656 74.5	2307.7 0.695 72.3	1935 0.696 63.7	1473 0.691 52.1	1078.6 0.658 39.9	814.8 0.64 31.1	
Las Vegas Lat. 36°05'N El. 2162 ft	$\bar{H}$ $\bar{K}_T$ $\bar{T}_a$	1035.8 0.654 47.5	1438 0.697 53.9	1926.5 0.728 60.3	2322.8 0.719 69.5	2629.5 0.732 78.3	2799.2 0.746 88.2	2524 0.685 95.0	2342 0.697 92.9	2062 0.716 85.4	1602.6 0.704 71.7	1190 0.657 57.8	964.2 0.668 50.2	
NEW JERSEY														
Seabrook Lat. 39°30'N El. 100 ft	$\bar{H}$ $\bar{K}_T$ $\bar{T}_a$	591.9 0.426 39.5	854.2 0.453 37.6	1195.6 0.476 43.9	1518.8 0.481 54.7	1800.7 0.504 64.9	1964.6 0.522 74.1	1949.8 0.530 79.8	1715 0.517 77.7	1445.7 0.524 69.7	1071.9 0.508 61.2	721.8 0.449 48.5	522.5 0.416 39.3	
NEW MEXICO														
Albuquerque Lat. 35°03'N El. 5314 ft	$\bar{H}$ $\bar{K}_T$ $\bar{T}_a$	1150.9 0.704 37.3	1453.9 0.691 43.3	1925.4 0.719 50.1	2343.5 0.722 59.6	2560.9 0.713 69.4	2757.5 0.737 79.1	2561.2 0.695 82.8	2387.8 0.708 80.6	2120.3 0.728 73.6	1639.8 0.711 62.1	1274.2 0.684 47.8	1051.6 0.704 39.4	
NEW YORK														
Ithaca Lat. 42°27'N El. 950 ft	$\bar{H}$ $\bar{K}_T$ $\bar{T}_a$	434.3 0.351 27.2	755 0.435 26.5	1074.9 0.45 36	1322.9 0.428 48.4	1779.3 0.502 59.6	2025.8 0.538 68.9	2031.3 0.554 73.9	1736.9 0.530 71.9	1320.3 0.497 64.2	918.4 0.465 53.6	466.4 0.324 41.5	370.8 0.337 29.6	
New York Lat. 40°46'N El. 52 ft	$\bar{H}$ $\bar{K}_T$ $\bar{T}_a$	539.5 0.406 35.0	790.8 0.435 34.9	1180.4 0.480 43.1	1426.2 0.455 52.3	1738.4 0.488 63.3	1994.1 0.53 72.2	1938.7 0.528 76.9	1605.9 0.486 75.3	1349.4 0.500 69.5	977.8 0.475 59.3	598.1 0.397 48.3	476 0.403 37.7	
Sayville Lat. 40°30'N El. 20 ft	$\bar{H}$ $\bar{K}_T$ $\bar{T}_a$	602.9 0.453 35	936.2 0.511 34.9	1259.4 0.510 43.1	1560.5 0.498 52.3	1857.2 0.522 63.3	2123.2 0.564 72.2	2040.9 0.555 76.9	1734.7 0.525 75.3	1446.8 0.530 69.5	1087.4 0.527 59.3	697.8 0.450 48.3	533.9 0.447 37.7	
Schenectady Lat. 42°50'N El. 217 ft	$\bar{H}$ $\bar{K}_T$ $\bar{T}_a$	488.2 0.406 24.7	753.5 0.441 24.6	1026.6 0.433 34.9	1272.3 0.413 48.3	1553.1 0.438 61.7	1687.8 0.448 70.8	1662.3 0.454 76.9	1494.8 0.458 73.7	1124.7 0.426 64.6	820.6 0.420 53.1	436.2 0.309 40.1	356.8 0.331 28.0	
Upton Lat. 40°52'N El. 75 ft	$\bar{H}$ $\bar{K}_T$ $\bar{T}_a$	583 0.444 35.0	872.7 0.483 34.9	1280.4 0.522 43.1	1609.9 0.514 52.3	1891.5 0.532 63.3	2159 0.574 72.2	2044.6 0.557 76.9	1789.6 0.542 75.3	1472.7 0.542 69.5	1102.6 0.538 59.3	686.7 0.448 48.3	551.3 0.467 37.7	

Table B-2 (*Continued*)

		Jan.	Feb.	Mar.	Apr.	May	June	July	Aug.	Sept.	Oct.	Nov.	Dec.
NORTH CAROLINA													
Greensboro	$\bar{H}$	743.9	1031.7	1323.2	1755.3	1988.5	2111.4	2033.9	1810.3	1517.3	1202.6	908.1	690.8
Lat. 36°05'N	$\bar{K}_T$	0.469	0.499	0.499	0.543	0.554	0.563	0.552	0.538	0.527	0.531	0.501	0.479
El. 891 ft	$\bar{T}_a$	42.0	44.2	51.7	60.8	69.9	78.0	80.2	78.9	73.9	62.7	51.5	43.2
Hatteras	$\bar{H}$	891.9	1184.1	1590.4	2128	2376.4	2438	2334.3	2085.6	1758.3	1337.6	1053.5	798.1
Lat. 35°13'N	$\bar{K}_T$	0.546	0.563	0.593	0.655	0.661	0.652	0.634	0.619	0.605	0.58	0.566	0.535
El. 7 ft	$\bar{T}_a$	49.9	49.5	54.7	61.5	69.9	77.2	80.0	79.8	76.7	67.9	59.1	51.3
NORTH DAKOTA													
Bismarck	$\bar{H}$	587.4	934.3	1328.4	1668.2	2056.1	2173.8	2305.5	1929.1	1441.3	1018.1	600.4	464.2
Lat. 46°47'N	$\bar{K}_T$	0.594	0.628	0.605	0.565	0.588	0.579	0.634	0.606	0.581	0.584	0.510	0.547
El. 1660 ft	$\bar{T}_a$	12.4	15.9	29.7	46.6	58.6	67.9	76.1	73.5	61.6	49.6	31.4	18.4
OHIO													
Cleveland	$\bar{H}$	466.8	681.9	1207	1443.9	1928.4	2102.6	2094.4	1840.6	1410.3	997	526.6	427.3
Lat. 41°24'N	$\bar{K}_T$	0.361	0.383	0.497	0.464	0.543	0.559	0.571	0.559	0.524	0.491	0.351	0.371
El. 805 ft	$\bar{T}_a$	30.8	30.9	39.4	50.2	62.4	72.7	77.0	75.1	68.5	57.4	44.0	32.8
Columbus	$\bar{H}$	486.3	746.5	1112.5	1480.8	1839.1	(2111)	2041.3	1572.7	1189.3	919.5	479	430.2
Lat. 40°00'N	$\bar{K}_T$	0.356	0.401	0.447	0.470	0.515	(0.561)	0.555	0.475	0.433	0.441	0.302	0.351
El. 833 ft	$\bar{T}_a$	32.1	33.7	42.7	53.5	64.4	74.2	78	75.9	70.1	58	44.5	34.0
OKLAHOMA													
Oklahoma City	$\bar{H}$	938	1192.6	1534.3	1849.4	2005.1	2355	2273.8	2211	1819.2	1409.6	1085.6	897.4
Lat. 35°24'N	$\bar{K}_T$	0.580	0.571	0.576	0.570	0.558	0.629	0.618	0.565	0.628	0.614	0.588	0.608
El. 1304 ft	$\bar{T}_a$	40.1	45.0	53.2	63.6	71.2	80.6	85.5	85.4	77.4	66.5	52.2	43.1
Stillwater	$\bar{H}$	763.8	1081.5	1463.8	1702.6	1879.3	2235.8	2224.3	2039.1	1724.3	1314	991.5	783
Lat. 36°09'N	$\bar{K}_T$	0.484	0.527	0.555	0.528	0.523	0.596	0.604	0.607	0.599	0.581	0.548	0.544
El. 910 ft	$\bar{T}_a$	41.2	45.6	53.8	64.2	71.6	81.1	85.9	85.9	77.5	67.6	52.6	43.9
ONTARIO													
Ottawa	$\bar{H}$	539.1	852.4	1250.5	1506.6	1857.2	2084.5	2045.4	1752.4	1326.6	826.9	458.7	408.5
Lat. 45°20'N	$\bar{K}_T$	0.499	0.540	0.554	0.502	0.529	0.554	0.560	0.546	0.521	0.450	0.359	0.436
El. 339 ft	$\bar{T}_a$	14.6	15.6	27.7	43.3	57.5	67.5	71.9	69.8	61.5	48.9	35	19.6

Location		Jan	Feb	Mar	Apr	May	Jun	Jul	Aug	Sep	Oct	Nov	Dec	Ann
Toronto Lat. 43°41'N El. 379 ft	$\bar{H}$ $\bar{K}_T$ $\bar{T}_a$	451.3 0.388 26.5	674.5 0.406 26.0	1088.9 0.467 34.2	1388.2 0.455 46.3	1785.2 0.506 58	1941.7 0.516 68.4	1968.6 0.539 73.8	1622.5 0.500 71.8	1284.1 0.493 64.3	835 0.438 52.6	458.3 0.336 40.9	352.8 0.346 30.2	
OREGON														
Astoria Lat. 46°12'N El. 8 ft	$\bar{H}$ $\bar{K}_T$ $\bar{T}_a$	338.4 0.330 41.3	607 0.397 44.7	1008.5 0.454 46.9	1401.5 0.471 51.3	1838.7 0.524 55.0	1753.5 0.466 59.3	2007.7 0.551 62.6	1721 0.538 63.6	1322.5 0.526 62.2	780.4 0.435 55.7	413.6 0.336 48.5	295.2 0.332 43.9	
Medford Lat. 42°23'N El. 1329 ft	$\bar{H}$ $\bar{K}_T$ $\bar{T}_a$	435.4 0.353 39.4	804.4 0.464 45.4	1259.8 0.527 50.8	1807.4 0.584 56.3	2216.2 0.625 63.1	2440.5 0.648 69.4	2607.4 0.710 76.9	2261.6 0.689 76.4	1672.3 0.628 69.6	1043.5 0.526 58.7	558.7 0.384 47.1	346.5 0.313 40.5	
PENNSYLVANIA														
State College Lat. 40°48'N El. 1175 ft	$\bar{H}$ $\bar{K}_T$ $\bar{T}_a$	501.8 0.381 31.3	749.1 0.413 31.4	1106.6 0.451 39.8	1399.2 0.448 51.3	1754.6 0.493 63.4	2027.6 0.539 71.8	1968.2 0.536 75.8	1690 0.512 73.4	1336.1 0.492 66.1	1017 0.496 55.6	580.1 0.379 43.2	4443.9 0.376 32.6	
RHODE ISLAND														
Newport Lat. 41°29'N El. 60 ft	$\bar{H}$ $\bar{K}_T$ $\bar{T}_a$	565.7 0.438 29.5	856.4 0.482 32.0	1231.7 0.507 39.6	1484.8 0.477 48.2	1849 0.520 58.6	2019.2 0.536 67.0	1942.8 0.529 73.2	1687.1 0.513 72.3	1411.4 0.524 66.7	1035.4 0.512 56.2	656.1 0.44 46.5	527.7 0.460 34.4	
SOUTH CAROLINA														
Charleston Lat. 32°54'N El. 46 ft	$\bar{H}$ $\bar{K}_T$ $\bar{T}_a$	946.1 0.541 53.6	1152.8 0.521 55.2	1352.4 0.491 60.6	1918.8 0.584 67.8	2063.4 0.574 74.8	2113.3 0.567 80.9	1649.4 0.454 82.9	1933.6 0.569 82.3	1557.2 0.525 79.1	1332.1 0.554 69.8	1073.8 0.539 59.8	952 0.586 54.0	
SOUTH DAKOTA														
Rapid City Lat. 44°09'N El. 3218 ft	$\bar{H}$ $\bar{K}_T$ $\bar{T}_a$	687.8 0.601 24.7	1032.5 0.627 27.4	1503.7 0.649 34.7	1807 0.594 48.2	2028 0.574 58.3	2193.7 0.583 67.3	2235.8 0.612 76.3	2019.9 0.622 75.0	1628 0.628 64.7	1179.3 0.624 52.9	763.1 0.566 38.7	590.4 0.588 29.2	
TEXAS														
Brownsville Lat. 25°55'N El. 20 ft	$\bar{H}$ $\bar{K}_T$ $\bar{T}_a$	1105.9 0.517 63.3	1262.7 0.500 66.7	1505.9 0.505 70.7	1714 0.509 76.2	2092.2 0.584 81.4	2288.5 0.627 85.1	2345 0.650 86.5	2124 0.617 86.9	1774.9 0.566 84.1	1536.5 0.570 78.9	1104.8 0.468 70.7	982.3 0.488 65.2	
El Paso Lat. 31°48'N El. 3916 ft	$\bar{H}$ $\bar{K}_T$ $\bar{T}_a$	1247.6 0.686 47.1	1612.9 0.714 53.1	2048.7 0.730 58.7	2447.2 0.741 67.3	2673 0.743 75.7	2731 0.733 84.2	2391.1 0.652 84.9	2350.5 0.669 83.4	2077.5 0.693 78.5	1704.8 0.695 69.0	1324.7 0.647 56.0	1051.6 0.626 48.5	

Table B-2 (*Continued*)

		Jan.	Feb.	Mar.	Apr.	May	June	July	Aug.	Sept.	Oct.	Nov.	Dec.
TEXAS (Cont'd.)													
Fort Worth	$\bar{H}$	936.2	1198.5	1597.8	1829.1	2105.1	2437.6	2293.3	2216.6	1880.8	1476	1147.6	913.6
Lat. 32°50'N	$\bar{K}_T$	0.530	0.541	0.577	0.556	0.585	0.654	0.624	0.653	0.634	0.612	0.576	0.563
El. 544 ft	$\bar{T}_a$	48.1	52.3	59.8	68.8	75.9	84.0	87.7	88.6	81.3	71.5	58.8	50.8
Midland	$\bar{H}$	1066.4	1345.7	1784.8	2036.1	2301.1	2317.7	2301.8	2193	1921.8	1470.8	1244.3	1023.2
Lat. 31°56'N	$\bar{K}_T$	0.587	0.596	0.638	0.617	0.639	0.622	0.628	0.643	0.642	0.600	0.609	0.611
El. 2854 ft	$\bar{T}_a$	47.9	52.8	60.0	68.8	77.2	83.9	85.7	85.0	78.9	70.3	56.6	49.1
San Antonio	$\bar{H}$	1045	1299.2	1560.1	1664.6	2024.7	2250	2364.2	2185.2	1844.6	1487.4	1104.4	954.6
Lat. 29°32'N	$\bar{K}_T$	0.541	0.550	0.542	0.500	0.563	0.62	0.647	0.637	0.603	0.584	0.507	0.528
El. 794 ft	$\bar{T}_a$	53.7	58.4	65.0	72.2	79.2	85.0	87.4	87.8	82.6	74.7	63.3	56.5
TENNESSEE													
Nashville	$\bar{H}$	589.7	907	1246.8	1662.3	1997	2149.4	2079.7	1862.7	1600.7	1223.6	823.2	614.4
Lat. 36°07'N	$\bar{K}_T$	0.373	0.440	0.472	0.514	0.556	0.573	0.565	0.554	0.556	0.540	0.454	0.426
El. 605 ft	$\bar{T}_a$	42.6	45.1	52.9	63.0	71.4	80.1	83.2	81.9	76.6	65.4	52.3	44.3
Oak Ridge	$\bar{H}$	604	895.9	1241.7	1689.6	1942.8	2066.4	1972.3	1795.6	1559.8	1194.8	796.3	610
Lat. 36°01'N	$\bar{K}_T$	0.382	0.435	0.471	0.524	0.541	0.551	0.536	0.534	0.542	0.527	0.438	0.422
El. 905 ft	$\bar{T}_a$	41.9	44.2	51.7	61.4	69.8	77.8	80.2	78.8	74.5	62.7	50.4	42.5
UTAH													
Salt Lake City	$\bar{H}$	622.1	986	1301.1	1813.3	...	...	...	...	1689.3	1250.2	...	552.8
Lat. 40°46'N	$\bar{K}_T$	0.468	0.909	0.529	0.579	...	...	...	...	0.621	0.610	...	0.467
El. 4227 ft	$\bar{T}_a$	29.4	36.2	44.4	53.9	63.1	71.7	81.3	79.0	68.7	57.0	42.5	34.0

WASHINGTON													
Seattle	$\bar{H}$	282.6	520.6	992.2	1507	1881.5	1909.9	2110.7	1688.5	1211.8	702.2	386.3	239.5
Lat. 47°27′N	$\bar{K}_T$	0.296	0.355	0.456	0.510	0.538	0.508	0.581	0.533	0.492	0.407	0.336	0.292
El. 386 ft	$\bar{T}_a$	42.1	45.0	48.9	54.1	59.8	64.4	68.4	67.9	63.3	56.3	48.4	44.4
Spokane	$\bar{H}$	446.1	837.6	1200	1864.6	2104.4	2226.5	2479.7	2076	1511	844.6	486.3	279
Lat. 47°40′N	$\bar{K}_T$	0.478	0.579	0.556	0.602	0.603	0.593	0.684	0.656	0.616	0.494	0.428	0.345
El. 1968 ft	$\bar{T}_a$	26.5	31.7	40.5	49.2	57.9	64.6	73.4	71.7	62.7	51.5	37.4	30.5
WISCONSIN													
Madison	$\bar{H}$	564.6	812.2	1232.1	1455.3	1745.4	2031.7	2046.5	1740.2	1443.9	993	555.7	495.9
Lat. 43°08′N	$\bar{K}_T$	0.40	0.478	0.522	0.474	0.493	0.540	0.559	0.534	0.549	0.510	0.396	0.467
El. 866 ft	$\bar{T}_a$	21.8	24.6	35.3	49.0	61.0	70.9	76.8	74.4	65.6	53.7	37.8	25.4
WYOMING													
Lander	$\bar{H}$	786.3	1146.1	1638	1988.5	2114	2492.2	2438.4	2120.6	1712.9	1301.8	837.3	694.8
Lat. 42°48′N	$\bar{K}_T$	0.65	0.672	0.691	0.647	0.597	0.662	0.665	0.649	0.647	0.666	0.589	0.643
El. 5370 ft	$\bar{T}_a$	20.2	26.3	34.7	45.5	56.0	65.4	74.6	72.5	61.4	48.3	33.4	23.8

Source: *Applications of Solar Energy for Heating and Cooling of Buildings*, ASHRAE, New York, 1977.

Table B-3 Ratio of monthly average daily global insolation on a tilted surface facing the equator to that on a horizontal surface[†]

Latitude (degrees)	Jan.	Feb.	Mar.	Apr.	May	Jun.	Jul.	Aug.	Sept.	Oct.	Nov.	Dec.
					$\bar{R}$ for $\bar{K}_T = 0.30$							
					Latitude − tilt = 15°							
25	1.09	1.06	1.03	1.00	0.98	0.98	0.98	0.99	1.02	1.05	1.08	1.09
30	1.15	1.10	1.05	1.01	0.98	0.97	0.97	0.99	1.03	1.08	1.13	1.16
35	1.23	1.15	1.07	1.01	0.97	0.96	0.96	1.00	1.05	1.12	1.20	1.25
40	1.34	1.22	1.11	1.02	0.97	0.95	0.96	1.00	1.07	1.18	1.30	1.38
45	1.51	1.31	1.15	1.03	0.97	0.94	0.95	1.00	1.10	1.25	1.45	1.58
50	1.77	1.44	1.21	1.05	0.97	0.93	0.95	1.01	1.13	1.35	1.67	1.91
55	2.24	1.65	1.29	1.07	0.96	0.93	0.94	1.02	1.18	1.50	2.04	2.53
					Latitude − tilt = 0							
25	1.17	1.11	1.04	0.97	0.93	0.91	0.92	0.95	1.01	1.08	1.16	1.19
30	1.24	1.15	1.05	0.97	0.92	0.90	0.91	0.95	1.02	1.11	1.21	1.27
35	1.33	1.20	1.08	0.97	0.91	0.89	0.90	0.95	1.03	1.16	1.29	1.38
40	1.46	1.27	1.11	0.98	0.90	0.87	0.89	0.94	1.05	1.21	1.41	1.53
45	1.65	1.37	1.15	0.99	0.90	0.86	0.88	0.94	1.08	1.29	1.57	1.76
50	1.96	1.52	1.21	1.00	0.89	0.85	0.87	0.95	1.11	1.40	1.82	2.14
55	2.51	1.75	1.29	1.01	0.89	0.84	0.86	0.95	1.16	1.56	2.25	2.88
					Latitude − tilt = −15°							
25	1.21	1.11	1.00	0.91	0.84	0.82	0.83	0.88	0.96	1.07	1.18	1.24
30	1.28	1.15	1.01	0.90	0.83	0.80	0.81	0.87	0.97	1.10	1.24	1.32
35	1.37	1.20	1.03	0.90	0.82	0.79	0.80	0.86	0.97	1.14	1.32	1.43
40	1.51	1.27	1.06	0.90	0.81	0.77	0.79	0.86	0.99	1.19	1.44	1.60
45	1.71	1.37	1.10	0.90	0.80	0.76	0.77	0.85	1.01	1.27	1.61	1.84
50	2.04	1.52	1.15	0.91	0.79	0.74	0.76	0.85	1.04	1.38	1.88	2.26
55	2.63	1.76	1.23	0.92	0.78	0.73	0.75	0.85	1.08	1.54	2.33	3.05
					Vertical surface							
25	0.94	0.78	0.62	0.48	0.42	0.40	0.41	0.45	0.56	0.73	0.90	0.99
30	1.04	0.85	0.67	0.52	0.44	0.42	0.43	0.48	0.60	0.79	0.99	1.10
35	1.17	0.94	0.72	0.55	0.47	0.44	0.45	0.51	0.65	0.86	1.10	1.24
40	1.33	1.04	0.78	0.59	0.50	0.47	0.48	0.55	0.70	0.95	1.25	1.44
45	1.57	1.18	0.86	0.64	0.53	0.49	0.51	0.59	0.76	1.06	1.45	1.72
50	1.93	1.36	0.95	0.68	0.56	0.52	0.54	0.63	0.82	1.20	1.75	2.17
55	2.55	1.62	1.06	0.74	0.60	0.55	0.57	0.67	0.91	1.40	2.24	3.00

[†] *Source: Intermediate Minimum Property Standards for Solar Heating and Domestic Hot Water Systems*, vol. 5, App. A, Table A.6, U.S. Dept. of Housing and Urban Development, Washington, D.C., 1977.

Table B-3 (*Continued*)

Latitude (degrees)	Jan.	Feb.	Mar.	Apr.	May	Jun.	Jul.	Aug.	Sept.	Oct.	Nov.	Dec.
				$\bar{R}$ for $\bar{K}_T = 0.40$								
				Latitude − tilt = 15°								
25	1.11	1.08	1.04	1.01	0.98	0.97	0.98	1.00	1.03	1.07	1.10	1.13
30	1.20	1.13	1.07	1.01	0.98	0.96	0.97	1.00	1.05	1.11	1.18	1.22
35	1.31	1.20	1.11	1.03	0.97	0.95	0.96	1.00	1.07	1.17	1.28	1.34
40	1.46	1.30	1.15	1.04	0.97	0.94	0.96	1.01	1.10	1.25	1.41	1.52
45	1.69	1.43	1.21	1.06	0.97	0.94	0.95	1.02	1.15	1.35	1.61	1.79
50	2.04	1.61	1.30	1.09	0.98	0.94	0.95	1.04	1.20	1.49	1.90	2.22
55	2.68	1.89	1.41	1.12	0.98	0.93	0.95	1.06	1.27	1.70	2.41	3.06
				Latitude − tilt = 0								
25	1.24	1.15	1.06	0.98	0.92	0.90	0.91	0.95	1.03	1.12	1.22	1.27
30	1.34	1.21	1.09	0.98	0.91	0.88	0.90	0.95	1.04	1.17	1.30	1.38
35	1.46	1.29	1.13	0.99	0.91	0.87	0.89	0.95	1.07	1.23	1.41	1.52
40	1.64	1.39	1.17	1.00	0.90	0.86	0.88	0.96	1.10	1.31	1.57	1.73
45	1.90	1.53	1.23	1.02	0.90	0.86	0.88	0.96	1.14	1.42	1.79	2.04
50	2.32	1.74	1.32	1.04	0.90	0.85	0.87	0.98	1.19	1.58	2.13	2.56
55	3.05	2.04	1.43	1.07	0.90	0.84	0.87	0.99	1.27	1.80	2.71	3.54
				Latitude − tilt = −15°								
25	1.31	1.17	1.03	0.91	0.82	0.79	0.80	0.87	0.98	1.12	1.27	1.35
30	1.41	1.23	1.06	0.91	0.81	0.77	0.79	0.86	0.99	1.17	1.36	1.46
35	1.54	1.31	1.09	0.91	0.80	0.76	0.78	0.86	1.01	1.23	1.47	1.62
40	1.73	1.41	1.13	0.92	0.80	0.75	0.77	0.86	1.04	1.31	1.64	1.84
45	2.01	1.56	1.19	0.93	0.79	0.74	0.76	0.87	1.08	1.42	1.87	2.18
50	2.45	1.77	1.27	0.95	0.79	0.73	0.76	0.88	1.12	1.58	2.23	2.74
55	3.24	2.08	1.39	0.98	0.79	0.72	0.75	0.89	1.19	1.81	2.85	3.80
				Vertical surface								
25	1.05	0.84	0.63	0.44	0.36	0.34	0.35	0.40	0.54	0.77	0.99	1.12
30	1.18	0.94	0.69	0.49	0.39	0.36	0.37	0.44	0.60	0.85	1.11	1.26
35	1.35	1.05	0.76	0.54	0.43	0.39	0.41	0.49	0.66	0.95	1.26	1.45
40	1.57	1.18	0.84	0.59	0.47	0.42	0.44	0.53	0.73	1.06	1.46	1.71
45	1.88	1.36	0.94	0.65	0.51	0.46	0.48	0.58	0.81	1.21	1.73	2.08
50	2.36	1.60	1.06	0.71	0.55	0.50	0.52	0.63	0.90	1.39	2.12	2.68
55	3.18	1.95	1.21	0.78	0.60	0.54	0.56	0.69	1.00	1.66	2.76	3.78

Table B-3 (Continued)

Latitude (degrees)	Jan.	Feb.	Mar.	Apr.	May	Jun.	Jul.	Aug.	Sept.	Oct.	Nov.	Dec.
$\bar{R}$ for $\bar{K}_T = 0.50$												
Latitude − tilt = 15°												
25	1.14	1.09	1.05	1.01	0.98	0.97	0.97	1.00	1.03	1.08	1.12	1.15
30	1.23	1.16	1.08	1.02	0.97	0.96	0.96	1.00	1.06	1.13	1.21	1.26
35	1.37	1.24	1.13	1.03	0.97	0.95	0.96	1.01	1.09	1.20	1.33	1.41
40	1.55	1.36	1.19	1.05	0.97	0.94	0.96	1.02	1.13	1.30	1.49	1.62
45	1.82	1.51	1.26	1.08	0.98	0.94	0.96	1.03	1.18	1.42	1.72	1.93
50	2.24	1.73	1.36	1.12	0.99	0.94	0.96	1.06	1.25	1.59	2.08	2.45
55	2.99	2.06	1.50	1.16	1.00	0.94	0.96	1.08	1.34	1.83	2.67	3.44
Latitude − tilt = 0												
25	1.29	1.19	1.08	0.98	0.91	0.88	0.90	0.95	1.04	1.15	1.26	1.32
30	1.40	1.26	1.11	0.99	0.91	0.87	0.89	0.95	1.06	1.21	1.36	1.45
35	1.56	1.35	1.16	1.00	0.90	0.86	0.88	0.96	1.09	1.28	1.50	1.63
40	1.77	1.48	1.22	1.02	0.90	0.86	0.88	0.97	1.13	1.38	1.68	1.87
45	2.08	1.65	1.30	1.04	0.90	0.85	0.87	0.98	1.18	1.52	1.95	2.25
50	2.57	1.89	1.40	1.08	0.91	0.85	0.87	1.00	1.25	1.70	2.36	2.86
55	3.44	2.26	1.54	1.12	0.92	0.85	0.88	1.02	1.34	1.97	3.04	4.02
Latitude − tilt = −15°												
25	1.38	1.22	1.05	0.91	0.81	0.77	0.79	0.86	0.99	1.16	1.33	1.43
30	1.50	1.29	1.09	0.91	0.80	0.76	0.78	0.86	1.01	1.22	1.44	1.57
35	1.66	1.39	1.13	0.92	0.80	0.75	0.77	0.86	1.04	1.30	1.58	1.75
40	1.89	1.52	1.19	0.94	0.79	0.74	0.76	0.87	1.08	1.40	1.78	2.02
45	2.22	1.69	1.26	0.96	0.79	0.73	0.76	0.88	1.12	1.53	2.06	2.43
50	2.75	1.94	1.36	0.98	0.79	0.73	0.76	0.89	1.19	1.72	2.49	3.09
55	3.68	2.32	1.50	1.02	0.80	0.72	0.75	0.91	1.27	1.99	3.22	4.34
Vertical surface												
25	1.13	0.89	0.63	0.42	0.32	0.29	0.30	0.37	0.53	0.80	1.06	1.21
30	1.29	1.00	0.71	0.47	0.35	0.32	0.33	0.41	0.60	0.89	1.20	1.38
35	1.48	1.13	0.79	0.53	0.40	0.35	0.37	0.47	0.67	1.01	1.38	1.60
40	1.74	1.29	0.89	0.59	0.44	0.39	0.41	0.52	0.75	1.14	1.61	1.91
45	2.11	1.50	1.00	0.66	0.49	0.44	0.46	0.58	0.84	1.31	1.92	2.34
50	2.67	1.78	1.14	0.73	0.54	0.48	0.51	0.64	0.95	1.54	2.39	3.04
55	3.64	2.19	1.32	0.81	0.60	0.53	0.56	0.71	1.08	1.84	3.15	4.34

Table B-3 (*Continued*)

Latitude (degrees)	Jan.	Feb.	Mar.	Apr.	May	Jun.	Jul.	Aug.	Sept.	Oct.	Nov.	Dec.
					$\bar{R}$ for $\bar{K}_T = 0.60$							
					Latitude − tilt = 15°							
25	1.15	1.11	1.06	1.01	0.98	0.96	0.97	1.00	1.04	1.09	1.14	1.17
30	1.27	1.18	1.10	1.02	0.97	0.95	0.96	1.00	1.07	1.15	1.24	1.29
35	1.41	1.28	1.15	1.04	0.97	0.94	0.96	1.01	1.10	1.23	1.37	1.46
40	1.62	1.40	1.21	1.07	0.98	0.94	0.95	1.02	1.15	1.34	1.56	1.70
45	1.92	1.58	1.30	1.10	0.98	0.94	0.96	1.04	1.21	1.48	1.82	2.05
50	2.40	1.83	1.41	1.14	0.99	0.94	0.96	1.07	1.29	1.67	2.22	2.64
55	3.24	2.20	1.57	1.19	1.01	0.94	0.97	1.10	1.39	1.95	2.89	3.75
					Latitude − tilt = 0							
25	1.33	1.21	1.09	0.98	0.91	0.87	0.89	0.95	1.05	1.17	1.30	1.37
30	1.46	1.30	1.13	0.99	0.90	0.86	0.88	0.95	1.08	1.24	1.41	1.51
35	1.63	1.40	1.19	1.01	0.90	0.85	0.87	0.96	1.11	1.33	1.57	1.71
40	1.88	1.55	1.26	1.03	0.90	0.85	0.87	0.97	1.16	1.44	1.78	1.99
45	2.23	1.74	1.35	1.06	0.91	0.85	0.87	0.99	1.22	1.59	2.08	2.41
50	2.78	2.02	1.47	1.10	0.92	0.85	0.88	1.01	1.30	1.81	2.54	3.10
55	3.76	2.43	1.63	1.15	0.93	0.85	0.88	1.05	1.40	2.11	3.31	4.41
					Latitude − tilt = −15°							
25	1.43	1.26	1.07	0.91	0.80	0.75	0.77	0.86	1.00	1.19	1.39	1.49
30	1.57	1.34	1.11	0.92	0.79	0.74	0.76	0.86	1.03	1.26	1.51	1.65
35	1.76	1.45	1.16	0.93	0.79	0.73	0.76	0.86	1.06	1.35	1.67	1.86
40	2.02	1.60	1.23	0.95	0.79	0.73	0.75	0.87	1.11	1.47	1.90	2.17
45	2.40	1.80	1.32	0.98	0.79	0.72	0.75	0.89	1.16	1.62	2.22	2.62
50	2.99	2.09	1.44	1.01	0.80	0.72	0.75	0.91	1.24	1.84	2.70	3.37
55	4.04	2.52	1.59	1.05	0.81	0.72	0.76	0.93	1.34	2.15	3.52	4.78
					Vertical surface							
25	1.20	0.92	0.63	0.39	0.28	0.25	0.26	0.34	0.53	0.82	1.12	1.28
30	1.37	1.04	0.72	0.46	0.32	0.28	0.30	0.39	0.60	0.93	1.28	1.48
35	1.59	1.19	0.81	0.52	0.37	0.32	0.34	0.45	0.68	1.06	1.48	1.73
40	1.88	1.37	0.92	0.59	0.42	0.37	0.39	0.51	0.77	1.21	1.73	2.07
45	2.30	1.61	1.05	0.66	0.48	0.42	0.44	0.58	0.87	1.40	2.09	2.56
50	2.93	1.93	1.21	0.75	0.54	0.47	0.50	0.65	0.99	1.65	2.61	3.34
55	4.01	2.39	1.41	0.84	0.60	0.52	0.55	0.72	1.13	2.00	3.46	4.80

Table B-3 (*Continued*)

Latitude (degrees)	Jan.	Feb.	Mar.	Apr.	May	Jun.	Jul.	Aug.	Sept.	Oct.	Nov.	Dec.
					$\bar{R}$ for $\bar{K}_T = 0.70$							
					Latitude − tilt = 15°							
25	1.17	1.12	1.06	1.01	0.98	0.96	0.97	1.00	1.04	1.10	1.16	1.19
30	1.30	1.20	1.11	1.03	0.97	0.95	0.96	1.00	1.07	1.17	1.27	1.33
35	1.46	1.31	1.17	1.05	0.97	0.94	0.95	1.01	1.12	1.26	1.42	1.51
40	1.69	1.45	1.24	1.08	0.98	0.94	0.95	1.03	1.17	1.38	1.62	1.78
45	2.03	1.65	1.34	1.11	0.99	0.94	0.96	1.06	1.24	1.53	1.92	2.18
50	2.56	1.93	1.47	1.16	1.00	0.94	0.97	1.09	1.33	1.75	2.36	2.83
55	3.50	2.34	1.64	1.22	1.02	0.94	0.98	1.13	1.45	2.06	3.11	4.06
					Latitude − tilt = 0							
25	1.37	1.24	1.11	0.98	0.90	0.86	0.88	0.95	1.06	1.20	1.34	1.41
30	1.52	1.34	1.16	1.00	0.90	0.85	0.87	0.95	1.09	1.27	1.47	1.58
35	1.71	1.46	1.22	1.02	0.90	0.85	0.87	0.96	1.13	1.37	1.64	1.80
40	1.98	1.62	1.30	1.05	0.90	0.84	0.87	0.98	1.19	1.50	1.88	2.11
45	2.38	1.84	1.40	1.08	0.91	0.84	0.87	1.00	1.26	1.67	2.21	2.58
50	3.00	2.15	1.53	1.13	0.92	0.85	0.88	1.03	1.35	1.91	2.73	3.35
55	4.09	2.61	1.72	1.19	0.94	0.85	0.89	1.07	1.47	2.25	3.58	4.80
					Latitude − tilt = −15°							
25	1.49	1.30	1.09	0.91	0.78	0.73	0.76	0.85	1.01	1.23	1.44	1.56
30	1.65	1.39	1.14	0.92	0.78	0.72	0.75	0.86	1.04	1.30	1.58	1.73
35	1.86	1.52	1.20	0.94	0.78	0.72	0.75	0.87	1.09	1.41	1.76	1.98
40	2.15	1.69	1.28	0.96	0.78	0.72	0.74	0.88	1.14	1.54	2.01	2.32
45	2.58	1.91	1.38	0.99	0.79	0.71	0.75	0.90	1.20	1.71	2.37	2.83
50	3.24	2.23	1.51	1.04	0.80	0.72	0.75	0.92	1.29	1.96	2.92	3.66
55	4.41	2.71	1.69	1.09	0.81	0.72	0.76	0.96	1.40	2.31	3.83	5.23
					Vertical surface							
25	1.26	0.96	0.64	0.37	0.25	0.21	0.23	0.31	0.52	0.85	1.18	1.36
30	1.46	1.09	0.73	0.44	0.29	0.25	0.27	0.37	0.60	0.97	1.35	1.57
35	1.70	1.26	0.84	0.51	0.35	0.29	0.32	0.43	0.69	1.11	1.57	1.85
40	2.03	1.46	0.96	0.59	0.41	0.34	0.37	0.50	0.79	1.28	1.86	2.23
45	2.48	1.72	1.10	0.67	0.47	0.40	0.43	0.57	0.90	1.49	2.25	2.77
50	3.18	2.07	1.27	0.76	0.53	0.45	0.48	0.65	1.03	1.77	2.83	3.65
55	4.39	2.59	1.50	0.87	0.60	0.51	0.55	0.73	1.19	2.15	3.78	5.26

Table B-4 Direct insolation†

	Pacific northwest (Tacoma)	Pacific southwest (L.A.)	North central (Bismarck)	Central (Dodge City)	Southwest (El Paso)	Gulf coast (Mobile)	Great Lakes (Madison)	Northeast (Boston)	Southeast (Charleston)
Jan.	1.8	6.0	4.4	6.0	6.6	4.0	4.0	4.0	4.0
Feb.	2.5	5.5	5.1	5.0	7.5	4.2	5.0	4.5	4.3
Mar.	3.8	8.0	6.0	6.0	8.5	5.0	6.0	5.0	5.0
Apr.	5.0	9.0	7.0	7.0	9.5	5.0	6.1	5.6	5.0
May	6.1	9.5	7.2	7.5	10.1	6.1	6.6	6.0	6.2
Jun.	7.9	9.3	9.0	8.0	9.9	6.3	8.0	6.0	6.0
Jul.	9.0	9.0	8.2	8.0	8.9	6.0	7.8	6.5	6.0
Aug.	6.5	7.8	8.0	8.0	8.8	5.5	7.2	5.2	5.2
Sept.	5.0	7.3	6.5	6.9	7.5	5.9	5.8	5.0	5.5
Oct.	3.1	7.4	5.5	7.0	7.6	4.8	4.8	4.0	4.7
Nov.	2.1	6.1	3.9	5.8	6.4	5.0	3.2	3.0	4.5
Dec.	1.8	6.0	3.4	5.0	6.0	4.0	3.2	3.5	4.1
Average	4.55	7.56	6.19	6.69	8.11	5.15	5.64	4.86	5.04
Mean Deviation	2.03	1.19	1.47	0.93	1.17	0.68	1.34	1.05	0.615
% Mean deviation (% deviation from mean)	44.7	15.7	23.8	13.9	14.4	13.1	23.8	21.6	12.2

Source: E. C. Boes, I. J. Hall, R. R. Prairie, R. P. Stromberg, and H. E. Anderson, "Distribution of Direct and Total Solar Radiation Available for the USA," *Proceedings of the 1976 Annual Meeting of the Am. Sec. of ISES, Sharing the Sun*, Winnipeg, 1976, pp. 238–263.
† All values in kWh/(m² · day).

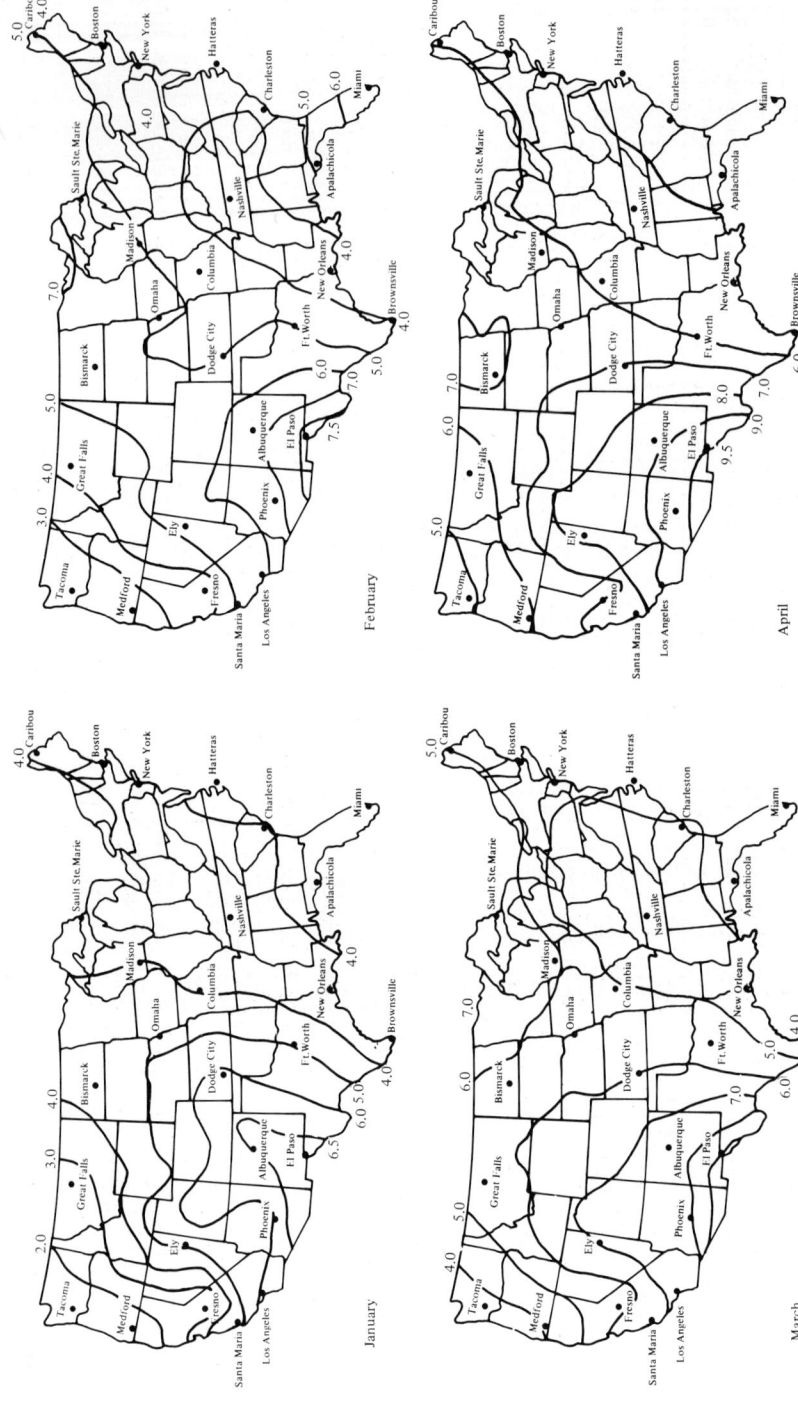

Figure B-2 Mean daily direct insolation, kWh/(m² · day). (*Note:* 1 kWh/m² = 320.5 Btu/ft².) (*Source:* E. C. Boes, I. J. Hall, R. R. Prairie, R. P. Stromberg, and H. E. Anderson, "Distribution of Direct and Total Solar Radiation Available for the USA," *Proceedings of the 1976 Annual Meeting of the Am. Sec. of ISES, Sharing the Sun, Winnipeg, 1976*, pp. 238–263.)

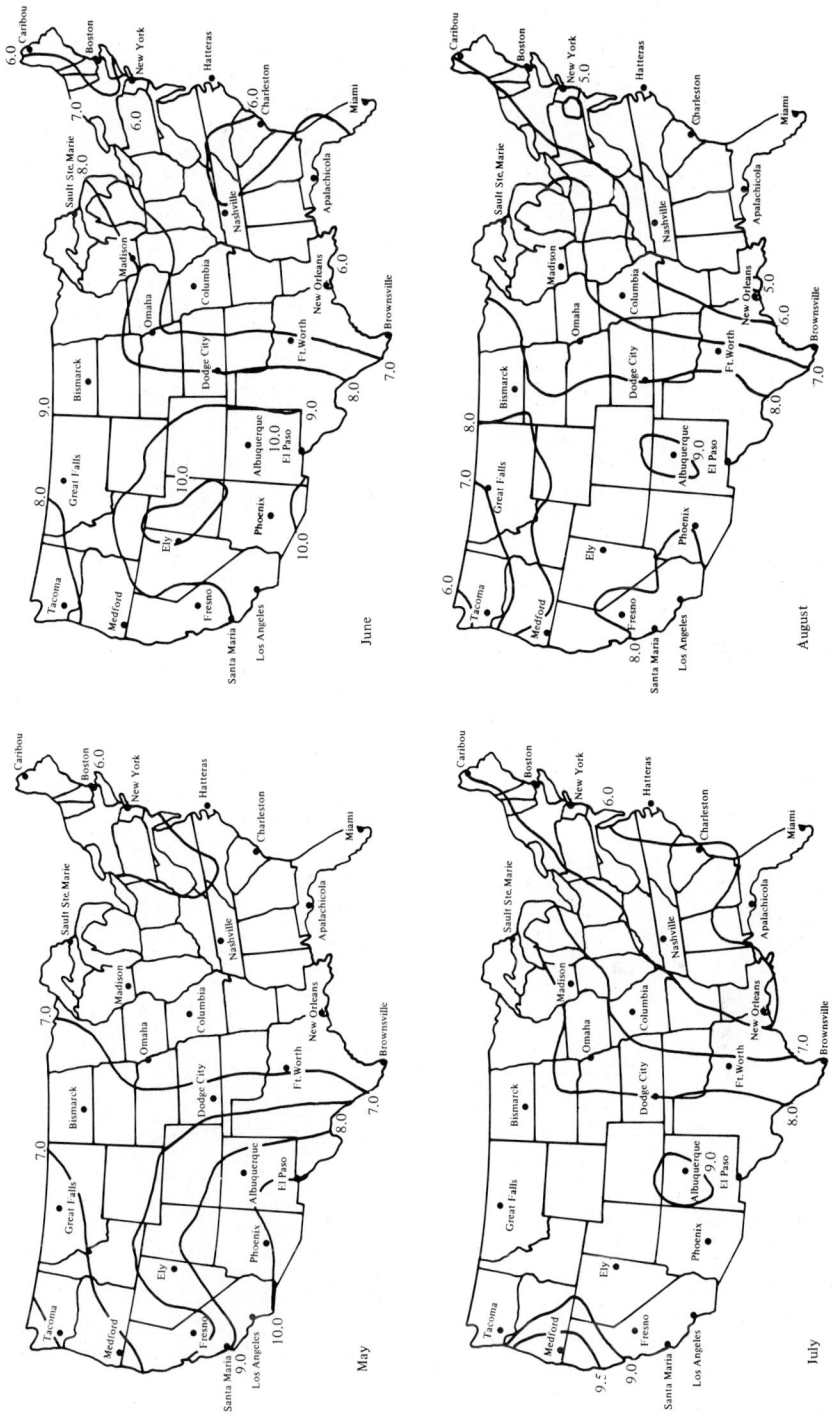

Figure B-2 (*Continued*)

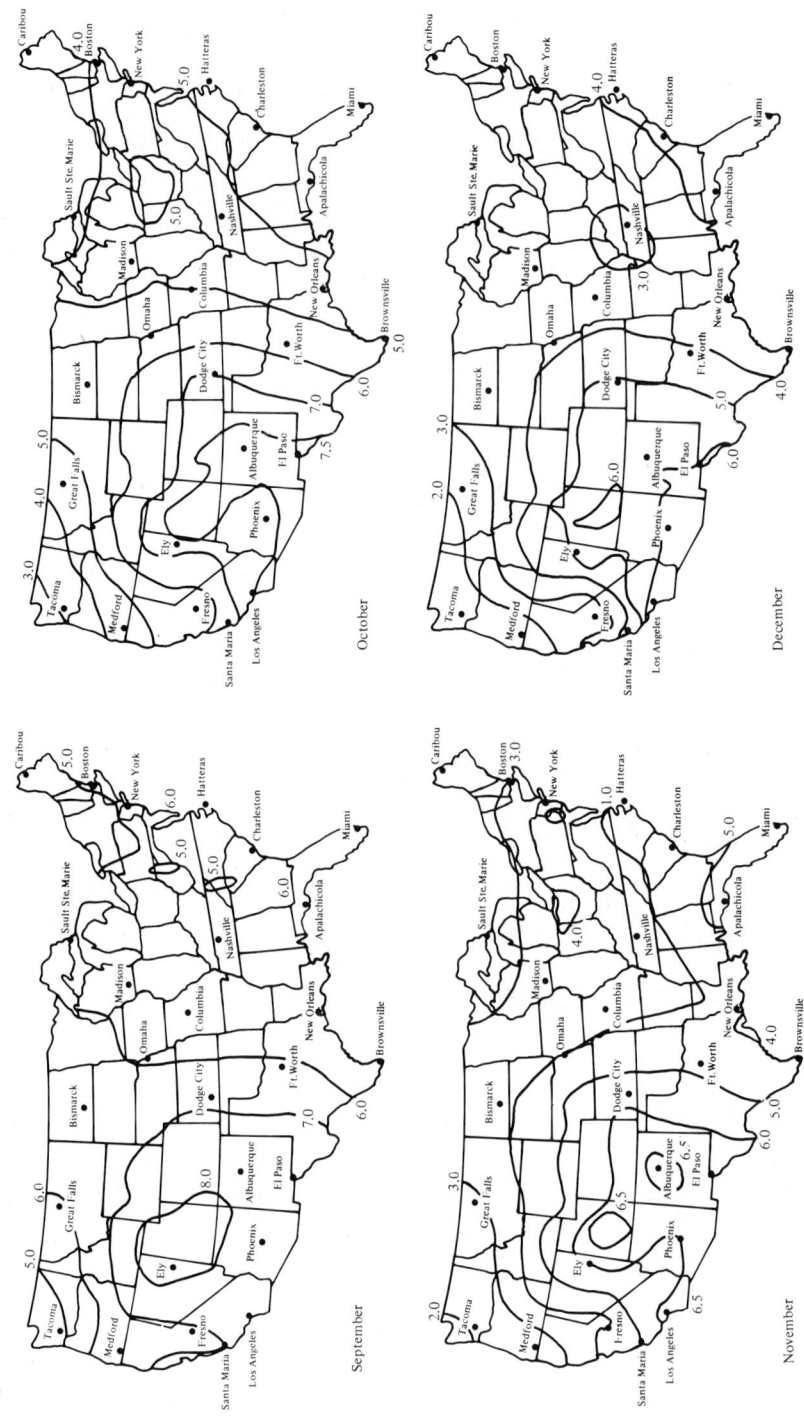

Figure B-2 (*Continued*)

Table B-5 Average monthly and yearly degree-days for heating (base 65°F) and 97½ percent winter design temperature (°F) for selected cities in the United States and Canada

Station†		97½% winter design temp.	Jul.	Aug.	Sept.	Oct.	Nov.	Dec.	Jan.	Feb.	Mar.	Apr.	May	Jun.	Yearly total
ALABAMA															
Montgomery	A	26	0	0	0	68	330	527	543	417	316	90	0	0	2291
ALASKA															
Fairbanks	A	−50	171	332	642	1203	1833	2254	2359	1901	1739	1068	555	222	14279
Juneau	A	−4	301	338	483	725	921	1135	1237	1070	1073	810	601	381	9075
Nome	A	−28	481	496	693	1094	1455	1820	1879	1666	1770	1314	930	573	14171
ARIZONA															
Flagstaff	A	5	46	68	201	558	867	1073	1169	991	911	651	437	180	7152
Tucson	A	32	0	0	0	25	231	406	471	344	242	75	6	0	1800
ARKANSAS															
Little Rock	A	23	0	0	9	127	465	716	756	577	434	126	9	0	3219
CALIFORNIA															
Eureka	C	35	270	257	258	329	414	499	546	470	505	438	372	285	4643
Los Angeles	A	43	28	28	42	78	180	291	372	302	288	219	158	81	2061
Sacramento	A	32	0	0	0	56	321	546	583	414	332	178	72	0	2502
San Francisco	C	42	192	174	102	118	231	388	443	336	319	279	239	180	3001
COLORADO															
Denver	A	3	6	9	117	428	819	1035	1132	938	887	558	288	66	6283
FLORIDA															
Miami	A	47	0	0	0	0	0	65	74	56	19	0	0	0	214
Tallahassee	A	29	0	0	0	28	198	360	375	286	202	36	0	0	1485

Source: ASHRAE Systems Handbook, ASHRAE, New York, 1976.
† A: Airport stations; C: city stations.

Table B-5 (*Continued*)

Station		97½% winter design temp.	Jul.	Aug.	Sept.	Oct.	Nov.	Dec.	Jan.	Feb.	Mar.	Apr.	May	Jun.	Yearly total
GEORGIA															
Atlanta	A	23	0	0	18	124	417	648	636	518	428	147	25	0	2961
Augusta	A	23	0	0	0	78	333	552	549	445	350	90	0	0	2397
IDAHO															
Boise	A	10	0	0	132	415	792	1017	1113	854	722	438	245	81	5809
ILLINOIS															
Chicago (Midway)	A	1	0	0	81	326	753	1113	1209	1044	890	480	211	48	6155
INDIANA															
Indianapolis	A	4	0	0	90	316	723	1051	1113	949	809	432	177	39	5699
IOWA															
Des Moines	A	−3	0	6	96	363	828	1225	1370	1137	915	438	180	30	6588
KANSAS															
Topeka	A	6	0	0	57	270	672	980	1122	893	722	330	124	12	5182
LOUISIANA															
Baton Rouge	A	30	0	0	0	31	216	369	409	294	208	33	0	0	1560
Shreveport	A	26	0	0	0	47	297	477	552	426	304	81	0	0	2184
MAINE															
Portland	A	0	12	53	195	508	807	1215	1339	1182	1042	675	372	111	7511
MARYLAND															
Baltimore	A	15	0	0	48	264	585	905	936	820	679	327	90	0	4654
MASSACHUSETTS															
Boston	A	10	0	9	60	316	603	983	1088	972	846	513	208	36	5634
MICHIGAN															
Lansing	A	6	6	22	138	431	813	1163	1262	1142	1011	579	273	69	6909
Sault Ste. Marie	A	−8	96	105	279	580	951	1367	1525	1380	1277	810	477	201	9048

MINNESOTA														
Duluth	A	−15	71	109	330	632	1131	1581	1745	1518	1355	840	490	198	10000
Minneapolis	A	−10	22	31	189	505	1014	1454	1631	1380	1166	621	288	81	8382
MISSOURI														
St. Louis	A	8	0	0	60	251	627	936	1026	848	704	312	121	15	4900
MONTANA														
Great Falls	A	−16	28	53	258	543	921	1169	1349	1154	1063	642	384	186	7750
NEVADA														
Las Vegas	A	26	0	0	0	78	387	617	688	487	335	111	6	0	2709
Reno	A	7	43	87	204	490	801	1026	1073	823	729	510	357	189	6332
NEW MEXICO														
Albuquerque	A	17	0	0	12	229	642	868	930	703	595	288	81	0	4348
NEW YORK														
New York (Cent. Park)	C	15	0	0	30	233	540	902	986	885	760	408	118	9	4871
Syracuse	A	2	6	28	132	415	744	1153	1271	1140	1004	570	248	45	6756
NORTH CAROLINA														
Greensboro	A	17	0	0	33	192	513	778	784	672	552	234	47	0	3805
NORTH DAKOTA														
Bismarck	A	−19	34	28	222	577	1083	1463	1708	1442	1203	645	329	117	8851
OHIO														
Toledo	A	5	0	16	117	406	792	1138	1200	1056	924	543	242	60	6494
OKLAHOMA														
Oklahoma City	A	15	0	0	15	164	498	766	868	664	527	189	34	0	3725
OREGON														
Medford	A	23	0	0	78	372	678	871	918	697	642	432	242	78	5008
Portland	A	24	25	28	114	335	597	735	825	644	586	396	245	105	4635
PENNSYLVANIA														
Pittsburgh	A	9	0	9	105	375	726	1063	1119	1002	874	480	195	39	5987
SOUTH CAROLINA														
Charleston	A	27	0	0	0	59	282	471	487	389	291	54	0	0	2033

Table B-5 (*Continued*)

Station		97½% winter design temp.	Jul.	Aug.	Sept.	Oct.	Nov.	Dec.	Jan.	Feb.	Mar.	Apr.	May	Jun.	Yearly total
SOUTH DAKOTA															
Sioux Falls	A	−10	19	25	168	462	972	1361	1544	1285	1082	573	270	78	7839
TENNESSEE															
Nashville	A	16	0	0	30	158	495	732	778	644	512	189	40	0	3578
TEXAS															
Brownsville	A	40	0	0	0	0	66	149	205	106	74	0	0	0	600
El Paso	A	25	0	0	0	84	414	648	685	445	319	105	0	0	2700
Fort Worth	A	24	0	0	0	65	324	536	614	448	319	99	0	0	2405
Houston	A	32	0	0	0	6	183	307	384	288	192	36	0	0	1396
Lubbock	A	15	0	0	18	174	513	744	800	613	484	201	31	0	3578
San Angelo	A	25	0	0	0	68	318	536	567	412	288	66	0	0	2255
UTAH															
Salt Lake City	A	9	0	0	81	419	849	1082	1172	910	763	459	233	84	6052
VIRGINIA															
Norfolk	A	23	0	0	0	136	408	698	738	655	533	216	37	0	3421
WASHINGTON															
Seattle-Tacoma	A	24	56	62	162	391	633	750	828	678	657	474	295	159	5145
Spokane	A	4	9	25	168	493	879	1082	1231	980	834	531	288	135	6655
WEST VIRGINIA															
Charleston	A	14	0	0	63	254	591	865	880	770	648	300	96	9	4476
WISCONSIN															
Madison	A	−5	25	40	174	474	930	1330	1473	1274	1113	618	310	102	7863
WYOMING															
Casper	A	−5	6	16	192	524	942	1169	1290	1084	1020	657	381	129	7410
ALBERTA															
Calgary	A	−25	109	186	402	719	1110	1389	1575	1379	1268	798	477	291	9703

BRITISH COLUMBIA Vancouver‡	A	19	81	87	219	456	657	787	862	723	676	501	310	156	5515
MANITOBA Winnipeg	A	−25	38	71	322	683	1251	1757	2008	1719	1465	813	405	147	10679
NEW BRUNSWICK Fredericton‡	A	−10	78	68	234	592	915	1392	1541	1379	1172	753	406	141	8671
NEWFOUNDLAND St. John's‡	A	6	186	180	342	651	831	1113	1262	1170	1187	927	710	432	8991
NORTHWEST TERRITORIES Fort Norman	C		164	341	666	1234	1959	2474	2592	2209	2058	1386	732	294	16109
NOVA SCOTIA Halifax	C	4	58	51	180	457	710	1074	1213	1122	1030	742	487	237	7361
ONTARIO Toronto	C	1	7	18	151	439	760	1111	1233	1119	1013	616	298	62	6827
QUEBEC Montreal‡	A	−10	9	43	165	521	882	1392	1566	1381	1175	684	316	69	8203
SASKATCHEWAN Regina	A	−29	78	93	360	741	1284	1711	1965	1687	1473	804	409	201	10806
YUKON TERRITORY Dawson	C		164	326	645	1197	1875	2415	2561	2150	1838	1068	570	258	15067

‡ The data for these normals were from the full 10-year period 1951–1960, adjusted to the standard normal period 1931–1960.

APPENDIX
C

HEAT TRANSFER IN SOLAR-THERMAL APPLICATIONS

Heat is transferred as a result of the temperature difference between surfaces or between a surface and a fluid. There are three primary modes of heat transfer: conduction, convection, and radiation. These are discussed in the following sections, with an emphasis on the applications of each in solar technology.

C-1 CONDUCTION

Conduction is the transfer of heat through materials due to molecular, electronic, and/or lattice effects. The conduction heat rate $\dot{Q}$ (measured in W) normal to a surface of area A (measured in m^2) can be expressed by Fourier's equation

$$\dot{Q} = -kA \frac{dT}{dx} \tag{C-1}$$

where dT/dx (°C/m) is the temperature gradient in the direction x, k is the material's thermal conductivity [W/(m · °C)], and the negative sign means the heat flow is down the temperature gradient.

The thermal conductivities of selected solids, liquids, and gases (as well as other properties to be discussed presently) are listed in Tables C-1, C-2, and C-3, respectively. One sees that, in general, the thermal conductivities increase from

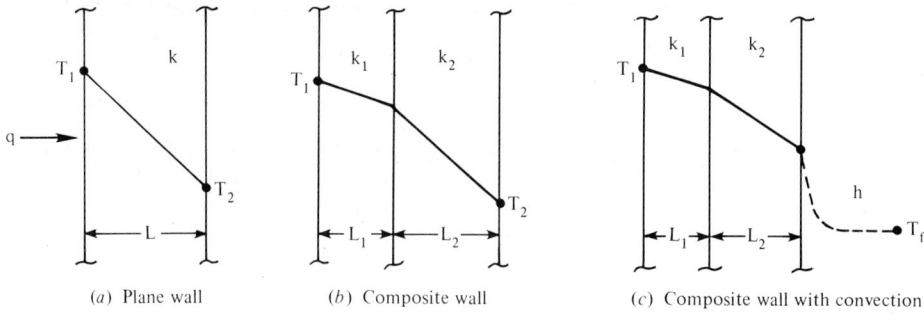

Figure C-1 One-dimensional heat transfer through single and composite walls.

gases to liquids to solids; however, there is a wide range within liquids and solids particularly and considerable overlap between them. Note that the thermal conductivities of foam and fiberglass insulation are similar to those for air.

The conduction heat flux q (measured in W/m²) through a plane wall of conductivity k and thickness L, with boundary temperatures of T_1 and T_2 (Fig. C-1a), is, by integrating Eq. (C-1), found to be

$$q = \frac{\dot{Q}}{A} = \frac{k(T_1 - T_2)}{L} = \frac{T_1 - T_2}{L/k} = \frac{T_1 - T_2}{R'} \tag{C-2}$$

where R' is the unit area wall resistance:

$$R' = L/k \quad \text{(in m}^2 \cdot {}^\circ\text{C/W)}$$

For heat flow through a composite plane wall as in Fig. C-1b, the resistances are additive: $\Sigma R' = R'_1 + R'_2$ and the conductive heat flux becomes

$$q = \frac{T_1 - T_2}{\Sigma R'} = \frac{T_1 - T_2}{L_1/k_1 + L_2/k_2} \tag{C-3}$$

If a surface of the composite wall is exposed to a fluid at T_f through a convective heat-transfer coefficient h,† then the unit area convective resistance is $1/h$. Again, the heat flow is through fluid and wall components in series, and the resistances are additive. Thus, the heat transfer (flux) through the composite wall with surface convection is

$$q = \frac{T_1 - T_f}{\Sigma R'} = \frac{T_1 - T_f}{L_1/k_1 + L_2/k_2 + 1/h} \tag{C-4}$$

† The convective heat-transfer coefficient is defined in Sec. C-2.

Table C-1 Properties of selected solids

Substance	Temperature °C	Thermal conductivity k W/(m·°C)	Density ρ kg/m³	Specific heat c kJ/(kg·°C)	Thermal diffusivity α m²/s × 10⁶	Reference†
Polyurethane foam, light	‡	0.016–0.033	32–48	…	…	3
Polyurethane foam, heavy	‡	0.022–0.040	64–112	…	…	3
Glass wool	23	0.038	24	0.7	2.26	2
Wood (pine)	20	0.10	497	2.8	0.075	1
Acrylic (general purpose)	‡	0.21	1190	1.47	0.12	3
Polyvinyl chloride	‡	0.15	1450	…	…	3
PVC (chlorinated)	‡	…	1530	…	…	3
Ethylene-propylene diene (EPDM)	‡	0.25	860	…	…	3
Soil (dry)	20	~0.35	~730	1.84	~0.26	1
(wet)	20	~2.36	…	…	~0.78	1
Ice	0	2.22	913	0.239	1.24	1
Concrete (1-2-4 mix)	20	1.37	2100	0.88	0.75	2
Building brick	20	0.66	1700	0.84	0.45	1
Granite	~20	~2.9	2640	0.82	~1.3	2
Glass (window)	20	0.78	2700	0.84	0.34	2
Glass (borosilicate)	~60	1.09	2200	…	…	2
Stainless steel (18-8)	20	16.3	7817	0.46	4.44	2
	100	17	…	…	…	2
Carbon steel (0.5%)	20	54	7833	0.465	14.74	2
	100	52	…	…	…	2
Nickel	20	90	8906	0.445	22.66	2
Brass (70 Cu/30 Zn)	20	111	8522	0.385	34.1	2
Aluminum	20	204	8418	0.896	84.2	2
	100	206	…	…	…	2
Copper	20	386	8954	0.383	112.3	2
	100	379	…	…	…	2

† See Refs. 1–3 at end of this appendix.
‡ Where temperature not stated, approximately room temperature.

Table C-2 Properties of selected liquids†

Liquid	Temperature °C	Density ρ kg/m³	Specific heat c_p kJ/(kg·°C)	Kinematic viscosity ν m²/s × 10⁶	Thermal conductivity k W/(m·°C)	Thermal diffusivity α m²/s × 10⁶	Thermal coefficient of volume expansion β K⁻¹ × 10³	Prandtl number Pr
Engine oil	20	888	1.88	900	0.145	0.087	0.70	10,400
	100	840	2.22	20.3	0.137	0.074	...	276
Ethylene glycol	0	1127	2.24	53.0	0.303	0.120	...	440
	20	1114	2.34	19.0	0.289	0.111	0.59	170
	60	1085	2.50	5.1	0.260	0.096	...	53
75 EG/25 H₂O‡	0	1106	2.68	19	0.356	0.120	...	160
	20	1094	2.80	7.3	0.346	0.113	0.63	65
	60	1065	3.03	2.4	0.327	0.101	...	24
50 EG/50 H₂O‡	0	1074	3.19	7.4	0.424	0.124	...	60
	20	1065	3.31	3.7	0.419	0.119	0.54	31
	60	1040	3.52	1.3	0.415	0.113	...	11.5
Propylene glycol	0	1050	2.35	240	0.227	0.092	...	2,600
	20	1036	2.47	54	0.216	0.084	0.68	640
	60	1006	2.70	8.2	0.194	0.071	...	115
75 PG/25 H₂O‡	0	1059	2.94	57	0.301	0.097	...	590
	20	1045	3.03	17	0.292	0.092	0.67	185
	60	1015	3.24	3.3	0.273	0.083	...	40
50 PG/50 H₂O‡	0	1052	3.52	16	0.387	0.105	...	150
	20	1034	3.56	5.7	0.382	0.104	~0.69	55
	60	1010	3.70	1.7	0.377	0.101	...	16.8
Triethylene glycol	20	1126	2.09	47	0.237	0.101	...	470
	60	1095	2.28	13.7	0.219	0.088	0.69	155
75 TEG/25 H₂O‡	20	1112	2.75	19	0.308	0.101	...	190
	60	1080	2.93	4.5	0.291	0.092	0.61	49

Freon	−20	1461	0.907	0.235	0.071	0.0539	...	4.4
	0	1397	0.935	0.214	0.073	0.0557	...	3.8
	20	1330	0.966	0.198	0.073	0.0560	...	3.5
Water	0	1000	4.225	1.79	0.566	0.134	...	13.25
	20	997	4.180	1.01	0.602	0.144	...	7.0
	60	983	4.179	0.479	0.654	0.159	0.49	3.01
	100	958	4.211	0.295	0.682	0.169	...	1.76
	260	785	4.731	0.136	0.616	0.166	...	0.83
NaK (56/44)	93	890	1.13	0.652	25.6	25.5	...	0.026
Dow Therm A	100	999	...	0.991	0.132	...	0.84	...
	200	910	1.65	0.431	0.119	0.079	1.02	5.4
	300	809	1.89	0.263	0.107	0.070	1.37	3.8
	400	681	2.10	0.203	0.094	0.066	...	3.1

† Information for all substances taken from Ref. 2 at end of this appendix, except for the following: glycol data taken from Ref. 4, and Dow Therm A data taken from Ref. 5.

‡ By weight.

Table C-3 Properties of selected gases† (at atmospheric pressure)

Gas	Temperature, K	Density ρ kg/m^3	Specific heat c_p kJ/(kg · °C)	Kinematic viscosity v m^2/s × 10^6	Thermal conductivity k W/(m · °C)	Thermal diffusivity α m^2/s × 10^6	Prandtl number Pr
Helium	250	0.1944	5.2	91	0.134	132.6	0.70
	300	0.1620	5.2	124	0.149	176.9	0.70
	400	0.1215	5.2	200	0.178	281.7	0.71
	500	0.0972	5.2	290	0.203	401.6	0.72
Nitrogen	300	1.142	1.041	15.63	0.0262	22.04	0.713
	400	0.854	1.046	25.74	0.0334	37.34	0.691
	500	0.682	1.056	37.66	0.0398	55.3	0.684
Air	250	1.143	1.005	9.49	0.0223	13.16	0.722
	300	1.177	1.006	15.68	0.0262	22.16	0.708
	400	0.883	1.014	25.9	0.0337	37.6	0.689
	500	0.705	1.030	37.9	0.0404	55.6	0.680
	600	0.588	1.055	51.3	0.0466	75.1	0.680
	700	0.503	1.075	66.3	0.0523	96.7	0.684
	800	0.441	1.098	82.3	0.0578	119.5	0.689
	900	0.393	1.121	99.3	0.0628	142.7	0.696
	1000	0.352	1.142	117.8	0.0675	167.8	0.702
Carbon dioxide	250	2.166	0.804	5.81	0.0129	7.40	0.793
	300	1.797	0.871	8.32	0.0166	10.59	0.770
	400	1.342	0.942	14.39	0.0246	19.46	0.738
	500	1.073	1.013	21.67	0.0335	30.84	0.702
Water vapor	400	0.554	2.014	24.2	0.0261	23.4	1.040
	500	0.441	1.99	38.6	0.0339	38.7	0.996
	600	0.365	2.03	56.6	0.0422	57.3	0.986
	700	0.314	2.09	77.2	0.0505	77.2	1.00
	800	0.274	2.15	102	0.0592	100.1	1.010

† All information taken from Ref. 2 at end of this appendix.

Note further that, if there is a contact resistance R'_{co} at the interface between the composite solids or a fouling resistance R'_f such as caused by scale on the surface in contact with the fluid, they are also included in the same manner as other resistances. R'_{co} and R'_f are not readily available except by experiment in many cases; however, they can be very significant.

Frequently a term "U factor" (or overall heat-transfer coefficient) is used in the building industry. It is the reciprocal of $\Sigma R'$ and has units of W/(m^2 · °C). For case c of Fig. C-1, then

$$U = 1/(L_1/k_1 + L_2/k_2 + 1/h) \tag{C-5}$$

Also, the term "R factor" is frequently used to quantify the thermal resistance of insulation. It is the numerical value of $\Sigma R'$ and is usually quoted in relation to the English engineering system of units, e.g., R-22 implies 22 (h · ft^2 · °F)/Btu.

HEAT TRANSFER IN SOLAR-THERMAL APPLICATIONS

Example C-1 What is the conductive heat flux (i.e., per unit area) and total heat transfer through a solid brick wall 8 in thick, having an area of 8 × 12 ft, and with boundary temperatures of 120 and 70°F?

SOLUTION

$$q = k \frac{T_1 - T_2}{L} = [0.66 \text{ W/(m} \cdot {}°\text{C)}]\{0.5778 \text{ [Btu/(h} \cdot \text{ft} \cdot {}°\text{F)][(m} \cdot {}°\text{C)/W]}\}$$

$$\times \left[\frac{(120 - 70)°\text{F}}{8/12 \text{ ft}}\right] = 28.6 \text{ Btu/(h} \cdot \text{ft}^2)$$

$$= 90.2 \text{ W/m}^2$$

$$\dot{Q} = qA = 28.6 \text{ Btu/(h} \cdot \text{ft}^2) \, (8 \times 12)\text{ft}^2 = 2750 \text{ Btu/h}$$

$$= 805 \text{ W}$$

Example C-2 What is the total heat transfer through the back of a 1 × 2-m collector which has 5 cm of glass wool insulation, 2 cm of low-density polyurethane foam insulation, a 0.08-cm-thick galvanized steel enclosure, and a convective heat-transfer coefficient of 5 W/(m² · °C). The absorber panel is at 70°C and the ambient air is at 10°C.

SOLUTION

$$\dot{Q} = A \frac{T_1 - T_f}{\Sigma R'} = (2 \text{ m}^2) \left[\frac{(70 - 10)°\text{C}}{\left(\frac{0.05}{0.038} + \frac{0.02}{0.025} + \frac{0.0008}{54} + \frac{1}{5}\right)(\text{m}^2 \cdot {}°\text{C})/\text{W}}\right]$$

$$= \frac{120 \text{ (m}^2 \cdot {}°\text{C)}}{2.316 (\text{m}^2 \cdot {}°\text{C})/\text{W}} = 51.8 \text{ W}$$

$$= 177 \text{ Btu/h}$$

Note: The U factor for this composite is 0.432 W/(m² · °C) = [0.76 Btu/(h · ft² · °F)].

Example C-3 What is the R factor for 10 in of glass wool insulation?

SOLUTION

$$R' = \frac{L}{k} = \frac{(10/12)\text{ft}}{0.038 \text{ Btu/(h} \cdot \text{ft} \cdot {}°\text{F)}} = 21.9 \text{ (ft}^2 \cdot \text{h} \cdot {}°\text{F)/Btu}$$

Therefore the R factor is approximately 22. (It is currently not quoted in a metric unit.)

Heat transfer by conduction through annular geometries (pipes with or without insulation) is also common. The conduction through an annular pipe wall of

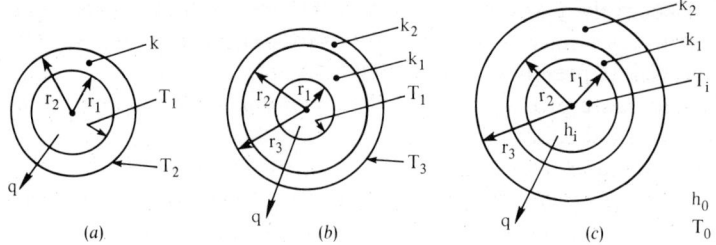

Figure C-2 One-dimensional heat transfer through single and composite annulas.

thermal conductivity k, radiuses r_1 and r_2, length L, and surface temperatures T_1 and T_2 (Fig. C-2a) is

$$\dot{Q} = \frac{T_1 - T_2}{(1/2\pi k L) \ln (r_2/r_1)} \qquad \text{(C-6)}$$

In the case of composite annular geometries, such as an insulated pipe (Fig. C-2b), the heat flow is in series through the composite, thus:

$$\dot{Q} = \frac{(T_1 - T_3)}{\dfrac{1}{2\pi L}\left(\dfrac{1}{k}\ln\dfrac{r_2}{r_1} + \dfrac{1}{k_2}\ln\dfrac{r_3}{r_2}\right)} \qquad \text{(C-7)}$$

If, as in Fig. C-2c, the composite annulus is exposed to convective heat-transfer coefficients h_i and h_o on the inside and outside surfaces, respectively, then

$$\dot{Q} = \frac{T_i - T_o}{\dfrac{1}{2\pi L}\left(\dfrac{1}{h_i r_1} + \dfrac{1}{k_1}\ln\dfrac{r_2}{r_1} + \dfrac{1}{k_2}\ln\dfrac{r_3}{r_2} + \dfrac{1}{h_o r_3}\right)} \qquad \text{(C-8)}$$

C-2 CONVECTION

Convection is that heat-transfer mode occurring in fluids as a result of the combined macroscopic and molecular motions of the fluid. The conduction effect (molecular) is thus lumped into what is referred to as convection. The macroscopic (gross) fluid motions may result from body forces such as gravity acting on a fluid with varying density (natural convection) or from fluid circulation created by an active device or effect such as a pump or the wind (forced convection). The subject of convection heat transfer is complicated and prediction of the convection heat-transfer rate is usually based on empirical equations. The convection heat-transfer rate $\dot{Q}$ (in W) is written as

$$\dot{Q} = hA(T_f - T_s)$$

where h, A, and T are the convective heat-transfer coefficient [W/(m² · °C)], area (m²), and temperature (°C), respectively. The subscripts f and s refer to the fluid and solid surface, respectively. Convection usually is between a fluid (f) and solid surface (s) but in some cases may be between two surfaces through a fluid. The equations and/or graphical presentations particularly useful to solar design are discussed below. These relate primarily to prediction of convection heat transfer in the design of solar-thermal collectors.

The convective heat-transfer correlations are in most cases written in terms of dimensionless quantities to make them more general. The convection heat-transfer coefficient h is nondimensionalized as the Nusselt number $\mathrm{Nu} \equiv hd/k$, where k is the fluid thermal conductivity [W/(m · °C)] and d is some characteristic dimension of the surface (m).

Forced Convection

In forced convection, the fluid velocity plays a dominant role and is normally a specified quantity. The Reynolds number $\mathrm{Re} \equiv dV\rho/\mu$ (V, ρ, μ, and d are the fluid velocity, density, dynamic viscosity, and some characteristic length dimension, respectively) is the dimensionless quantity defining the nature of the fluid flow. For flow in passages (pipes and ducts), d is the passage diameter and V is the average velocity, while for flow such as wind over a solar collector, $d = L$ is the length of the surface in the flow direction and V is the average wind velocity. For flow in passages, laminar to turbulent transition occurs near $\mathrm{Re}_d \simeq 2300$, while for flow over surfaces, transition occurs near $\mathrm{Re}_L \simeq 3 \times 10^5$.

For flow in passages, the convection heat transfer under turbulent conditions ($\mathrm{Re} \gg 2300$) is correlated as [6]

$$\mathrm{Nu} = 0.027 \ \mathrm{Re}^{0.8} \ \mathrm{Pr}^{1/3} \left(\frac{\mu}{\mu_w}\right)^{0.14}$$

where $\mathrm{Pr} \equiv c_p \mu/k$ is a dimensionless fluid parameter where c_p and μ are the fluid specific heat [kJ/(kg · °C)] and dynamic viscosity [kg/(m · s)], respectively, and μ/μ_w is a viscosity ratio with μ_w evaluated at the wall temperature and all other fluid properties evaluated at the average or bulk temperature. For flow in passages under laminar conditions ($\mathrm{Re} < 2300$), the average convection heat transfer coefficient over the length L is correlated as [6]

$$\mathrm{Nu} = 1.86 \ (\mathrm{Re} \ \mathrm{Pr})^{1/3} \left(\frac{d}{L}\right)^{1/3} \left(\frac{\mu}{\mu_w}\right)^{0.14}$$

where d/L is the ratio of passage diameter to length and the other quantities are defined as above. Note that the laminar heat-transfer coefficient is affected by the passage length.

For air flow over a cylinder such as wind blowing across the receiver tube of a linear concentrating collector, the convection heat-transfer correlation recommended by Ref. 1 is

$$\mathrm{Nu} = 0.25 \ \mathrm{Re}^{0.6} \ \mathrm{Pr}^{0.38}$$

for $10^3 < \mathrm{Re} < 2 \times 10^5$ where all properties are evaluated at the film temperature which is the average between the fluid and surface and the length dimension in Nu and Re is the cylinder diameter.

For air flow over flat surfaces such as the glazing of a flat-plate solar collector, the expression for the heat-transfer coefficient recommended by Ref. 7 is

$$h = 5.7 + 3.8V$$

where h and V are in W/(m² · °C) and m/s, respectively.

More recently, Ref. 8 recommended the correlation

$$h = 0.931(k/L) \, \text{Re}^{1/2} \, \text{Pr}^{1/3}$$

for $\text{Re} < 4 \times 10^5$.

Natural Convection

Natural convection is driven by gravity acting on fluid having varying density and the dimensionless parameter salient to natural convection is the Grashof number $\text{Gr} = g\beta \, \Delta T \, d^3/v^2$, where g, β, ΔT, d, and v are the local gravitational acceleration, fluid coefficient of volumetric expansion, temperature difference between surface and fluid, surface characteristic dimension, and kinematic viscosity, respectively. The Nusselt number Nu is usually correlated in terms of the Grashof and Prandtl numbers:

$$\text{Nu} = f_n \, (\text{Gr}, \text{Pr})$$

The most common natural convection problem associated with solar energy collection is that occurring between the cover plates or between the absorber and cover plate of a flat-plate collector. Tabor [9] has presented these convection coefficients for air between parallel inclined surfaces in the graphical form presented in Figs. C-3 and C-4. The symbols U, H, and D refer to upward, horizontal, and downward directions of heat flow, respectively—i.e., H means the heated surface is vertical and thus the heat transfer is horizontal. Figure C-3 provides Nu as a function of Gr with tilt a parameter. To facilitate the determination of the convection coefficient for air, Fig. C-4 allows determination of the quantities F_1 and F_2 as functions of the average air temperature. For specified values of the plate spacing l and temperature difference between surfaces, ΔT, computation of the quantity $F_1 \, \Delta T \, l^3$ allows determination of $F_2 hl$ for a given orientation, and thus the heat-transfer coefficient h can be calculated.

Other geometries of interest to solar design involving natural convection include vertical, inclined and horizontal flat surfaces, horizontal cylinders, and horizontal concentric annulas, the latter two being applicable to the absorber tube and glazed absorber tube of a linear concentrating collector. Table C-4, partially adapted from Ref. 10, lists recommended general heat-transfer correlations as well as simplified equations for air near 25°C and 1 atm pressure. For the horizontal annulus the effective thermal conductivity k_e (to be used in the conduction equation) is presented for convenience; otherwise, the heat-transfer coefficient h is presented.

HEAT TRANSFER IN SOLAR-THERMAL APPLICATIONS 319

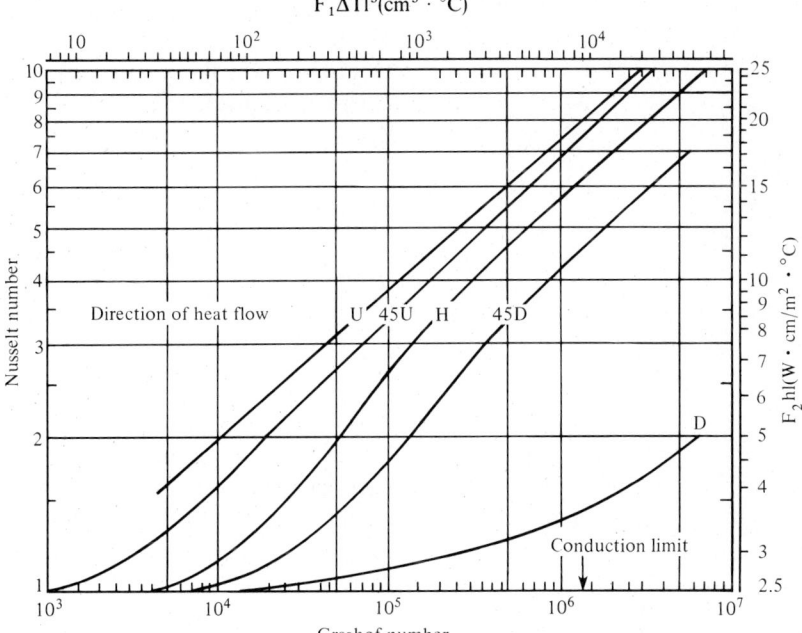

Figure C-3 Nusselt number as a function of Grashof number for free-convection heat transfer between parallel planes. See Fig. C-4 for values of F_1 and F_2. (*Adapted from Ref. 9.*)

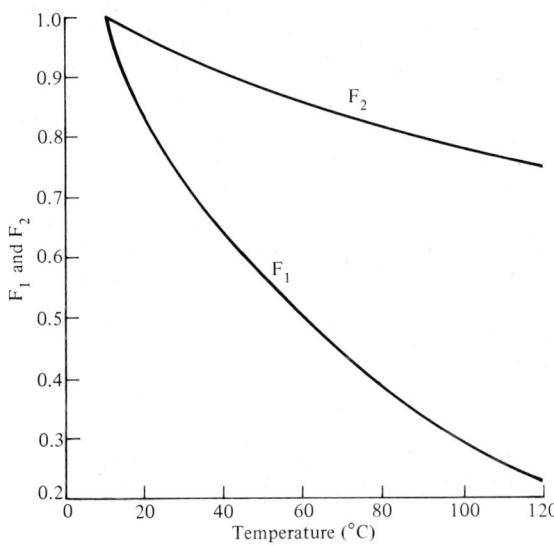

Figure C-4 Air property corrections F_1 and F_2 for Fig. C-3.

Table C-4 Natural convection correlations for several geometries

Geometry	Gr Pr	Correlation	Reference	Simplified equation† for air at 25°C and 1 atm
Vertical plates or cylinder	10^4–10^9	$\mathrm{Nu}_L = 0.59(\mathrm{Gr}_L\,\mathrm{Pr})^{1/4}$	11	$h = 1.42\left(\dfrac{\Delta T}{L}\right)^{1/4}$
	10^9–10^{13}	$\mathrm{Nu}_L = 0.10(\mathrm{Gr}_L\,\mathrm{Pr})^{1/3}$	12	$h = 0.95(\Delta T)^{1/3}$
Horizontal cylinders	10^4–10^9	$\mathrm{Nu}_d = 0.53(\mathrm{Gr}_d\,\mathrm{Pr})^{1/4}$	11	$h = 1.32\left(\dfrac{\Delta T}{d}\right)^{1/4}$
	10^9–10^{12}	$\mathrm{Nu}_d = 0.13(\mathrm{Gr}_d\,\mathrm{Pr})^{1/3}$	11	$h = 1.24(\Delta T)^{1/3}$
Upper surface of heated plates	2×10^4 – 10^6	$\mathrm{Nu}_L = 0.54(\mathrm{Gr}_L\,\mathrm{Pr})^{1/4}$	13, 14	$h = 1.32\left(\dfrac{\Delta T}{L}\right)^{1/4}$
	8×10^6 – 10^{11}	$\mathrm{Nu}_L = 0.15(\mathrm{Gr}_L\,\mathrm{Pr})^{1/3}$	13, 14	$h = 1.43(\Delta T)^{1/3}$
Inclined heated plates facing upward	10^5–10^{11}	$\mathrm{Nu}_L = 0.13[(\mathrm{Gr}_L\,\mathrm{Pr})^{1/3} - (\mathrm{Gr}_{c,L}\,\mathrm{Pr})^{1/3}]$ $+ 0.56(\mathrm{Gr}_{c,L}\,\mathrm{Pr}\cos\theta)^{1/4}$‡	13	
Horizontal annulus ($\delta = r_o - r_i$)	6×10^3–6×10^6	$\dfrac{k_e}{k} = 0.11(\mathrm{Gr}_\delta\,\mathrm{Pr})^{0.29}$	15	$k_e = 0.59(\Delta T\,\delta^3)^{0.29}$ $\left.\begin{array}{c}\\ \\ \\ \\ \end{array}\right\} q = \dfrac{2\pi k_e L\,\Delta T}{\ln(r_o/r_i)}$
	10^6–10^8	$\dfrac{k_e}{k} = 0.40(\mathrm{Gr}_\delta\,\mathrm{Pr})^{0.20}$	15	$k_e = 0.41(\Delta T\,\delta^3)^{0.20}$

† h in W/(m²·°C), k_e in W/(m·°C), ΔT in °C; L, d, and δ in m.

‡ *Note:* θ measured from vertical. Critical Grashof number, $\mathrm{Gr}_c \simeq 2\times 10^9$, 2×10^8, and 2×10^7 for $\theta = -30°$, $-45°$, and $-60°$, respectively, and the first term omitted when $\mathrm{Gr}_L < \mathrm{Gr}_{c,L}$.

C-3 RADIATION

Thermal radiation is that portion of the electromagnetic spectrum extending from approximately 0.1 to 1000 μm.† The regions of approximately 0.1–0.4 μm, 0.4–0.7 μm and 0.7–1000 μm are the ultraviolet, visible, and infrared regions, respectively. (Sometimes the regions of approximately 0.7–3.0 μm and 3.0–1000 μm are referred to as the "near-infrared" and "far-infrared" regions, respectively.)

Radiation exhibits both particlelike and wavelike behavior, and is quantized as discrete photons, the energy of a photon being given as its frequency v times Planck's constant:

$$e = hv$$

Since photons travel at the speed of light ($c = v\lambda$), then

$$e = \frac{hc}{\lambda}$$

where e, h, c, v, and λ are the energy, Planck's constant ($=6.63 \times 10^{-27}$ erg · s), speed of light ($=3.0 \times 10^8$ m/s), frequency, and wavelength, respectively. Because thermal radiation extends over a spectrum of wavelengths and the radiation properties of some materials may be strongly wavelength-dependent, property variations with wavelength can be important factors in potential applications of solar energy.

When radiation from one medium is incident upon a different material, it may be reflected, transmitted, and/or absorbed, the terms reflectivity (ρ), transmissivity (τ), and absorptivity (α) referring to the fractional amount of each (compared to incident). The sum of these is 1:

$$\rho + \tau + \alpha = 1$$

for any surface, and for an opaque surface where $\tau = 0$, therefore,

$$\rho + \alpha = 1$$

The term "opaque" really is relative to thickness since the radiation-interface interaction is not just a surface phenomenon but extends to some depth into a material. Many surfaces can be considered opaque, however.

Blackbody Radiation

A "blackbody" designates a body that at a particular temperature absorbs all incident radiation, and as a consequence, emits the maximum possible amount of radiation for that temperature. The Stefan-Boltzmann equation states that for a blackbody the emitted radiative flux in W/m² is

$$W_b = \sigma T^4$$

---

† 1 μm equals 10^{-6} m and 10,000 Å.

where σ is the Stefan-Boltzmann constant 5.669×10^{-8} W/(m$^2 \cdot$ K^4), and T is the absolute temperature, in K, of the body.

The blackbody has a spectral emissive power [measured in W/(m$^2 \cdot \mu$m)] as indicated in Fig. C-5 for different temperatures. The spectral emissive power is the emissive power per unit wavelength band at a particular wavelength. Note that the spectral emissive power is a function of wavelength and of temperature and exhibits a peak which occurs at a wavelength that varies with temperature. This peak in the spectral emissive power is related to the temperature by Wien's displacement law:

$$\lambda_{\max} T = 2897.6 \; \mu\text{m} \cdot \text{K}$$

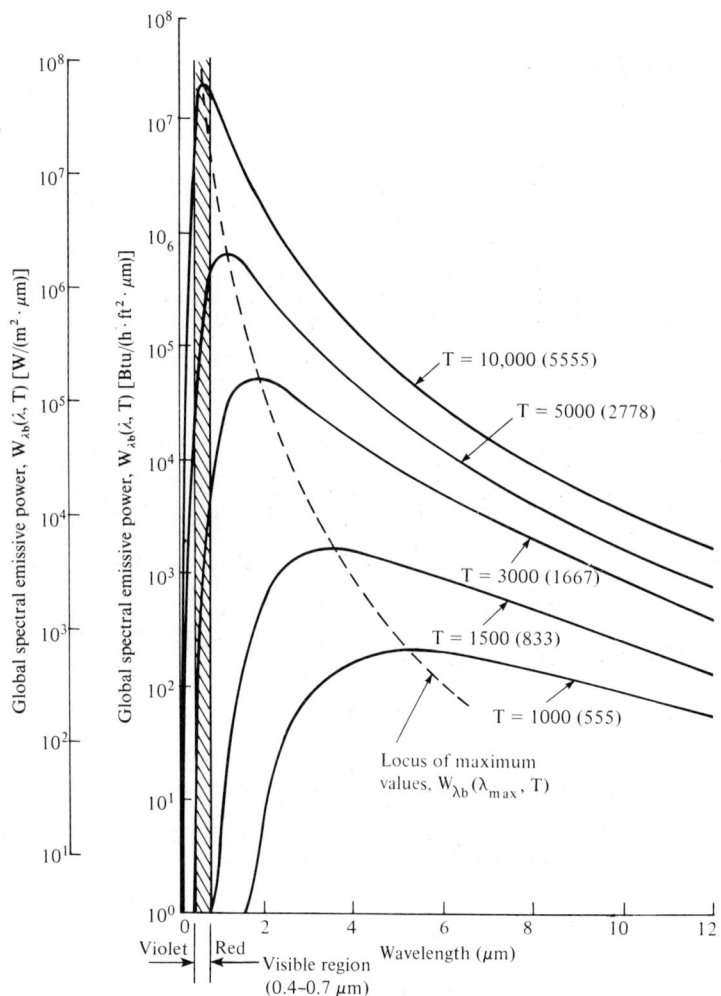

Figure C-5 Global spectral emissive power of a blackbody for several different temperatures. Blackbody temperature T given in °R (K). (*From Ref. 16.*)

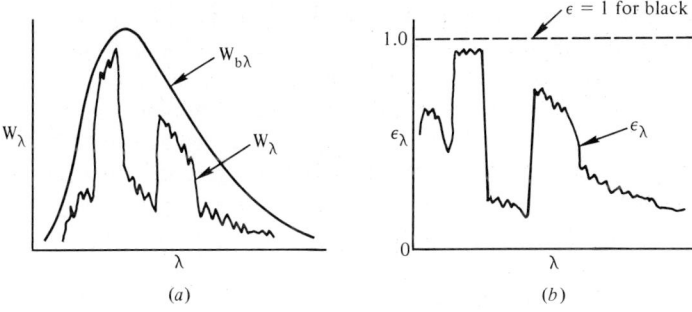

Figure C-6 Spectral emissive power and emissivity.

That is, the product of the wavelength at maximum spectral emissive power and absolute temperature is a constant.

The normalized spectral emissive power $W_{\lambda b}/\sigma T^5$ as a function of λT is presented in Fig. C-7a as well as in Table C-5. Note that the maximum in $W_{\lambda b}$ occurs at $\lambda T \cong 2900$ μm · K (Fig. C-7a). The fraction of total energy that is emitted up to λT, i.e., $W_{(0-\lambda T)}/\sigma T^4$, is also presented in Fig. C-7b and Table C-5.

Real-Body Radiation

A real body exhibits an emitted radiative flux (total global emissive power) of some fraction of the blackbody emissive power:

$$W = \epsilon \sigma T^4$$

where ϵ is the effective emissivity (total global emissivity) of the surface, is less than or equal to 1, and is normally a function of the body temperature. Surfaces also exhibit spectrally dependent emissivity ϵ_λ, as depicted in Fig. C-6b. If at a particular temperature the spectral emissive power of a body compared to a blackbody is as shown in Fig. C-6a, then the ratio of $W_\lambda/W_{b\lambda}$ is the spectral emissivity ϵ_λ in Fig. C-6b. The total global emissivity ϵ is simply the "effective average" emissivity used to multiply σT^4 to provide the correct emissive power for the real body.

Grey Body

A grey body is a body for which the spectral emissivity ϵ_λ is a constant (i.e., not a function of λ). Many surfaces can be approximated as being grey, while for others it is a poor approximation. For a grey body the emissive power is as above:

$$W = \epsilon \sigma T^4$$

Emissivity and Absorptivity

Just as the emissivity may be spectrally dependent, so may the absorptivity. It can be shown that the spectral emissivity ϵ_λ and spectral absorptivity α_λ of a

Table C-5 Blackbody radiation functions

λT		$W_{b\lambda}/T^5$		$W_{bO-\lambda T}$	λT		$W_{b\lambda}/T^5$		$W_{bO-\lambda T}$
		Btu	W				Btu	W	
$\mu m \cdot °R$	$\mu m \cdot K$	$h \cdot ft^2 \cdot °R^5 \cdot \mu m \times 10^{15}$	$m^2 \cdot K^5 \cdot \mu m \times 10^{11}$	$\dfrac{W_{bO-\lambda T}}{\sigma T^4}$	$\mu m \cdot °R$	$\mu m \cdot K$	$h \cdot ft^2 \cdot °R^5 \cdot \mu m \times 10^{15}$	$m^2 \cdot K^5 \cdot \mu m \times 10^{11}$	$\dfrac{W_{bO-\lambda T}}{\sigma T^4}$
1,000	555.6	0.000671	0.400×10^{-5}	0.170×10^{-7}	10,200	5,666.7	92.145	0.54877	0.70754
1,200	666.7	0.0202	0.120×10^{-3}	0.756×10^{-6}	10,400	5,777.8	88.181	0.52517	0.71806
1,400	777.8	0.204	0.00122	0.106×10^{-4}	10,600	5,888.9	84.394	0.50261	0.72813
1,600	888.9	1.057	0.00630	0.738×10^{-4}	10,800	6,000.0	80.777	0.48107	0.73777
1,800	1,000.0	3.544	0.02111	0.321×10^{-3}	11,000	6,111.1	77.325	0.46051	0.74700
2,000	1,111.1	8.822	0.05254	0.00101	11,200	6,222.2	74.031	0.44089	0.75583
2,200	1,222.2	17.776	0.10587	0.00252	11,400	6,333.3	70.889	0.42218	0.76429
2,400	1,333.3	30.686	0.18275	0.00531	11,600	6,444.4	67.892	0.40434	0.77238
2,600	1,444.4	47.167	0.28091	0.00983	11,800	6,555.6	65.036	0.38732	0.78014
2,800	1,555.6	66.334	0.39505	0.01643	12,000	6,666.7	62.313	0.37111	0.78757
3,000	1,666.7	87.047	0.51841	0.02537	12,200	6,777.8	59.717	0.35565	0.79469
3,200	1,777.8	108.14	0.64404	0.03677	12,400	6,888.9	57.242	0.34091	0.80152
3,400	1,888.9	128.58	0.76578	0.05059	12,600	7,000.0	54.884	0.32687	0.80806
3,600	2,000.0	147.56	0.87878	0.06672	12,800	7,111.1	52.636	0.31348	0.81433
3,800	2,111.1	164.49	0.97963	0.08496	13,000	7,222.2	50.493	0.30071	0.82035
4,000	2,222.2	179.04	1.0663	0.10503	13,200	7,333.3	48.450	0.28855	0.82612
4,200	2,333.3	191.05	1.1378	0.12665	13,400	7,444.4	46.502	0.27695	0.83166
4,400	2,444.4	200.51	1.1942	0.14953	13,600	7,555.6	44.645	0.26589	0.83698
4,600	2,555.6	207.55	1.2361	0.17337	13,800	7,666.7	42.874	0.25534	0.84209
4,800	2,666.7	212.32	1.2645	0.19789	14,000	7,777.8	41.184	0.24527	0.84699
5,000	2,777.8	215.06	1.2808	0.22285	14,200	7,888.9	39.572	0.23567	0.85171
5,200	2,888.9	216.00	1.2864	0.24803	14,400	8,000.0	38.033	0.22651	0.85624
5,400	3,000.0	215.39	1.2827	0.27322	14,600	8,111.1	36.565	0.21777	0.86059
5,600	3,111.1	213.46	1.2713	0.29825	14,800	8,222.2	35.163	0.20942	0.86477
5,800	3,222.2	210.43	1.2532	0.32300	15,000	8,333.3	33.825	0.20145	0.86880
6,000	3,333.3	206.51	1.2299	0.34734	16,000	8,888.9	27.977	0.16662	0.88677
6,200	3,444.4	201.88	1.2023	0.37118	17,000	9,444.4	23.301	0.13877	0.90168
6,400	3,555.6	196.69	1.1714	0.39445	18,000	10,000.0	19.536	0.11635	0.91414

6,600	3,666.7	191.09	1.1380	0.41708	19,000	10,555.6	16.484	0.09817	0.92462
6,800	3,777.8	185.18	1.1029	0.43905	20,000	11,111.1	13.994	0.08334	0.93349
7,000	3,888.9	179.08	1.0665	0.46031	21,000	11,666.7	11.949	0.07116	0.94104
7,200	4,000.0	172.86	1.0295	0.48085	22,000	12,222.2	10.258	0.06109	0.94751
7,400	4,111.1	166.60	0.99221	0.50066	23,000	12,777.8	8.852	0.05272	0.95307
7,600	4,222.2	160.35	0.95499	0.51974	24,000	13,333.3	7.676	0.04572	0.95788
7,800	4,333.3	154.16	0.91813	0.53809	25,000	13,888.9	6.687	0.03982	0.96207
8,000	4,444.4	148.07	0.88184	0.55573	26,000	14,444.4	5.850	0.03484	0.96572
8,200	4,555.5	142.10	0.84629	0.57267	27,000	15,000.0	5.139	0.03061	0.96892
8,400	4,666.7	136.28	0.81163	0.58891	28,000	15,555.6	4.532	0.02699	0.97174
8,600	4,777.8	130.63	0.77796	0.60449	29,000	16,111.1	4.012	0.02389	0.97423
8,800	4,888.9	125.15	0.74534	0.61941	30,000	16,666.7	3.563	0.02122	0.97644
9,000	5,000.0	119.86	0.71383	0.63371	40,000	22,222.2	1.273	0.00758	0.98915
9,200	5,111.1	114.76	0.68346	0.64740	50,000	27,777.8	0.560	0.00333	0.99414
9,400	5,222.2	109.85	0.65423	0.66051	60,000	33,333.3	0.283	0.00168	0.99649
9,600	5,333.3	105.14	0.62617	0.67305	70,000	38,888.9	0.158	0.940×10^{-3}	0.99773
9,800	5,444.4	100.62	0.59925	0.68506	80,000	44,444.4	0.0948	0.564×10^{-3}	0.99845
10,000	5,555.6	96.289	0.57346	0.69655	90,000	50,000.0	0.0603	0.359×10^{-3}	0.99889
					100,000	55,555.6	0.0402	0.239×10^{-3}	0.99918

Source: Ref. 10.

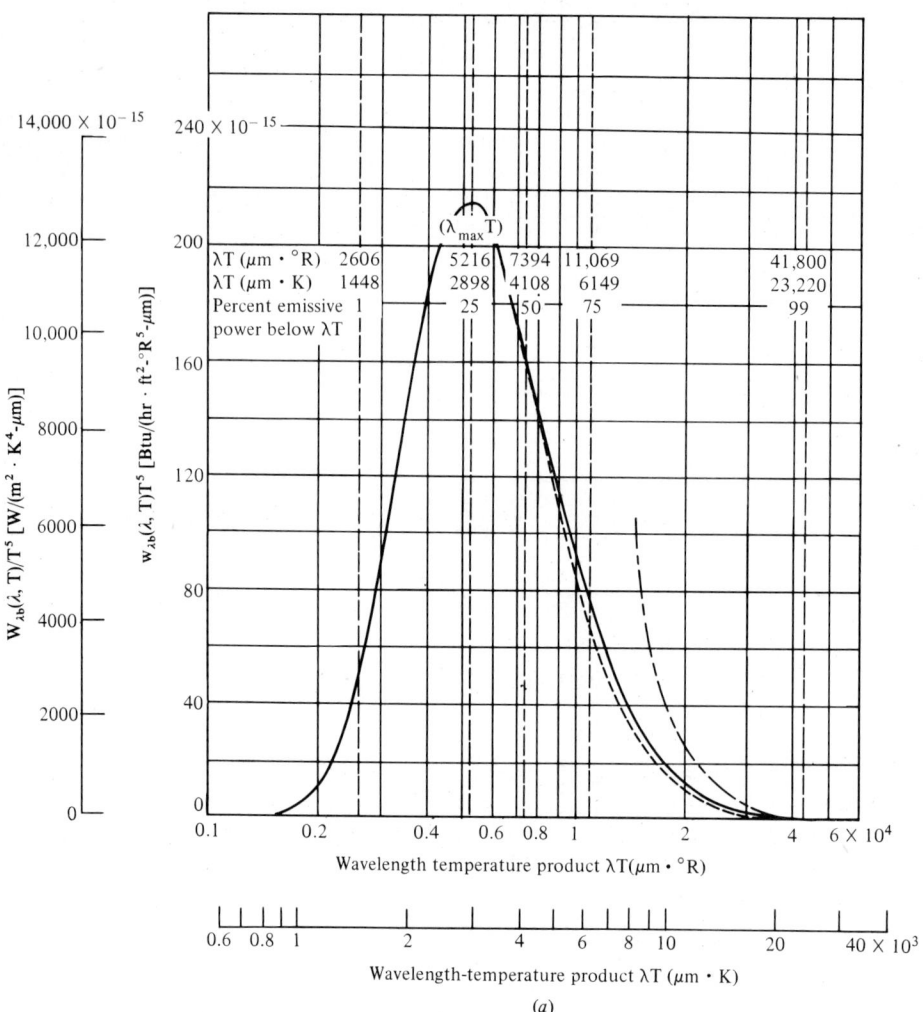

Figure C-7 (a) Spectral distribution of blackbody hemispherical emissive power. ——— Planck's law; --- Wien's distribution; — — — Rayleigh-Jeans' distribution. (b) Fractional blackbody emissive power in the range 0 to T. (*From Ref. 16.*)

body at the same wavelength are equal (Kirchhoff's law), i.e.,

$$\epsilon_\lambda = \alpha_\lambda$$

If the body has grey properties it can further be shown that

$$\epsilon = \alpha$$

Since most of the energy in the solar spectrum occurs below 1 μm and since the peak in the spectral emissive power for surfaces in many solar applications

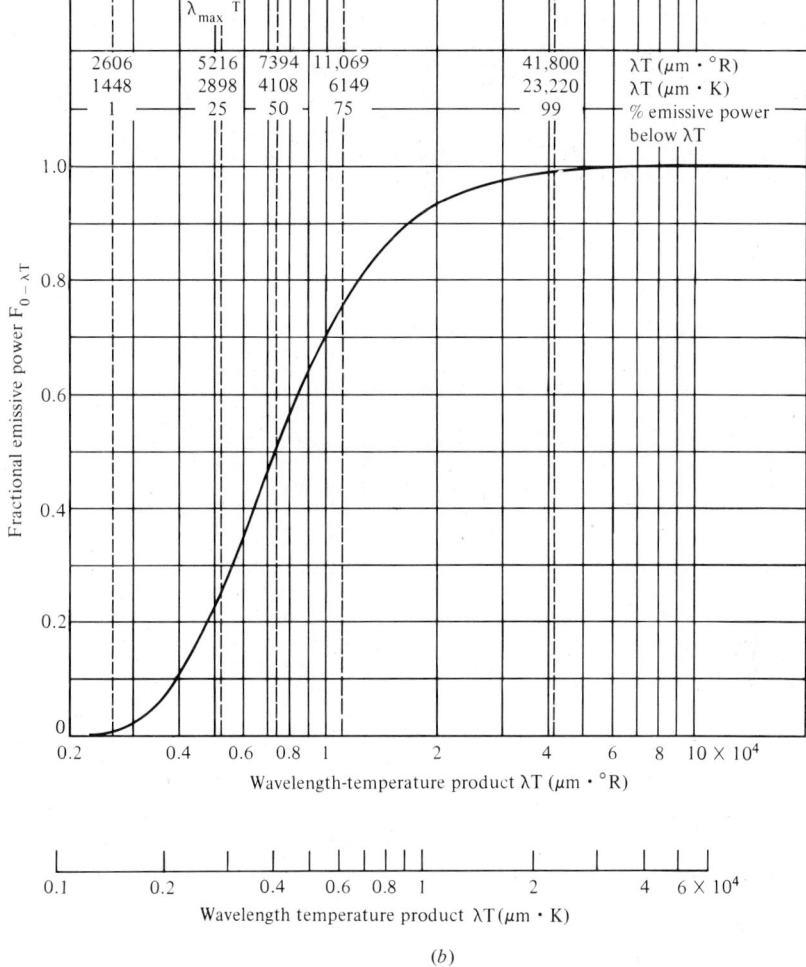

(b)

occurs at much larger wavelengths (i.e., the blackbody peaks occur at approximately 10, 5, and 3.3 μm for 25, 300 and 600°C surfaces, respectively), the spectral qualities of the surface are very important.

Table D-1 lists the infrared emittance ϵ for a variety of materials. Note that the emittances are generally functions of temperature. Table D-2 lists the solar absorptivities α for several surfaces. Note that the ratio of α/ϵ is one measure of the quality of the surface for efficient collection of solar energy. Figure C-8 presents the wavelength-dependent reflectivity of different surfaces. White epoxy paint (D) has a low α/ϵ and represents a good coating for reflection of solar energy, i.e., good for roofs or walls of buildings to reduce heat loads in summer or for spacecraft thermal control. Stainless steel (C) is typical of many metals. Curves A and B show the spectral properties of chrome-oxide and nickel-oxide,

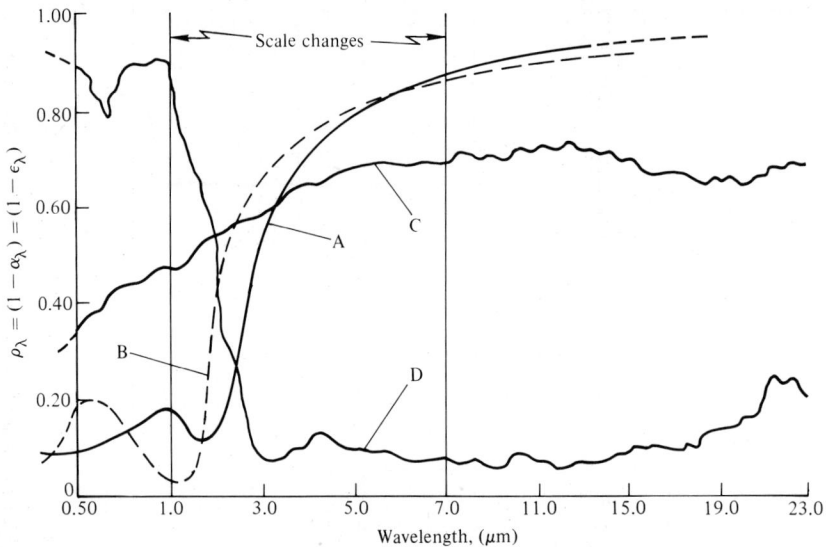

Figure C-8 Monochromatic reflectivity of various surfaces. A, black chrome; B, black nickel; C, stainless steel, type 301; D, white epoxy paint on aluminum. *(From D. K. Edwards, K. E. Nelson, R. D. Roddick, and J. T. Gier, "Basic Studies on the Use and Control of Solar Energy," Department of Engineering Report No. 60-93, University of California, Los Angeles, 1960.)*

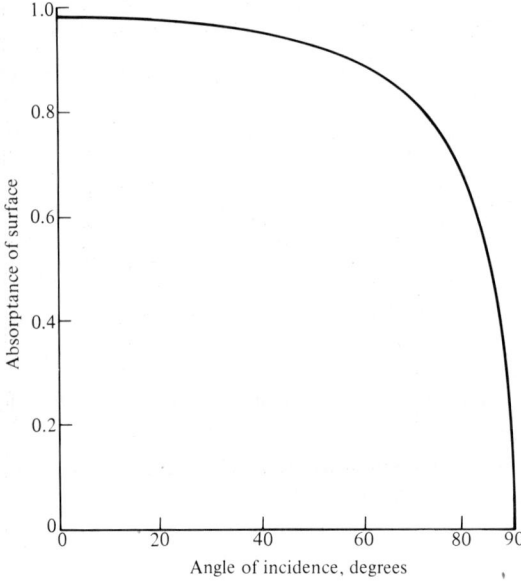

Figure C-9 Directional absorptance of a blackened surface for artificial sunlight transmitted through glass.

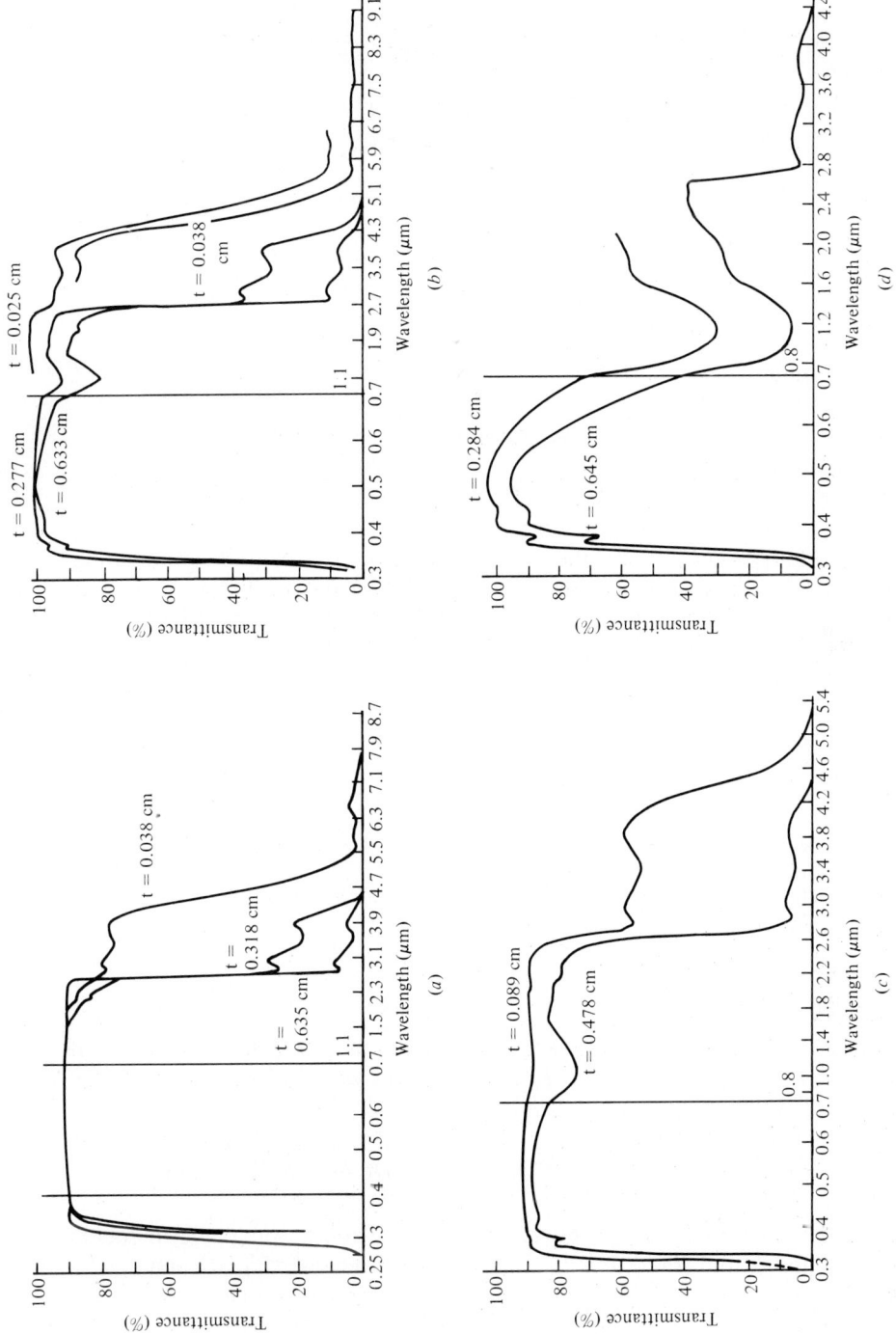

Figure C-10 Spectral transmittance of glass for varying amounts of Fe_2O_3. (a) 0.02; (b) 0.10; (c) 0.15; (d) 0.50. [*From A. G. H. Dietz "Diathermanous Materials and Properties of Surfaces" in R. W. Hamilton (ed.), Space Heating with Solar Energy, MIT, 1954.*]

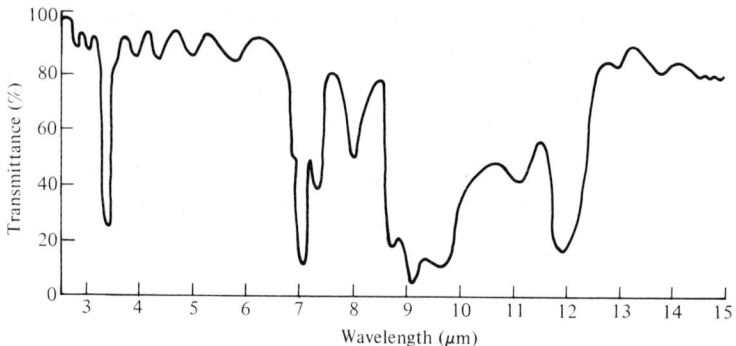

Figure C-11 Spectral transmittance of Tedlar, a polyvinyl fluoride film. *(Courtesy duPont.)*

respectively, which exhibit good selective properties for solar energy collection.

The absorptivity of surfaces is also a function of the incident angle. Figure C-9 presents the absorptivity of a blackened surface as a function of angle from normal for a simulated solar spectrum.

Figures C-10 and C-11 give the spectral transmittances of glasses of different thicknesses for varying amounts of iron-oxide (Fe_2O_3) and for Tedlar film (polyvinyl fluoride), respectively. Figure 5-2 earlier in the book indicated the effects of incidence angle and number of cover glasses on transmittance.

The reflectivity of a surface to the solar spectrum is very important in reflective concentrating collectors. Since the reflectivity $\rho = 1 - \alpha$ for opaque surfaces, the solar reflectivity for several surfaces can be determined from the absorptivity data of Table D-2.

Radiation Exchange

The net radiation lost by a small object of area A at temperature T_1 and emissivity ϵ_1 surrounded by a large enclosure at T_∞ is

$$\dot{Q}_{net} = \sigma \epsilon_1 A \left(T_1^4 - T_\infty^4 \right)$$

The net radiation exchange between two "large" parallel plates of area A and with temperatures and emissivities T_1 and T_2 and ϵ_1 and ϵ_2 for the case of grey surfaces is

$$\dot{Q}_{net} = \frac{\sigma A}{1/\epsilon_1 + 1/\epsilon_2 - 1} (T_1^4 - T_2^4) \qquad \text{(if positive then net loss from 1)}$$

The net radiation exchange between long, diffuse concentric grey cylinders of radiuses r_1 and r_2 and length L with temperatures and emissivities T_1 and T_2 and ϵ_1 and ϵ_2 is

$$\dot{Q}_{net} = \frac{\sigma 2\pi r_1 L (T_1^4 - T_2^4)}{1/\epsilon_1 + (r_1/r_2)(1/\epsilon_2 - 1)} \qquad \text{(if positive then net loss from 1)}$$

The net radiation exchange between surface 1 of emissivity ϵ and a large enclosure at T_2 is

$$\dot{Q}_{net} = \epsilon \sigma A_1 (T_1^4 - T_2^4)$$

It is frequently convenient in heat-transfer analysis to "linearize" the radiation heat-transfer equation by factoring out the absolute temperature difference, which for the general exchange equation above becomes

$$\dot{Q}_{net} = \epsilon \sigma A_1 (T_1^2 + T_2^2)(T_1 + T_2)(T_1 - T_2)$$
$$\equiv h_R A_1 (T_1 - T_2)$$

where $h_R \equiv \epsilon \sigma (T_1^2 + T_2^2)(T_1 + T_2)$ is defined as the radiation heat-transfer coefficient. While it is temperature-dependent, it facilitates determination of the overall thermal loss coefficients when both convection and radiation heat transfer are important.

REFERENCES

1. Frank Kreith, *Principles of Heat Transfer*, 3d ed., Harper & Row, New York, 1973.
2. A. I. Brown, and S. M. Marco, *Introduction to Heat Transfer*, 3d ed., McGraw Hill Book Co., New York, 1958.
3. *Materials Engineering—Materials Selector '78*, Reinhold Publishing Company, New York, 1977.
4. *Glycols*, Union Carbide Bulletin, 1971.
5. *Dow Therm Heat Transfer Fluids*, Dow Chemical Company, 1967.
6. E. N. Sieder and C. E. Tate, "Heat Transfer and Pressure Drop of Liquids in Tubes," *Ind. Engr. Chem.*, vol. 28, p. 1429, 1936.
7. J. A. Duffie and W. A. Beckman, *Solar Energy Thermal Processes*, John Wiley & Sons, New York, 1974.
8. E. M. Sparrow and K. K. Tien, "Forced Convection Heat Transfer at an Inclined and Yawed Square Plate-Application to Solar Collector," *J. Heat Transfer*, vol. 99, pp. 507–512, 1977.
9. H. Tabor, "Radiation, Convection and Conduction Coefficients in Solar Collectors," *Bull. Res. Counc., Israel*, vol. 6C, p. 155, 1958.
10. J. P. Holman, *Heat Transfer*, McGraw-Hill Book Co., New York, 1976.
11. W. H. McAdams, *Heat Transmission*, 3d ed., McGraw-Hill Book Co., New York, 1954.
12. C. Y. Warner and V. S. Arpaci, "An Experimental Investigation of Turbulent Natural Convection in Air at Low Pressure Along a Vertical Heated Flat Plate," *Int. J. Heat Mass Transfer*, vol. 11, p. 397, 1968.
13. T. Fujii and H. Imura, "Natural Convection Heat Transfer from a Plate with Arbitrary Inclinations," *Int. J. Heat Mass Transfer*, vol. 15, p. 755, 1972.
14. J. R. Lloyd and W. R. Moran, "Natural Convection Adjacent to Horizontal Surface of Various Planforms," ASME Paper 74-WA/HT-66, ASME Winter Annual Meeting, New York, November 1974.
15. C. Y. Lui, W. K. Mueller, and F. Landis, "Natural Convection Heat Transfer in Long Horizontal Cylindrical Annuli," *Int. Developments in Heat Transfer, Proc. Second Int. Heat Transfer Conf.*, part 5, p. 976, Boulder, CO and Westminster, Eng., 1961.
16. R. Siegel and J. R. Howell, *Thermal Radiation Heat Transfer*, 2d ed., McGraw-Hill Book Co., New York, 1980.

APPENDIX D

SELECTED RADIATIVE PROPERTIES OF MATERIALS

Tables of absorptivities for insolation and of total emissivities are provided here for convenience in working problems and to give the reader an indication of the magnitudes to be expected. Many factors such as roughness and oxidation can strongly affect the radiative properties. No attempt is made here to describe in detail the conditions of the material samples; hence the values given here are only reasonable approximations in some instances. For detailed information on radiative properties including sample descriptions and results from many sources, the reader is referred to Refs. 1–3. Data in Ref. 3 is very extensive. Reference 4 also provides a limited amount of additional information. Reference 5 gives information particularly useful in solar collector design. As will be seen from these references, there can sometimes be considerable differences in the properties for the same material measured by different investigators. Reference 6 discusses radiative properties in detail and presents equations for prediction of properties and for the extrapolation of limited experimental data. Reference 7 provides reflectivity values for common surfaces.

Table D-1 Total emissivity

Material	Surface temperature,† °F (K)	ϵ
Metals		
Aluminum:		
Highly polished plate	400–1100 (480–870)	0.038–0.06
Bright foil	70 (295)	0.04
Polished plate	212 (373)	0.095
Heavily oxidized	200–1000 (370–810)	0.20–0.33
Brass:		
Highly polished	500–700 (530–640)	0.028–0.031
Polished	200 (370)	0.09
Dull	120–660 (320–620)	0.22
Oxidized	400–1000 (480–810)	0.60
Chromium, polished	100–2000 (310–1370)	0.08–0.40
Copper:		
Highly polished	100 (310)	0.02
Polished	100–500 (310–530)	0.04–0.05
Scraped, shiny	100 (310)	0.07
Slightly polished	100 (310)	0.15
Black oxidized	100 (310)	0.78
Gold:		
Highly polished	200–1100 (370–870)	0.018–0.135
Polished	266 (400)	0.018
Inconel X and B, polished (see stainless steel)		
Iron:		
Highly polished, electrolytic	100–500 (310–530)	0.05–0.07
Polished	800–900 (700–760)	0.14–0.38
Freshly rubbed with emery	100 (310)	0.24
Wrought iron, polished	100–500 (310–530)	0.28
Cast iron, freshly turned	100 (310)	0.44
Iron plate, pickled, then rusted red	68 (293)	0.61
Cast iron, oxidized at 1100°F	400–1100 (480–870)	0.64–0.78
Cast iron, rough, strongly oxidized	100–500 (310–530)	0.95
Magnesium, polished	100–500 (310–530)	0.07–0.13
Monel:		
Polished	100 (310)	0.17
Oxidized at 1100°F	1000 (810)	0.45
Nickel:		
Electrolytic	100–500 (310–530)	0.04–0.06
Technically pure, polished	440–710 (500–650)	0.07–0.087
Electroplated on iron, not polished	68 (293)	0.11

† When temperatures and emissivities both have ranges, linear interpolation can be used over these values.

Table D-1 Total emissivity (*Continued*)

Material	Surface temperature,† °F (K)	ε
Plate oxidized at 1100°F	390–1110 (470–870)	0.37–0.48
Nickel oxide	1200–2300 (920–1530)	0.59–0.86
Platinum:		
Electrolytic	500–1000 (530–810)	0.06–0.10
Polished plate	440–1160 (500–900)	0.054–0.104
Silver polished	100–1000 (310–810)	0.01–0.03
Stainless steel:		
Inconel X, polished	−300–900 (90–760)	0.19–0.20
Inconel B, polished	−300–900 (90–760)	0.19–0.22
Type 301, polished	75 (297)	0.16
Type 310, smooth	1500 (1090)	0.39
Type 316, polished	400–1900 (480–1310)	0.24–0.31
Steel:		
Polished sheet	−300–0 (90–273)	0.07–0.08
Polished sheet	0–300 (273–420)	0.08–0.14
Mild steel, polished	500–1200 (530–920)	0.27–0.31
Sheet with skin due to rolling	70 (295)	0.66
Sheet with rough oxide layer	70 (295)	0.81
Tin:		
Polished sheet	93 (310)	0.05
Bright tinned iron	76 (298)	0.043–0.064
Zinc:		
Polished	100–1000 (310–810)	0.02–0.05
Galvanized sheet, fairly bright	100 (310)	0.23
gray oxidized	70 (295)	0.23–0.28
Nonmetals		
Alumina on Inconel	1000–2000 (810–1370)	0.65–0.45
Asbestos:		
Paper	100 (310)	0.93
Board	100 (310)	0.96
Black chrome	800 (700)	0.06–0.15
Black acrylic paint	325–375 (436–464)	0.95–0.967
Brick:		
White refractory	2000 (1370)	0.29
Fireclay	1800 (1260)	0.75
Rough red	100 (310)	0.93
Carbon, lampsoot	100 (310)	0.95

† When temperatures and emissivities both have ranges, linear interpolation can be used over these values.

Table D-1 Total emissivity (*Continued*)

Material	Surface temperature,† °F (K)	ϵ
Concrete, rough	100 (310)	0.94
Copper oxide	375 (464)	0.12–0.15
Corundum, emery rough	200 (370)	0.86
Ice:		
Smooth	32 (273)	0.966
Rough crystals	32 (273)	0.985
Lead sulfide crystals	350 (506)	0.19–0.37
Magnesium oxide, refractory	300–900 (420–760)	0.69–0.55
Marble, white	100 (310)	0.95
Nickel black	400 (478)	0.11–0.18
Paint:		
Oil, all colors	212 (373)	0.92–0.96
Lacquer, flat black	100–200 (310–370)	0.96–0.98
Red lead	200 (370)	0.93
Paper:		
White	100 (310)	0.96
Roofing	100 (310)	0.91
Plaster	100 (310)	0.91
Porcelain, glazed	70 (295)	0.92
Sandstone	100–500 (310–530)	0.83–0.90
Silicon carbide	300–1200 (420–920)	0.83–0.96
Slate	100 (310)	0.67–0.80
Snow	20 (270)	0.82
Soot, candle	200–500 (370–530)	0.95
Water, deep	32–212 (273–373)	0.96
Wood:		
Oak, planed	70 (295)	0.90
Beech	158 (340)	0.94
Sawdust	100 (310)	0.75

† When temperatures and emissivities both have ranges, linear interpolation can be used over these values.

Table D-2 Absorptivity for insolation receiving material at 295K (70° F)

Material	α
Metals	
Aluminum:	
Highly polished	0.10
Polished	0.20
Chromium, electroplated	0.40
Copper:	
Highly polished	0.18
Clean	0.25
Tarnished	0.64
Oxidized	0.70
Galvanized iron	0.38
Gold, bright foil	0.29
Iron:	
Ground with fine grit	0.36
Blued	0.55
Sandblasted	0.75
Magnesium, polished	0.19
Nickel:	
Highly polished	0.15
Polished	0.36
Electrolytic	0.40
Platinum, bright	0.31
Silver:	
Highly polished	0.07
Polished	0.13
Commercial sheet	0.30
Stainless steel No. 301, polished	0.37
Tungsten, highly polished	0.37
Nonmetals	
Aluminum-oxide (Al_2O_3)	0.06–0.23
Asphalt pavement, dust-free	0.93
Black acrylic paint	0.95–0.967
Black chrome	0.90–0.95

Table D-2 (Continued)

Material	α
Metals	
Brick, red	0.75
Concrete roofing tile:	
Uncolored	0.73
Brown	0.91
Black	0.91
Copper oxide	0.81–0.93
Earth, plowed field	0.75
Felt, black	0.82
Graphite	0.88
Grass	0.75–0.80
Gravel	0.29
Lead sulfide, crystals	0.89
Leaves, green	0.71–0.79
Magnesium-oxide (MgO)	0.15
Marble, white	0.46
Nickel, black	0.89–0.94
Paint:	
Aluminum	0.55
Oil, zinc, white	0.30
Oil, light green	0.50
Oil, light gray	0.75
Oil, black on galvanized iron	0.90
Paper, white	0.28
Slate, blue gray	0.88
Snow, clean	0.2–0.35
Soot, coal	0.95
Titanium dioxide (TiO_2)	0.12
Zinc oxide	0.15
Zinc sulfide (ZnS)	0.21

Table D-3 Effective reflectances (integrated over the solar spectrum and angle of incidence)

Characteristic flat surfaces	ρ_r
Snow (fresh)	0.75
Water	0.07
Soils	0.14
Earth roads	0.04
Coniferous forest (winter)	0.07
Forest (autumn), plants	0.26
Weathered blacktop	0.10
Weathered concrete	0.22
Dead leaves	0.30
Dry grass	0.20
Green leaves	0.26
Bituminous and gravel roof	0.13
Crushed rock surface	0.20
Building surfaces (dark)	0.27
Building surfaces (light)	0.60

Characteristic landscapes	ρ_r
Fields with snow cover	~0.6
Open water	0.16
Urban commercial	0.16
Urban institutional	0.38
Residential areas	0.2–0.4

Source: Ref. 7.

REFERENCES

1. G. G. Gubareff, J. E. Janssen, and R. H. Torborg, *Thermal Radiation Properties Survey*, 2d ed., Minneapolis-Honeywell Regulator Co., Minneapolis, 1960.
2. W. D. Wood, H. W. Deem, and C. F. Lucks, *Thermal Radiative Properties*, Plenum Press, New York, 1964.
3. Y. S. Touloukian et al., *Metallic Elements and Alloys*, vol. 7, *Nonmetallic Solids*, vol. 8, and *Coatings*, vol. 9, in *Thermal Radiative Properties*, Thermophysical Properties Research Center of Purdue University, Data Series, Plenum Press, New York, 1970.
4. Darii Yakovlevich Svet, *Thermal Radiation, Metals, Semiconductors, Ceramics, Partly Transparent Bodies, and Films*, Plenum Press, New York, 1965.
5. Arthur Ratzel, "Design of Flat Plate Solar Energy Collectors for Space Cooling Applications Utilizing Commercially Available Materials," M.S. thesis, University of Houston, Houston, May 1976.
6. Robert Siegel and John R. Howell, *Thermal Radiation Heat Transfer*, 2d ed., McGraw-Hill Book Co., New York, 1980.
7. B. D. Hunn and D. O. Calafell, II, "Determination of Average Ground Reflectivity for Solar Collectors," *Solar Energy*, vol. 19, no. 1, pp. 87–89, 1977.

APPENDIX

E

SOLSIM

E-1 GENERAL INFORMATION

The solar-thermal energy system simulation program SOLSIM is designed as a flexible learning tool. It provides the user with outputs that show system conditions at selected time intervals as well as system efficiencies calculated over the complete simulation period.

SOLSIM is available in both a FORTRAN and BASIC version. Both are programmed for interactive use with remote terminals, and model the system described in Chap. 2. Options allow for user to simulate a Carnot heat engine driven by solar energy, a domestic hot-water system, and others.

Insolation can be chosen as a repeating sinusoidally varying function, in which case the user must specify the time of sunrise, day length, and peak insolation. Insolation may also be specified as a repeating arbitrary daily profile, as a repeating daily profile specified at each time increment, or as a profile specified at each time increment throughout simulation.

Loads may be specified as constant, given at each time increment, given as a repeating arbitrary daily profile, or given as a repeating daily profile specified at each time increment. The Carnot engine option requires the user to specify the work output of the engine, and SOLSIM calculates the thermal energy provided to the engine by the solar collection and storage system.

Storage capacity is specified, and the storage can be chosen as completely mixed or as stratified using the model discussed in Chap. 6 for liquid systems. Heat losses from storage are accounted for by specifying the overall thermal loss coefficient.

Ambient temperature is modeled as a daily sinusoidal variation around an average ambient temperature. The average ambient and the swing, which is the deviation from average ambient, are user-specified. The sine function has a phase shift built in so that maximum ambient occurs at 12 h after sunrise, and minimum ambient at sunrise.

The collector parameters are those described in Chaps. 2, 4, and 5 and include F_R, b, $(\tau\alpha)_{\text{eff}}$, $q_{s,\text{ref}}$ and, for the Carnot engine system, a.

Finally, the program will calculate the allowable capital investment for the system modeled if the user inputs the interest rate, fuel escalation rate, system life, and cost of conventional fuel (Chap. 10).

Users may choose either SI or English engineering units for data input as well as output, and the input and output units may be different, i.e., inputs in English and output in SI, or vice versa.

Two methods of input are available. The user may choose the "base case," in which case the default values of data are input. Data may then be altered piece by piece. Otherwise, *all* data can be input by the user. Data can always be displayed and/or altered before the simulation is run.

The user should have all data at hand before beginning input. Rough hand calculation of sizes, flow rates, etc., will save a lot of useless computer runs.

Accuracy of the Simulation

The simulation is limited to a maximum of 10 days. Because the insolation data does not easily account for daily, hourly, or even seasonal variations from day to day, the simulation is not good for most engineering design except for comparing systems.

Because the runs are over relatively short time periods, startup transients can strongly influence the accuracy of the calculated average system quantities such as system collection efficiency, fraction of load supplied, and allowable system cost. This problem can be overcome by a careful choice of initial storage temperature.

Common Problems

Choice of storage capacity that is far too small or far too large will cause the simulation to appear in error by causing immediate excess storage temperature when the collector is operating, or no load to be supplied, respectively.

Excessive mass flow rate in the collector loop will cause on-off cycling of the collector pump at each time increment (as would happen in a real system), and the system collection efficiency will be poor.

Conclusions

SOLSIM is written as a teaching/learning tool. It is written in a block structure so that it can be easily modified by the addition of subroutines, and teachers and students are encouraged to modify the program for their needs. For example, local insolation data for typical summer and winter periods at the user's location could be made available as a user option.

The program is interactive so that it can be used in the classroom or for easy homework application.

Table E-1 SOLSIM required inputs

Inputs	Default value for base case
General:	
Specific heat capacity of working fluid, J/(kg · °C)[Btu/(lb$_m$ · °R)]	4183 (1)
Length of simulation, min	2880
Δt between calculations, min	15
Number of Δt's between printouts	4
Collector information:	
Aperture area, m^2 (ft^2)	30 (323)
Value of $(\tau\alpha)_{\text{eff}}$	0.8
Value of a	0
Value of b	1.2
Value of F_R	0.9
Collector mass flow rate, kg/min (lb$_m$/min)	20 (44)
Value of $q_{s,\text{ref}}$ used in b, W/m^2 [Btu/(h · ft^2)]	800 (254)
Load information:	
Is load a DHW system? If so, input supply main temperature, K (°R).	—
Is load a Carnot engine? If so, specify Watts output and collector constant a.	—
Is the load constant? Input J/min (Btu/h).	2.5 × 10^5 (1.42 × 10^5)
If not, input at each Δt or repeated daily profile.	—
Load mass flow rate (unless DHW system), kg/min (lb$_m$/min).	10 (22)
Storage information:	
Thermal capacity of storage, J/K (Btu/°R)	5 × 10^6 (8530)
Initial temperature, K (°R)	300 (540)
Minimum temperature to load, K (°R)	300 (540)
Maximum allowed storage temperature, K (°R)	373 (671)
Tank overall loss coefficient, W/(m^2 · °C)[Btu/(h · ft^2 · °F)]	0.284 (0.050)
Tank L/D ratio	3
If stratified storage, number of tank levels to be treated and initial temperature of each level	—
Ambient temperature:	
Treated as sinusoidal around ambient average, K (°R)	285 (513)
With a positive maximum swing, K (°R)	10 (18)
Insolation information:	
Is a simple sine curve to be used?	Yes
If so, maximum insolation, W/m^2 [Btu/h · ft^2)]	800 (254)
Day length (sunrise-sunset), min	600
Time of sunrise, h (local time plus decimal fraction)	7.0
If not, input insolation at each time, or as a profile repeated each day	—
Economic analysis:	
Expected system life, years	10
Interest rate, %	12
Expected fuel cost escalation rate, %	15
Cost of competing fuel, $/kWh	$0.05

Table E-1 (Continued)

Outputs
At each selected time interval:
Time, min
Insolation, W/m^2 [Btu/(h · ft^2)]
Load supplied, J/min (Btu/h)
Load outlet temperature, K (°R)
Storage temperature (if mixed storage), K (°R)
Collector outlet (or stagnation) temperature, K (°R)
Net rate of energy added to storage, J/min (Btu/h)
Collector pump status (on, off)
Load pump status (on, off)
Collector efficiency, %
Temperature distribution in storage (if stratified storage), K(°R)
Carnot engine efficiency (if load is Carnot engine)
Hot water supplied (if DHW system), kg/min (gal/h)
At completion of simulation; for the simulation period:
Percentage of load supplied by solar energy
Percentage of time load pump operated
Collection efficiency of system, percentage
Allowable capital cost of system (if desired), $

Table E-2 Symbol correspondence for basic version of SOLSIM

Program symbol	Text symbol	Definition
A$	...	Index to denote altered data wanted
A, A(M), A(N)	...	Index on various loops
A1	$(\dot{m}_c c \, \Delta t/[(mc)_{st} n])\delta_c$	Checks on allowable Δt for stratified storage subroutine
A2	$(\dot{m}_L c \, \Delta t/[(mc)_{st} n])\delta_L$	
A7	$A_{st,m}$	Area of tank layer for heat loss
A8	a	Collector radiation constant
A9	A_c	Collector aperture area
B$	...	Index to denote sinusoidal insolation
B	...	Time increment index
B9	b	Collector convective loss constant
C$	...	Index to denote constant load
C4	$\int_{t=0}^{t_{sim}} q_s(t) \, dt$	Cumulative insolation over simulation
C5	...	Control function to start insolation input
C7	...	Cost of competing fuel
C8	$(Mc)_{st}$	Heat capacity of storage
C9	c	Specific heat of fluid
D$	...	Index to denote heat engine
D1	t_d	Length of day from sunrise to sunset, min

Table E-2 (Continued)

Program symbol	Text symbol	Definition
D2	t_{rise}	Time of sunrise, min
D4	m	Index of tank division number
D5	n	Number of divisions in stratified tank
D7	D	Diameter of storage tank
D9	$(\Delta T)_a$	Ambient temperature swing
E\$	...	Index to denote economics wanted
E1	$\int_{t=0}^{t_{sim}} \eta(t) \, dt / t_{sim}$	System collection efficiency over simulation
E5	$\eta(t)$	Instantaneous collector efficiency
E6	η_C	Instantaneous Carnot efficiency
F\$	...	Index to denote altered data is being used internally
F1	...	Value of fuel saved
F9	F_R	Collector heat removal factor
G6	$a/b\bar{T}_a^3$	
G9	$\dot{m}_c c \bar{T}_a / (60 A_c b q_{s,ref})$	
H4	...	Δt required to assure convergence in stratified storage subroutine
H5		Time of starting stratified liquid calculation
I\$	...	Index to denote SI input data
I1	$q_{s,max}$	Maximum insolation
I4	$(1 + i_{eff})^n$	
I5	j	Fuel cost escalation rate
I6	i	Interest rate
I7	i_{eff}	Effective interest rate
I8	$q_{s,ref}$	Reference insolation used in a and b
I9(B)	$q_s(t)$	Time-varying insolation
K\$	...	Index to specify base case
L	...	Index to denote method of load input
L8	L/D	Length-to-diameter ratio of storage tank
L7	L	Length of storage tank
M	...	Dummy index on tank layers
M8	$\dot{m}_L$	Mass flow rate to load
M9	$\dot{m}_c$	Mass flow rate to collector
N1	n	Years of expected system life
O\$	...	Index to denote SI output data
O1	...	Number of Δt intervals between prints
P1	...	Number of Δt intervals load pump is on
P4	...	Percent of load supplied
P5	$P, (PV)_0$	Allowable system capital cost, \$
P6	...	Percent of time load pump on
P7	...	Load supplied by solar in Δt
Q(B)	$(\dot{Q}_c - \dot{Q}_L)$ − loss rate from tank	Net rate of stored energy at Δt
Q1	$\int_{t=0}^{t_{sim}} \dot{Q}_c(t) \, dt$	Cumulative useful energy collected

Table E-2 (Continued)

Program symbol	Text symbol	Definition
Q5	...	Cumulative solar energy supplied to load over simulation period
Q6	...	Annual energy supplied to load by solar
Q7	$\int_{t=0}^{t_{sim}} \dot{Q}_L(t)\,dt$	Cumulative load over period of simulation
Q9	$\dot{Q}_L$	Load at time t, if constant
Q9(B)	$\dot{Q}_L(t)$	Load at time t, if varying
R8	...	Mass flow through storage that sets time increment necessary for stratified system
R9	ρ_{H_2O}	Density of water
S$	...	Index to denote input units in units in use
S8	δ_L	Switch on load pump
S9	δ_c	Switch on collector pump
S	...	Dummy switch index
S1	...	Dummy switch index
S5(A)	t	Time at which load is specified
S6(A)	$\dot{Q}_L(t)$	Load at time t for daily profile
T	...	Time increment index
T0(T)	$T_{L,out}(t)$	Load outlet temperature
T1(T)	$T_{st}(t)$	Storage temperature (mixed)
T2(T)	$T_{f,out}(t)$	Collector outlet temperature
T3	$\bar{T}_a$	Average ambient temperature
T3(M)	$T(m)$	Stratified tank temperature at level m
T4	$T_{st,max}$	Maximum allowed storage temperature
T5		Index to denote mixed (1) or stratified (2) storage
T5(B)	$T_a(t)$	Instantaneous ambient temperature
T6	$T_{st}(0)$	Initial storage temperature
T7	Δt	Time increment used in calculations
T8	t_{sim}	Length of simulation
T9	$(\tau\alpha)_{eff}$	Collector property
U8	$\bar{U}_{st}$	Storage loss coefficient
V8	$(mc)_{st}/c\rho_{H_2O}$	Volume of water storage tank
W3	...	Largest load required throughout simulation
W9	$\dot{W}_C$	Power output of Carnot engine
X1	$T_{st,bottom}$	Stratified storage bottom and
X2	$T_{st,top}$	top temperatures
Y$	...	Index to denote DHW system
Y1	T_{sup}	City main supply temperature for DHW system
Z1	...	Index for choice of data alteration
Z9	$T_{st,min}$	Minimum allowed storage temperature, or temperature needed by load for DHW systems

E-2 SAMPLE BASIC PROGRAM

```
00010    DIM I9(1000),Q9(1000),T5(1000),S5(100),S6(100),T3(10),T4(10)
00020    DIM A(10),B(10)
00030    R9=1000
00040      REM    THIS IS THE DENSITY OF WATER USED TO CALCULATE
00050      REM    VOLUME OF STORAGE. IF OTHER FLUIDS ARE USED,
00060      REM    SUCH AS ETHYLENE GLYCOL, APPROPRIATE DATA
00070      REM    SHOULD BE USED INSTEAD.
00080      PRINT
00090    PRINT "THIS PROGRAM WORKS IN SI OR ENG. UNITS"
00100      PRINT
00110    PRINT "DO YOU WANT TO RUN A CASE MUCH DIFFERENT FROM THE BASE?"
00120      INPUT K$
00130        IF K$="YES" THEN 660
00140        IF K$="NO" THEN 230
00150          REM    IF THE USER WANTS TO RUN A DIFFERENT CASE THAN
00160          REM    THE BASE CASE, GO TO THE INPUT DATA SECTION.
00170          REM    OTHERWISE, GO TO THE BASE CASE DATA SECTION.
00180        GOSUB 11110
00190        GO TO 110
00200    REM
00210    REM                BASE CASE DATA (IN SI UNITS)
00220    REM
00230      T8=2880
00240      T7=15
00250      O1=4
00260      B$="YES"
00270      I1=800
00280      D1=600
00290         D2=420
00300      I8=800
00310      T3=285
00320      D9=10
00330      T9=.8
00340      B9=1.2
00350      A8=0
00360         F9=0.95
00370      M9=20
00380      C9=4183
00390      A9=30
00400      C8=5E6
00410      U8=.28385
00420      L8=3
00430      T5=1
00440      T6=300
00450      Z9=300
00460      T4=373
00470      D$="NO"
00480      C$="YES"
00490      Q9=2.5E5
00500      M8=10
00510      E$="YES"
00520      N1=10
00530      I6=12
00540      I5=15
```

```
00550         C7=.05
00560         I$="SI"
00570         S$="SI"
00580         O$="SI"
00590         F$="INPUT"
00600         Y$="NO"
00610         GO TO 3350
00620            REM   GO TO THE DISPLAY DATA SECTION
00630   REM
00640   REM                  INPUT DATA SECTION
00650   REM
00660         F$="INPUT"
00670         PRINT
00680      PRINT "THIS PROGRAM REQUIRES SYSTEM INFORMATION"
00690         PRINT
00700      PRINT "ENTER THE SYSTEM OF UNITS TO BE USED FOR INPUT"
00710         PRINT "(ENTER 'SI' OR 'ENG' IN CAPS)"
00720         INPUT I$
00730         S$=I$
00740            IF I$="SI" THEN 770
00750            IF I$="ENG" THEN 770
00760         GO TO 700
00770         PRINT
00780      PRINT "ENTER THE SYSTEM OF UNITS TO BE USED FOR OUTPUT"
00790         PRINT "(ENTER 'SI' OR 'ENG' IN CAPS)"
00800         INPUT O$
00810            IF O$="SI" THEN 840
00820            IF O$="ENG" THEN 840
00830         GO TO 780
00840         PRINT
00850      PRINT "ENTER THE TOTAL TIME OF THE SIM";
00860      PRINT "ULATION PERIOD, IN MINUTES"
00870         INPUT T8
00880         PRINT
00890      PRINT "ENTER THE TIME INTERVAL BETWEEN CALCULATON ";
00900      PRINT "POINTS, IN MINUTES"
00910         INPUT T7
00920            R=1440/T7
00930            IF R=INT(R) THEN 960
00940         PRINT "TIME INTERVAL MUS
00940         PRINT "TIME INTERVAL MUST DIVIDE 1440 EVENLY"
00950         GO TO 890
00960         PRINT
00970      PRINT "ENTER THE FREQUENCY OF OUTPUT IN NUMBER OF INTERVALS"
00980         INPUT O1
00990         PRINT
01000      PRINT "IS THE INSOLATION SINUSOIDAL?"
01010         INPUT B$
01020         PRINT
01030            IF B$="YES" THEN 1110
01040            IF B$="NO" THEN 1240
01050               REM   IF THE INSOLATION IS SINUSOIDAL, ENTER THE MAX.
01060               REM   INSOLATION AND THE LENGTH OF THE SOLAR DAY.
01070               REM   IF THE INSOLATION IS NOT SINUSOIDAL, ENTER THE
01080               REM   INSOLATION FOR EACH CALCULATION INTERVAL
01090         GOSUB 11110
```

```
01100            GO TO 1000
01110         PRINT"ENTER THE PEAK INSOLATION IN WATTS/SQ.M.  (BTU/HR-SQ.FT.)"
01120            INPUT I1
01130            PRINT
01140         PRINT"ENTER THE LENGTH OF THE SOLAR DAY IN MINUTES"
01150            INPUT D1
01160            PRINT
01170          PRINT "ENTER THE TIME OF SUNRISE AS LOCAL HOUR PLUS"
01180          PRINT "DECIMAL FRACTION ON 24-HOUR CLOCK."
01190            INPUT D2
01200            D2=D2*60
01210            PRINT
01220            GO TO 1320
01230              REM   SKIP THE VARIABLE INSOLATION INPUT
01240          B=-1
01250         PRINT "ENTER THE INSOLATION DATA IN WATTS/SQ.M.  (BTU/HR-SQ.FT.)"
01260         FOR A=0 TO T8 STEP T7
01270            B=B+1
01280            PRINT "ENTER THE INSOLATION AT ";A;" MINUTES"
01290              INPUT I9(B)
01300          NEXT A
01310            PRINT
01320         PRINT "ENTER THE VALUE OF QS,REF USED IN EVALUATING A AND B"
01330            PRINT "IN WATTS/SQ.M.(BTU/HR-SQ.FT.)"
01340            INPUT I8
01350            PRINT
01360         PRINT "ENTER THE DAILY MEAN T THAT WILL EXIST OVER THE "
01370            PRINT "PERIOD OF THE SIMULATION, IN KELVINS (RANKINE)"
01380            INPUT T3
01390            PRINT
01400         PRINT "ENTER THE DELTA-T SWING ABOVE THE DAILY MEAN"
01410            PRINT "IN KELVINS(RANKINE)"
01420            INPUT D9
01430            PRINT
01440         PRINT "ENTER THE VALUE OF TAU-ALPHA FOR THE COLLECTOR"
01450            INPUT T9
01460            PRINT
01470         PRINT "ENTER THE VALUE OF B FOR THE COLLECTOR"
01480            INPUT B9
01490            PRINT
01500           PRINT "ENTER THE VALUE OF F SUB-R"
01510            INPUT F9
01520            PRINT
01530         PRINT " ENTER THE MASS FLOW RATE TO BE RUN THROUGH THE COLLE";
01540            PRINT "CTOR BANK, KG/MIN (LBM/MIN.)"
01550            INPUT M9
01560            PRINT
01570         PRINT "ENTER THE SPECIFIC HEAT CAPACITY OF THE FLUID IN ";
01580            PRINT "JOULES/KG-DEG.C (BTU/LBM-DEG.F.)"
01590            INPUT C9
01600            PRINT
01610         PRINT "ENTER THE COLLECTOR AREA IN SQ.M.  (SQ.FT.)"
01620            INPUT A9
01630            PRINT
01640         PRINT "ENTER THE HEAT CAPACITY OF THE STORAGE SYSTEM,";
01650            PRINT " JOULES/DEG.C (BTU/DEG.F)"
```

```
01660        INPUT C8
01670        PRINT
01680      PRINT "ENTER THE U-VALUE OF THE STORAGE TANK IN ";
01690        PRINT "W/SQ.M.-DEG.C.(BTU/HR-SQ.FT.-DEG.F.)"
01700        INPUT U8
01710        PRINT
01720      PRINT "ENTER THE LENGTH-DIAMETER RATIO OF THE ";
01730        PRINT "STORAGE TANK"
01740        INPUT L8
01750        PRINT
01760      PRINT "ENTER THE APPROPRIATE NUMBER FOR THE STOR";
01770        PRINT "AGE SYSTEM USED"
01780        PRINT "1) LUMPED LIQUID STORAGE"
01790        PRINT "2) STRATIFIED LIQUID STORAGE"
01800        INPUT T5
01810          IF T5=1 THEN 1970
01820     IF T5=2 THEN 1890
01830     PRINT "INPUT EITHER 1 OR 2"
01840     GO TO 1800
01850            REM   IF THE STORAGE IS LUMPED, SKIP THE STRATIFIED
01860            REM   STORAGE INPUTS. OTHERWISE, ENTER THE NUMBER
01870            REM   OF DIVISIONS AND THE INITIAL STORAGE TEMPS.
01880            REM   FOR EACH DIVISION OF STORAGE.
01890     PRINT
01900          PRINT "ENTER THE NUMBER OF DIVISIONS IN THE TANK"
01910            INPUT D5
01920            D4=D5-1
01930            GOSUB 11420
01940            REM   GO TO THE STRATIFIED STORAGE INIT. SUBROUTINE
01950          IF T5=2 THEN 2010
01960            REM   IF STRATIFIED STORAGE, SKIP LUMPED STORAGE INPUT
01970          PRINT
01980       PRINT "ENTER THE INITIAL VALUE OF THE STORAGE ";
01990         PRINT "TEMPERATURE, IN KELVINS (RANKINE)"
02000         INPUT T6
02010         PRINT
02020       PRINT "ENTER THE MINIMUM ALLOWABLE STORAGE TEMPERATURE";
02030         PRINT " THAT WILL GIVE"
02040         PRINT "USEFUL ENERGY TO THE LOAD, IN KELVINS (RANKINE)"
02050         INPUT Z9
02060         PRINT
02070       PRINT "ENTER THE MAXIMUM SAFE STORAGE TEMPERATURE, ";
02080         PRINT "IN KELVINS (RANKINE)"
02090         INPUT T4
02100         PRINT
02110       PRINT "IS THE LOAD A HEAT INPUT TO A CARNOT CYCLE?  "
02120         INPUT D$
02130         PRINT
02140          IF D$="YES" THEN 2220
02150          IF D$="NO" THEN 2320
02160            REM   IF THE LOAD IS A CARNOT CYCLE, ENTER THE WORK
02170            REM   OUTPUT EXPECTED AND THE VALUE OF A FOR THE
02180            REM   HIGH TEMP. COLL. OTHERWISE, ASK IF THE LOAD
02190            REM   IS CONSTANT
02200         GO SUB 11110
02210         GO TO 2110
```

```
02220          PRINT "ENTER THE WORK OUTPUT YOU EXPECT FROM THE HEAT ";
02230            PRINT "ENGINE, IN WATTS"
02240            INPUT W9
02250            PRINT
02260          PRINT "ENTER THE VALUE OF A THAT APPLIES TO THE HIGH ";
02270            PRINT "TEMPERATURE COLLECTORS BEING USED"
02280            INPUT A8
02290            PRINT
02300            GO TO 3060
02310              REM    SKIP THE OTHER LOAD INPUTS
02320      PRINT "IS THE LOAD DOMESTIC HOT WATER?"
02330      INPUT Y$
02340      PRINT
02350      IF Y$="YES" THEN 2410
02360      IF Y$="NO" THEN 2440
02370      REM   IF THE LOAD IS DOMESTIC HOT WATER, INPUT SUPPLY
02380      REM   WATER TEMP. OTHERWISE, SKIP IT.
02390      GOSUB 11110
02400      GO TO 2320
02410      PRINT "ENTER THE CITY MAIN SUPPLY TEMP IN KELVINS(RANKINE)"
02420      INPUT Y1
02430      PRINT
02440        PRINT "IS THE LOAD CONSTANT?   "
02450          INPUT C$
02460          PRINT
02470          D$="NO"
02480          A8=0
02490            IF C$="YES" THEN 2560
02500            IF C$="NO" THEN 2610
02510              REM   IF THE LOAD IS CONSTANT, ENTER THE CONSTANT LOAD
02520              REM    OTHERWISE, ENTER THE LOAD FOR EACH CALCULATION
02530              REM    INTERVAL
02540            GO SUB 11110
02550            GO TO 2440
02560          PRINT "ENTER THE LOAD, IN JOULES/MIN. (BTU/HR)"
02570            INPUT Q9
02580      PRINT
02590            GO TO 3040
02600              REM   SKIP THE VARIABLE LOAD INPUT
02610      PRINT "ENTER THE METHOD OF INPUT FOR VARIABLE LOAD"
02620      PRINT "1) LOAD AT EACH TIME INCREMENT"
02630      PRINT "2) TYPICAL DAY'S LOAD AT EACH INCREMENT"
02640      PRINT "3) TYPICAL DAY'S LOAD PROFILE"
02650      INPUT L
02660      IF L=1 THEN 2810
02670      IF L=2 THEN 2710
02680      IF L=3 THEN 2910
02690      PRINT "INVALID ENTRY, ENTER A NUMBER (1-3)"
02700      GO TO 2610
02710          B=-1
02720          PRINT "ENTER THE LOAD DATA IN JOULES/MIN. (BTU/HR)"
02730          FOR A=0 TO 1440-T7 STEP T7
02740            B=B+1
02750            PRINT "ENTER THE LOAD AT ";A;" MINUTES"
02760              INPUT Q9(B)
02770          NEXT A
```

```
02780          PRINT
02790          GO TO 3040
02800          REM    SKIP THE OTHER VARIABLE LOAD INPUTS
02810          B=-1
02820          PRINT "ENTER THE LOAD DATA IN JOULES/MIN. (BTU/HR)"
02830          FOR A=0 TO T8 STEP T7
02840             B=B+1
02850             PRINT "ENTER THE LOAD AT ";A;" MINUTES"
02860                INPUT Q9(B)
02870          NEXT A
02880          PRINT
02890          GO TO 3040
02900          REM    SKIP THE OTHER VARIABLE LOAD INPUTS
02910           PRINT "ENTER THE TIME IN HOURS FOLLOWED BY"
02920          PRINT"A COMMA (,) FOLLOWED BY THE LOAD AT THAT TIME IN"
02930          PRINT"JOULES/MIN.(BTU/HR)"
02940          PRINT"YOU MUST ENTER A LOAD AT T=0 AND T=24 HOURS"
02950          PRINT"ENTER DATA POINTS CONSECUTIVELY"
02960          A=-1
02970          A=A+1
02980          INPUT S5(A),S6(A)
02990          IF S5(A)=24 THEN 3040
03000          IF S5(A)<24 THEN 2970
03010          PRINT "TIME MUST BE IN HOURS (0-24)"
03020          PRINT "REENTER LAST DATA POINT"
03030          GO TO 2980
03040          IF Y$="YES" THEN 3090
03050          PRINT
03060            PRINT "ENTER THE MASS FLOW RATE TO BE RUN THROUGH THE LOAD H";
03070               PRINT "EAT EXCHANGER, KG/MIN. (LBM/MIN.)"
03080               INPUT M8
03090               PRINT
03100           PRINT "DO YOU WANT ECONOMIC ANALYSIS?"
03110              INPUT E$
03120              PRINT
03130                 IF E$="YES" THEN 3200
03140                 IF E$="NO" THEN 3350
03150                    REM    IF THE USER WANTS ECONOMIC ANALYSIS, ENTER
03160                    REM    THE REQUIRED ECONOMIC DATA. OTHERWISE,
03170                    REM    GO TO THE DISPLAY DATA SECTION
03180                 GOSUB 11110
03190                 GO TO 3100
03200           PRINT "ENTER EXPECTED SYSTEM LIFE IN YEARS"
03210              INPUT N1
03220              PRINT
03230            PRINT "ENTER INTEREST RATE,%"
03240               INPUT I6
03250               PRINT
03260            PRINT "ENTER FUEL ESCALATION RATE,%"
03270               INPUT I5
03280               PRINT
03290            PRINT "ENTER THE COST OF COMPETING FUEL,DOLLARS/KWH"
03300               INPUT C7
03310               PRINT
03320          REM
03330          REM                DISPLAY DATA SECTION
```

```
03340       REM
03350         PRINT
03360       IF I$=S$ THEN 3440
03370         REM    IF INPUT DATA HAS CORRECT UNITS, SKIP CONVERSION
03380       IF I$="SI" THEN 3420
03390       GOSUB 12070
03400         REM    CHANGE INPUT DATA FROM SI TO ENG. UNITS
03410       GO TO 3440
03420       GOSUB 11610
03430         REM    CHANGE INPUT DATA FROM ENG. TO SI UNITS
03440       PRINT,"INPUT  DATA  IN  ";S$;"  UNITS"
03450         PRINT
03460     IF Y$="NO" THEN 3500
03470     PRINT "THIS IS A SIMULATION OF A DHW SYSTEM"
03480     PRINT "CITY WATER SUPPLY IS AT ";Y1;" KELVINS(RANKINE)"
03490     PRINT "HOT WATER IS SUPPLIED AT ";Z9;" KELVINS(RANKINE)"
03500       PRINT"TIME OF SIMULATION IS   ";T8;" MINS."
03510       PRINT"CALCULATION INTERVAL IS ";T7;" MINS."
03520       PRINT"FREQUENCY OF OUTPUT IS ";O1;" INTERVALS"
03530       IF B$="NO"   THEN 3600
03540         REM    IF THE INSOLATION IS NOT SINUSOIDAL, SKIP
03550         REM    THE SINUS. INSOL. DISPLAY AND SHOW THE
03560         REM    VARIABLE INSOL. VALUES IN A TABLE
03570        PRINT"QS(T)=";I1;"SIN(PI*(T-";D2;")/";D1;") W/SQ.M. (BTU/HR.-SQ.FT.)"
03580       GO TO 3690
03590         REM    SKIP THE VARIABLE INSOL. DISPLAY
03600       PRINT"INSOLATION IS VARIABLE WITH THE FOLLOWING VALUES"
03610         PRINT
03620       PRINT"TIME-MINUTES",,"INSOLATION-W/SQ.M.  (BTU/HR-SQ.FT.)"
03630       B=-1
03640       FOR A=0 TO T8 STEP T7
03650         B=B+1
03660         PRINT A,,I9(B)
03670       NEXT  A
03680         PRINT
03690       PRINT"QS,REF IS ";I8;" W/SQ.M.  (BTU/HR-SQ.FT.)"
03700       PRINT"DAILY MEAN TEMP.  IS ";T3;" KELVINS (RANKINE)"
03710       PRINT"DELTA-T SWING IS ";D9;" KELVINS (RANKINE)"
03720       PRINT"TAU-ALPHA IS ";T9
03730       PRINT"VALUE OF B IS ";B9
03740          PRINT"VALUE OF F SUB-R IS ";F9
03750       PRINT"COLLECTOR MASS FLOW IS ";M9;" KG/MIN.   (LBM/MIN.)"
03760       PRINT"SPECIFIC HEAT OF FLUID IS ";C9;" JOULES/KG-DEG.C."
03770          PRINT TAB(39) "(BTU/LBM-DEG.F.)"
03780       PRINT"COLLECTOR AREA IS ";A9;" SQ.M.   (SQ.FT.)"
03790       IF T5=2 THEN   3850
03800         REM    IF THE STORAGE IS STRATIFIED, SKIP THE LUMPED
03810         REM    STORAGE DISPLAY
03820       PRINT"LUMPED LIQUID STORAGE"
03830       GO TO 3860
03840         REM    SKIP THE STRATIFIED STORAGE DISPLAY
03850       PRINT"STRATIFIED LIQUID STORAGE WITH ";D5;" DIVISIONS"
03860       PRINT"STORAGE CAPACITY IS ";C8;" JOULES/DEG.C(BTU/DEG.F)"
03870       PRINT"U-VALUE OF STORAGE TANK IS ";U8;" W/SQ.M.-DEG.C."
03880          PRINT TAB(35) "(BTU/HR.-SQ.FT.-DEG.F.)"
03890       PRINT"LENGTH/DIAM. OF STORAGE IS ";L8
```

```
03900     IF T5=2 THEN 3940
03910     PRINT"INITIAL STORAGE TEMP.  IS ";T6;" KELVINS (RANKINE)"
03920     GO TO 3990
03930       REM    SKIP STRAT. STOR. TEMP. DISPLAY
03940     PRINT"INITIAL STORAGE TEMPERATURES(TOP TO BOTTOM)"
03950     FOR M=0 TO D4
03960       PRINT T3(M);
03970     NEXT M
03980     PRINT
03990     PRINT"MIN. USEABLE TEMP.  IS ";Z9;" KELVINS (RANKINE)"
04000     PRINT"MAX. SAFE TEMP.  IS ";T4;" KELVINS (RANKINE)"
04010     IF D$="NO"  THEN 4080
04020       REM   IF THE LOAD IS NOT A CARNOT CYCLE, SKIP THE
04030       REM   CARNOT CYCLE DISPLAYS
04040     PRINT"LOAD IS A CARNOT CYCLE TO DELIVER ";W9;" WATTS"
04050     PRINT"VALUE OF A IS ";A8
04060     GO TO 4500
04070       REM   SKIP THE REST OF THE LOAD DISPLAYS
04080     IF C$="NO"  THEN   4140
04090       REM   IF THE LOAD IS NOT CONSTANT, SKIP THE
04100       REM   CONSTANT LOAD DISPLAY
04110     PRINT"LOAD IS CONSTANT AT ";Q9;" JOULES/MIN.  (BTU/HR)"
04120     GO TO 4500
04130       REM   SKIP THE VARIABLE LOAD DISPLAY
04140     IF L=2 THEN 4270
04150     IF L=3 THEN 4380
04160       PRINT"LOAD IS VARIABLE WITH THE FOLLOWING VALUES"
04170       PRINT
04180       PRINT"TIME-MINUTES",,"LOAD-JOULES/MIN.  (BTU/HR)"
04190       B=-1
04200       FOR A=0 TO T8 STEP T7
04210         B=B+1
04220         PRINT A,,Q9(B)
04230       NEXT   A
04240       PRINT
04250     GO TO 4500
04260     REM    SKIP THE OTHER LOAD DISPLAYS
04270     PRINT "TYPICAL DAY'S LOAD"
04280       PRINT
04290       PRINT"TIME-MINUTES",,"LOAD-JOULES/MIN.  (BTU/HR)"
04300       B=-1
04310       FOR A=0 TO 1440-T7 STEP T7
04320         B=B+1
04330         PRINT A,,Q9(B)
04340       NEXT   A
04350       PRINT
04360     GO TO 4500
04370     REM    SKIP THE OTHER LOAD DISPLAYS
04380     PRINT "TYPICAL DAY'S LOAD PROFILE"
04390     PRINT "(LINEAR BETWEEN CONSECUTIVE DATA POINTS)"
04400     PRINT "IF TIMES ARE NOT CONSECUTIVE, THIS DATA ";
04410     PRINT "NEEDS TO BE CORRECTED"
04420     PRINT
04430     PRINT "TIME",,"LOAD"
04440     PRINT "HOURS",,"JOULES/MIN. (BTU/HR)"
04450     A=-1
```

```
04460      A=A+1
04470      PRINT S5(A),,S6(A)
04480      IF S5(A)<24 THEN 4460
04490      PRINT
04500      IF Y$="YES" THEN 4520
04510        PRINT"LOAD MASS FLOW IS ";M8;" KG/MIN.   (LBM/MIN.)"
04520      IF E$="NO" THEN 4590
04530         REM    IF NO ECONOMIC ANALYSIS IS TO BE DONE
04540         REM    SKIP THE ECONOMIC DISPLAYS
04550        PRINT"EXPECTED SYSTEM LIFE IS ";N1;" YEARS"
04560        PRINT"INTEREST RATE IS ";I6;" %"
04570        PRINT"FUEL ESCALATION RATE IS ";I5;" %"
04580        PRINT"FUEL COST IS NOW ";C7;" $/KWH"
04590          PRINT
04600      IF F$="ALTER" THEN 8660
04610        PRINT"DO YOU WANT TO CHANGE ANY OF THE INPUT DATA?"
04620          INPUT A$
04630            IF A$="YES" THEN 8060
04640            IF A$="NO" THEN 4740
04650               REM    IF THE USER WANTS TO ALTER ANY OF THE DATA
04660               REM    GO TO THE ALTER DATA SECTION.  OTHERWISE,
04670               REM    INITIALIZE THE SYSTEM PARAMETERS AND RUN
04680               REM    THE SIMULATION
04690             GO SUB 11110
04700             GO TO 4610
04710      REM
04720      REM                INITIALIZE SYSTEM PARAMETERS
04730      REM
04740      IF S$="SI" THEN 4760
04750      GOSUB 11610
04760      IF B$="NO" THEN 4790
04770      GOSUB 11220
04780      IF D$="YES" THEN 4850
04790      IF C$="NO" THEN 4820
04800      GOSUB 11540
04810      GO TO 4850
04820      ON L GO TO 4840,4840,4830
04830      GOSUB 12810
04840      GOSUB 12930
04850      GOSUB 11150
04860      IF T5=1 THEN 5000
04870      IF Y$="NO" THEN 4910
04880      GOSUB 13050
04890      R8=W3/(C9*(Z9-Y1))
04900      GO TO 4920
04910      R8=M8
04920      H4=.1*C8/(M9*C9*D5)
04930      IF H4<=.1*C8/(R8*C9*D5) THEN 4950
04940      H4=.1*C8/(R8*C9*D5)
04950      IF H4>=T7 THEN 4980
04960      H4=T7/(INT(T7/H4)+1)
04970      GO TO 5010
04980      H4=T7
04990      GO TO 5010
05000      T1=T6
05010        S9=0
```

```
05020      S8=0
05030      G9=(M9*C9*T3)/(A9*I8*60)
05040      G6=A8/(B9*(T3^3))
05050      V8=C8/(R9*C9)
05060      L7=(4*V8*L8^2/3.14159)^.333333
05070      D7=L7/L8
05080      A7=3.14159*D7*(D7/2+L7)
05090      U7=U8*60*A7
05100      Q7=0
05110      Q5=0
05120      P1=0
05130      C5=-1
05140      Q1=0
05150      C4=0
05160   REM
05170   REM    PRINT THE HEADINGS FOR THE OUTPUT DATA COLUMNS
05180   REM
05190      PRINT"TIME(MIN)","INSOLATION","LOAD","LOAD OUT",
05200      PRINT "STORAGE TEMP"
05210      IF O$="ENG" THEN 5270
05220         REM   IF OUTPUT IS IN ENG. UNITS SKIP SI HEADING
05230         REM   OTHERWISE PRINT SI HEADING
05240      PRINT,"WATTS/SQ M","JOULES/MIN.","   T(K)","   T(K)",
05250      GO TO 5280
05260         REM   SKIP ENG. HEADING
05270   PRINT,"BTU/HR-SQ.FT.","BTU/HR","   T(R)","   T(R)"
05280      PRINT
05290      PRINT "COLLECTOR OUT","NET STORED","COLLECTOR","LOAD",
05300      PRINT "COLLECTOR"
05310      IF O$="ENG" THEN 5370
05320         REM   IF OUTPUT IS IN ENG. UNITS SKIP SI HEADING
05330         REM   OTHERWISE PRINT SI HEADING
05340      PRINT "   T(K)","JOULES/MIN","PUMP","PUMP","EFFICIENCY"
05350      GO TO 5380
05360         REM   SKIP ENG. HEADING
05370      PRINT "   T(R)","BTU/HR","PUMP","PUMP","EFFICIENCY"
05380      PRINT
05390   REM
05400   REM              MAIN CALCULATION LOOP
05410   REM
05420      B=-1
05430      FOR  T=0 TO T8 STEP T7
05440         B=B+1
05450         S1=S9
05460         S=S8
05470         Q=0
05480         IF S3=0 THEN 5520
05490            REM   IF THE LOAD PUMP IS OFF, SKIP THE
05500            REM   LOAD PUMP COUNTER
05510         P1=P1+1
05520         Q7=Q7+Q9(B)*T7
05530         P7=Q9(B)*S3
05540      IF Y$="YES" THEN 5560
05550         Q5=Q5+P7*T7
05560         C5=C5+1
05570            REM   INCREASE THE OUTPUT COUNTER
```

```
05580        IF T5=1 THEN 5750
05590           REM    IF THE STORAGE IS LUMPED, SKIP THE
05600           REM    STRATIFIED STORAGE INITIALIZATIONS
05610        H5=T
05620           X1=T3(D4)
05630           X2=T3(0)
05640        IF Y$="NO" THEN 5710
05650         IF X2<=Z9 THEN 5680
05660        M8=Q9(B)/(C9*(X2-Y1))
05670        GO TO 5690
05680        M8=Q9(B)/(C9*(Z9-Y1))
05690           IF M8>0 THEN 5710
05700           M8=0
05710        A1=M9*H4*C9/(C8/D5)*S9
05720        A2=M8*H4*C9/(C8/D5)*S8
05730        GO TO 5840
05740           REM    SKIP THE LUMPED STORAGE OUTLET TEMP. EQUATIONS
05750        X1=T1
05760        X2=T1
05770        IF Y$="NO" THEN 5840
05780         IF X2<=Z9 THEN 5810
05790        M8=Q9(B)/(C9*(X2-Y1))
05800         GO TO 5820
05810        M8=Q9(B)/(C9*(Z9-Y1))
05820           IF M8>0 THEN 5840
05830           M8=0
05840         IF S9=0 THEN 5950
05850         IF I9(B)=0 THEN 5950
05860           REM    IF THE COLL. PUMP IS OFF, OR THE SUN
05870           REM    IS NOT SHINING, CALCULATE THE STAGNATION
05880           REM    TEMP. OF THE COLLECTOR. OTHERWISE,
05890           REM    CALCULATE THE OPERATING TEMP.
05900        D7=(A8/T3^4)*(X1^4-T5(B)^4)
05910        T2=X1*(1-((B9*F9)/G9))
05920          T2=T2+(F9*T3/G9)*((T9*I9(B)/I8)+B9*(T5(B)/T3))-D7*T3/G9
05930        GO TO 5970
05940           REM    SKIP THE STAGNATION TEMP. EQUATIONS
05950        D8=G6*(X1^4-T5(B)^4)
05960        T2=T5(B)+((T9*T3*I9(B))/(I8*B9))-D8
05970        IF T2>=T5(B) THEN 6010
05980        T2=T5(B)
05990           REM    THESE STEPS ENSURE THAT THE COLLECTOR TEMP.
06000           REM    STAYS ABOVE AMBIENT
06010        IF Y$="YES" THEN 6040
06020           T0=X2-((S8*Q9(B))/(M8*C9))
06030        GO TO 6090
06040        T0=Y1
06050         IF X2<Z9 THEN 6080
06060        Q5=Q5+S8*T7*Q9(B)
06070        GO TO 6090
06080        Q5=Q5+Q9(B)*T7*(X2-T0)/(Z9-T0)
06090           IF T0<=0 THEN 13140
06100           REM    IF THE ABSOLUTE TEMP. OF THE LOAD RETURN
06110           REM    FALLS BELOW ABSOLUTE ZERO, PRINT AN ERROR
06120           REM    MESSAGE AND END THE RUN.
06130        IF T5=2 THEN 6260
```

```
06140          REM    IF STRATIFIED STORAGE, SKIP LUMPED STORAGE CALC.
06150          REM
06160          REM    LUMPED STORAGE CALCULATIONS
06170          REM
06180          Q=(S9*M9*C9*(T2-X1))-(S8*M8*C9*(X2-T0))-U7*(T1-T5(B))
06190          T1=T1+Q*T7/C8
06200          GO TO 6580
06210          REM    IF THE STORAGE IS LUMPED, SKIP THE
06220          REM    STRATIFIED STORAGE CALCULATIONS
06230          REM
06240          REM    STRATIFIED STORAGE CALCULATIONS
06250          REM
06260          FOR M=0 TO D4
06270            IF T2>=T3(M)  THEN 6300
06280            A(M)=0
06290          NEXT M
06300          A(M)=T2-T3(M)
06310          FOR N=M+1 TO  D4
06320            A(N)=T3(N-1)-T3(N)
06330          NEXT N
06340          FOR M=D4 TO 0 STEP -1
06350            IF T3(M)>=T0 THEN 6380
06360            B(M)=0
06370          NEXT M
06380          B(M)=T0-T3(M)
06390          FOR N=0 TO  M-1
06400            B(N)=T3(N+1)-T3(N)
06410          NEXT N
06420          FOR M=0 TO D4
06430            T4(M)=T3(M)+A1*A(M)+A2*B(M)-U7/C8*(T3(M)-T5(B))*H4
06440            Q=Q+C8/D5*(T4(M)-T3(M))
06450            T3(M)=T4(M)
06460          NEXT M
06470          H5=H5+H4
06480          IF H5<T+T7 THEN 5620
06490          REM    IF THE STRATIFIED STORAGE CALCULATIONS
06500          REM    HAVE NOT BEEN REPEATED OFTEN ENOUGH TO
06510          REM    FILL UP THE GIVEN CALCULATION INTERVAL,
06520          REM    REPEAT THE STRATIFIED STORAGE CALCULATIONS
06530          REM    OTHERWISE, CONTINUE WITH THE PUMP SWITCH
06540          REM    CALCULATIONS
06550          REM
06560          REM    PUMP SWITCH CALCULATIONS
06570          REM
06580          IF X2>T4 THEN 6770
06590          REM    IF THE STORAGE IS TOO HOT, TURN OFF THE
06600          REM    COLLECTOR PUMP.  OTHERWISE, USE TWO DIFFERENT
06610          REM    CRITERIA TO DECIDE THE NEW STATUS OF THE
06620          REM    COLLECTOR PUMP.  THIS HELPS REDUCE THE ON-
06630          REM    OFF PROBLEM AT THE START OF A NEW DAY.
06640          IF S9=0 THEN 6710
06650          IF (T2-X1)<2 THEN 6770
06660          REM    IF THE COLLECTOR IS OPERATING, TURN OFF THE
06670          REM    COLLECTOR IF THE TEMP.  RISE IS LESS THAN
06680          REM    2 DEG.  K.  OTHERWISE LEAVE IT ON.
06690          S9=1
```

```
06700           GO TO 6780
06710           IF (T2-X1)<6 THEN 6770
06720              REM   IF THE COLLECTOR IS STAGNANT, TURN ON THE
06730              REM   COLLECTOR IF THE TEMP. RISE IS MORE THAN
06740              REM   6 DEG. K. OTHERWISE, LEAVE IT OFF.
06750           S9=1
06760           GO TO 6780
06770           S9=0
06780           IF Q9(B)=0 THEN 6900
06790           IF Y$="YES" THEN 6840
06800           IF X2<=Z9 THEN 6900
06810              REM   IF THE STORAGE TEMP IS TOO LOW, OR THERE
06820              REM   IS NO LOAD, THEN TURN OFF THE LOAD PUMP.
06830              REM   OTHERWISE, LEAVE IT ON.
06840           S8=1
06850            IF Y$="YES" THEN 6930
06860           GO TO 6960
06870                 REM   IF DHW,THEN FIND PORTION OF DHW SUPPLIED
06880                 REM   BELOW SUPPLY TEMPERATURE BUT ABOVE
06890                 REM   MAIN TEMPERATURE
06900            IF Y$="YES" THEN 6930
06910            S8=0
06920           GO TO 6960
06930            IF X2<=Y1 THEN 6950
06940           GO TO 6960
06950            S8=0
06960           IF D$="NO" THEN 7000
06970              REM   IF THE LOAD IS NOT A CARNOT CYCLE, SKIP
06980              REM   THE CARNOT CYCLE LOAD EQUATION.
06990           Q9(B+1)=(60*W9)/(1-T5(B)/X2)
07000           Q1=Q1+S1*M9*C9*(T2-X1)*T7
07010           C4=I9(B)*A9*60*T7+C4
07020           IF C5=01 THEN 7120
07030              REM   IF THE OUTPUT COUNTER REACHES THE DESIRED
07040              REM   OUTPUT FREQUENCY, GO TO THE OUTPUT SECTION
07050           IF T=0 GO TO 7120
07060              REM   OUTPUT THE FIRST SET OF CALCULATIONS
07070           GO TO 7620
07080              REM   GO TO NEXT T IF THERE IS NO OUTPUT.
07090        REM
07100        REM                OUTPUT SECTION
07110        REM
07120           C5=0
07130           IF S1=0 THEN 7200
07140           IF I9(B)=0 THEN 7200
07150              REM   IF THE COLLECTOR PUMP IS OFF, OR THE SUN IS
07160              REM   NOT SHINING, SET THE COLLECTOR EFFICIENCY
07170              REM   E5=0. OTHERWISE, CALCULATE IT DIRECTLY.
07180           E5=M9*C9*(T2-X1)/(I9(B)*A9*60)*100
07190           GO TO 7210
07200           E5=0
07210           M$="OFF"
07220           N$="OFF"
07230           IF S1=0 THEN 7270
07240              REM   IF THE COLL. PUMP IS OFF, M$="OFF", OTHERWISE
07250              REM   SET M$="ON".
```

```
07260           M$="ON"
07270           IF S=0 THEN 7310
07280              REM   IF THE LOAD PUMP IS OFF, N$="OFF",  OTHERWISE
07290              REM   SET N$="ON".
07300           N$="ON"
07310           PRINT
07320           IF O$="SI" THEN 7370
07330              REM   IF OUTPUT IS IN SI UNITS, SKIP THE
07340              REM   UNITS CONVERSION SUBROUTINE
07350           GOSUB 12530
07360              REM   CHANGE OUTPUT UNITS FROM SI TO ENG.
07370           IF T5=2 THEN 7400
07380           PRINT T,I9(B),Q9(B),T0,T1,T2,Q,M$,N$,E5;"%"
07390           GO TO 7410
07400           PRINT T,I9(B),Q9(B),T0,"N/A",T2,Q,M$,N$,E5;"%"
07410           IF D$="YES" THEN 7450
07420           GO TO 7470
07430              REM   IF LOAD IS A CARNOT CYCLE, CALCULATE ENGINE
07440              REM   EFFICIENCY. OTHERWISE, SKIP E6
07450           E6=(1-(T5(B)/X2))*(S8*100)
07460           PRINT "ENGINE EFF=";E6;"%"
07470           IF T5=1 THEN 7530
07480              REM   IF LUMPED STORAGE, SKIP STRAT. STORAGE
07490           PRINT "STRATIFIED STORAGE TEMPERATURES,TOP TO BOTTOM"
07500           FOR M=0 TO D4
07510              PRINT T3(M);
07520           NEXT M
07530        IF Y$="NO" THEN 7560
07540           PRINT
07550         PRINT "HOT WATER FROM STORAGE AT ";M8;" KG/MIN.(GPH)"
07560           PRINT
07570           IF O$="SI" THEN 7620
07580              REM   IF OUTPUT IS IN SI UNITS, SKIP THE
07590              REM   UNITS CONVERSION SUBROUTINE
07600           GOSUB 12670
07610              REM   CHANGE OUTPUT UNITS BACK TO SI
07620        NEXT T
07630        IF  Q7<>0 THEN 7660
07640           P4=0
07650           GO TO 7690
07660           P4=(Q5/Q7)*100
07670              REM   IF LOAD REQUIRED IS 0 SET % LOAD SUPPLIED TO 0
07680              REM   OTHERWISE, CALCULATE % LOAD DIRECTLY
07690           PRINT "PERCENTAGE OF LOAD SUPPLIED BY SOLAR IS ";P4;"%"
07700           P6=(P1/(B+1))*100
07710        PRINT "LOAD PUMP WAS ON ";P6;"% OF THE TIME"
07720           E1=Q1/C4*100
07730        PRINT "SYSTEM COLLECTION EFFICIENCY IS ";E1;"%"
07740           IF E$="NO" THEN 7860
07750              REM   IF THERE IS NO ECONOMIC ANALYSIS, SKIP THE
07760              REM   ECONOMICS EQUATIONS
07770           Q6=(0.146*Q7)/T8
07780           I7=(I6-I5)/(100+I5)
07790           I4=(1+I7)^N1
07800           F1=((P4*Q6)/100)*C7
07810           P5=F1*(I4-1)/(I7*I4)
```

```
07820     PRINT "ALLOWABLE SYSTEM CAPITAL COST IS $";P5
07830     REM
07840     REM   CHOICE OF CONTINUATIONS SECTION
07850     REM
07860       PRINT" TYPE THE NUMBER OF WHAT YOU WANT TO DO "
07870         PRINT" 1) RUN A DIFFERENT SIMULATION"
07880         PRINT" 2) RUN ALMOST THE SAME SIMULATION"
07890         PRINT" 3) STOP THE PROGRAM "
07900       PRINT "NOTE: IF YOU CHOOSE 2) AND ARE USING STRATIFIED"
07910       PRINT "    STORAGE, THE PROGRAM WILL RESTART WITH THE"
07920       PRINT "    STORAGE TEMPERATURES FROM THE END OF THE"
07930       PRINT "    LAST RUN UNLESS YOU INPUT OTHER VALUES."
07940           INPUT O7
07950             IF O7=3 THEN 13170
07960               REM   STOP THE PROGRAM
07970             IF O7=1 THEN 660
07980               REM   GO TO THE INPUT DATA SECTION
07990             IF O7=2 THEN 8060
08000               REM   GO TO THE ALTER DATA SECTION
08010             PRINT" NOT A VALID NUMBER"
08020             GO TO 7860
08030     REM
08040     REM                 ALTER DATA SECTION
08050     REM
08060       IF I$=S$ THEN 8140
08070         REM   IF INPUT DATA IS IN CORRECT UNITS, SKIP CONVERSION
08080       IF I$="SI" THEN 8120
08090       GOSUB 12070
08100         REM   CHANGE INPUT DATA FROM SI TO ENG.
08110       GO TO 8140
08120       GOSUB 11610
08130         REM   CHANGE INPUT DATA FROM ENG. TO SI UNITS
08140       F$="ALTER"
08150       PRINT" TYPE THE NUMBER OF THE ITEM WHICH YOU WISH"
08160         PRINT" TO CHANGE, FOLLOWED BY ITS NEW VALUE"
08170         PRINT;" ON THE NEXT LINE"
08180         PRINT "WHEN YOU WISH TO RUN,TYPE'36'"
08190         PRINT" 1) SIMULATION TIME"
08200           PRINT"    NOTE:IF YOU INCREASE 1) AND WANT VARIABLE"
08210           PRINT"    INSOLATION OR VARIABLE LOAD, YOU MUST ENTER THE"
08220           PRINT"    ENTIRE ARRAY AGAIN.  "
08230         PRINT" 2) CALCULATION INTERVAL"
08240           PRINT"    NOTE:IF YOU CHANGE 2) AND WANT EITHER VARIABLE"
08250           PRINT"    INSOLATION OR VARIABLE LOAD, YOU MUST ENTER THE"
08260           PRINT"    ENTIRE ARRAY AGAIN.  "
08270         PRINT" 3) FREQUENCY OF OUTPUT"
08280         PRINT" 4) OPTION FOR SINUSOIDAL INSOLATION"
08290         PRINT" 5) OPTION FOR VARIABLE INSOLATION"
08300         PRINT" 6) VALUE OF QS,REF"
08310         PRINT" 7) DAILY MEAN AMBIENT TEMP."
08320         PRINT" 8) DAILY TEMPERATURE SWING ABOVE MEAN"
08330         PRINT" 9) VALUE OF TAU-ALPHA"
08340         PRINT"10) VALUE OF B"
08350         PRINT"11) VALUE OF F SUB-R"
08360         PRINT"12) MASS FLOW THROUGH THE COLLECTOR"
08370         PRINT"13) SPECIFIC HEAT OF THE FLUID"
```

```
08380          PRINT"14) COLLECTOR AREA"
08390          PRINT"15) HEAT CAPACITY OF STORAGE"
08400          PRINT"16) U-VALUE OF STORAGE"
08410          PRINT"17) LENGTH/DIAM. OF STORAGE"
08420          PRINT"18) TYPE OF STORAGE USED"
08430            PRINT"     1) LUMPED LIQUID STORAGE"
08440            PRINT"     2) STRATIFIED LIQUID STORAGE"
08450          PRINT"19) NUMBER OF DIVISIONS IN STRATIFIED STORAGE"
08460          PRINT"20) INITIAL STORAGE TEMPERATURE(S)"
08470          PRINT"21) MIN.  USEABLE STORAGE TEMPERATURE"
08480          PRINT"22) MAX.  SAFE STORAGE TEMPERATURE"
08490          PRINT"23) OPTION FOR CARNOT CYCLE LOAD"
08500     PRINT"24) OPTION FOR DOMESTIC HOT WATER LOAD"
08510          PRINT"25) OPTION FOR CONSTANT LOAD"
08520          PRINT"26) OPTION FOR VARIABLE LOAD"
08530          PRINT"27) MASS FLOW THROUGH THE LOAD"
08540          PRINT"28) OPTION FOR ECONOMIC ANALYSIS"
08550          PRINT"29) EXPECTED SYSTEM LIFE"
08560          PRINT"30) INTEREST RATE"
08570          PRINT"31) FUEL ESCALATION RATE"
08580          PRINT"32) FUEL COST"
08590          PRINT"33) OPTION TO CHANGE INPUT DATA UNITS"
08600            PRINT"    NOTE:EXISTING DATA CONVERTED INTERNALLY"
08610          PRINT"34) OPTION TO CHANGE OUTPUT DATA UNITS"
08620          PRINT"35) DISPLAY DATA"
08630          PRINT"36) STOP NEW INPUT AND RUN"
08640          PRINT"TYPE THE NUMBER OF THE PARAMETER TO BE CHANGED"
08650          PRINT"ALTER NUMBERS 1) AND 2) FIRST, IF NECESSARY"
08660          PRINT"ITEM NUMBER?"
08670          INPUT Z1
08680            IF Z1=36 THEN 4740
08690            IF Z1=35 THEN 3350
08700            IF Z1<>1 THEN 8730
08710              INPUT T3
08720              GO TO 8660
08730            IF Z1<>2 THEN 8800
08740              INPUT T7
08750     R=1440/T7
08760     IF R=INT(R) THEN 8660
08770     PRINT "TIME INTERVAL MUST DIVIDE 1440 EVENLY"
08780     PRINT "REENTER CALCULATION INTERVAL"
08790     GO TO 8740
08800            IF Z1<>3 THEN 8830
08810              INPUT O1
08820              GO TO 8660
08830            IF Z1<>4 THEN 8930
08840              PRINT"ENTER THE PEAK INSOLATION IN W/SQ.M."
08850                INPUT I1
08860              PRINT"ENTER THE LENGTH OF THE SOLAR DAY IN MIN."
08870                INPUT D1
08880              PRINT "ENTER SUNRISE TIME AS HOUR PLUS DECIMAL FRACTION."
08890                INPUT D2
08900                D2=D2*60
08910            B$="YES"
08920            GO TO 8660
08930            IF Z1<>5  THEN 9010
```

```
08940              B=-1
08950              FOR  A=0 TO T8 STEP T7
08960                PRINT"ENTER THE INSOLATION AT";A;"MIN. IN W/SQ.M."
08970                  INPUT   I9(B)
08980              NEXT  A
08990              B$="NO"
09000              GO TO 8660
09010          IF Z1<>6 THEN 9040
09020              INPUT I8
09030              GO TO 8660
09040          IF Z1<>7 THEN 9070
09050              INPUT T3
09060              GO TO 8660
09070          IF Z1<>8 THEN 9100
09080              INPUT D9
09090              GO TO 8660
09100          IF Z1<>9  THEN   9130
09110              INPUT T9
09120              GO TO 8660
09130          IF Z1<>10 THEN 9160
09140              INPUT B9
09150              GO TO 8660
09160          IF Z1<>11 THEN 9190
09170              INPUT F9
09180              GO TO 8660
09190          IF Z1<>12 THEN 9220
09200              INPUT M9
09210              GO TO 8660
09220          IF Z1<>13 THEN 9250
09230              INPUT C9
09240              GO TO 8660
09250          IF Z1<>14 THEN 9280
09260              INPUT A9
09270              GO TO 8660
09280          IF Z1<>15 THEN 9310
09290              INPUT C8
09300              GO TO 8660
09310          IF Z1<>16 THEN 9340
09320              INPUT U8
09330              GO TO 8660
09340          IF Z1<>17 THEN 9370
09350              INPUT L8
09360              GO TO 8660
09370          IF Z1<>18 THEN 9490
09380              INPUT T5
09390                IF T5=1 THEN 8660
09400     IF T5=2 THEN 9430
09410     PRINT "INPUT EITHER 1 OR 2"
09420     GO TO 9380
09430              PRINT"ENTER THE NUMBER OF DIVISIONS IN STORAGE"
09440              INPUT D5
09450              D4=D5-1
09460              GOSUB 11420
09470                REM   GO TO STRAT. STOR. INIT. SUBROUTINE
09480              GO TO 8660
09490          IF Z1<>19 THEN 9550
```

```
09500              INPUT D5
09510              D4=D5-1
09520              GOSUB 11420
09530                REM   GO TO STRAT.STOR.INIT. SUBROUTINE
09540              GO TO 8660
09550            IF Z1<>20 THEN 9620
09560              IF T5=2 THEN 9590
09570                INPUT T6
09580                GO TO 8660
09590              GOSUB 11420
09600                REM   GO TO THE STRAT. STOR. INIT. SUBROUTINE
09610              GO TO 8660
09620            IF Z1<>21 THEN 9650
09630              INPUT Z9
09640              GO TO 8660
09650            IF Z1<>22 THEN 9680
09660              INPUT T4
09670              GO TO 8660
09680            IF Z1<>23 THEN 9780
09690              PRINT"ENTER THE WORK OUTPUT YOU EXPECT FROM THE";
09700                PRINT" HEAT ENGINE, IN WATTS"
09710              INPUT W9
09720              PRINT"ENTER THE VALUE OF A THAT APPLIES TO THE ";
09730                PRINT"HIGH TEMPERATURE COLLECTORS BEING USED"
09740              INPUT A8
09750              D$="YES"
09760              D$="YES"
09770              GO TO 8660
09780          IF Z1<>24 THEN 9930
09790          PRINT "IS THE LOAD DOMESTIC HOT WATER?"
09800          INPUT Y$
09810          PRINT
09820          IF Y$="YES" THEN 9880
09830          IF Y$="NO" THEN 3660
09840          REM  IF THE LOAD IS DOMESTIC HOT WATER, INPUT SUPPLY
09850          REM  WATER TEMP. OTHERWISE, SKIP IT.
09860          GOSUB 11110
09870          GO TO 9790
09880          PRINT "ENTER THE CITY MAIN SUPPLY TEMP IN KELVINS(RANKINE)"
09890          INPUT Y1
09900          D$="NO"
09910          PRINT
09920          GO TO 8660
09930            IF Z1<>25 THEN 9990
09940              INPUT Q9
09950              D$="NO"
09960              A8=0
09970              C$="YES"
09980              GO TO 8660
09990            IF Z1<>26 THEN 10440
10000          C$="NO"
10010          PRINT "ENTER THE METHOD OF INPUT FOR VARIABLE LOAD"
10020          PRINT "1) LOAD AT EACH TIME INCREMENT"
10030          PRINT "2) TYPICAL DAY'S LOAD AT EACH INCREMENT"
10040          PRINT "3) TYPICAL DAY'S LOAD PROFILE"
10050          INPUT L
```

```
10060     IF L=1 THEN 10210
10070     IF L=2 THEN 10110
10080     IF L=3 THEN 10310
10090     PRINT "INVALID ENTRY, ENTER A NUMBER (1-3)"
10100     GO TO 10010
10110        B=-1
10120        PRINT "ENTER THE LOAD DATA IN JOULES/MIN.  (BTU/HR)"
10130        FOR A=0 TO 1440-T7 STEP T7
10140          B=B+1
10150          PRINT "ENTER THE LOAD AT ";A;" MINUTES"
10160            INPUT Q9(B)
10170        NEXT A
10180        PRINT
10190     GO TO 8660
10200     REM    SKIP THE OTHER VARIABLE LOAD INPUTS
10210        B=-1
10220        PRINT "ENTER THE LOAD DATA IN JOULES/MIN.  (BTU/HR)"
10230        FOR A=0 TO T8 STEP T7
10240          B=B+1
10250          PRINT "ENTER THE LOAD AT ";A;" MINUTES"
10260            INPUT Q9(B)
10270        NEXT A
10280        PRINT
10290     GO TO 8660
10300     REM    SKIP THE OTHER VARIABLE LOAD INPUTS
10310     PRINT "ENTER THE TIME IN HOURS FOLLOWED BY"
10320     PRINT"A COMMA (,) FOLLOWED BY THE LOAD AT THAT TIME IN"
10330     PRINT"JOULES/MIN.(BTU/HR)"
10340     PRINT"YOU MUST ENTER A LOAD AT T=0 AND T=24 HOURS"
10350     PRINT"ENTER DATA POINTS CONSECUTIVELY"
10360     A=-1
10370     A=A+1
10380     INPUT S5(A),S6(A)
10390     IF S5(A)=24 THEN 8660
10400     IF S5(A)<24 THEN 10370
10410     PRINT "TIME MUST BE IN HOURS (0-24)"
10420     PRINT "REENTER LAST DATA POINT"
10430     GO TO 10380
10440          IF Z1<>27 THEN 10470
10450            INPUT M3
10460          GO TO 8660
10470          IF Z1<>28 THEN 10630
10480            PRINT"DO YOU WANT ECONOMIC ANALYSIS?"
10490              INPUT E$
10500                IF E$="YES" THEN 10540
10510                IF E$="NO" THEN 8660
10520                GOSUB 11110
10530                GO TO 10480
10540                PRINT"ENTER THE EXPECTED SYSTEM LIFE IN YEARS"
10550                  INPUT N1
10560                PRINT"ENTER THE INTEREST RATE,%"
10570                  INPUT I6
10580                PRINT"ENTER THE FUEL ESCALATION RATE,%"
10590                  INPUT I5
10600                PRINT"ENTER THE FUEL COST,$/KWH"
10610                  INPUT C7
```

```
10620            GO TO 8660
10630            IF Z1<>29 THEN 10660
10640              INPUT N1
10650              GO TO 8660
10660            IF Z1<>30 THEN 10690
10670              INPUT I6
10680              GO TO 8660
10690            IF Z1<>31 THEN 10720
10700              INPUT I5
10710              GO TO 8660
10720            IF Z1<>32 THEN 10750
10730              INPUT C7
10740              GO TO 8660
10750            IF Z1<>33 THEN 10970
10760              PRINT"ENTER THE SYSTEM OF UNITS TO BE USED FOR INPUT"
10770                PRINT"(ENTER 'SI' OR 'ENG' IN CAPS)"
10780                PRINT"NOTE:   EXISTING DATA CONVERTED INTERNALLY"
10790                INPUT J$
10800                  IF J$="SI" THEN 10830
10810                  IF J$="ENG" THEN 10830
10820                  GO TO 10760
10830                  IF J$=I$ THEN 10950
10840                    REM   IF INPUT DATA IS ALREADY IN NEW UNITS
10850                    REM   SKIP CONVERSION AND PRINT MESSAGE
10860                  IF S$="SI" THEN 10910
10870                  GOSUB 11610
10880                    REM   CHANGE INPUT DATA FROM ENG. TO SI UNITS
10890                  I$=J$
10900                  GO TO 8660
10910                  GOSUB 12070
10920                    REM   CHANGE INPUT DATA FROM SI TO ENG. UNITS
10930                  I$=J$
10940                  GO TO 8660
10950                  PRINT"NO CHANGES MADE, DATA ALREADY IN ";J$;" UNITS"
10960                  GO TO 8660
10970            IF Z1<>34 THEN 11040
10980              PRINT"ENTER THE SYSTEM OF UNITS TO BE USED FOR OUTPUT"
10990                PRINT"(ENTER 'SI' OR 'ENG' IN CAPS)"
11000                INPUT O$
11010                  IF O$="SI" THEN 8660
11020                  IF O$="ENG" THEN 8660
11030                  GO TO 10980
11040            PRINT"ENTER NUMBER OF PARAMETER TO BE CHANGED, 1-36, ";
11050              PRINT"FOLLOWED BY ITS NEW VALUE ON THE NEXT LINE"
11060              GO TO 8660
11070 REM
11080 REM                SUBROUTINES
11090 REM
11100    REM   YES-NO ANSWER CORRECTION
11110      PRINT "PLEASE ANSWER WITH A 'YES' OR 'NO' IN CAPS"
11120      PRINT
11130      RETURN
11140    REM   SINUSOIDAL AMBIENT TEMPERATURE
11150      B=-1
11160      FOR A=0 TO T8 STEP T7
11170        B=B+1
```

```
11180          T5(B)=T3+D9*SIN(6.28318*((B*T7)-(D2+360))/1440)
11190       NEXT A
11200       RETURN
11210   REM    SINUSOIDAL INSOLATION
11220       B=-1
11230       FOR A=0 TO T8 STEP T7
11240          B=B+1
11250          FOR D=0 TO 10
11260             IF A<=24*60*D THEN 11280
11270             C=A-24*60*D
11280          NEXT D
11290          IF C<=D2 THEN 11360
11300             REM    IF TIME OF DAY IS BEFORE SUNRISE,
11310             REM    SET INSOLATION TO ZERO
11320          IF C<=(D1+D2) THEN 11380
11330             REM    IF THE TIME OF DAY, C, IS LESS THAN THE
11340             REM    SOLAR DAY LENGTH, CALCULATE THE INSOLATION.
11350             REM    OTHERWISE, SET THE INSOLATION TO 0
11360          I9(B)=0
11370          GO TO 11390
11380          I9(B)=I1*SIN(3.14159*(C-D2)/D1)
11390       NEXT A
11400       RETURN
11410   REM    STRATIFIED STORAGE TEMPERATURE
11410   REM    STRATIFIED STORAGE TEMPERATURE INIT. SUBROUTINE
11420       PRINT "ENTER THE INITIAL STORAGE TEMPERATURE IN ";
11430          PRINT "EACH DIVISION OF STORAGE,"
11440          PRINT " IN KELVINS(RANKINE)"
11450          PRINT "DIVISIONS ARE NUMBERED FROM TOP (1) TO ";
11460          PRINT "BOTTOM (";D5;")"
11470          PRINT
11480          FOR M=1 TO D5
11490             PRINT "DIVISION #";M;"?"
11500                INPUT T3(M-1)
11510          NEXT M
11520       RETURN
11530   REM    CONSTANT LOAD
11540       B=-1
11550       FOR A=0 TO T8 STEP T7
11560          B=B+1
11570          Q9(B)=Q9
11580       NEXT A
11590       RETURN
11600   REM    ALTER UNITS OF INPUT DATA FROM ENG. TO SI UNITS
11610       IF B$="NO" THEN 11640
11620       I1=I1*(3.15393)
11630       GO TO 11670
11640       FOR B=0 TO T8/T7
11650          I9(B)=I9(B)*(3.15393)
11660       NEXT B
11670   IF Y$="NO" THEN 11690
11680   Y1=Y1*(5/9)
11690       I8=I8*(3.15393)
11700       T3=T3*(5/9)
11710       D9=D9*(5/9)
11720       M9=M9*(.45359)
```

```
11730          C9=C9*(4185.8)
11740          A9=A9*(.0929)
11750          C8=C8*(1898.64)
11760          U8=U8*(5.677)
11770     IF T5=2 THEN 11800
11780          T6=T6*(5/9)
11790     GO TO 11830
11800     FOR M=0 TO D4
11810     T3(M)=T3(M)*(5/9)
11820     NEXT M
11830          Z9=Z9*(5/9)
11840          T4=T4*(5/9)
11850          IF D$="YES" THEN 12030
11860          IF C$="NO" THEN 11890
11870          Q9=Q9*(17.576)
11880          GO TO 12030
11890     IF L=2 THEN 11950
11900     IF L=3 THEN 11990
11910     FOR B=0 TO T8/T7
11920     Q9(B)=Q9(B)*(17.576)
11930     NEXT B
11940     GO TO 12030
11950     FOR B=0 TO 1440/T7
11960     Q9(B)=Q9(B)*(17.576)
11970     NEXT B
11980     GO TO 12030
11990     A=-1
12000     A=A+1
12010     S6(A)=S6(A)*(17.576)
12020     IF S5(A)<24 THEN 12000
12030          M8=M8*(.45359)
12040          S$="SI"
12050          RETURN
12060       REM   ALTER INPUT DATA UNITS FROM SI TO ENG.
12070          IF B$="NO" THEN 12100
12080          I1=I1/(3.15393)
12090          GO TO 12130
12100          FOR B=0 TO T8/T7
12110             I9(B)=I9(B)/(3.15393)
12120          NEXT B
12130     IF Y$="NO" THEN 12150
12140     Y1=Y1/(5/9)
12150          I8=I8/(3.15393)
12160          T3=T3/(5/9)
12170          D9=D9/(5/9)
12180          M9=M9/(.45359)
12190          C9=C9/(4185.8)
12200          A9=A9/(.0929)
12210          C8=C8/(1898.64)
12220          U8=U8/(5.677)
12230     IF T5=2 THEN 12260
12240          T6=T6/(5/9)
12250     GO TO 12290
12260     FOR M=0 TO D4
12270     T3(M)=T3(M)/(5/9)
12280     NEXT M
```

```
12290      Z9=Z9/(5/9)
12300      T4=T4/(5/9)
12310      IF D$="YES" THEN 12490
12320      IF C$="NO" THEN 12350
12330      Q9=Q9/(17.576)
12340      GO TO 12490
12350  IF L=2 THEN 12410
12360  IF L=3 THEN 12450
12370  FOR B=0 TO T8/T7
12380      Q9(B)=Q9(B)/(17.576)
12390  NEXT B
12400  GO TO 12490
12410  FOR B=0 TO 1440/T7
12420      Q9(B)=Q9(B)/(17.576)
12430  NEXT B
12440  GO TO 12490
12450  A=-1
12460  A=A+1
12470  S6(A)=S6(A)/(17.576)
12480  IF S5(A)<24 THEN 12460
12490      M8=M8/(.45359)
12500      S$="ENG"
12510      RETURN
12520    REM   OUTPUT DATA FROM SI TO ENG. UNITS
12530      I9(B)=I9(B)/(3.15393)
12540      Q9(B)=Q9(B)/(17.576)
12550      T0=T0/(5/9)
12560      IF  T5=2 THEN 12590
12570      T1=T1/(5/9)
12580      GO TO 12620
12590      FOR M=0 TO D4
12600         T3(M)=T3(M)/(5/9)
12610      NEXT M
12620      T2=T2/(5/9)
12630      Q=Q/(17.575)
12640  IF Y$="NO" THEN 12650
12650      RETURN
12660    REM   OUTPUT DATA FROM ENG. TO SI UNITS
12670      I9(B)=I9(B)*(3.15393)
12680      Q9(B)=Q9(B)*(17.576)
12690      T0=T0*(5/9)
12700      IF  T5=2 THEN 12730
12710      T1=T1*(5/9)
12720      GO TO 12760
12730      FOR M=0 TO D4
12740         T3(M)=T3(M)*(5/9)
12750      NEXT M
12760      T2=T2*(5/9)
12770      Q=Q*(17.576)
12780  M8=M8*(.0631313)
12790      RETURN
12800    REM   TRANSLATE DAILY LOAD PROFILE INTO DAILY LOAD ARRAY
12810  A=0
12820  B=-1
12830  FOR T=0 TO 1440-T7 STEP T7
12840     B=B+1
```

```
12850      IF T<S5(A+1)*60 THEN 12880
12860      A=A+1
12870      GO TO 12850
12880      W5=(T/60-S5(A))/(S5(A+1)-S5(A))
12890      Q9(B)=S6(A)+W5*(S6(A+1)-S6(A))
12900   NEXT T
12910   RETURN
12920   REM   REPEAT DAILY LOAD ARRAY TO FILL SIMULATION TIME
12930   W6=1440/T7
12940   W7=-1
12950   B=-1
12960   FOR T=0 TO T8 STEP T7
12970      B=B+1
12980      W7=W7+1
12990      IF W7<W6 THEN 13010
13000         W7=0
13010      Q9(B)=Q9(W7)
13020   NEXT T
13030   RETURN
13040   REM   FIND THE LARGEST LOAD OF THE SIMULATION
13050   B=-1
13060   W3=Q9(0)
13070   FOR T=0 TO T8 STEP T7
13080      B=B+1
13090      IF Q9(B)<=W3 THEN 13110
13100      W3=Q9(B)
13110   NEXT T
13120   RETURN
13130      REM   LOAD OUTLET ERROR MESSAGE
13140         PRINT"REQUIRED TEMP. DROP ACROSS LOAD WOULD RESULT IN";
13150         PRINT" SUB-ZERO ABSOLUTE TEMP."
13160         PRINT"RUN ABORTED BY PROGRAM"
13170      STOP
13180      END
```

Table E-3 Symbol correspondence for FORTRAN version of SOLSIM

Program symbol	Text symbol	Definition
ACOLL	a	Collector a value
AREAC	A_c	Collector aperture area
ASTOR	...	Surface area of storage tank
BCOLL	b	Collector b value
BIGLD		Maximum load over period of simulation
EAVAIL	$A_c \int q_s(t)\, dt$	Cumulative insolation integrated over simulation period
ESTNET	...	Net energy added to storage
ETAC	η	Collector efficiency
CAPST	$(Mc)_{st}$	Thermal capacity of storage
CCEFF	η_C	Carnot engine efficiency
CLDELT	...	ΔT between collector outlet and stratified tank layer
CONLD	$\dot{Q}_L$	Constant specified load
DELTA 1	...	Values used to determine Δt for use
DELTA 2	...	in stratified storage calculations
DSTOR	D	Diameter of storage tank
ECOLL	$\int \dot{Q}_u\, dt$	Useful energy collected during simulation
FCOST	F	Cost of competing fuel
FLWMAX	...	Maximum flow rate in load loop
FRATE	j	Fuel cost escalation rate
FRLLST	...	Values of fraction Δt in stratified
FRLDST	...	storage calculations
FSUBR	F_R	Collector heat removal parameter
IALTER	...	Index to note whether data are being altered
IBASE	...	Index to denote base case
IECONO	...	Index to denote economics desired
IEND	...	Time index value at end of simulation
IFLAG	...	Time index
IFRQ	...	Index showing frequency of desired output
ILDCC	...	Index to denote load is a Carnot cycle
ILDCON	...	Index to denote load is constant
ILDDHW	...	Index to denote load is DHW system
INPVI	...	Index to denote nonsinusoidal insolation
INPVLD	...	Index to denote type of load profile
INORFX	...	Index to denote whether original data input or altered data
INSOL	$\dot{q}_s(t)$	Insolation at time t
INUNIT	...	Index to denote system of input units
ISTAT	...	Index to denote whether units are in correct system for SOLSIM equation
ISWTCL	δ_c	Switch on collector pump
ISWTLD	δ_L	Switch on load pump
ITYPST	...	Index to denote mixed ($=1$) or stratified ($=2$) storage

Table E-3 (Continued)

Program symbol	Text symbol	Definition
LDDELT	...	Δt between load outlet and stratified tank layer
LDPMP	...	Time load pump was on during simulation
LDPTM	...	Time of data point arbitrary load profile
LDPVAL	...	Value of load, arbitrary load profile
LOAD	$\dot{Q}_L$	Energy load on system
LSTOR	L	Length of storage tank
MDOTC	$\dot{m}_c$	Mass flow rate in collector loop
MDOTLD	$\dot{m}_L$	Mass flow rate in load loop
NODIV	N	Number of stratified tank layers
OUNIT	...	Index to denote system of output units
OUTPUT	...	Index to control printing of output data
PERLD	...	Percentage of load supplied over simulation period
PMPLD	...	Percentage of time load pump was on during simulation
PWRCC	$\dot{W}_C$	Power output of Carnot engine
QSREF	$q_{s,\text{ref}}$	Reference insolation
RATE	i	Interest rate
REQLD	$\int \dot{Q}_L(t)\, dt$	Cumulative required load over simulation period
RHO	ρ_{H_2O}	Density of water
RLNDIA	L/D	Length-to-diameter ratio of storage tank
SOLMAX	$q_{s,\text{max}}$	Maximum insolation
SPHT	c	Specific heat of working fluid
SSPTM	...	Time of input point, arbitrary insolation
SSPVAL	...	Value of insolation in arbitrary profile
SUNSIN	...	Index to denote sinusoidal
SUPLD	...	Load supplied over period of simulation
SWING	$(\Delta T)_a$	Ambient temperature swing
SYSCST	P	Justifiable cost of system
SYSEFF	...	Collection efficiency of system
SYSLIFE	n	System life
SUNDAY	t_d	Length of solar day
SUNRIS	t_{rise}	Time of sunrise
TAAVG	$\bar{T}_a$	Daily average ambient temperature
TAMB	T_a	Ambient temperature
TAUA	$(\tau\alpha)_{\text{eff}}$	Collector property
TCALC	Δt	Calculation interval
TCOUT	$T_{f,\text{out}}$	Collector fluid outlet temperature
TIME	t	Instantaneous time
TLDIN	$T_{L,\text{in}}$	Load inlet temperature
TSIM	...	Length of simulation period
TSPLY	T_{sup}	City main supply temperature
TSTMAX	$T_{\text{st,max}}$	Maximum safe storage temperature
TSTMIN	$T_{\text{st,min}}$	Minimum storage temperature that supplied useful energy
TSTOR	T_{st}	Mixed storage temperature

Table E-3 (*Continued*)

Program symbol	Text symbol	Definition
TSTRAT(M)	$T_{st}(m)$	Temperature of stratified storage layer m
TSTTMP	...	Intermediate stratified storage temperatures

E-3 SAMPLE FORTRAN PROGRAM

```
      PROGRAM SOLSIM(TTY=101B,OUTPUT,TAPE5=TTY,TAPE6=OUTPUT)
      COMMON//TSIM,TCALC,SUNSIN,SOLMAX,INPVI,INSOL(1000),QSREF,TAAVG,
     1SWING,MDOTC,SPHT,AREAC,CAPST,USTOR,ITYPST,NODIV,TSTRAT(10),TSTOR
     2TSTMIN,TSTMAX,ILDCC,ILDDHW,ILDCON,CONLD,INPVLD,LOAD(1000),MDOTLD
     3UAREA,UCAPST,UINP,UINSOL(2),ULOAD,UMSFLW,USPHT(2),UTEMP,UUVAL(2)
     4LDPTM(100),LDPVAL(100),SSPTM(100),SSPVAL(100),ISTAT,RHO,TSPLY,
     5SUNRIS
      DIMENSION TAMB(1000),TSTTMP(10),CLRET(10),CLDELT(10),LDRET(10)
      DIMENSION LDDELT(10)
      REAL INSOL,LOAD,LDPTM,LDPVAL,LDDELT,MDOTC,MDOTLD,LSTOR
      INTEGER UINP,UINSOL,UTEMP,UMSFLW,USPHT,UAREA,UCAPST
      INTEGER UUVAL,ULOAD,CLRET,SUNSIN,OUNIT,OUTPUT
      DATA PI/3.1415927/
      RHO=1000.
      WRITE(5,7000)
7000  FORMAT(21X,'S O L S I M')
      WRITE(5,1000)
1000  FORMAT(' THIS PROGRAM SIMULATES SOLAR ENERGY SYSTEMS')
100   WRITE(5,1001)
1001  FORMAT(' DO YOU WANT TO USE THE BASE CASE? TYPE 1 FOR YES, 2 FOR
     1 NO')
      READ(5,*)IBASE
      IF(IBASE.EQ.1)GO TO 101
      IF(IBASE.EQ.2)GO TO 101
      GO TO 100
C*******************************************
C********    BASE CASE DATA SECTION    ****
C*******************************************
101   INORFX=2
      TSPLY=300.
      IUNIT=1
      OUNIT=1
      TSIM=2880.
      TCALC=15
      IFRQ=4
      SUNSIN=1
      SOLMAX=800
      SUNDAY=600
      SUNRIS=420
      QSREF=800
      TAAVG=285
      SWING=10
      TAUA=.8
```

```
      BCOLL=1.2
      FSUBR=0.95
      ACOLL=0.0
      MDOTC=20.
      SPHT=4183.
      AREAC=30.
      CAPST=5000000.
      USTOR=.28385
      RLNDIA=3.
      ITYPST=1
      TSTOR=300
      TSTMIN=300.
      TSTMAX=373.
      ILDCC=2
      ILDDHW=2
      ILDCON=1
      CONLD=250000.
      MDOTLD=10.
      IECONO=1
      SYSLIF=10.
      RATE=.12
      FRATE=.15
      FCOST=.05
      UINP=2HSI
      UINSOL(1)=7HW/SQ.M.
      UINSOL(2)=1H
      UTEMP=7HKELVINS
      UMSFLW=8HKG./MIN.
      USPHT(1)=8HJ/KG.-DE
      USPHT(2)=4HG.C.
      UAREA=5HSQ.M.
      UCAPST=8HJ/DEG.C.
      UUVAL(1)=10HW/SQ.M.-DE
      UUVAL(2)=4HG.C.
      ULOAD=6HJ/MIN.
      ISTAT=1
      IF(IBASE.EQ.2)GO TO 171
      GO TO 200
C****************************************
C*********    INPUT DATA SECTION    ****
C****************************************
  171 INORFX=2
C*********    ENTER INPUT DATA UNITS    ****
  102 WRITE(5,1002)
 1002 FORMAT(' ENTER THE SYSTEM OF UNITS FOR INPUT (1 OR 2)'/
     1' 1) SI'/' 2) ENGLISH')
      READ(5,*)IUNIT
      IF(IUNIT.EQ.1)GO TO 103
      IF(IUNIT.EQ.2)GO TO 104
      GO TO 102
  103 CONTINUE
      UINP=2HSI
      UINSOL(1)=7HW/SQ.M.
      UINSOL(2)=1H
      UTEMP=7HKELVINS
      UMSFLW=8HKG./MIN.
```

```
      USPHT(1)=8HJ/KG.-DE
      USPHT(2)=4HG.C.
      UAREA=5HSQ.M.
      UCAPST=8HJ/DEG.C.
      UUVAL(1)=10HW/SQ.M.-DE
      UUVAL(2)=4HG.C.
      ULOAD=6HJ/MIN.
      IF((INORFX.EQ.1).AND.(ISTAT.NE.IUNIT))CALL IETOSI
      ISTAT=1
      IF(INORFX.EQ.1)GO TO 351
      GO TO 105
  104 CONTINUE
      UINP=3HENG
      UINSOL(1)=7HBTU/HR.
      UINSOL(2)=7H-SQ.FT.
      UTEMP=6HDEG.R.
      UMSFLW=9HLBM./MIN.
      USPHT(1)=8HBTU/LBM.
      USPHT(2)=7H-DEG.F.
      UAREA=6HSQ.FT.
      UCAPST=10HBTU/DEG.F.
      UUVAL(1)=10HBTU/H-SQ.F
      UUVAL(2)=9HT.-DEG.F.
      ULOAD=7HBTU/HR.
      IF((INORFX.EQ.1).AND.(ISTAT.NE.IUNIT))CALL ISITOE
      ISTAT=2
      IF(INORFX.EQ.1)GO TO 351
C********     ENTER OUTPUT DATA UNITS     ****
  105 WRITE(5,1003)
 1003 FORMAT(' ENTER THE SYSTEM OF UNITS FOR OUTPUT (1 OR 2)'/
     1' 1) SI'/' 2) ENGLISH')
      READ(5,*)OUNIT
      IF((OUNIT.EQ.1).OR.(OUNIT.EQ.2))GO TO 106
      GO TO 105
  106 IF(INORFX.EQ.1)GO TO 351
C********     ENTER SIMULATION TIME     ****
  108 WRITE(5,1004)
 1004 FORMAT(' ENTER THE TOTAL TIME OF THE SIMULATION IN MINUTES')
      READ(5,*)TSIM
      IF(INORFX.EQ.1)GO TO 351
C********     ENTER CALCULATION INTERVAL     ****
  109 WRITE(5,1005)
 1005 FORMAT(' ENTER THE TIME INTERVAL BETWEEN CALCULATION POINTS IN',
     1' MINUTES')
      READ(5,*)TCALC
      TEST=1440./TCALC
      IF(TEST.EQ.INT(TEST))GO TO 110
      WRITE(5,1006)
 1006 FORMAT(' CALCULATION INTERVAL MUST DIVIDE 1440 EVENLY')
      GO TO 109
  110 TEST=TSIM/TCALC
      IF(TEST.EQ.INT(TEST))GO TO 170
      WRITE(5,1007)
 1007 FORMAT(' CALCULATION INTERVAL MUST DIVIDE TOTAL TIME EVENLY')
      GO TO 109
  170 IF(INORFX.EQ.1)GO TO 351
```

```
C*********      ENTER FREQUENCY OF OUTPUT      ****
  111 WRITE(5,1008)
 1008 FORMAT(' ENTER THE FREQUENCY OF OUTPUT IN NUMBER OF INTERVALS')
      READ(5,*)IFRQ
      IF(INORFX.EQ.1)GO TO 351
C*********      OFFER SINUSOIDAL INSOLATION      ****
  112 WRITE(5,1009)
 1009 FORMAT(' IS THE INSOLATION SINUSOIDAL? TYPE 1 FOR YES, 2 FOR NO')
      READ(5,*)SUNSIN
      IF(SUNSIN.EQ.1)GO TO 113
      IF(SUNSIN.EQ.2)GO TO 115
      GO TO 112
C*********      ENTER SINUSOIDAL INSOLATION DATA      ****
  113 WRITE(5,1010)UINSOL(1),UINSOL(2)
 1010 FORMAT(' ENTER THE MAXIMUM INSOLATION IN ',2A7)
      SUNSIN=1
      READ(5,*)SOLMAX
      WRITE(5,1011)
 1011 FORMAT(' ENTER THE LENGTH OF THE SOLAR DAY IN MINUTES')
      READ(5,*)SUNDAY
      WRITE(5,2024)
 2024 FORMAT(' ENTER TIME OF SUNRISE AS LOCAL HOUR,',
     1'(WITH DECIMAL FRACTION) ON 24-HOUR CLOCK')
      READ(5,*)SUNRIS
      SUNRIS=60*SUNRIS
      IF(INORFX.EQ.1)GO TO 351
      GO TO 123
C*********      ENTER VARIABLE INSOLATION DATA      ****
  115 WRITE(5,1012)
 1012 FORMAT(' ENTER THE METHOD OF INPUT FOR VARIABLE INSOLATION',
     1'(1,2 OR 3)'/' 1) INSOLATION AT EACH TIME INTERVAL'/
     2' 2) TYPICAL DAY INSOLATION AT EACH TIME INTERVAL'/
     3' 3) TYPICAL DAY INSOLATION PROFILE')
      SUNSIN=2
      READ(5,*)INPVI
      IF(INPVI.EQ.1)GO TO 116
      IF(INPVI.EQ.2)GO TO 117
      IF(INPVI.EQ.3)GO TO 118
C*********      INSOLATION AT EACH TIME INTERVAL      ****
  116 IEND=INT(TSIM/TCALC)+1
      TIME=0.
      DO 1 IFLAG=1,IEND
      WRITE(5,1015)TIME,UINSOL(1),UINSOL(2)
 1015 FORMAT(' ENTER THE INSOLATION AT ',F8.1,' MINUTES IN ',2A7)
      READ(5,*)INSOL(IFLAG)
      TIME=TIME+TCALC
    1 CONTINUE
      IF(INORFX.EQ.1)GO TO 351
      GO TO 123
C*********      TYPICAL DAY INSOLATION DATA      ****
  117 IEND=INT(1440./TCALC)+1
      TIME=0.
      DO 2 IFLAG=1,IEND-1
      WRITE(5,1015)TIME,UINSOL(1),UINSOL(2)
      READ(5,*)INSOL(IFLAG)
      TIME=TIME+TCALC
```

```
    2 CONTINUE
      IF(INORFX.EQ.1)GO TO 351
      GO TO 123
C********   TYPICAL DAY INSOLATION PROFILE    ****
  118 IFLAG=0
      WRITE(5,1016)UINSOL(1),UINSOL(2)
 1016 FORMAT(' ENTER THE TIME IN HOURS, FOLLOWED BY A COMMA,',
     1' FOLLOWED BY THE'/'INSOLATION IN ',2A7/' YOU MUST ENTER DATA ',
     2'POINTS AT T=0 AND T=24 HOURS'/' ENTER DATA POINTS ',
     3' CONSECUTIVELY')
  119 IFLAG=IFLAG+1
  120 WRITE(5,1017)UINSOL(1),UINSOL(2)
 1017 FORMAT(' TIME(HRS.),INSOLATION(',2A7,')')
      READ(5,*)SSPTM(IFLAG),SSPVAL(IFLAG)
      IF(IFLAG.EQ.1)GO TO 119
      IF(SSPTM(IFLAG).GT.SSPTM(IFLAG-1))GO TO 121
      WRITE(5,1018)
 1018 FORMAT(' DATA POINTS NOT CONSECUTIVE'/' REENTER LAST DATA ',
     1'POINT')
      GO TO 120
  121 IF(SSPTM(IFLAG).LT.24.)GO TO 119
      IF(INORFX.EQ.1)GO TO 351
C********        ENTER QS,REF     ****
  123 WRITE(5,1021)UINSOL(1),UINSOL(2)
 1021 FORMAT(' ENTER THE VALUE OF QS,REF USED IN A AND B IN ',2A7)
      READ(5,*)QSREF
      IF(INORFX.EQ.1)GO TO 351
C********   ENTER DAILY MEAN AMBIENT TEMP.    ****
  124 WRITE(5,1022)UTEMP
 1022 FORMAT(' ENTER THE DAILY MEAN AMBIENT TEMP. IN ',A7)
      READ(5,*)TAAVG
      IF(INORFX.EQ.1)GO TO 351
C********    ENTER DAILY TEMP. SWING    ****
  125 WRITE(5,1023)UTEMP
 1023 FORMAT(' ENTER THE DAILY TEMP. SWING IN ',A7,/
     1' SWING IS MAXIMUM ABSOLUTE DIFFERENCE FROM AVG. AMBIENT.')
      READ(5,*)SWING
      IF(INORFX.EQ.1)GO TO 351
C********      ENTER TAU-ALPHA      ****
  126 WRITE(5,1024)
 1024 FORMAT(' ENTER THE VALUE OF TAU-ALPHA FOR THE COLLECTOR')
      READ(5,*)TAUA
      IF(INORFX.EQ.1)GO TO 351
C********      ENTER VALUE OF B     ****
  127 WRITE(5,1025)
 1025 FORMAT(' ENTER THE VALUE OF B FOR THE COLLECTOR (B>0)')
      READ(5,*)BCOLL
      IF(BCOLL.LE.0.)GO TO 127
      IF(INORFX.EQ.1)GO TO 351
C********     ENTER VALUE OF FSUB-R      ******
  159 WRITE(5,2023)
 2023 FORMAT(* ENTE
 2023 FORMAT(* ENTER THE VALUE OF F SUB-R FOR THE COLLECTOR*)
      READ(5,*),FSUBR
      IF(INORFX.EQ.1) GO TO 351
```

```
C*********      ENTER VALUE OF A     ****
  128 WRITE(5,1026)
 1026 FORMAT(' ENTER THE VALUE OF A FOR THE COLLECTOR')
      READ(5,*)ACOLL
      IF(INORFX.EQ.1)GO TO 351
C*********      ENTER COLLECTOR MASS FLOW RATE    ****
  130 WRITE(5,1028)UMSFLW
 1028 FORMAT(' ENTER THE MASS FLOW RATE THROUGH THE COLLECTOR IN ',A9)
      READ(5,*)MDOTC
      IF(INORFX.EQ.1)GO TO 351
C*********      ENTER SPECIFIC HEAT OF FLUID    ****
  131 WRITE(5,1029)USPHT(1),USPHT(2)
 1029 FORMAT(' ENTER THE SPECIFIC HEAT OF THE FLUID IN ',2A8)
      READ(5,*)SPHT
      IF(INORFX.EQ.1)GO TO 351
C*********      ENTER COLLECTOR AREA    ****
  132 WRITE(5,1030)UAREA
 1030 FORMAT(' ENTER THE COLLECTOR AREA IN ',A6)
      READ(5,*)AREAC
      IF(INORFX.EQ.1)GO TO 351
C*********      ENTER HEAT CAPACITY OF STORAGE    ****
  133 WRITE(5,1031)UCAPST
 1031 FORMAT(' ENTER THE HEAT CAPACITY OF STORAGE IN ',A10)
      READ(5,*)CAPST
      IF(INORFX.EQ.1)GO TO 351
C*********      ENTER U-VALUE OF STORAGE TANK    ****
  134 WRITE(5,1032)UUVAL(1),UUVAL(2)
 1032 FORMAT(' ENTER THE U-VALUE OF THE STORAGE INSULATION IN ',2A10)
      READ(5,*)USTOR
      IF(INORFX.EQ.1)GO TO 351
C*********      ENTER LENGTH-DIAMETER RATIO OF TANK    ****
  135 WRITE(5,1033)
 1033 FORMAT(' ENTER THE LENGTH-DIAMETER RATIO OF THE STORAGE TANK')
      READ(5,*)RLNDIA
      IF(INORFX.EQ.1)GO TO 351
C*********      ENTER TYPE OF STORAGE    ****
  136 WRITE(5,1034)
 1034 FORMAT(' ENTER THE TYPE OF STORAGE SYSTEM USED (1 OR 2)'/
     1' 1) LUMPED LIQUID STORAGE'/' 2) STRATIFIED LIQUID STORAGE')
      READ(5,*)ITYPST
      IF(ITYPST.EQ.1)GO TO 139
      IF(ITYPST.EQ.2)GO TO 137
      GO TO 136
C*********      ENTER NUMBER OF DIVISIONS (STRATIFIED)    ****
  137 IF(ITYPST.EQ.1)WRITE(5,1063)
 1063 FORMAT(' STORAGE IS LUMPED, ALTER ITEM NUMBER 20 TO GET ',
     1'STRATIFIED STORAGE').
      IF(ITYPST.EQ.1)GO TO 351
      WRITE(5,1035)
 1035 FORMAT(' ENTER THE NUMBER OF DIVISIONS IN STORAGE'/
     1' (MAXIMUM 10, SUGGESTED 3-5)')
      READ(5,*)NODIV
      IF(NODIV.GT.10)GO TO 137
C*********      ENTER INITIAL STORAGE TEMPS. (STRATIFIED)    ****
  138 IF(ITYPST.EQ.1)WRITE(5,1063)
      IF(ITYPST.EQ.1)GO TO 351
```

```
      WRITE(5,1036)UTEMP,NODIV
 1036 FORMAT(' ENTER THE INITIAL STORAGE TEMP. AT EACH DIVISION IN ',
     1A7/' DIVISIONS ARE NUMBERED FROM TOP (1) TO BOTTOM (',I2,')')
      DO 3 IFLAG=1,NODIV
      WRITE(5,1037)IFLAG
 1037 FORMAT(' DIVISION #',I2)
      READ(5,*)TSTRAT(IFLAG)
    3 CONTINUE
      IF(INORFX.EQ.1)GO TO 351
      GO TO 140
C*********     ENTER INITIAL STORAGE TEMP. (LUMPED)     ****
  139 IF(ITYPST.EQ.2)WRITE(5,1064)
 1064 FORMAT(' STORAGE IS STRATIFIED, ALTER ITEM NUMBER 20 TO GET',
     1' LUMPED STORAGE')
      IF(ITYPST.EQ.2)GO TO 351
      WRITE(5,1038)UTEMP
 1038 FORMAT(' ENTER THE INITIAL STORAGE TEMPERATURE IN ',A7)
      READ(5,*)TSTOR
      IF(INORFX.EQ.1)GO TO 351
C*********     ENTER MIN. USEFUL STORAGE TEMP.     ****
  140 WRITE(5,1039)UTEMP
 1039 FORMAT(' ENTER THE MINIMUM STORAGE TEMPERATURE THAT WILL GIVE'/
     1' USEFUL ENERGY TO THE LOAD IN ',A7)
      READ(5,*)TSTMIN
      IF(INORFX.EQ.1)GO TO 351
C*********     ENTER MAX. SAFE STORAGE TEMP.     ****
  141 WRITE(5,1040)UTEMP
 1040 FORMAT(' ENTER THE MAXIMUM SAFE STORAGE TEMPERATURE IN ',A7)
      READ(5,*)TSTMAX
      IF(INORFX.EQ.1)GO TO 351
C*********     OFFER CARNOT CYCLE     ****
  142 WRITE(5,1041)
 1041 FORMAT(' IS THE LOAD A HEAT INPUT TO A CARNOT CYCLE? TYPE 1 FOR ',
     1'YES, 2 FOR NO')
      READ(5,*)ILDCC
      IF(ILDCC.EQ.1)GO TO 143
      IF(ILDCC.EQ.2)GO TO 173
      GO TO 142
C*********     ENTER POWER OUTPUT OF CARNOT CYCLE     ****
  143 WRITE(5,1042)
 1042 FORMAT(' ENTER THE EXPECTED POWER OUTPUT FROM THE HEAT ENGINE IN ',
     1'WATTS')
      ILDDHW=2
      READ(5,*)PWRCC
      IF(INORFX.EQ.1)GO TO 351
      GO TO 156
  173 IF(INORFX.EQ.1)GO TO 146
C*********     OFFER DOMESTIC HOT WATER     ****
  144 WRITE(5,1043)
 1043 FORMAT(' IS THE LOAD DOMESTIC HOT WATER (DHW)? TYPE 1 FOR YES, 2',
     1' FOR NO')
      READ(5,*)ILDDHW
      IF(ILDDHW.EQ.1)GO TO 145
      IF(ILDDHW.EQ.2)GO TO 174
      GO TO 144
```

```
C*********     ENTER CITY MAIN SUPPLY TEMP.     ****
  145 WRITE(5,1044)UTEMP
 1044 FORMAT(' ENTER THE CITY MAIN SUPPLY TEMP. IN ',A7)
      ILDCC=2
      READ(5,*)TSPLY
  700 WRITE(5,7001)
 7001 FORMAT(' NOTE: CITY MAIN SUPPLY TEMP.'/
     1' MUST BE LESS THAN THE MINIMUM TEMP.'/
     2' THAT PROVIDES USEFUL ENERGY TO LOAD.')
      IF(INORFX.EQ.1)GO TO 351
  174 IF(INORFX.EQ.1)GO TO 156
C*********     OFFER CONSTANT LOAD     ****
  146 WRITE(5,1045)
 1045 FORMAT(' IS THE LOAD CONSTANT? TYPE 1 FOR YES, ',
     1'2 FOR NO')
      READ(5,*)ILDCON
      IF(ILDCON.EQ.1)GO TO 147
      IF(ILDCON.EQ.2)GO TO 148
      GO TO 146
C*********     ENTER CONSTANT LOAD     ****
  147 IF(ILDCC.EQ.1)WRITE(5,1065)
 1065 FORMAT(' CARNOT CYCLE LOAD IS CALCULATED FROM CARNOT CYCLE'/
     1' EFFICIENCY EQUATION'/' LOAD IS NOT AN INPUT FOR CARNOT CYCLES'
     2)
      IF(ILDCC.EQ.1)GO TO 351
      WRITE(5,1046)ULOAD
 1046 FORMAT(' ENTER THE CONSTANT LOAD IN ',A7)
      ILDCON=1
      READ(5,*)CONLD
      IF(INORFX.EQ.1)GO TO 351
      GO TO 156
C*********     ENTER VARIABLE LOAD     ****
  148 IF(ILDCC.EQ.1)WRITE(5,1065)
      IF(ILDCC.EQ.1)GO TO 351
      WRITE(5,1047)
 1047 FORMAT(' ENTER THE METHOD OF INPUT FOR VARIABLE LOAD '
     1' (1,2 OR 3)'/' 1) LOAD AT EACH TIME INTERVAL'/
     2' 2) TYPICAL DAY LOAD AT EACH TIME INTERVAL'/
     3' 3) TYPICAL DAY LOAD PROFILE')
      ILDCON=2
      READ(5,*)INPVLD
      IF(INPVLD.EQ.1)GO TO 149
      IF(INPVLD.EQ.2)GO TO 150
      IF(INPVLD.EQ.3)GO TO 151
      GO TO 148
C*********     LOAD AT EACH TIME INTERVAL     ****
  149 IEND=INT(TSIM/TCALC)+1
      TIME=0.
      DO 4 IFLAG=1,IEND
      WRITE(5,1048)TIME,ULOAD
 1048 FORMAT(' ENTER THE LOAD AT',F8.1,' MINUTES IN ',A7)
      READ(5,*)LOAD(IFLAG)
      TIME=TIME+TCALC
    4 CONTINUE
      IF(INORFX.EQ.1)GO TO 351
      GO TO 156
```

```
C*********      TYPICAL DAY LOAD DATA      ****
  150 IEND=INT(1440./TCALC)+1
      TIME=0.
      DO 5 IFLAG=1,IEND
      WRITE(5,1048)TIME,ULOAD
      READ(5,*)LOAD(IFLAG)
      TIME=TIME+TCALC
    5 CONTINUE
      IF(INORFX.EQ.1)GO TO 351
      GO TO 156
C*********      TYPICAL DAY LOAD PROFILE      ****
  151 IFLAG=0
      WRITE(5,1049)ULOAD
 1049 FORMAT(' ENTER THE TIME IN HOURS, FOLLOWED BY A COMMA, ',
     1'FOLLOWED BY'/' THE LOAD IN ',A7/' YOU MUST ENTER DATA POINTS',
     2' AT T=0 AND T=24 HOURS'/' ENTER DATA POINTS CONSECUTIVELY')
  152 IFLAG=IFLAG+1
  153 WRITE(5,1050)ULOAD
 1050 FORMAT(' TIME(HRS.),LOAD(',A7,')')
      READ(5,*)LDPTM(IFLAG),LDPVAL(IFLAG)
      IF(IFLAG.EQ.1)GO TO 152
      IF(LDPTM(IFLAG).GT.LDPTM(IFLAG-1))GO TO 154
      WRITE(5,1018)
      GO TO 153
  154 IF(LDPTM(IFLAG).LT.24.)GO TO 152
      IF(INORFX.EQ.1)GO TO 351
C*********      ENTER LOAD MASS FLOW RATE      ****
  156 IF((ILDDHW.EQ.1).AND.(INORFX.EQ.1))WRITE(5,1051)
 1051 FORMAT(' LOAD MASS FLOW DEPENDS ON LOAD AND STORAGE TEMP. ',/
     1' AND IS NOT AN INPUT FOR DHW SYSTEMS')
      IF(ILDDHW.EQ.1)GO TO 172
      WRITE(5,1052)UMSFLW
 1052 FORMAT(' ENTER THE MASS FLOW RATE THROUGH THE LOAD HEAT ',
     1'EXCHANGER IN ',A9)
      READ(5,*)MDOTLD
  172 IF(INORFX.EQ.1)GO TO 351
C*********      OFFER ECONOMIC ANALYSIS      ****
  157 WRITE(5,1053)
 1053 FORMAT(' DO YOU WANT ECONOMIC ANALYSIS? TYPE 1 FOR YES, ',
     1'2 FOR NO')
      READ(5,*)IECONO
      IF(IECONO.EQ.1)GO TO 158
      IF(IECONO.EQ.2)GO TO 200
      GO TO 157
C*********      ENTER ECONOMIC DATA      ****
  158 WRITE(5,1054)
 1054 FORMAT(' ENTER THE EXPECTED SYSTEM LIFE IN YEARS')
      READ(5,*)SYSLIF
      WRITE(5,1055)
 1055 FORMAT(' ENTER THE INTEREST RATE FOR FINANCING AS A DECIMAL')
      READ(5,*)RATE
      WRITE(5,1056)
 1056 FORMAT(' ENTER THE FUEL ESCALATION RATE AS A DECIMAL')
      READ(5,*)FRATE
      WRITE(5,1057)
 1057 FORMAT(' ENTER THE COST OF THE COMPETING FUEL IN $/KWH')
```

```
      READ(5,*)FCOST
      IF(INORFX.EQ.1)GO TO 351
C*******************************************
C*********    DISPLAY DATA SECTION    ****
C*******************************************
  200 WRITE(5,2000)
 2000 FORMAT(' TO PAUSE THE DISPLAY OF INPUT DATA, TYPE <CTL-S>'/
     1' TO CONTINUE THE DISPLAY, TYPE <CTL-Q>')
      WRITE(5,2001)UINP
 2001 FORMAT(/' INPUT DATA IN ',A3,' UNITS')
      WRITE(5,2002)TSIM,TCALC,IFRQ
 2002 FORMAT(' SIMULATION TIME = ',F8.2,' MINUTES'/
     1' CALCULATION INTERVAL = ',F8.2,' MINUTES'/
     2' FREQUENCY OF OUTPUT = ',I5,' INTERVALS')
      IF(SUNSIN.EQ.2)GO TO 202
      WRITE(5,2003)SOLMAX,UINSOL(1),UINSOL(2),SUNDAY,SUNRIS
 2003 FORMAT(' MAX. INSOLATION = ',F9.2,1X,2A7/
     1' LENGTH OF SOLAR DAY = ',F7.2,' MINUTES'/
     2' TIME OF SUNRISE=',F7.2,' MINUTES')
      GO TO 207
  202 IF(INPVI.EQ.1)GO TO 203
      IF(INPVI.EQ.2)GO TO 204
      IF(INPVI.EQ.3)GO TO 205
  203 IEND=INT(TSIM/TCALC)
      WRITE(5,2004)UINSOL(1),UINSOL(2)
 2004 FORMAT(' VARIABLE INSOLATION'/' TIME(MIN.)',3X,'INSOLATION',
     1'(',2A7,')'/)
      WRITE(5,2005)((I-1)*TCALC,INSOL(I),I=1,(IEND+1))
 2005 FORMAT(1000(1X,F7.2,6X,F9.2/))
      GO TO 207
  204 IEND=INT(1440./TCALC)
      WRITE(5,2006)UINSOL(1),UINSOL(2)
 2006 FORMAT(' TYPICAL DAY INSOLATION'/' TIME(MIN.)',3X,'INSOLATION',
     1'(',2A7,')'/)
      WRITE(5,2005)((I-1)*TCALC,INSOL(I),I=1,(IEND+1))
      GO TO 207
  205 IFLAG=0
      WRITE(5,2007)UINSOL(1),UINSOL(2)
 2007 FORMAT(' TYPICAL DAY INSOLATION PROFILE'/' TIME(HRS.)',3X,
     1'INSOLATION(',2A7,')'/)
  206 IFLAG=IFLAG+1
      WRITE(5,2005)SSPTM(IFLAG),SSPVAL(IFLAG)
      IF(SSPTM(IFLAG).LT.24.)GO TO 206
  207 WRITE(5,2008)QSREF,UINSOL(1),UINSOL(2),TAAVG,UTEMP,SWING,UTEMP
 2008 FORMAT(' QS,REF = ',F9.2,1X,2A7/' AVG. AMBIENT TEMP. = ',
     1F5.1,1X,A7/' DAILY TEMP. SWING = ',F7.1,1X,A7)
      WRITE(5,2009)TAUA,BCOLL,ACOLL,FSUBR,MDOTC,UMSFLW,SPHT,USPHT(1),
     1USPHT(2),AREAC,UAREA,
 2009 FORMAT(' TAU-ALPHA = ',F6.4/' VALUE OF B = ',F6.4/' VALUE OF A',
     1' =',E9.2/' VALUE OF F SUB-R=',F5.3/' COLL. MASS FLOW ='
     2,E9.2,1X,A9/' SPEC. HEAT OF FLUID = ',E9.2,1X,2A8/
     3' COLL.AREA = ',E9.2,1X,A6)
      IF(ITYPST.EQ.1)GO TO 208
      IF(ITYPST.EQ.2)GO TO 209
  208 WRITE(5,2010)CAPST,UCAPST,TSTOR,UTEMP
 2010 FORMAT(' LUMPED LIQUID STORAGE CAPACITY = ',E9.2,1X,A10/
```

```
          1' INITIAL STORAGE TEMP. = ',F6.2,1X,A7)
          GO TO 210
  209 WRITE(5,2011)CAPST,UCAPST,NODIV,(I,TSTRAT(I),I=1,NODIV)
 2011 FORMAT(' STRATIFIED LIQUID STORAGE CAPACITY = ',E9.2,1X,A10/
          1' NUMBER OF DIVISIONS = ',I2/' INITIAL STORAGE TEMPS (TOP=1)'/
          2(5X,'DIVISION #',I2,5X,F6.1))
  210 WRITE(5,2012)TSTMIN,UTEMP,TSTMAX,UTEMP,USTOR,UUVAL(1),UUVAL(2)
 2012 FORMAT(' MIN. USEABLE STORAGE TEMP. = ',F6.1,1X,A7/
          1' MAX. SAFE STORAGE TEMP. = ',F6.1,1X,A7/
          2' U-VALUE OF STORAGE TANK=',F6.4,1X,2A10)
          IF(ILDCC.EQ.2)GO TO 211
          WRITE(5,2013)PWRCC
 2013 FORMAT(' LOAD IS A CARNOT CYCLE TO DELIVER ',E9.2,' WATTS')
          GO TO 218
  211 IF(ILDDHW.EQ.2)GO TO 212
          WRITE(5,2014)TSTMIN,UTEMP,TSPLY,UTEMP
 2014 FORMAT(' LOAD IS DHW PROVIDED AT ',F6.1,1X,A7/
          1' CITY MAIN SUPPLY TEMP. = ',F6.1,1X,A7)
  212 IF(ILDCON.EQ.2)GO TO 213
          WRITE(5,2015)CONLD,ULOAD
 2015 FORMAT(' LOAD IS CONSTANT, LOAD = ',E9.2,1X,A7)
          GO TO 218
  213 IF(INPVLD.EQ.1)GO TO 214
          IF(INPVLD.EQ.2)GO TO 215
          IF(INPVLD.EQ.3)GO TO 216
  214 WRITE(5,2016)ULOAD,((I-1)*TCALC,LOAD(I),I=1,(INT(TSIM/TCALC)+1))
 2016 FORMAT(' VARIABLE LOAD'/' TIME(MIN.)',3X,'LOAD(',A7,')'/
          11000(1X,F7.2,6X,E12.5/))
          GO TO 218
  215 WRITE(5,2017)ULOAD,((I-1)*TCALC,LOAD(I),I=1,
          1(INT(1440./TCALC)+1))
 2017 FORMAT(' TYPICAL DAY LOAD'/' TIME(MIN.)',3X,'LOAD(',A7,
          1')'/1000(1X,F7.2,6X,E12.5/))
          GO TO 218
  216 I=0
          WRITE(5,2018)ULOAD
 2018 FORMAT(' TYPICAL DAY LOAD PROFILE'/' TIME(HRS.)',3X,'LOAD(',
          1A7,')')
  217 I=I+1
          WRITE(5,2019)LDPTM(I),LDPVAL(I)
 2019 FORMAT(1X,F7.2,6X,E12.5)
          IF(LDPTM(I).LT.24.)GO TO 217
  218 IF(ILDDHW.EQ.1)WRITE(5,1051)
          IF(ILDDHW.EQ.1)GO TO 219
          WRITE(5,2020)MDOTLD,UMSFLW
 2020 FORMAT(' LOAD MASS FLOW RATE = ',E9.2,1X,A9)
  219 IF(IECONO.EQ.2)GO TO 220
          WRITE(5,2021)SYSLIF,RATE,FRATE,FCOST
 2021 FORMAT(' EXPECTED SYSTEM LIFE = ',F5.2,' YEARS'/
          1' INTEREST RATE = ',F5.2/' FUEL ESC. RATE = ',F5.2/
          2' FUEL COST = $',F6.3,'/KWH')
  220 IF(INORFX.EQ.1)GO TO 351
  221 WRITE(5,2022)
 2022 FORMAT(' DO YOU WISH TO ALTER ANY OF THE INPUT DATA?',
          1' TYPE 1 FOR YES, 2 FOR NO')
          READ(5,*)IALTER
```

```
      IF(IALTER.EQ.1)GO TO 300
      IF(IALTER.EQ.2)GO TO 400
      GO TO 221
C****************************************
C*********     ALTER DATA SECTION      ****
C****************************************
  300 WRITE(5,3000)
      INORFX=1
 3000 FORMAT(' TO ALTER THE INPUT DATA, TYPE THE NUMBER OF THE ITEM'/
     1' YOU WISH  TO CHANGE (1-33)'/' TO REDISPLAY THE INPUT DATA,',
     2* TYPE 34*/* WHEN YOU WISH TO RUN THE SIMULATION, TYPE 35*/
     3'  1) INPUT DATA UNITS'/4X,'NOTE: EXISTING DATA CONVERTED',
     3' INTERNALLY'/'  2) OUTPUT DATA UNITS'/
     4'  3) SIMULATION TIME'/'  4) CALCULATION INTERVAL'/
     5'  5) FREQUENCY OF OUTPUT'/'  6) OPTION FOR SINUSOIDAL',
     6' INSOLATION'/'  7) OPTION FOR VARIABLE INSOLATION'/
     7'  8) QS,REF'/'  9) DAILY MEAN TEMP.'/ 10) DAILY TEMP. ',
     8*SWING*/* 11) TAU-ALPHA*/* 12) VALUE OF B*/* 13) VALUE OF*,
     8* F SUB-R*/* 14) VALUE OF*,
     9' A'/' 15) COLLECTOR MASS FLOW RATE'/' 16) SPECIFIC HEAT ',
     1'OF FLUID'/' 17) COLLECTOR AREA'/' 18) HEAT CAPACITY OF ',
     2'STORAGE'/' 19) U-VALUE OF STORAGE TANK'/' 20) LENGTH-DIAM',
     3'ETER OF STORAGE TANK'/' 21) TYPE OF STORAGE'/' 22) NUMBER',
     4' OF DIVISIONS IN STRATIFIED STORAGE'/' 23) INITIAL STRATI',
     5'FIED STORAGE TEMPS.'/' 24) INITIAL LUMPED STORAGE TEMP.'/
     6' 25) MIN. USEABLE STORAGE TEMP.'/' 26) MAX. SAFE STORAGE ',
     7'TEMP.'/' 27) OPTION FOR CARNOT CYCLE'/' 28) OPTION FOR DOM',
     8'ESTIC HOT WATER SYSTEM'/' 29) OPTION FOR CONSTANT LOAD'/
     9' 30) OPTION FOR VARIABLE LOAD'/' 31) LOAD MASS FLOW RATE'/
     1' 32) OPTION FOR ECONOMIC ANALYSIS'/' 33) OPTION FOR BASE ',
     2'CASE'/' 34) DISPLAY INPUT DATA'/' 35) RUN SIMULATION'/
     3' NOTE: IF YOU WISH TO CHANGE ANY OF 1-4, DO SO FIRST')
  351 WRITE(5,3001)
 3001 FORMAT(' ITEM NUMBER?')
      READ(5,*)ITEM
      IF((ITEM.GE.1).OR.(ITEM.LE.35))GO TO 352
      WRITE(5,3002)
 3002 FORMAT(' ENTER AN INTEGER BETWEEN 1 AND 35')
      GO TO 351
  352 GO TO(102,105,108,109,111,113,115,123,124,125,126,127,159,128,
     *130,131,
      1132,133,134,135,136,137,138,139,140,141,142,144,147,148,156,157,10
      11,200,400),ITEM
C*******************************************************
C*********     INITIALIZE SYSTEM PARAMETERS    ****
C*******************************************************
C**************WRITE INPUT LIST ON OUTPUT FILE********************
  400 WRITE(6,2001)UINP
      WRITE(6,2002)TSIM,TCALC,IFRQ
      IF(SUNSIN.EQ.2)GO TO 602
      WRITE(6,2003)SOLMAX,UINSOL(1),UINSOL(2),SUNDAY,SUNRIS
      GO TO 607
  602 IF(INPVI.EQ.1)GO TO 603
      IF(INPVI.EQ.2)GO TO 604
      IF(INPVI.EQ.3)GO TO 605
```

```
    603 IEND=INT(TSIM/TCALC)
        WRITE(6,2004)UINSOL(1),UINSOL(2)
        WRITE(6,2005)((I-1)*TCALC,INSOL(I),I=1,(IEND+1))
        GO TO 607
    604 IEND =INT(1440./TCALC)
        WRITE(6,2006)UINSOL(1),UINSOL(2)
        WRITE(6,2005)((I-1)*TCALC,INSOL(I),I=1,(IEND+1))
        GO TO 607
    605 IFLAG=0
        WRITE(6,2007)UINSOL(1),UINSOL(2)
    606 IFLAG=IFLAG+1
        WRITE(6,2005)SSPTM(IFLAG),SSPVAL(IFLAG)
        IF(SSPTM(IFLAG).LT.24.)GO TO 606
    607 WRITE(6,2008)QSREF,UINSOL(1),UINSOL(2),TAAVG,UTEMP,SWING,UTEMP
        WRITE(6,2009)TAUA,BCOLL,ACOLL,FSUBR,MDOTC,UMSFLW,SPHT,USPHT(1),
       1USPHT(2),AREAC,UAREA
        IF(ITYPST.EQ.1)GO TO 608
        IF(ITYPST.EQ.2)GO TO 609
    608 WRITE(6,2010)CAPST,UCAPST,TSTOR,UTEMP
        GO TO 610
    609 WRITE(6,2011)CAPST,UCAPST,NODIV,(I,TSTRAT(I),I=1,NODIV)
    610 WRITE(6,2012)TSTMIN,UTEMP,TSTMAX,UTEMP,USTOR,UUVAL(1),UUVAL(2)
        IF(ILDCC.EQ.2)GO TO 611
        WRITE(6,2013)PWR
        WRITE(6,2013)PWRCC
        GO TO 618
    611 IF(ILDDHW.EQ.2)GO TO 612
        WRITE(6,2014)TSTMIN,UTEMP,TSPLY,UTEMP
    612 IF(ILDCON.EQ.2)GO TO 613
        WRITE(6,2015)CONLD,ULOAD
        GO TO 618
    613 IF(INPVLD.EQ.1)GO TO 614
        IF(INPVLD.EQ.2)GO TO 615
        IF(INPVLD.EQ.3)GO TO 616
    614 WRITE(6,2016)ULOAD,((I-1)*TCALC,LOAD(I),I=1,(INT(TSIM/TCALC)+1))
        GO TO 618
    615 WRITE(6,2017)ULOAD,((I-1)*TCALC,LOAD(I),I=1,
       1(INT(1440./TCALC)+1))
        GO TO 618
    616 I=0
        WRITE(6,2018)ULOAD
    617 I=I+1
        WRITE(6,2019)LDPTM(I),LDPVAL(I)
        IF(LDPTM(I).LT.24.)GO TO 617
    618 IF(ILDDHW.EQ.1)WRITE(6,1051)
        IF(ILDDHW.EQ.1)GO TO 619
        WRITE(6,2020)MDOTLD,UMSFLW
    619 IF(IECONO.EQ.2)GO TO 620
        WRITE(6,2021)SYSLIF,RATE,FRATE,FCOST
    620 IF(ISTAT.EQ.2)CALL IETOSI
        IF(SUNSIN.EQ.1)CALL SINSOL(TSIM,TCALC,SOLMAX,SUNDAY,INSOL,
       1SUNRIS)
        IF(SUNSIN.EQ.1)GO TO 401
        GO TO (401,402,403)INPVI
    403 CALL PRODAY(TCALC,SSPTM,SSPVAL,INSOL)
    402 CALL DAYSIM(TSIM,TCALC,INSOL)
```

```
    401 IF((ILDCC.EQ.1).AND.(ITYPST.EQ.1))GO TO 414
        IF((ILDCC.EQ.1).AND.(ITYPST.EQ.2))GO TO 410
        IF(ILDCON.EQ.2)GO TO 404
        DO 405 IFLAG=1,1+INT(TSIM/TCALC)
        LOAD(IFLAG)=CONLD
    405 CONTINUE
        GO TO 408
    404 GO TO (408,407,406)INPVLD
    406 CALL PRODAY(TCALC,LDPTM,LDPVAL,LOAD)
    407 CALL DAYSIM(TSIM,TCALC,LOAD)
    408 IF(ITYPST.EQ.1)GO TO 414
        IF(ILDDHW.EQ.2)GO TO 410
        BIGLD=LOAD(1)
        DO 409 IFLAG=2,INT(TSIM/TCALC)
        IF(BIGLD.GE.LOAD(IFLAG))GO TO 409
        BIGLD=LOAD(IFLAG)
    409 CONTINUE
        IF((TSTMIN-TSPLY).LE.1.) GO TO 480
        FLWMAX=BIGLD/(SPHT*(TSTMIN-TSPLY))
        IF(FLWMAX.GE.0.) GO TO 411
        FLWMAX=0.001
        GO TO 411
    410 FLWMAX=MDOTLD
        IF(FLWMAX.GE.0.) GO TO 411
        FLWMAX=0.001
        GO TO 411
    480 FLWMAX=500.
    411 DELTA1=.1*CAPST/(MDOTC*SPHT*NODIV)
        DELTA2=.1*CAPST/(FLWMAX*SPHT*NODIV)
        DELTA=AMIN1(DELTA1,DELTA2)
        IF(DELTA.LT.TCALC)DELTA=TCALC/(INT(TCALC/DELTA)+1)
        IF(DELTA.GE.TCALC)DELTA=TCALC
    414 ISWTCL=0
        ISWTLD=0
        IF(ITYPST.EQ.1)TLDIN=TSTOR
        IF(ITYPST.EQ.2)TLDIN=TSTRAT(1)
        CALL TAMB1(TSIM,TCALC,TAAVG,SWING,TAMB,SUNRIS)
        IF(ILDCC.EQ.1)LOAD(1)=(60.*PWRCC)/(1.-TAMB(1)/TLDIN)
        G9=(MDOTC*SPHT*TAAVG)/(QSREF*AREAC*60.)
        G6=ACOLL/(BCOLL*TAAVG**3)
        VSTOR=CAPST/(RHO*SPHT)
        LSTOR=(4.*VSTOR*RLNDIA**2/PI)**.333333
        DSTOR=LSTOR/RLNDIA
        ASTOR=PI*DSTOR*(DSTOR/2.+LSTOR)
        UTOT=USTOR*ASTOR*60.
        REQLD=0.
        SUPLD=0.
        LDPMP=0
        OUTPUT=-1
        ECOLL=0.
        EAVAIL=0.
   C********    WRITE HEADINGS TO FILE OUTPUT     ****
        IF(ITYPST.EQ.2)GO TO 416
        WRITE(6,4000)
   4000 FORMAT(/ ' TIME',9X,'INSOLATION',3X,'COLLECTOR',4X,'COLLECTOR',
       14X,'COLLECTOR',4X,'NET STORED',3X,'STORAGE',6X,'LOAD',9X,
```

```
     2'LOAD',9X,'LOAD')
      IF(OUNIT.EQ.2)GO TO 417
      WRITE(6,4001)
 4001 FORMAT(' (MIN.)',7X,'(W/SQ.M.)',4X,'OUTLET (K)',4X,'EFFICIENCY',
     13X,'PUMP',9X,'(J/MIN.)',5X,'TEMP. (K)',4X,'(J/MIN.)',5X,
     2'OUTLET (K)',3X,'PUMP'/)
      GO TO 450
  417 WRITE(6,4002)
 4002 FORMAT(' (MIN.)',7X,'(BTU/HR-SQ.FT)',2X,'OUTLET (R)',2X,
     1'EFFICIENCY',3X,'PUMP',9X,'(BTU/HR.)',4X,'TEMP. (R)',4X,
     2'(BTU/HR.)',4X,'OUTLET (R)',3X,'PUMP'/)
      GO TO 450
  416 WRITE(6,4003)
 4003 FORMAT(/ ' TIME',9X,'INSOLATION',3X,'COLLECTOR',4X,'COLLECTOR',
     14X,'COLLECTOR',4X,'NET STORED',3X,'LOAD',9X,'LOAD',9X,'LOAD')
      IF(OUNIT.EQ.2)GO TO 418
      WRITE(6,4004)
 4004 FORMAT(' (MIN.)',7X,'(W/SQ.M.)',4X,'TEMP. (K)',4X,'EFFICIENCY',
     13X,'PUMP',9X,'(J/MIN.)',5X,'(J/MIN.)',5X,'OUTLET (K)',3X,'PUMP'/)
      GO TO 450
  418 WRITE(6,4005)
 4005 FORMAT(' (MIN.)',7X,'(BTU/HR-SQ.FT)',2X,'TEMP. (R)',2X,
     1'EFFICIENCY',3X,'PUMP',9X,'(BTU/HR.)',4X,'(BTU/HR.)',4X,
     2'OUTLET (R)',3X,'PUMP'/)
C********       WRITE HEADINGS TO TTY      ****
  450 WRITE(5,4050)
 4050 FORMAT(/ 'TIME',9X,'INSOLATION',3X,'COLLECTOR',4X,'COLLECTOR',
     14X,'COLLECTOR')
      IF(OUNIT.EQ.2)GO TO 452
      WRITE(5,4051)
 4051 FORMAT(' (MIN.)',7X,'(W/SQ.M.)',4X,'TEMP. (K)',4X,'EFFICIENCY',
     13X,'PUMP'/)
      GO TO 453
  452 WRITE(5,4052)
 4052 FORMAT(' (MIN.)',7X,'(BTU/HR-SQFT)',2X,'TEMP. (R)',2X,'EFFICIENCY',
     13X,'PUMP'/)
  453 IF(ITYPST.EQ.2)GO TO 4
   53 IF(ITYPST.EQ.2)GO TO 451
      WRITE(5,4053)
 4053 FORMAT(' NET STORED',3X,'STORAGE',6X,3('LOAD',9X))
      IF(OUNIT.EQ.2)GO TO 454
      WRITE(5,4054)
 4054 FORMAT(' (J/MIN.)',5X,'TEMP. (K)',4X,'(J/MIN.)',5X,'OUTLET (K)',
     13X,'PUMP'/)
      GO TO 500
  454 WRITE(5,4055)
 4055 FORMAT(' (BTU/HR.)',4X,'TEMP. (R)',4X,'(BTU/HR.)',4X,'OUTLET (R)',
     13X,'PUMP'/)
      GO TO 500
  451 WRITE(5,4056)
 4056 FORMAT(' NET STORED',3X,3('LOAD',9X))
      IF(OUNIT.EQ.2)GO TO 455
      WRITE(5,4057)
 4057 FORMAT(' (J/MIN.)',5X,'(J/MIN.)',5X,'OUTLET (K)',3X,'PUMP')
      GO TO 500
  455 WRITE(5,4058)
```

```
      4058 FORMAT(' (BTU/HR.)',4X,'(BTU/HR.)',4X,'OUTLET (R)',3X,'PUMP')
C****************************************************
C*********     MAIN CALCULATION SECTION     ****
C****************************************************
   500 DO 599 IFLAG=1,1+INT(TSIM/TCALC)
C*********     SET VARIABLES FOR CALCULATIONS     ****
       JSWTCL=ISWTCL
       JSWTLD=ISWTLD
       ESTNET=0.
       IF(ISWTLD¬EQ.
       IF(ISWTLD.EQ.1)LDPMP=LDPMP+1
       OUTPUT=OUTPUT+1
       REQLD=REQLD+LOAD(IFLAG)*TCALC
       IF((TLDIN.GE.TSTMIN).AND.(ILDDHW.EQ.1))SUPLD=LOAD(IFLAG)*
      1TCALC*ISWTLD+SUPLD
       IF((TLDIN.LT.TSTMIN).AND.(ILDDHW.EQ.1))SUPLD=LOAD(IFLAG)*
      1TCALC*ISWTLD*((TLDIN-TSPLY)/(TSTMIN-TSPLY))+SUPLD
       IF(ILDDHW.EQ.2)SUPLD=SUPLD+ISWTLD*LOAD(IFLAG)*TCALC
       IF(ITYPST.EQ.1)GO TO 501
       TIME=TCALC*(IFLAG-1)
   512 TCIN=TSTRAT(NODIV)
       TLDIN=TSTRAT(1)
       IF(((ILDDHW.EQ.1).AND.(TLDIN.GE.TSTMIN))MDOTLD=LOAD(IFLAG)/
      1(SPHT*(TLDIN-TSPLY))
       IF(((ILDDHW.EQ.1).AND.(TLDIN.LT.TSTMIN))MDOTLD=LOAD(IFLAG)/
      1(SPHT*(TSTMIN-TSPLY))
       IF(MDOTLD.LE.0.)MDOTLD=0.
       FRCLST=MDOTC*SPHT*DELTA*ISWTCL/(CAPST/NODIV)
       FRLDST=MDOTLD*SPHT*DELTA*ISWTLD/(CAPST/NODIV)
       GO TO 502
   501 TCIN=TLDIN=TSTOR
       IF((ILDDHW.EQ.1).AND.(TLDIN.GE.TSTMIN))MDOTLD=LOAD(IFLAG)/
      1(SPHT*(TLDIN-TSPLY))
       IF((ILDDHW.EQ.1).AND.(TLDIN.LT.TSTMIN))MDOTLD=LOAD(IFLAG)/
      1(SPHT*(TSTMIN-TSPLY))
       IF(MDOTLD.LE.0.)MDOTLD=0.
C*********     COLLECTOR CALCULATIONS     ****
   502 IF(ISWTCL.EQ.0)GO TO 503
       IF(INSOL(IFLAG).LE.0.)GO TO 503
       D7=(ACOLL/TAAVG**4)*(TCIN**4-TAMB(IFLAG)**4)
       TCOUT=TCIN*(1.-BCOLL*FSUBR/G9)+(TAAVG*FSUBR/G9)*(TAUA*INSOL(IFLAG)
      1/QSREF+BCOLL*TAMB(IFLAG)/TAAVG)-D7*TAAVG/G9
       GO TO 504
   503 D8=G6*(TCIN**4-TAMB(IFLAG)**4)
       TCOUT=TAMB(IFLAG)+TAUA*TAAVG*INSOL(IFLAG)/(BCOLL*QSREF)-D8
   504 IF(TCOUT.LT.TAMB(IFLAG))TCOUT=TAMB(IFLAG)
C*********     LOAD OUTLET CALCULATION     ****
       IF(ILDDHW.EQ.1)TLDOUT=TSPLY
       IF(ILDDHW.EQ.2)TLDOUT=TLDIN-(ISWTLD*LOAD(IFLAG))/(MDOTLD*SPHT)
C*********     STORAGE CALCULATIONS     ****
       IF(ITYPST.EQ.2)GO TO 513
C*********     LUMPED STORAGE     *****
       ESTNET=(ISWTCL*MDOTC*SPHT*(TCOUT-TCIN))-(ISWTLD*MDOTLD*SPHT*
      1(TLDIN-TLDOUT))-UTOT*(TSTOR-TAMB(IFLAG))
       TSTOR=TSTOR+ESTNET*TCALC/CAPST
       GO TO 514
```

```
C*********       STRATIFIED STORAGE        ****
   513 DO 505 M=1,NODIV
       IF(TCOUT.GE.TSTRAT(M))GO TO 506
       CLDELT(M)=0.
   505 CONTINUE
   506 CLDELT(M)=TCOUT-TSTRAT(M)
       DO 507 N=M+1,NODIV
       CLDELT(N)=TSTRAT(N-1)-TSTRAT(N)
   507 CONTINUE
       DO 508 M=1,NODIV
       MM=NODIV+1-M
       IF(TLDOUT.LE.TSTRAT(MM))GO TO 509
       LDDELT(MM)=0.
   508 CONTINUE
   509 LDDELT(MM)=TLDOUT-TSTRAT(MM)
       DO 510 M=1,MM-1
       LDDELT(M)=TSTRAT(M+1)-TSTRAT(M)
   510 CONTINUE
       DO 511 M=1,NODIV
       TSTTMP(M)=TSTRAT(M)+FRCLST*CLDELT(M)+FRLDST*LDDELT(M)-
      1(UTOT/CAPST)*(TSTRAT(M)-TAMB(IFLAG))*DELTA
       ESTNET=ESTNET+CAPST/NODIV*(TSTTMP(M)-TSTRAT(M))
       TSTRAT(M)=TSTTMP(M)
   511 CONTINUE
       TIME=TIME+DELTA
C*********    REPEAT STRATIFIED CALCULATIONS TO FILL INTERVAL    ****
       IF(TIME.LT.(TCALC*IFLAG))GO TO 512
       ESTNET=ESTNET/TCALC
C*********       COLLECTOR AND LOAD PUMP SWITCHES    ****
   514 IF(((ISWTCL.EQ.1).AND.((TCOUT.GE.TSTMAX).OR.((TCOUT-TCIN)
      1.LT.2.))).OR.((ISWTCL.EQ.0).AND.((TCOUT-TCIN).LT.6.)))ISWTCL=0
       IF(((ISWTCL.EQ.1).AND.(TCOUT.LT.TSTMAX).AND.((TCOUT-TCIN)
      1.GE.2.)).OR.((ISWTCL.EQ.0).AND.((TCOUT-TCIN).GE.6.)))
      2ISWTCL=1
       IF((LOAD(IFLAG).LE.0.).OR.(TLDIN.LE.TSTMIN))ISWTLD=0
       IF((LOAD(IFLAG).GT.0.).AND.(TLDIN.GT.TSTMIN))ISWTLD=1
       IF((ILDDHW.EQ.1).AND.(TLDIN.GT.TSPLY)) ISWTLD=1
       IF(ILDCC.EQ.1)LOAD(IFLAG)=(60.*PWRCC)/(1.+TAMB(IFLAG)/TLDIN)
C*********    ADD UP COLLECTED ENERGY AND AVAILABLE ENERGY    ****
       ECOLL=ECOLL+JSWTCL*MDOTC*SPHT*(TCOUT-TCIN)*TCALC
       EAVAIL=EAVAIL+INSOL(IFLAG)*AREAC*60.*TCALC
       IF(OUTPUT.EQ.IFRQ)GO TO 515
       GO TO 599
C*********       OUTPUT SECTION    ****
   515 OUTPUT=0
       IF((JSWTCL.EQ.0).OR.(INSOL(IFLAG).LE.0.))ETAC=0.
       IF((JSWTCL.EQ.1).AND.(INSOL(IFLAG).GT.0.))ETAC=(MDOTC*SPHT*
      1(TCOUT-TCIN))/(INSOL(IFLAG)*AREAC*60.)*100.
       IF(JSWTCL.EQ.0)UCLPMP=3HOFF
       IF(JSWTCL.EQ.1)UCLPMP=2HON
       IF(JSWTLD.EQ.0)ULDPMP=3HOFF
       IF(JSWTLD.EQ.1)ULDPMP=2HON
       IF(OUNIT.EQ.1)GO TO 530
       INSOL(IFLAG)=INSOL(IFLAG)/(3.15393)
       TCOUT=TCOUT/(5./9.)
       ESTNET=ESTNET/(17.576)
```

```
            IF(ITYPST.EQ.2)GO TO 531
            TSTOR=TSTOR/(5./9.)
            GO TO 532
        531 DO 8 M=1,NODIV
            TSTRAT(M)=TSTRAT(M)/(5./9.)
          8 CONTINUE
        532 LOAD(IFLAG)=LOAD(IFLAG)/(17.576)
            TLDOUT=TLDOUT/(5./9.)
            IF(ILDDHW.EQ.1)MDOTLD=MDOTLD/(.0631313)
            UMSFLW=3HGPH
        530 IF(ITYPST.EQ.2)GO TO 516
            WRITE(6,5000)TCALC*(IFLAG-1),INSOL(IFLAG),TCOUT,ETAC,UCLPMP,
           1ESTNET,TSTOR,LOAD(IFLAG),TLDOUT,ULDPMP
       5000 FORMAT(' ',F8.1,5X,F8.1,5X,F7.1,6X,F7.1,6X,A3,10X,E9.2,2X,
           1F7.1,6X,E9.2,5X,F7.1,6X,A3)
            GO TO 517
        516 WRITE(6,5001)TCALC*(IFLAG-1),INSOL(IFLAG),TCOUT,ETAC,UCLPMP,
           1ESTNET,LOAD(IFLAG),TLDOUT,ULDPMP
       5001 FORMAT(/ ' ',F8.1,5X,F8.1,5X,F7.1,6X,F7.1,6X,A3,10X,E9.2,2X,
           1E9.2,5X,F7.1,6X,A3)
            WRITE(6,5002)(TSTRAT(M),M=1,NODIV)
       5002 FORMAT(' STRATIFIED STORAGE TEMPERATURES (TOP TO BOTTOM)'/
           1(3X,F7.1))
        517 IF(ILDDHW.EQ.1)WRITE(6,5003)MDOTLD,UMSFLW
       5003 FORMAT(' HOT WATER FLOW RATE FROM STORAGE IS ',F6.2,1X,A9)
            IF(ILDCC.EQ.2)GO TO 518
            CCEFF=(1.-TAMB(IFLAG)/TLDIN)*100.
            WRITE(6,5004)CCEFF
       5004 FORMAT(' CARNOT EFFICIENCY IS ',F7.1,' %')
        518 IF(ITYPST.EQ.2)GO TO 519
            WRITE(5,5005)TCALC*(IFLAG-1),INSOL(IFLAG),TCOUT,ETAC,UCLPMP,
           1ESTNET,TSTOR,LOAD(IFLAG),TLDOUT,ULDPMP
       5005 FORMAT(1X,F8.1,5X,F8.1,5X,F7.1,6X,F8.3,6X,A3/1X,E9.2,2X,
           1F7.1,6X,E9.2,2X,F9.1,8X,A3)
            GO TO 520
        519 WRITE(5,5006)TCALC*(IFLAG-1),INSOL(IFLAG),TCOUT,ETAC,UCLPMP,
           1ESTNET,LOAD(IFLAG),TLDOUT,ULDPMP
       5006 FORMAT(/ ' ',F8.1,5X,F7.1,5X,F7.1,6X,F8.3,6X,A3/1X,E9.2,2X,
           1E9.2,2X,F9.1,8X,A3)
            WRITE(5,5002)(TSTRAT(M),M=1,NODIV)
        520 IF(ILDDHW.EQ.1)WRITE(5,5003)MDOTLD,UMSFLW
            IF(ILDCC.EQ.2)GO TO 521
            WRITE(5,5004)CCEFF
        521 WRITE(5,5007)
       5007 FORMAT(' ')
            IF(OUNIT.EQ.1)GO TO 599
            INSOL(IFLAG)=INSOL(IFLAG)*(3.15393)
            TCOUT=TCOUT*(5./9.)
            ESTNET=ESTNET*(17.576)
            IF(ITYPST.EQ.2)GO TO 541
            TSTOR=TSTOR*(5./9.)
            GO TO 542
        541 DO 9 M=1,NODIV
            TSTRAT(M)=TSTRAT(M)*(5./9.)
          9 CONTINUE
        542 LOAD(IFLAG)=LOAD(IFLAG)*(17.576)
```

```
      TLDOUT=TLDOUT*(5./9.)
      IF(ILDDHW.EQ.1)MDOTLD=MDOTLD*(.0631313)
      UMSFLW=8HKG./MIN.
  599 CONTINUE
C**************************************************
C*********     OUTPUT SUMMARY OF RESULTS     ****
C**************************************************
      IF(REQLD.LE.0.001)PERLD=0.
      IF(REQLD.GT.0.001)PERLD=(SUPLD/REQLD)*100.
      WRITE(5,5008)PERLD
      WRITE(6,5008)PERLD
 5008 FORMAT(/ ' PERCENTAGE OF LOAD SUPPLIED BY SOLAR IS ',F5.1,' %')
      PMPLD=(FLOAT(LDPMP)/FLOAT(INT(TSIM/TCALC)))*100.
      WRITE(5,5009)PMPLD
      WRITE(6,5009)PMPLD
 5009 FORMAT('   LOAD PUMP WAS ON ',F5.1,' % OF THE TIME')
      SYSEFF=(ECOLL/EAVAIL)*100.
      WRITE(5,5010)SYSEFF
      WRITE(6,5010)SYSEFF
 5010 FORMAT('   SYSTEM COLLECTION EFFICIENCY IS ',F5.1,' %')
      IF(IECONO.EQ.2)GO TO 600
      Q6=(.146*REQLD)/TSIM
      R7=(RATE-FRATE)/(1.+FRATE)
      R4=(1+R7)**SYSLIF
      F1=((PERLD*Q6)/100.)*FCOST
      SYSCST=F1*(R4-1.)/(R7*R4)
      WRITE(5,5011)SYSCST
      WRITE(6,5011)SYSCST
 5011 FORMAT('   ALLOWABLE SYSTEM CAPITAL COST IS $',F9.2)
C**********************************************************
C*********     CHOICE OF CONTINUATIONS SECTION     ****
C**********************************************************
  600 IF(IUNIT.EQ.2)CALL ISITOE
  601 WRITE(5,6000)
 6000 FORMAT(' ENTER THE NUMBER OF WHAT YOU WANT TO DO'/
     1'1) RUN A DIFFERENT SIMULATION'/'2) RUN A SIMILAR ',
     2'SIMULATION'/'3) STOP THE PROGRAM'/'NOTE: IF YOU CHOOSE 2),'
     3,'THE PROGRAM WILL RESTART WITH',
     4' THE'/'STORAGE TEMPERATURES FROM THE END OF THE LAST'/'RUN '
     5,'UNLESS YOU INPUT OTHER VALUES')
      READ(5,*)NEXT
      IF(NEXT.EQ.1)GO TO 171
      IF(NEXT.EQ.2)GO TO 300
      IF(NEXT.EQ.3)GO TO 702
      GO TO 601
  702 STOP
      END
      SUBROUTINE DAYSIM(TSIM,TCALC,ARRAY)
      DIMENSION ARRAY(1000)
      NCALC=INT(1440./TCALC)
      ISUB1=0
      DO 1 IFLAG=1,1+INT(TSIM/TCALC)
      ISUB1=ISUB1+1
      IF(ISUB1.GT.NCALC)ISUB1=1
      ARRAY(IFLAG)=ARRAY(ISUB1)
    1 CONTINUE
```

```
      RETURN
      END
      SUBROUTINE IETOSI
      COMMON//TSIM,TCALC,SUNSIN,SOLMAX,INPVI,INSOL(1000),QSREF,TAAVG,
     1SWING,MDOTC,SPHT,AREAC,CAPST,USTOR,ITYPST,NODIV,TSTRAT(10),TSTOR,
     2TSTMIN,TSTMAX,ILDCC,ILDDHW,ILDCON,CONLD,INPVLD,LOAD(1000),MDOTLD,
     3UAREA,UCAPST,UINP,UINSOL(2),ULOAD,UMSFLW,USPHT(2),UTEMP,UUVAL(2),
     4LDPTM(100),LDPVAL(100),SSPTM(100),SSPVAL(100),ISTAT,RHO,TSPLY
      REAL INSOL,LOAD,LDPTM,LDPVAL,MDOTC,MDOTLD
      INTEGER SUNSIN,UAREA,UCAPST,UINP,UINSOL,ULOAD,UMSFLW,USPHT,UTEMP,
     1UUVAL
      IF(SUNSIN.EQ.2)GO TO 1
      SOLMAX=SOLMAX*(3.15393)
      GO TO 7
    1 IF(INPVI.EQ.2)GO TO 3
      IF(INPVI.EQ.3)GO TO 5
      DO 2 IFLAG=1,1+INT(TSIM/TCALC)
      INSOL(IFLAG)=INSOL(IFLAG)*(3.15393)
    2 CONTINUE
      GO TO 7
    3 DO 4 IFLAG=1,1+INT(1440./TCALC)
      INSOL(IFLAG)=INSOL(IFLAG)*(3.15393)
    4 CONTINUE
      GO TO 7
    5 IFLAG=0
    6 IFLAG=IFLAG+1
      SSPVAL(IFLAG)=SSPVAL(IFLAG)*(3.15393)
      IF(SSPTM(IFLAG).LT.24.)GO TO 6
    7 IF(ILDDHW.EQ.2)GO TO 8
      TSPLY=TSPLY*(5./9.)
    8 QSREF=QSREF*(3.15393)
      RHO=RHO*(77.8597)
      TAAVG=TAAVG*(5./9.)
      SWING=SWING*(5./9.)
      MDOTC=MDOTC*(.45359)
      SPHT=SPHT*(4185.8)
      AREAC=AREAC*(.0929)
      CAPST=CAPST*(1898.64)
      USTOR=USTOR*(5.677)
      IF(ITYPST.EQ.2)GO TO 9
      TSTOR=TSTOR*(5./9.)
      GO TO 11
    9 DO 10 IFLAG=1,NODIV
      TSTRAT(IFLAG)=TSTRAT(IFLAG)*(5./9.)
   10 CONTINUE
   11 TSTMIN=TSTMIN*(5./9.)
      TSTMAX=TSTMAX*(5./9.)
      IF(ILDCC.EQ.1)GO TO 18
      IF(ILDCON.EQ.2)GO TO 12
      CONLD=CONLD*(17.576)
      GO TO 18
   12 IF(INPVLD.EQ.2)GO TO 14
      IF(INPVLD.EQ.3)GO TO 16
      DO 13 IFLAG=1,1+INT(TSIM/TCALC)
      LOAD(IFLAG)=LOAD(IFLAG)*(17.576)
   13 CONTINUE
```

```
      GO TO 18
14 DO 15 IFLAG=1,1+INT(1440./TCALC)
   LOAD(IFLAG)=LOAD(IFLAG)*(17.576)
15 CONTINUE
   GO TO 18
16 IFLAG=0
17 IFLAG=IFLAG+1
   LDPVAL(IFLAG)=LDPVAL(IFLAG)*(17.576)
   IF(LDPTM(IFLAG).LT.24.)GO TO 17
18 IF(ILDDHW.EQ.1)GO TO 19
   MDOTLD=MDOTLD*(.45359)
19 ISTAT=1
   UINP=2HSI
   UINSOL(1)=7HW/SQ.M.
   UINSOL(2)=1H
   UTEMP=7HKELVINS
   UMSFLW=8HKG./MIN.
   USPHT(1)=8HJ/KG.-DE
   USPHT(2)=4HG.C.
   UAREA=5HSQ.M.
   UCAPST=8HJ/DEG.C.
   UUVAL(1)=10HW/SQ.M.-DE
   UUVAL(2)=4HG.C.
   ULOAD=6HJ/MIN.
   RETURN
   END
   SUBROUTINE ISITOE
   COMMON//TSIM,TCALC,SUNSIN,SOLMAX,INPVI,INSOL(1000),QSREF,TAAVG,
  1SWING,MDOTC,SPHT,AREAC,CAPST,USTOR,ITYPST,NODIV,TSTRAT(10),TSTOR,
  2TSTMIN,TSTMAX,ILDCC,ILDDHW,ILDCON,CONLD,INPVLD,LOAD(1000),MDOTLD,
  3UAREA,UCAPST,UINP,UINSOL(2),ULOAD,UMSFLW,USPHT(2),UTEMP,UUVAL(2),
  4LDPTM(100),LDPVAL(100),SSPTM(100),SSPVAL(100),ISTAT,RHO,TSPLY
   REAL INSOL,LOAD,LDPTM,LDPVAL,MDOTC,MDOTLD
   INTEGER SUNSIN,UAREA,UCAPST,UINP,UINSOL,ULOAD,UMSFLW,USPHT,UTEMP,
  1UUVAL
   IF(SUNSIN.EQ.2)GO TO 1
   SOLMAX=SOLMAX/(3.15393)
   GO TO 7
 1 IF(INPVI.EQ.2)GO TO 3
   IF(INPVI.EQ.3)GO TO 5
   DO 2 IFLAG=1,1+INT(TSIM/TCALC)
   INSOL(IFLAG)=INSOL(IFLAG)/(3.15393)
 2 CONTINUE
   GO TO 7
 3 DO 4 IFLAG=1,1+INT(1440./TCALC)
   INSOL(IFLAG)=INSOL(IFLAG)/(3.15393)
 4 CONTINUE
   GO TO 7
 5 IFLAG=0
 6 IFLAG=IFLAG+1
   SSPVAL(IFLAG)=SSPVAL(IFLAG)/(3.15393)
   IF(SSPTM(IFLAG).LT.24.)GO TO 6
 7 IF(ILDDHW.EQ.2)GO TO 8
   TSPLY=TSPLY/(5./9.)
 8 QSREF=QSREF/(3.15393)
   RHO=RHO/(77.8597)
```

```
      TAAVG=TAAVG/(5./9.)
      SWING=SWING/(5./9.)
      MDOTC=MDOTC/(.45359)
      SPHT=SPHT/(4185.8)
      AREAC=AREAC/(.0929)
      CAPST=CAPST/(1898.64)
      USTOR=USTOR/(5.677)
      IF(ITYPST.EQ.2)GO TO 9
      TSTOR=TSTOR/(5./9.)
      GO TO 11
    9 DO 10 IFLAG=1,NODIV
      TSTRAT(IFLAG)=TSTRAT(IFLAG)/(5./9.)
   10 CONTINUE
   11 TSTMIN=TSTMIN/(5./9.)
      TSTMAX=TSTMAX/(5./9.)
      IF(ILDCC.EQ.1)GO TO 18
      IF(ILDCON.EQ.2)GO TO 12
      CONLD=CONLD/(17.576)
      GO TO 18
   12 IF(INPVLD.EQ.2)GO TO 14
      IF(INPVLD.EQ.3)GO TO 16
      DO 13 IFLAG=1,1+INT(TSIM/TCALC)
      LOAD(IFLAG)=LOAD(IFLAG)/(17.576)
   13 CONTINUE
      GO TO 18
   14 DO 15 IFLAG=1,1+INT(1440./TCALC)
      LOAD(IFLAG)=LOAD(IFLAG)/(17.576)
   15 CONTINUE
      GO TO 18
   16 IFLAG=0
   17 IFLAG=IFLAG+1
      LDPVAL(IFLAG)=LDPVAL(IFLAG)/(17.576)
      IF(LDPTM(IFLAG).LT.24.)GO TO 17
   18 IF(ILDDHW.EQ.1)GO TO 19
      MDOTLD=MDOTLD/(.45359)
   19 ISTAT=2
      UINP=3HENG
      UINSOL(1)=7HBTU/HR.
      UINSOL(2)=7H-SQ.FT.
      UTEMP=6HDEG.R.
      UMSFLW=9HLBM./MIN.
      USPHT(1)=8HBTU/LBM.
      USPHT(2)=7H-DEG.F.
      UAREA=6HSQ.FT.
      UCAPST=10HBTU/DEG.F.
      UUVAL(1)=10HBTU/H-SQ.F
      UUVAL(2)=9HT.-DEG.F.
      ULOAD=7HBTU/HR.
      RETURN
      END
      SUBROUTINE PRODAY(TCALC,TM,VAL,ARRAY)
      DIMENSION TM(100),VAL(100),ARRAY(1000)
      ISUB=1
      DO 1 IFLAG=1,1+INT(1440./TCALC)
      TIME=TCALC*(IFLAG-1)
    2 IF(TIME.LT.TM(ISUB+1)*60.) GO TO 3
```

```
      ISUB=ISUB+1
      GO TO 2
    3 FRAC=(TIME/60.-TM(ISUB))/(TM(ISUB+1)-TM(ISUB))
      ARRAY(IFLAG)=VAL(ISUB)+FRAC*(VAL(ISUB+1)-VAL(ISUB))
    1 CONTINUE
      IFLAG=INT(1440./TCALC)+1
      ARRAY(IFLAG)=VAL(24)
      RETURN
      END
      SUBROUTINE SINSOL(TSIM,TCALC,SOLMAX,SUNDAY,INSOL,
     1SUNRIS)
      REAL INSOL(1000)
      SUNSET=SUNRIS+SUNDAY
      DO 1 IFLAG=1,1+INT(TSIM/TCALC)
      TIME=TCALC*(IFLAG-1)
      DO 2 IDAY=1,10
      IF(TIME.LT.(24.*60.*IDAY))GO TO 3
    2 CONTINUE
    3 TIME=TIME-(24.*60.*(IDAY-1))
      IF(TIME.LT.SUNRIS)GO TO 4
      IF(TIME.GT.SUNSET)GO TO 4
      INSOL(IFLAG)=SOLMAX*SIN(3.1415927*(TIME-SUNRIS)/SUNDAY)
      GO TO 1
    4 INSOL(IFLAG)=0.
    1 CONTINUE
      RETURN
      END
      SUBROUTINE TAMB1(TSIM,TCALC,TAAVG,SWING,TAMB,SUNRIS)
      DIMENSION TAMB(1000)
      DO 1 IFLAG=1,1+INT(TSIM/TCALC)
      TIME=TCALC*(IFLAG-1)
      TAMB(IFLAG)=TAAVG+SWING*SIN(3.1415927*(TIME-SUNRIS-180)/720.)
    1 CONTINUE
      RETURN
      END
```

APPENDIX F
GLOSSARY

Absolute humidity Mass of water vapor per mass of dry air.

Absorber plate or absorber Component of a solar collector that absorbs solar energy and transfers it to a flowing fluid. The absorber plate is usually made of metal coated with a "black" or highly absorbing coating.

Absorption air conditioner Air conditioning unit that uses heat to evaporate a volatile component (refrigerant) from a solvent in a regenerator. Condensation of the volatile component and then expansion through a throttling device and subsequent evaporation produces the cooling effect in an evaporator. The resulting vapor is then absorbed again into the solvent in an absorber and pumped back to the regenerator. Common fluid combinations are ammonia/water and water/lithium bromide where the first is the refrigerant and the second the solvent.

Absorption refrigerator Similar to an absorption air conditioner, but refrigeration usually refers to freezing and thus the refrigerant must be capable of producing cooling below 0°C (32°F). A common fluid combination is ammonia/water.

Absorptivity, absorptance Property of a material that specifies the fraction of incident radiant energy that is absorbed. Ranges between 0 and 1.

Active system Solar heating or cooling system using an active device (pump or blower) to circulate the fluid between collector, storage, and load, usually with a control system to turn on or off depending on the environmental conditions.

Adiabatic Refers to no heat transfer to or from region; perfectly thermally insulated.

Air collector Solar collector using air rather than a liquid as the circulated heat-transfer fluid.

Air mass Effective mass of air that direct beam radiation penetrates relative to the air mass in the vertical direction. Equal to $1/\cos \zeta_s$ where ζ_s is the angle between the sun's rays and the vertical.

Alpha/epsilon ratio or α/ϵ Ratio of solar absorptivity to infrared emissivity of a given material. A good collector absorber using a selective surface should have a large value of α/ϵ, but must also have a large value of α (absorptivity).

Altitude, solar Angular elevation of the sun above the horizon.

Ambient temperature Temperature of the surroundings; for a collector, usually the outdoor dry bulb air temperature.

AS/ISES The American Section of the International Solar Energy Society.

ASHRAE The American Society of Heating, Refrigerating, and Air-Conditioning Engineers.

ASME The American Society of Mechanical Engineers.

ASTM The American Society for Testing Materials.

Azimuth angle (solar) Angular direction of the sun relative to the direction of the equator.

Back-loss coefficient Measure of the heat losses through the insulation and structure behind the absorber plate of a flat-plate solar collector.

Base load Load on a utility system that stays fairly constant over a full day. See *Peak load*.

Beam radiation, beam insolation That portion of the insolation that comes directly from the sun without scattering by the atmosphere or clouds.

Biomass Plant or animal material produced by photochemical reaction, and which may be used to produce useful energy.

Black Perfectly absorbing to all incident radiant energy. Many practical surfaces for solar collectors are nearly black, and absorb 90 percent or more of the incident solar energy.

Blackbody Ideal absorber for radiant energy. Real surfaces always reflect some radiant energy, and are therefore not true blackbodies.

Btu British thermal unit. The amount of heat required to raise the temperature of 1 lb_m of water by 1°F.

Calorie Amount of heat necessary to raise the temperature of 1 g of water by 1°C.

Capital-intensive Having a high initial cost, but possibly low operating and maintenance costs. Solar heating and cooling systems are capital-intensive.

Chemical storage Storage of heat by transforming it into chemical energy, which can be then recovered by reversing the chemical reaction.

Coefficient of performance (COP) Ratio of cooling rate to energy input rate (same units) for a refrigerator or air conditioner.

Collector efficiency For a flat-plate collector, the fraction of the global insolation incident on the collector aperture that is transferred as thermal energy to the working fluid. For a concentrating collector, the fraction of the beam insolation incident on the collector aperture that is transferred as thermal energy to the working fluid.

Collector plate See *Absorber plate*.

Comfort zone Range of combined ambient air temperature, relative humidity, air movement, and radiation in which a majority of people find comfort.

Concentrator Lens (refractor) or mirror (reflector) which directs the intercepted solar radiation onto an absorber surface which is smaller than the aperture. The idea is to reduce the absorber area and therefore losses so that high efficiency can be maintained at higher absorber (collection) temperature.

Cover plate Transparent cover(s) or glazing(s) on a collector; purpose is to admit short-wavelength radiation and reduce radiation and convection losses from the absorber.

Declination Sun angle relative to the plane of equator.

Desiccant Material which has affinity for water vapor such that it will absorb that vapor out of air, e.g., silica gel, calcium chloride. Desiccant cooling uses desiccant drying in conjunction with cooling and humidification to produce air conditioning.

Diffuse insolation Portion of the global insolation reaching a collector or building surface after scattering from clouds, atmospheric particles or any other materials (i.e., that portion whose direction is not from the sun).

Direct radiation, direct insolation See *Beam insolation*.

Double-wall heat exchanger Heat exchanger having two metal surfaces between hot and cold fluids, required when a toxic antifreeze solution used in collector loop could contaminate potable water.

Drainback Method of draining a vented solar system; when pump is off, the fluid in the collector automatically drains back to storage.

Draindown Solar system that is automatically drained down for freeze protection by actuated solenoid valves.

Drawdown Removal of all useful heat from the heat storage system (see *Swing temperature*).

Dry bulb temperature Commonly quoted temperature which is measured by a dry sensor (differs from *Wet bulb temperature*).

Edge losses Heat losses from a solar collector through the sides or edges of the absorber-plate enclosure.

Emissivity, emittance Property of a surface that determines its ability to emit radiant energy. The ratio of the radiation emitted by a surface at a particular temperature to that emitted by a blackbody at the same temperature.

Energy efficiency ratio Ratio of cooling rate (Btu/h) to electrical power input (W), in units of Btu/Wh (Btu/Watt-hr). Normally specified for vapor-compression air-conditioners or refrigerators.

Enthalpy Property that describes the energy content of a flowing material.

Entropy Term in thermodynamics useful in determining the efficiency with which processes occur.

Equinox One of two dates in the year when the sun declination is zero. Spring equinox occurs on March 21 and autumn equinox occurs on September 21.

Evaporative cooler Device which humidifies air, i.e., increases its humidity, and in the process decreases its dry bulb temperature.

Fin-tube absorber Standard absorber-plate design consisting of parallel tubes connected by heat-conducting fins.

Fixed collector Permanently oriented collector that has no provision for seasonal adjustment or tracking of the sun.

Flat-plate collector Collector without external concentrators or focusing devices, usually consisting of an absorber plate, transparent cover(s), back and side insulation, and a container.

Focusing collector Collector using some type of focusing device (parabolic mirror, Fresnel lens, etc.) to concentrate the beam insolation onto an absorbing element.

Forced circulation Circulation of the collecting fluid by a pump or blower.

Glazing Transparent cover material (glass or plastic) for a solar collector (see *Cover plate*).

Global insolation Includes the insolation striking the surface from all directions. Includes the diffuse plus beam insolation.

Gray Surface with emissivity and absorptivity that are independent of the wavelength of radiation. Such a surface will have equal values for absorptivity and emissivity.

Greenhouse effect Effective trapping of energy in an enclosure with a transparent cover when exposed to insolation. The cover is transparent to a large fraction of the insolation, but is fairly opaque to infrared radiation emitted by materials within the enclosure and acts as a barrier to convection losses.

Heat capacity storage See *Sensible heat storage*.

Heat exchanger Device that transfers heat from one fluid to another because of a temperature difference between the two.

Heat pump Active heating and cooling device that can provide both heating and air conditioning from the same unit.

Heat sink Material, location, or reservoir to which waste heat is discharged.

Heliostat One of an array of reflectors, its mount and two-axis tracking system used for directing beam insolation onto a central receiver.

Honeycomb Egg crate structure which is sometimes used between the absorber plate and cover plates of some collectors to reduce convection losses and thus produce greater collector efficiency.

Humidify To increase the moisture (water) content in the air.

Humidity Measure of the moisture content in the air (see also *Absolute humidity* and *Relative humidity*).

HVAC Acronym for the *h*eating, *v*entilating, *a*ir-*c*onditioning system.

Hydrated salt Salt which associates (bonds) with a number of molecules of water and in the process releases energy.

Incidence, angle of The angle relative to the surface normal at which insolation strikes a surface.

Infiltration Volumetric rate of exchange of air into a building from the surrounding environment.

Infrared That portion of the radiation spectrum where wavelengths are longer than those visible to the human eye, i.e., above about 0.78 μm (7800 Å).

Insolation Amount of solar energy reaching a surface per unit of time.

Insulation Material used to retard the flow of thermal energy from one space to another (material which has a low thermal conductivity).

ISES International Solar Energy Society.

Langley Unit of insolation; 1 Langley = 1 cal/cm^2.

Latent heat Heat used to change the phase of a material without changing its temperature, i.e., solid to liquid or liquid to vapor.

Latent heat storage Energy storage in a material by absorbing the latent heat as the material changes phases, such as in melting, freezing, or hydration.

Life-cycle cost Total cost of a system over its entire usable life, including capital costs, operating costs, fuel costs, maintenance costs, and financing costs.

Loss coefficient Measure of the rate of thermal loss from a collector per unit area and per unit of temperature difference by the combined processes of conduction, radiation, and convection.

Natural circulation Circulation of a fluid due to density differences resulting from temperature variations.

Natural convection Convection (motion) of a fluid due to density differences resulting from temperature variations.

Net present value (NPV) Value in present dollars of money or equipment at a future date taking into account an appropriate interest rate.

Nocturnal radiation Loss of energy by radiation to the sky at night when the surface (collector) is warmer than the effective sky temperature.

Nondirect insolation See *Diffuse insolation*.

Normal Line perpendicular to the plane of the surface under consideration.

Ocean thermal energy conversion (OTEC) Production of electrical energy from the ocean by using the temperature variation with depth to run a heat engine.

Orientation Rotation and/or position of a building or collector relative to the influences of natural environment, mainly the effects of sun and wind. Also refers frequently simply to direction, i.e., the long axis of a building is east-west.

Overhang Architectural device such as a roof extension placed over a window or vertical wall surface to intentionally shield the sun's direct radiation from a surface.

Passive design Use of design features such as shading, orientation, insulation, thermal mass, etc., to reduce or eliminate heating and cooling requirements.

Payback, payout Time required for savings in the operation and maintenance costs to make up for the greater capital cost of one system relative to another.

Peak load Maximum load that a utility system must provide.

Pebble-bed storage See *Rock storage*.

Phase change storage See *Latent heat storage*.

Photochemical conversion Conversion of photoenergy directly to a chemical energy form in a material. Plants use such reactions in photosynthesis.

Photovoltaic conversion Use of semiconductors or other photovoltaic devices that convert solar radiation (photons) directly to electricity.

Power tower (central receiver) A solar thermal electric power concept where many mirrors (heliostats) surrounding a central absorber on a tower, are tracked to reflect the sun's rays to the absorber to vaporize the fluid in the boiler (usually of a Rankine cycle plant).

Pyranometer Measuring device to determine local values of global (direct and diffuse) insolation.

Pyrheliometer Measuring device to determine local values of direct (beam) insolation.

Radiative heat transfer Transfer of energy from a body or between two bodies by electromagnetic radiation, usually restricted to consideration of visible and infrared radiation.

Rankine cycle Power cycle using a fluid which changes phase during the cycle and which consists of a boiler, turbine, condenser, and pump.

Reflected radiation Portion of the incident radiation on a surface (window, wall, collector) reflected by the surface.

Reflectivity Property of a material that specifies the fraction of incident radiant energy reflected. Ranges between 0 and 1.

Relative humidity Measure of water vapor in air, expressed as a percentage of the maximum water vapor which air can hold at that same temperature.

Retrofit Addition of solar heating, cooling, or hot-water-heating equipment to an existing structure to aid or replace the original nonsolar system.

Rock storage Energy storage in a bed of packed rock, usually used with an air system.

R&D Research and development.

Selective surface Surface which responds differently to different wavelengths of radiation (i.e., having wavelength-dependent properties). In solar energy applications, refers to a surface with a large value of absorptivity for solar energy and a small value of emissivity for infrared wavelengths. A solar collector with such a surface absorbs energy well, but has low radiation energy losses. See *Gray* for a contrast in types of surfaces.

Sensible heat Energy absorbed by a material which results in a temperature change.

Sensible heat storage Storage of energy by raising the temperature of a storage medium, usually a tank of water or a bed of rocks.

SERI Solar Energy Research Institute in Golden, CO.

SHAC Acronym for *s*olar *h*eating *a*nd *c*ooling.

Shading loss Loss of collector efficiency caused by the shading of the absorber plate by collector edges or components, or other obstructions such as trees or buildings.

Solar-assisted heat pumps Using solar collectors to provide energy for the evaporator coil of a heat pump to permit it to operate more effectively. The collector will also operate more effectively if required to collect at a temperature lower than that at which energy can be delivered directly to the heated space.

Solar cell A device for direct conversion of solar energy to electricity.

Solar constant Insolation on a surface in space at the earth's mean distance from the sun. Presently accepted as about 1353 W/m^2 [1.94 Langleys/min; 428 Btu/(h · ft^2)].

Solar irradiation, solar radiation See *Insolation*.

Solar spectrum Distribution of the sun's energy with wavelength. About 40 percent is in the visible wavelengths, with most of the remainder in the long-wavelength (infrared) part of the spectrum and a small fraction in the ultraviolet part.

Solar time Hour of the day as reckoned by the apparent position of the sun. Solar noon is that time on any day that the sun reaches its highest altitude angle.

Solstice One of the two dates during the year (summer solstice on June 21 and winter solstice on December 21) when the sun's declination to the plane of the equator is a maximum ($23\frac{1}{2}°$ and $-23\frac{1}{2}°$, respectively).

Specific heat Energy required to raise a mass of substance one degree in temperature [kJ/(kg · K) or Btu/(lb$_m$ · °F)].

Spectral distribution Distribution of some quantity (such as solar energy, emissivity, or absorptivity) with wavelength.

Stagnation temperature Temperature a collector absorber plate will reach in steady state with no coolant flow when exposed to insolation.

Stratification (thermal) Increase in temperature in the vertical direction in a storage fluid.

Sun rights If you install and pay for a solar heating and cooling system and your neighbor then builds a two-story house on the lot next door that shades your collectors, you may argue the question of "sun rights" in the courts.

Swamp cooler See *Evaporative cooler*.

Swing (temperature) Variation in temperature over a specific time interval such as the difference between maximum and minimum storage temperature or between the maximum and minimum ambient temperature.

Thermal capacity Ability of a medium to store energy.

Thermal lag Time delay in the temperature rise of the fluid leaving the collector during startup because part of the insolation goes to heat the collector itself. In a building, the time delay in the transmission of heat from an external to an internal surface.

Thermosiphon Use of natural circulation for transferring energy from a collector to storage.

Tilt Angle of a collector surface to the local horizontal.

Total radiation The integral of radiation over all wavelengths.

Tracking collector Collector using some device, usually electromechanical, to keep it aligned with the sun during daily and/or seasonal movement. May be one- or two-axis tracking and usually applied only to concentrating collectors.

Transmissivity, transmittance Property of a material that specifies the fraction of incident radiation that is transmitted through a given thickness. Reduction in transmittance is due both to absorption and reflection. Varies between 0 and 1.

Trombe wall Passive heating device consisting of a vertical building wall with glazing and massive masonry wall behind it to absorb the incoming solar energy and transfer it to the building by natural circulation and radiation.

Ultraviolet radiation Short-wavelength portion of the solar spectrum (less than about 0.38 μm or 3800 Å) which is largely absorbed by the atmosphere.

Visible radiation Portion of the solar spectrum sensed by the human eye, generally between about 0.38 and 0.78 μm (3800–7800 Å); accounts for about 40 percent of solar radiation.

Wet bulb temperature Temperature obtained if a wet wick is wrapped around a thermometer and the thermometer moved rapidly in the air. Used in connection with dry bulb temperature to determine absolute or relative humidity.

INDEX

Absorber, 9, 397
Absorption:
  atmospheric, 5–7
  by collectors, 126
  properties, 127, 328, 337–338
Absorptivity, 321, 323–328, 397
  data, 337–338
Acceptance angle, 66, 92, 131
Air heaters, 10, 97–98, 100, 397
Air mass, 6, 397
Altitude angle, solar, 47, 49–52, 55, 398
Ambient temperature, 398
  data, 284–293
  simulation, 31
Aperture, 10
ASHRAE (American Society of Heating, Refrigeration, and Air Conditioning Engineers) collector test, 151–152
Astronomical unit (AU), 3
Azimuth angle, solar, 47, 49–50, 55, 398
  surface, 57

Beam insolation, 6, 62–63, 65–66, 398
  data, 299–302
  extraterrestrial, 59
  on a tilted surface, 73, 77–78
Biot number, 179
Blackbody radiation, 321–327, 398
Blackbody spectral emissive power, 322

Central receiver, 13, 37–40, 115, 271
Circumsolar insolation, 66
Clear sky radiation, 8

Clearness index:
  data, 284–298
  hourly, 73–74
  monthly average, 76
Collector (*see* Concentrating collector; Flat-plate collector)
Compound parabolic concentrator (CPC), 102–103
Concentrating collector, 9–11, 101–116, 398
  concentration ratio, 89, 92, 131, 133
  design, 101–102
  efficiency, 92, 116, 398
  focal length, 131–132
  image, 132
  intercept factor, 93, 133
  linear focus, 101–112, 116, 130–133
  optical design, 130–133
  point focus, 107–115
  thermal analysis, 142–146
  thermal performance, 116
Concentration ratio, 133
  geometric, 89, 92, 131
Conduction heat transfer, 309–316
Conductivity, thermal, 309
  data, 311–314
Control, 28–30, 201–213
Convective heat transfer, 316–320
  forced, 147, 317–318
  free, 135, 318–320
Conversion of units, table, 276–277
Cooling, 191–192, 212, 222–224
Cost-benefit comparison, 245–247
Covers, 10, 95–97, 398
  optical properties, 125–129, 157, 329
CPC (compound parabolic concentrator), 102–103

Cylindrical collector, 103

Day length, 51
Declination, 47–49, 398
Degree-day, 189, 191, 221
   data, 189, 191, 303–307
Density, data, 311–314
Diffuse insolation, 6, 63, 68–69, 399
   on a tilted surface, 73, 77
Direct insolation (*see* Beam insolation)
Discounting, 243–244
Domestic hot water systems, 34–35, 251–255
Drainback, 204–207
Draindown, 204–206

Economics, 10–11, 40–42, 116–118, 241–249, 267–268
Effective interest rate, 41–42
Effectiveness, heat exchanger, 179–180
Efficiency, collector, 22–26, 88, 90–93, 100, 116, 136–141, 148–154, 398
   all-day, 88, 149
Elevation, data for selected cities, 284–293
Emissivity, 323–328, 399
   data, 333–336
Energy use patterns, data, 2, 192–193, 197–199
Equation of time, 48, 49, 56
Equatorial mount, 66
Equinox, 47
Evacuated tubular collector, 98–100
Evaporative loss, 194–196
Extraterrestrial insolation, 3–5
   on a horizontal surface, 59, 62–64
   spectral distribution, 4–5

f-chart method, 227–232, 257, 260
   annual, 233–236, 252–253, 257–260
Flat-plate collector, 9–11, 399
   covers (or glazing), 10, 95–97, 398
   design, 93–98
   efficiency, 22–26, 88, 90–93, 100, 136–141, 148–154, 398
   heat removal factor, 24–25, 144–146
   optical design, 124–129

Flat-plate collector (*Cont.*):
   orientation, 101, 237
   overall thermal loss coefficient, 90, 137, 146–147
   performance, 100, 117, 138–141
   shading, 237–239
   stagnation temperature, 25–26, 402
   temperature distribution, 142–145
   thermal analysis, 133–148
   tilt, 73, 76–77, 237
Focal length, 131–132
Focusing collector (*see* Concentrating collector)
Freeze protection, 166–167, 204–208
Fresnel lens, 105, 112
Fresnel mirror, 104, 106–107

Glazing (*see* Covers)
Global insolation, 6, 63, 67–69, 399
   data, 280–293
   on tilted surfaces, 71–82
Grashof number, 135, 318–320
Grey body, 323, 399
Ground-reflected insolation, 73
   data, 339

Heat engine, simulation, 36, 268–271
Heat exchanger effectiveness, 179–180
Heat removal factor, collector, 23–24, 87, 89, 144
Heat transfer coefficients:
   forced convection, 317–318
   free convection, 135, 319–320
   radiation, 331
   volumetric (pebble bed), 174
   wind, 147, 157
Heating system:
   costs, 242–245
   design, 217–221, 227–236, 255–257
   loads, 187–191, 221
   performance, 224–226, 257–262
Heliostat, 109–110, 115
History of solar devices, 12–17
Hottel-Whillier-Woertz-Bliss method, 142–146
Hour angle, 49
Hourly clearness index, 73–74
Hourly insolation, 71–76, 79

Hydrated salt, 162–164

Incident angle to a surface, 57–58, 60–61, 400
Industrial process heat, 262–268
Infrared insolation, 4
Insolation, 400
   atmospheric attenuation, 5–8
   data, 69–71, 280–293, 299–302
   on a tilted surface, 71–82, 294–298
   (*See also* Beam insolation; Diffuse insolation; Extraterrestrial insolation; Global insolation; Ground-reflected insolation; Terrestrial insolation)
Intercept factor, 93, 133

Kinematic viscosity, data, 312–314
Kirchhoff's law, 326

Latent heat storage, 162–165, 400
   analysis, 178–185
   materials data, 162–164
Latitude, 47, 49
   data for selected cities, 284–293
Life-cycle cost, 241–245, 400
Liquid heater, 10, 93–97, 100
Loads:
   cooling, 191–192
   heating, 187–191, 221
   industrial, 196–199
   swimming pool, 193–196
   water heating, 192–193

Modeling (*see* Simulation)
Monthly average insolation, 76–77
   data for a horizontal surface, 280–293, 299–302

Net present value (*see* Present value)
No flow temperature (*see* Stagnation temperature)
Nondirect insolation (*see* Diffuse insolation)
Nontracking concentrators, 101–103
NTUs (number of transfer units), 179–181

Nusselt number, 317–319

Optical loss, concentrator, 92–93, 130–133
   flat plate, 124–129
Optical thickness, spectral, 6–8
Overall thermal loss coefficient, 22, 87, 146–148

Parabolic collector, 103–105
Payback period, 248, 401
Pebble-bed storage, 173–177
Phase change storage (*see* Latent heat storage)
Power tower (*see* Central receiver)
Prandtl number data, 312–314
Present value, 244–245
Pyranometer, 65, 67–69, 401
Pyrheliometer, 65–66, 401

$\overline{R}$, 73, 77
   data, 294–298
Radiation:
   extraterrestrial, 3–5
   terrestrial, 5
Radiation heat transfer, 321–331
Radiative properties, 125–127, 129, 328–330, 333–339
Rayleigh scattering, 6–8
Receiver, 10
Reflectivity, 321, 330
   data, 339
Resistance, thermal, 310
Rim angle, 131–132
Rock storage (*see* Pebble-bed storage)
Rules of thumb for sizing, 217

Scattering, atmospheric, 5–6
Selective surface, 10, 129, 401
Sensible heat storage, 161–165, 401
   materials data, 162, 169
   model, 26, 168, 173–174
Shadow band, 68–69
Simulation:
   collector, 21–26
   heat engine, 36, 268–271
   heating, 255–261

Simulation (*Cont.*):
  hot water, 251–255
  load, 27
  process heat, 262–268
  storage, 26–27, 34–37, 39, 168–173
  system, 27–40, 227
Solar altitude angle, 47, 49–52, 55, 398
Solar azimuth angle, 47, 49–50, 55, 398
Solar constant, 3
Solar day, 56
Solar history, 12–17
Solar radiation (*see* Insolation)
Solar spectrum, 3–7
Solar time, 56
Solar zenith angle, 6, 47
SOLSIM, 30–34, 148, 247, 341–346, 371
  examples, 34–41, 253–254, 260–261, 271
  sample BASIC program, 347–370
  sample FORTRAN program, 373–395
Solstice, 47, 52
Specific heat, data, 311–314
Stagnation temperature, 25–26, 402
Storage, 160–185
  (*See also* Latent heat storage; Pebble-bed storage; Sensible heat storage)
Storage model, 26–27, 34–37, 39, 168–173
Stratification, 167–173, 402
Sun, 1, 3
Sunrise-sunset hour angle, 51
  relative to a tilted surface, 59
Swimming pool heating, 193–196
  collectors, 10, 99–100
  control, 210
  design, 217
  design data, 194–195

Terrestrial insolation, 5
  data, 280–293, 299–302
Testing collectors, 87–91, 149–154
Thermal conductivity, 309
  data, 311–314

Thermal diffusivity, data, 311–314
Thermal loss coefficients:
  overall, 22, 87, 146–148
  residential, 190, 221
Thermal resistance, 310
Thermosiphon, 201–202
Tilt angle, surface, 57–59
Time, equation of, 48–49, 56
  local, 56–57
  sunrise, 51
  sunset, 51
Time zones, 56
Top loss coefficient, 146
Tracking, 103–116, 209–210
Transmission:
  atmospheric, 6
  through cover-plate assembly, 125–129
  data, 125–127
  spectral transmittance, 125
Transmissivity, 321, 329–330
Transmittance-absorptance product, 22, 24, 81, 128, 152–153
Turbidity, 6–8

Ultraviolet insolation, 4, 6–7

Viscosity, kinematic, data, 312–314
Visible insolation, 4, 6

Water heating, 13, 30–35
  control, 211–212
  design, 217–221, 227–232
  forced circulation, 202–208
  loads, 192–193
  performance, 224–226
  thermosiphon, 201–202
Water storage, 166–168
  (*See also* Sensible heat storage)
Wien's displacement law, 322

Zenith angle, solar, 6, 47